Essentials of Conservation Biology

Essentials of

Richard B. Primack
BOSTON UNIVERSITY

Conservation
Biology

Sinauer Associates Inc. • Publishers
Sunderland, Massachusetts U.S.A.

THE COVER

It is often the case in conservation biology that threatened species, even if they are very similar, have different conservation needs. An illustration of this principle is provided by two American crane species, the whooping crane, *Grus americana,* and the sandhill crane, *Grus canadiensis.* The two species are closely related and both were threatened by human encroachment on their habitat; today, however, the overall population of sandhill cranes is estimated at well over 500,000 birds, while the total whooping crane population is estimated at only 155 individuals. Box 22 in Chapter 13 (page 311) discusses conservation efforts aimed at saving these cranes.

The cover shows a flock of gray sandhill cranes in the Bosque del Apache Refuge, New Mexico. The single white whooping crane in their midst was raised by the sandhills after scientists placed "extra" whooping crane eggs in sandhill nests in an attempt to increase the whooping crane population. (Photograph by Art Wolfe.)

ESSENTIALS OF CONSERVATION BIOLOGY

The images of John Muir, Gifford Pinchot, and Aldo Leopold on pages 13–16 appear courtesy of the State Historical Society of Wisconsin, 816 State Street, Madison 53706.

Library of Congress Cataloging-in-Publication Data

Primack, Richard B., 1950–
    Essentials of conservation biology / Richard B. Primack.
        p.   cm.   Includes bibliographical references and index.
    ISBN 0-87893-722-6
    1. Biological diversity conservation.   I. Title.
QH75.P752   1993
333.95'11—dc20   93-6933
                                                                        CIP

Printed in U.S.A.   6   5   4

To my wife, Margaret, my parents, Shirley and Saul, and my children, Daniel and William, who gave me the gifts of time and love.

Contents

Preface

Conservation biology has emerged during the last ten years as a major new synthetic discipline addressing the alarming loss of biological diversity throughout the world. This biodiversity crisis has attracted increasing scientific, government, and popular attention. Conservation biology represents a fusion of theory, basic research, applied research, and public education that is evolving at a rapid pace. Evidence of the explosive increase of interest in this new scientific field is shown by the rapidly increasing membership of the Society for Conservation Biology, the great intellectual excitement displayed in many newsletters and journals, and the large number of new edited books and advanced texts that appear almost weekly.

University students have enrolled enthusiastically and in large numbers in introductory conservation biology courses. However, there has been no suitable textbook that covers the breadth of subjects included in modern conservation biology; its biological core alone incorporates population biology, community ecology, evolution, genetics, taxonomy, paleontology, zoo management, wildlife ecology, agriculture, and forestry, and the subject also contains elements of history, philosophy, economics, anthropology, and public policy. Providing reading materials for such a wide-ranging course has been problematic.

My purpose in writing this book has been to provide a modern, up-to-date textbook of the basics of conservation biology for use by university undergraduates, beginning graduate students, and others wanting to know more about the subject. I have not presumed that people who use this book will have taken specific science courses prior to reading it, though some exposure to biology and ecology would be advantageous. Students with a strong background in ecology and evolution may find much of the material in Chapters 2 and 3 familiar. Suggested readings at the end of each chapter, as well as the references within the text, will allow interested students and researchers to delve deeper into any given subject than this introductory text allows.

Keeping this book current has been a challenge because of the rapid rate of change in the field and expansion of its horizons. I would certainly enjoy hearing from readers who could provide updated information and relevant examples that I could use in the next edition.

Acknowledgments

Special thanks are due to Kamaljit Bawa and Bob Tamarin for their constant encouragement. I would also like to express my thanks to Boston University for providing me with the excellent facilities and environment that made this project possible. Elizabeth Platt was the principal research assistant for the project, in particular helping to write the boxes and providing valuable criticism of the early drafts. Ernesto Hayn assisted in putting together the bibliography. Andy Sinauer, Carol Wigg, Norma Roche, Kathaleen Emerson, and the rest of the Sinauer Associates staff did a great job turning the manuscript into a book.

The manuscript was reviewed in its entirety by April Algaier, Kamaljit Bawa, Phil Cafaro, Brian Drayton, Les Kaufman, Mark Primack, and David Woodruff. Individual chapters or groups of chapters were reviewed by Michael Bean, Eric Dinerstein, Nicholas Gotelli, Gary Hartshorn, Sam McNaughton, Eric Menges, Christine Padoch, Mary Pearl, Alan Randall, Jamie Ressor, Holmes Rolston III, Sahotra Sarkar, Christine Schonewald-Cox, and Bob Tamarin. The boxes, with their case studies of specific topics, are crucial to the book, and many people offered valuable specialized input that helped in making them current: Michael Balick, Vadim Birstein, Jay Blakeney, Russell Burke, Tom Cade, Kevin Chu, Tim Clark, James Dietz, James Estes, D. A. Etnier, Donald Falk, Richard Forman, Robert Goodland, J. Frederick Grassle, Samuel Gruber, David Janos, Paul Kerlinger, Lloyd Kiff, Devra Kleiman, Scott Kraus, Lloyd Loope, Francis Howarth, Thomas Michael Power, Scott Swengel, George Schaller, Phil Tabas, John Terborgh, Tom Thorne, Dagmar Werner, David Wilcove, and Garrison Wilkes.

Richard B. Primack

PART I

Major Issues that Define the Discipline

CHAPTER 1

What Is Conservation Biology?

Popular interest in protecting the world's plant and animal species has intensified during the last 20 years. Both scientists and the general public have realized that we are living in a time of unprecedented mass extinction. Around the globe, biological communities that took millions of years to develop are being devastated by human actions. Today's mass extinctions can be compared to mass extinctions in the geological past, in which tens of thousands of species died out following some massive, unknown catastrophe, possibly a collision with an asteroid. Unless something is done to reverse the trend, the wonderful species that symbolize the essence of wildlife, such as elephants, tigers, and grizzly bears, will soon no longer be found in the wild. Thousands—possibly even millions—of less conspicuous plant and invertebrate species will join them in extinction unless their habitats and populations are protected—and their loss may prove even more devastating to the planet and its human inhabitants.

The main cause of the present extinctions is habitat destruction stemming from human activities, such as the clear-cutting of old-growth forests in the temperate zone and rain forests in the tropics, overgrazing grasslands, draining wetlands, and polluting freshwater and marine ecosystems. A second major cause of extinctions is the overharvesting of animals and plants, especially when it is done using modern technology, to supply national and international markets. Even when parcels of natural habitat are preserved as national parks and nature reserves, extreme vigilance is required to prevent extinction of the remaining species there; many of these species have been so dramatically reduced in numbers that they are now extremely vulnerable to extinction. Also, the environment in these habitat fragments may be so altered from its original condition that the site may no longer be suitable for the continued existence of certain species. Island communities, such as the native biota of the Hawaiian Islands and Madagascar, have been devastated by human activity, especially by the introduction of species such as cattle, pigs, and goats.

The powerful technologies developed by Western societies have allowed us to alter our environment on an unprecedented scale. Some of these transformations have been deliberate, resulting from the creation of dams or the development of new agricultural land; other changes, such as air pollution and the overgrazing of grasslands, are the accidental by-products of our activities. Unregulated dumping of chemicals and sewage into streams, rivers, and lakes has polluted major freshwater and coastal marine systems throughout the world and has driven significant numbers of species to extinction. Pollution has reached such levels that even large marine environments such as the Mediterranean Sea and the Arabian Gulf are threatened with biological death. Inland water bodies, such as the Aral Sea, have been completely destroyed, along with many unique fish species. Air pollution from factories and cars has turned rainwater into an acidic solution that weakens and kills plant life and, in turn, the animals that depend on the plants. Recently scientists have recognized that levels of air pollution have become severe enough to alter global climate patterns and strain the capacity of the atmosphere to filter out harmful ultraviolet radiation. The implications of these events for biological communities are enormous and ominous; they are the stimuli for the growth of conservation biology.

Concern for Biological Diversity

The concern over today's mass extinction spasm is based on four factors (Soulé 1985; Wilson 1985, 1989). First, the present threats to biological diversity are unprecedented. Never before in the history of life have so many species been threatened with extinction in so short a period of time. Second, the threat to biological diversity is accelerating due to the demands of a rapidly increasing human population, as well as continued advances in technology. This dire situation is exacerbated by the unequal distribution of the world's wealth and the crushing poverty in many tropical countries that have an abundance of species. Third, scientists now realize that many of the threats to biological diversity are synergistic; that is, several independent factors—such as acid rain, logging, and overhunting—combine additively or even multiplicatively to make a situation worse (Myers 1987). Finally, people are realizing that what is bad for biological diversity will almost certainly be bad for human populations, since humans are dependent on the natural environment for raw materials, food, medicines, water, and other goods and services.

Some people feel discouraged by the avalanche of species extinctions occurring in the world today, but it is also possible to feel challenged by the need to do something to stop the destruction. The next few decades will determine how many of the world's species will survive. The massive efforts now being made to protect new conservation areas and national parks will determine which of the world's species are preserved for the future. In the future, people may look back on these closing years of the twentieth century as a time when a relative handful of determined people saved numerous species.

Conservation biology is the new, multidisciplinary science that has developed to deal with the crisis confronting biological diversity. Conservation biology has two goals: first, to investigate human impacts on biological diversity and, second, to develop practical approaches to prevent the extinction of species (Soulé 1985, 1986; Wilson 1992). In many cases this means developing compromises between conservation priorities and human needs. Conservation biology arose because none of the traditional applied disciplines were comprehensive enough by themselves to address the critical threats to biological diversity. Agriculture, forestry, wildlife management, and fisheries biology have been primarily concerned with developing methods to manage a small range of species for the market economy and for recreation. These disciplines generally have not addressed the protection of the full range of species found in biological communities, or have regarded it as a secondary issue. In particular, these disciplines in the past often overlooked rare species with little direct economic value, species that are of considerable concern to conservation biology. Conservation biology complements the applied disciplines and provides a more general theoretical approach to the protection of biological diversity (Figure 1.1); it differs from these disciplines by having the long-term preservation of the entire biological community as its primary consideration, with economic factors often being only a secondary consideration.

The academic disciplines of population biology, taxonomy, and ecology are central to conservation biology, and many of the new conservation biologists have been drawn from these ranks. These disciplines traditionally emphasize the understanding of species in their

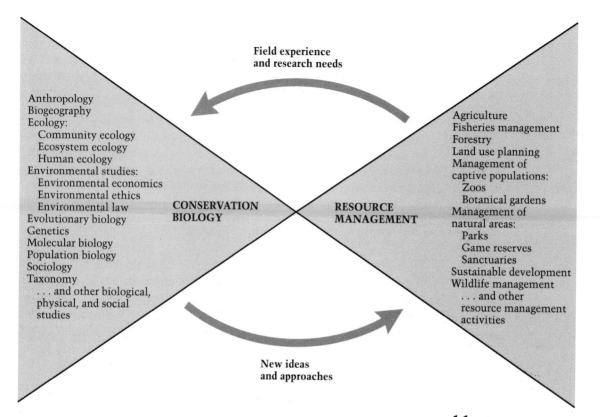

Field experience
and research needs

Anthropology
Biogeography
Ecology:
    Community ecology
    Ecosystem ecology
    Human ecology
Environmental studies:
    Environmental economics
    Environmental ethics
    Environmental law
Evolutionary biology
Genetics
Molecular biology
Population biology
Sociology
Taxonomy
    . . . and other biological,
    physical, and social
    studies

**CONSERVATION
BIOLOGY**

**RESOURCE
MANAGEMENT**

Agriculture
Fisheries management
Forestry
Land use planning
Management of
captive populations:
    Zoos
    Botanical gardens
Management of
natural areas:
    Parks
    Game reserves
    Sanctuaries
Sustainable development
Wildlife management
    . . . and other
    resource management
    activities

New ideas
and approaches

1.1 **Conservation biology represents a new synthesis of many basic sciences (left) that provide principles and new approaches for the applied fields of resource management. The experiences gained in the field in turn influence the direction of the basic sciences. (After Temple 1991.)**

natural environments; human activity is generally not included in their research models. Researchers often prefer to operate in remote, pristine field sites where human activity does not dominate or even affect their results. Yet many scientists now recognize that human activity is so prevalent today that ignoring it is now not only almost impossible, but would result in excluding from consideration one of the dominant components in modern biological communities (Primack 1992).

Many of the leaders in conservation biology have also come from zoos and botanical gardens, where they have gained experience in maintaining and propagating species in captivity. These zoologists and botanists have made first-hand observations of extinctions and habitat destruction in the wild and understand the urgency of the biodiversity crisis.

Modern conservation biology not only provides a linkage between academic and applied biological disciplines, but also draws on ideas and expertise from a broad range of fields. For example, environmental law provides the legal basis for protecting endangered species and critical habitats. Environmental ethics provides a rationale for preserving species. Social sciences, such as anthropology, sociology, and geography, provide insight into how local people can protect and manage the species found in their immediate environment. Environmental economists provide analyses of the economic values of biological diversity, which often support arguments for preservation. Ecosystem ecologists and climatologists monitor the physical characteristics of the environment and develop models of how the environment will change in the future.

A crucial difference between conservation biology and the traditional academic disciplines is that conservation biology is trying to provide answers to specific questions that can be applied in actual field situations. These questions revolve around determining the best strategies for protecting rare species, designing nature reserves, initiating breeding programs to maintain genetic variability in small populations, and reconciling conservation concerns with the needs of local people. The critical test for conservation biology is whether its methods succeed in preserving species and biological communities. If species continue to live and evolve in the wild, conservation biology will have succeeded; if its methods fail to preserve species, then conservation biology can justifiably be considered a failure.

Conservation biology should be considered a crisis discipline. Decisions on conservation questions are being made every day under severe time pressures. Conservation biologists and scientists in related fields are well suited to provide the advice that governments, businesses, and the general public need in order to make crucial decisions. However, because of time constraints, decisions on such matters as park design and species management often have to be made without the thorough investigations scientists normally require. Conservation biologists have to be willing to express an opinion based on available evidence, accepted theory, comparable examples, and informed judgment. If scientists are unwilling to offer such advice, the decision to act or not act will be made by someone with less training and in-depth knowledge of the needs of biological communities and endangered species.

Origins of Conservation Biology

The origins of conservation biology can be traced to religious and philosophical beliefs concerning the relationship between human society and the natural world (Hargrove 1986a,b; Zaidi 1986, 1989). In many of the world's religions, people are seen as both physically and spiritually connected to the plants and animals in the surrounding environment. In the Chinese Taoist and Japanese Shinto philosophies, wilderness areas are valued and protected because of their capacity to provide intense spiritual experiences (Waley 1934; Needham 1962; Davies 1987; Dwiveddi and Tiwari 1987; Badiner 1990). In other words, a direct connection is seen between the natural world and the spiritual world (Figure 1.2). This connection is broken when the natural world is altered and destroyed by human activity. Strict adherents to the Jainist and Hindu religions in India believe that all killing of animal life is wrong (Gadgil and Guha 1992). Descriptions of environmentally devastating events such as droughts, fires, and earthquakes that occur when people act against the wishes of the gods are prominent in the religious traditions of many societies.

Biological diversity often has immediate significance to traditional societies where the people live close to the land and water. In Native American tribes of the Pacific Northwest, hunters had to undergo purification rituals in order to be considered worthy of hunting animals (Pascua 1991). Hunting and gathering societies, such as the Penan of Borneo, give thousands of names to individual trees, animals, and places of their surroundings to create a cultural landscape that is vital to the well-being of the tribe (Brosius 1990). This relationship to the natural world was described eloquently at the Fourth World Wilderness Congress in 1987 by the delegate from the Kuna people of Panama (Gregg 1991):

> For the Kuna culture, the land is our mother and all living things that we live on are her brothers in such a manner that we must take care of her and live in a harmonious manner on her, because the extinction of one thing is also the end of another . . . .

In an original synthesis of the ecology and history of the Indian subcontinent, Gadgil and Guha (1992) argue that the belief systems, religions, and myths of hunter–gatherer societies and stable agricultural societies tend to emphasize conservation themes and the wise use of natural resources because these groups have learned over time to live within the constraints of a fixed resource base. In contrast, the belief systems of pastoralists and rapidly expanding agricultural and industrial societies emphasize the rapid consumption and destruction of natural resources as a way of maximizing growth and asserting control over other groups. These groups move to new localities when the resources of any one place are exhausted. The rituals found in many ancient religions that involve burning wood and sacrificing animals are seen by Gadgil and Guha as an attempt to dominate and subdue the natural world. Modern industrial states represent the ex-

1.2 A Shinto spirit gate in Japan. The Shinto philosophy teaches respect for wilderness areas and natural settings, emphasizing a direct connection between the natural and spiritual worlds. (Photograph by Lara Hartley/TERRAPHOTOGRAPHICS.)

treme culture of excessive and wasteful consumption, in which resources are transported to urban centers in ever-widening circles of resource depletion.

The religious responsibility of people to save species from extinction is well illustrated by the Biblical story of Noah's Ark (Hargrove 1989; Kay 1988). In this story, God punished the wicked ways of humans by sending a great flood, saving only Noah and his family. God ordered Noah to build an ark and fill it with a pair of each animal species, saying "Keep them alive with you." After the flood subsided, these animals were released to populate the Earth again. This story can be viewed as an early awareness of the importance of biological diversity and the initiation of a species preservation plan. The prophet Muhammad, founder of Islam, continued this theme of human responsibility, saying: "The world is green and beautiful and God has appointed you as His stewards over it. He sees how you acquit yourselves."

European Origins

In the Bible of the western Judeo-Christian tradition, wilderness was often seen as a place where an individual or group could go to strive in isolation for spiritual purity—not because it was soothing, but because it was harsh and unforgiving. Examples include Moses, Elijah, John the Baptist, and the ancient Israelites. In the New Testament, Jesus went into the wilderness to test himself against the temptations of the devil. In medieval Europe, the wilderness was perceived to be useless land and was often believed to be inhabited by evil spirits or monsters, in contrast to the pleasant qualities and appearance of agricultural landscapes (Nash 1982).

In European thinking, the prevalent view has been that God put nature here for humans to use, and that any use could be justified as long as humans benefitted. The dominant tenet of Western philosophy has been that nature should be converted into wealth as rapidly as possible and used for the benefit of people (Locke 1690). This anthropocentric (human-centered) view of nature led to the exploitation and degradation of the vast resources of the regions colonized by European countries from the sixteenth century onward. In practice, the wealth and benefits that came from this policy accrued primarily to the citizens of the colonial powers, while the needs of non-European native peoples were largely overlooked. The long-term ramifications of this philosophy for natural resources were not considered; the unexplored territories of the Americas, Asia, Africa, and Australia seemed so vast and rich that it was inconceivable to the colonial powers that natural resources would ever run out.

However, one important element of the conservation movement did develop in Europe out of the experience of scientific officers, often imbued with a Romantic idealism, who were sent to assist in the development of colonies during the nineteenth century (Grove 1990, 1992). These scientists were trained to make detailed observations on the biology, natural history, geography, and anthropology of the colonial regions. Many of them expected to find the indigenous people living in a wonderful harmony with nature; instead they found devastated forests, damaged watersheds, and poverty. In colonies throughout the world, certain scientists came to see that protection of forests was necessary to prevent soil erosion, maintain wood supplies, and prevent famine. Some colonial administrators also argued that certain intact forests should remain uncut because of their necessary role in ensuring a steady supply of rainfall in adjacent agricultural areas—foreshadowing modern concerns about global climate change. Such arguments led directly to conservation ordinances. On the Indian Ocean island of Mauritius, for example, the French colonial administration stipulated in 1769 that 25% of land holdings must remain in forest to prevent erosion, degraded areas were to be planted with trees, and forests growing within 200 meters of water were to be protected. Laws passed in the late eighteenth century regulated the pollutants being discharged by indigo and sugar mills, in order to prevent water pollution and the destruction of fish supplies. On the Caribbean island of Tobago, British officers set aside 20% of the island as "reserved in wood for rain."

These experiences and experiments on small tropical islands had considerable influence on British scientists working in India, who issued a report in 1852 urging the establishment of forest reserves throughout the vast subcontinent in order to avert environmental calamities and economic losses (Grove 1992). In particular, the report linked deforestation to lowered rainfall and water supplies, leading to famine among the local people. The report was embraced by the leadership of the British East India Company, who could see that conservation made good economic sense. During the mid-nineteenth century, Indian state governments established an extensive system of forest reserves protected and managed by professional foresters. This system of managed forest reserves in India was widely adopted in other parts of the colonial world, such as Southeast Asia, Australia, and Africa, and it influenced forestry in North America as well. The irony is that prior to colonization, indigenous peoples in these regions often had well-developed systems of natural resource management that were swept aside by the colonial governments (Poffenberger 1990; Gadgil and Guha 1992.)

Many of the themes of contemporary conservation biology were established in European scientific writings of a century or more ago (Grove 1992). The possibility of species extinction was demonstrated by the loss of wild cattle (*Bos primigenius*, also known as the aurochs) from Europe in 1627 and the extinction of the endemic dodo bird in Mauritius in the 1670s (Szafer 1968). To deal with the problem of the decline and possible extinction of wild cattle, Polish authorities established a nature reserve in 1564 in which hunting was prohibited. This nature reserve represents one of the earliest deliberate European efforts at conserving a species. While this action failed to preserve wild cattle (the progenitor of modern cattle), the nature reserve did protect the sole remaining population of the wisent, the European bison.

In Europe, concern for the protection of wildlife began to be widely expressed in the late nineteenth century (Ratcliffe 1984; Moore 1987; Green 1989). The combination of increasing area of land

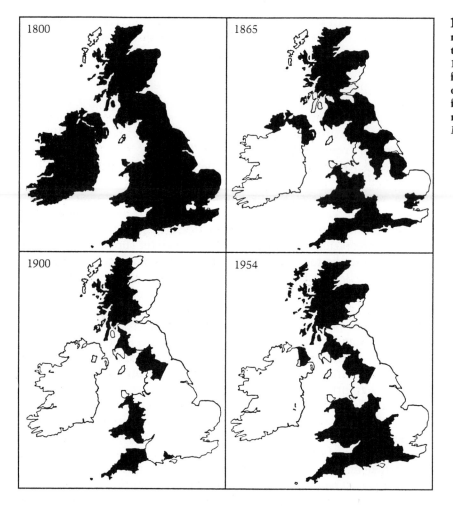

1.3 Changes in the range of the buzzard in the British Isles since 1800. Conservation efforts have led to a return of buzzard populations from the low levels reached in 1900. (From Moore 1987.)

under cultivation and more widespread use of firearms for hunting was leading to a marked reduction in wild animals. In Britain, many culturally and ecologically significant species became extinct in the wild at about this time: storks, cranes, great bustards, ospreys, sea eagles, wild boar, and wolves. Other species were showing rapid declines; one such example, that of the buzzard, is shown in Figure 1.3. These dramatic changes stimulated the formation of the British conservation movement leading to the founding of the Commons, Open Spaces, and Footpaths Preservation Society in 1865, the National Trust in 1895, and the Royal Society for the Preservation of Birds in 1899. Because of the intensive human use of the British landscape, conservation efforts in Britain have emphasized the preservation and management of relatively small fragments of land. Even today, less than 1% of the land in Great Britain is in nature reserves.

American Origins

Many of the themes of the North American conservation movement can be seen in the novels of James Fenimore Cooper, written in the early nineteenth century. In novels such as *The Pioneers, The Prairie*, and *The Deerslayer*, Cooper described the moral, spiritual, and aesthetic value of wilderness and deplored its thoughtless destruction. His frontier hero, Leatherstocking, condemns the exploiters with these words: "They scourge the very 'arth with their axes. Such hills and hunting grounds have I seen stripped of the gifts of the Lord, without remorse or shame!"

Among the early major intellectual figures in the United States arguing for the protection of natural areas were the nineteenth century philosophers Ralph Waldo Emerson and Henry David Thoreau (Nash 1982; Callicott 1990). Emerson, in his writings on the philosphical movement known as Transcendentalism, argued that nature could be viewed as a temple in which people can commune with the spiritual world (Emerson 1836). He was influenced in his writing by Eastern religions, which emphasize the importance of natural beauty as an aid to spiritual enlightenment.

John Muir (1838–1914)

Thoreau was both an advocate for nature and an opponent of materialistic society, believing that people needed far fewer possessions than they sought. To prove his point, he lived simply in a shack near Walden Pond, writing up his ideas and experiences in an essay that has had great impact on many generations of students (Thoreau 1863). Thoreau believed that the experience of nature was a necessary counterweight to the overrefining tendencies of civilization. He argued that

> [in] wilderness is the preservation of the world. Every tree sends its fibers forth in search of the Wild. The cities import it at any price. Men plow and sail for it. From the forest and wilderness come the tonics and barks which brace mankind. Our ancestors were savages. The story of Romulus and Remus being suckled by a wolf is not a meaningless fable. The founders of every state which has risen to eminence have drawn their nourishment and vigor from a similar wild source. (Thoreau, quoted in Homan 1991)

The eminent American wilderness advocate John Muir used the transcendental themes of Emerson and Thoreau in his campaigns to

preserve natural areas. In Muir's view, also known as the **Preservationist Ethic**, beautiful natural areas such as forest groves, mountaintops, and waterfalls had great value in fostering religious and spiritual experiences and for emotional refreshment (Muir 1901, 1916).
Muir believed that the aesthetic and spiritual values of nature were
comparable to the monetary values of commercial exploitation and
were often superior to the more tangible materialistic gains obtained
by exploitation (Callicott 1990). It has been argued that the Preservationist Ethic has undemocratic or even aristocratic overtones, since it
emphasizes the needs of the philosophers, poets, artists, and spiritual
seekers who require the aesthetic properties of nature for their development over the needs of the mass of society, who require jobs and
material goods from the natural environment in order to survive.
(Nations 1988; Guha 1989).

In addition to Muir's advocacy for the preservation of nature on
the grounds of human spiritual needs, he was among the first American conservationists to explicitly state that nature had an intrinsic
value in itself, apart from its value to humanity. Muir argued on
Biblical grounds that since God had created species and nature, to destroy them was undoing God's work. In Muir's viewpoint, people
have an equal place with all other species in God's scheme of nature:

> Why should man value himself as more than a small part of the one
> great unit of creation? And what creature of all that the Lord has
> taken the pains to make is not essential to the completeness of that
> unit—the cosmos? The universe would be incomplete without the
> smallest transmicroscopic creature that dwells beyond our conceitful
> eyes and knowledge. (Muir 1916)

Muir also viewed biological communities as composed of species
evolving together and dependent on one another. This ecological–
evolutionary perspective foreshadowed later theories of ecologists
such as Cowles, Forbes, and Clements (McIntosh 1985), as well as
those of the premier American conservationist Aldo Leopold. Biological communities have even been viewed as superorganisms in which
each species plays a vital role. The most recent development of this
argument is the Gaia hypothesis, in which the biological, physical,
and chemical components of the Earth interact to regulate the characteristics of the climate and atmosphere (Lovelock 1988).

An alternative view of nature, known as the **Resource Conservation Ethic**, was developed by Gifford Pinchot, the first director of the
U.S. Forest Service (Nash 1982; Callicott 1990). According to Pinchot, the qualities found in nature could be considered "natural re-

**Gifford Pinchot
(1865–1946)**

sources" for humans (Pinchot 1947). The goal of proper use of natural resources, according to the Resource Conservation Ethic, is "the greatest good of the greatest number [of people] for the longest time." Its first principle is that resources should be fairly distributed among present users and consumers, as well as future consumers. In this principle, we see the origins of modern sustainable development doctrines and attempts by ecological economists to put a monetary value on natural resources (Costanza 1991). As defined by the World Commission on the Environment and Development (1987), "**sustainable development** is development that meets the needs of the present without compromising the ability of future generations to meet their own needs." From the perspective of conservation biology, sustainable development is development that best meets present and future human needs without damaging the environment and biological diversity (Lubchenco et al. 1991).

The second principle of the Resource Conservation Ethic is that resources should be used with efficiency—that is, put to the best possible use and not wasted. The efficiency principle implies that there can be an ordering of uses with some favored over others, or possibly a "multiple use" of resources. Valuing natural beauty and aesthetic experiences can be considered one such use, which in some situations will take precedence over material uses.

Although the Resource Conservation Ethic can be linked to economic analyses that attempt to determine the "best" or most profitable use of the land, such methods involve pure market forces and have a tendency to minimize or even disregard the costs of environmental degradation and also to discount the future value of natural resources. Consequently, Pinchot argued that government bodies are needed to regulate and control natural resources with a long-term perspective to prevent their destruction. The Resource Conservation Ethic came to dominate American thinking in the twentieth century because of its democratic social philosophy and its emphasis on meeting the needs of society (Norton 1991). Government bodies that manage natural resources for multiple use, such as the Bureau of Land Management and the U.S. Forest Service, are the legacy of this Resource Conservationist approach, in contrast to the general preservationist philosophy of groups such as the Wilderness Society and the Sierra Club.

The Resource Conservation Ethic was the philosophy initially embraced by Aldo Leopold in his early years as a government forester. However, Leopold eventually came to believe that the Resource Conservation Ethic was inadequate, since it viewed the land merely as a collection of individual goods that can be used in different ways. Leopold began to consider nature as a landscape organized as a system of interrelated processes:

> Ecology is a new fusion point for all of the sciences.... The emergence of ecology has placed the economic biologist in a peculiar dilemma: with one hand he points out the accumulated findings of his search for utility, or lack of utility, in this or that species; with the other he lifts the veil from a biota so complex, so conditioned by interwoven cooperations and competitions, that no man can say where utility begins or ends. (Leopold 1939a)

**Aldo Leopold
(1886–1948)**

Leopold eventually came to the conclusion that the most important goal of land management is to maintain the health of natural ecosystems and ecological processes. Maintaining these ecological processes will ultimately give greater long-term value to humans than managing natural areas only for particular resources. As a result, he and many others lobbied successfully for certain parts of national forests to be set aside as wilderness areas (Nash 1982). He also considered humans to be part of the ecological community rather than standing apart from nature and exploiting it, as the proponents of the Resource Conservation Ethic argued. Despite Leopold's philosophical shift, he remained committed to the idea that humans should be involved in land management while avoiding both the dangers of over-exploitation and the inactivity of preservation. Leopold's synthesis has been termed the **Evolutionary–Ecological Land Ethic** (Callicott 1989, 1990). In his writings and his own practice at his family farm, Leopold advocated a land use policy in which human use of natural resources was compatible with, or even enhanced, biological diversity (Leopold 1939b, 1949). He believed that developing woodlots, fields, and ponds could in many cases create an environment more complex and biologically richer than a completely natural environment. Leopold's vision of an integration and complementation of nature and human activities is in line with modern research suggesting that humans have been an integral part of most ecosystems of the world for thousands of years, even in the tropical rain forests of the world (Gomez-Pompa and Kaus 1988, 1992; Dufour 1990). Integrating human activity into preservationist philosophy also makes practical sense since completely eliminating human impact on natural reserves has always been very difficult, and is now becoming impossible due to increasing human populations, air pollution, and global climate change.

Development of these philosophies has taken place alongside the growth of the many American conservation organizations, such as the Wilderness Society, the Audubon Society, Ducks Unlimited, and the Sierra Club, the development of national and provincial park systems, and the passing of numerous environmental laws (Nash 1982; Norton 1991). Elements of these philosophies are still present in contemporary writings, conservation organizations, and government policy (Norton 1991). Many government departments charged with land management continue to follow the Resource Conservation Ethic; established and new conservation organizations still tend to follow the Preservation Ethic; and many academic conservation biologists advocate either the Evolutionary–Ecological Land Ethic or the Preservation Ethic. Disagreements over policy and practice among conservation organizations and among individual conservationists continue to reflect these long-term philosophical differences (Nash 1982). Bitter differences of opinion can occur even within organizations. The continuing dialogue over elements of conservation philosophy and ethics is useful in deciding how to balance the long-term needs of protecting biological diversity with the more immediate needs of modern society for natural resources. However, Norton (1991) has alerted the environmental community on the need to put aside these differences in order to work together to protect as much of the natural world as possible.

Within the American conservation movement there has also been a series of writers warning in a prophetic fashion about the rising destruction of biological diversity and the natural environment. These authors have found a receptive audience among the general public, and have galvanized citizens by the millions to join efforts to protect birds and other wildlife, to protect mountains, seashores, wetlands, and other habitats, and to stop environmental pollution. One early author was G. P. Marsh (1864), with his *Man and Nature, Or, Physical Geography as Modified by Human Action*, followed by Fairfield Osborn's *Our Plundered Planet* (1948). In more recent years, four influential books documenting environmental destruction are *Silent Spring* by Rachel Carson (1962), *The Population Bomb* by Paul R. and Anne H. Ehrlich (1968), *The Closing Circle* by Barry Commoner (1971), and *The Arrogance of Humanism* by David Ehrenfeld (1975). Two recent books in this vein are *Earth in the Balance: Ecology and the Human Spirit* (1992) by U.S. Vice-President Albert Gore and *The Diversity of Life* by E. O. Wilson (1992).

The Current Status of Conservation Biology

By the early 1970s, scientists were aware of the impending biological diversity crisis, but there was no central forum or organization to deal with the issue. The growing number of people thinking about and conducting research on conservation issues needed to communicate with one another to develop new ideas and approaches. To discuss their common interests, ecologist Michael Soulé organized the First International Conference on Conservation Biology in 1978. It was a meeting that brought together wildlife conservationists, zoo managers, and academics at the San Diego Wild Animal Park (Jacobson 1990; Gibbons 1992). At that meeting, Soulé proposed a new interdisciplinary approach that could help save plants and animals from a wave of mass extinctions caused by humans. Following that meeting, Soulé, along with colleagues including Paul Ehrlich of Stanford University and Jared Diamond of the University of California at Los Angeles, began to develop conservation biology as a discipline that would combine the practical experience of wildlife management, forestry, and fisheries biology with the theories of population biology and biogeography to develop new approaches and methods for preserving species.

In 1985, this core of scientists founded the Society for Conservation Biology, which has become one of the fastest-growing societies in biology. This society has already enrolled 3600 professional members, approaching in size the 6500-member Ecological Society of America, founded over 70 years ago. There are other indicators of the growing success of conservation biology. First, the U.S. National Science Foundation has established a special program to support research in conservation biology and restoration ecology, and is funding conservation research through its regular panels as well. Second, major foundations have designated conservation biology as a primary target for funding. For example, the MacArthur Foundation spent roughly $17 million in 1992 on the conservation of biological diversity, a dramatic increase over previous years. Hundreds of millions of dollars per year may be spent on biological diversity by the new Global Environment Facility established recently by the United Nations and the World Bank. Third, at least 16 American universities have established graduate programs in conservation biology in the last three years alone (Jacobson 1990). This development is being driven by the interests of students, by the changing research interests of professors, and by foundations interested in supporting new programs. In 1992, the Pew Charitable Trusts spent $15.5 million on both biological diversity research and graduate programs to train students in "conservation and sustainable development." Finally, the journal *Conservation Biology* has emerged as an original and highly readable academic journal, further stimulating interest in the subject.

By these objective criteria—a rapidly expanding professional society, increased funding, active graduate programs, and a lively journal—conservation biology can certainly be judged as a relevant and popular new field. Ultimately, however, conservation biology must be judged by its ability to preserve biological diversity. When conservation biologists can confidently point to successful examples of species preserved using approaches developed by conservation biology, and not simply by its parent disciplines of wildlife management, forestry, fisheries biology, and ecology, then the practitioners of this new field will consider conservation biology to be a success.

Statement of Ethical Principles

Conservation biology rests on certain underlying assumptions that are generally agreed upon by members of the discipline (Soulé 1985). These statements cannot be proved or disproved, and accepting all of them is not a requirement for conservation biologists. Nonetheless, these assumptions represent a set of ethical and ideological statements that form the basis of the discipline and suggest research approaches and practical applications. As long as one or two are accepted, there is a rationale for conservation efforts.

1. *The diversity of organisms is good.* In general, humans enjoy seeing biological diversity. The hundreds of millions of visitors each year to zoos, national parks, botanical gardens, and aquaria are testament to the general public's interest in biological diversity. Genetic variation within species also has popular appeal, as shown by dog shows, cat shows, agricultural expositions, flower exhibitions, and large numbers of specialty clubs (African violet societies, rose societies, etc.) At a local level, home gardeners pride themselves on how many types of plants they have in their gardens, while birdwatchers compete to see how many species they can see in one day or in their lifetimes. It has even been speculated that humans may have a genetic predisposition to like biological diversity, called *biophilia* (Orians 1980; Wilson 1984). At the earliest stages of human society, increased biological diversity would have been advantageous for the hunting-and-gathering lifestyle that humans had for hundreds of thousands of years before the invention of agriculture. Greater biological diversity would have provided a greater variety of foods and other resources, and also buffered humans against environmental catastrophes and starvation.

2. *The untimely extinction of populations and species is bad.* The extinction of species and populations as a result of natural processes is a neutral event. Through the millennia of geological time, the natural extinction of certain species has tended to be balanced by the evolution of new species. The local loss of a

population of a species likewise is usually offset by the establishment of a new population through dispersal. However, as a result of human activity the rate of extinction has increased a thousandfold. Virtually all of the hundreds of extinctions of vertebrate species, and the presumed extinctions of thousands of invertebrate species, in the last century have been caused by humans.

3. *Ecological complexity is good.* Many of the most interesting properties of biological diversity are only expressed in natural environments. For example, complex co-evolutionary and ecological relationships exist among tropical flowers, hummingbirds that visit the flowers to drink nectar, and the mites that live in the flowers and use the hummingbirds' beaks as a "bus" to go from flower to flower (Colwell 1973, 1986). These relationships would never be suspected if the animals and plants were housed separately and in isolation at zoos and botanical gardens. While the biological diversity of species may be preserved in zoos and gardens, the ecological complexity that exists in natural communities will be largely lost.

4. *Evolution is good.* Evolutionary adaptation is the process that eventually leads to new species and increased biological diversity. Therefore, allowing populations to continue to evolve in nature is good. Human processes that limit or even destroy the ability of populations to evolve, such as severe reductions in population size, are bad. Preserving species in captivity when they are no longer able to survive in the wild is important, but the species is then cut off from the natural evolutionary process. In such cases the species may no longer be able to survive in the wild if it is released.

5. *Biological diversity has intrinsic value.* Species have a value all their own, regardless of their material value to human society. This value is in part conferred by their evolutionary history and unique ecological role, and also by their very existence. This biological valuation stands in contrast to an economic viewpoint that assigns monetary value to each species on the basis of the goods and services that it provides or could potentially provide.

Summary

1. Thousands of species are going extinct as a result of human activites. Conservation biology is a new synthetic discipline to deal with this unprecedented crisis. Conservation biologists combine basic and applied research ap-

proaches to prevent the extinction of species, the loss of genetic variation, and the destruction of biological communities.

2. Conservation biology draws on religous and philosophical traditions. European scientists in the eighteenth and nineteenth centuries reacted to the destruction of forests and water pollution in the colonies by proposing some of the first environmental legislation. The decline and extinction of species in Europe led to the establishment of the first nature reserves and an active popular interest in conservation. In the United States, Ralph Waldo Emerson, Henry David Thoreau, and John Muir argued for the preservation of wilderness and the intrinsic value of species; Gifford Pinchot proposed developing a balance among competing natural resource needs for present and future society; and Aldo Leopold advocated managing land to maintain ecological processes instead of just to satisfy human needs.

3. Conservation biology rests on a number of underlying assumptions that are accepted by most conservation biologists: The diversity of species is good; the extinction of species by human activities is bad; the complex interaction of species in natural communities is good; the evolution of new species is good; and biological diversity has value in and of itself.

Suggested Readings

Callicott, J. B. 1990. Whither conservation ethics? *Conservation Biology* 4: 15–20. A summary of major conservation themes.

Fiedler, P. L. and S. K. Jain (eds). 1992. *Conservation Biology: The Theory and Practice of Nature Conservation, Preservation, and Management.* Chapman and Hall, New York. A current look at selected topics.

Nash, R. 1982. *Wilderness and the American Mind.* Yale University Press, New Haven. An excellent review of the history and people of the environmental movement in the United States.

Norton, B. 1991. *Toward Unity among Environmentalists.* Oxford University Press, New York. An overview of the modern environmental movement, and strategies for accomplishing goals.

Soulé, M. 1985. What is conservation biology? *BioScience* 35: 727–734. Classic statement of the subject.

Soulé, M. (ed.) 1986. *Conservation Biology: The Science of Scarcity and Diversity.* Sinauer Associates, Sunderland, MA. A collection of essays, with a strong emphasis on sensitive habitats and the genetics of rare species.

Soulé, M. and B. A. Wilcox (eds.) 1980. *Conservation Biology: An Evolutionary-Ecological Perspective.* Sinauer Associates, Sunderland, MA. The first detailed book on conservation biology, with original papers by leading authors in the field.

Wilson, E. O. 1992. *The Diversity of Life.* Belknap Press of Harvard University Press, Cambridge, MA. An outstanding description of biological diversity, written for the general public.

Wilson, E. O., and F. M. Peter, eds. 1988. *Biodiversity.* National Academy Press, Washington, D.C. Short essays by leading authorities.

CHAPTER 2

What Is Biological Diversity?

Although protection of biological diversity is central to conservation biology, the phrase "biological diversity" has different meanings to different people. The definition given by the U.S. Office of Technology Assessment (1987) is "the variety and variability among living organisms and the ecological complexes in which they occur." This concept can be subdivided into three levels. Biological diversity at its most basic level includes the full range of species on Earth, from species such as bacteria, viruses, and protists through the multicellular kingdoms of plants, animals, and fungi. At finer levels of organization, biological diversity includes the genetic variation within species, both among geographically separated populations and among individuals within single populations. On a wider scale, biological diversity includes variations in the biological communities in which species live, the ecosystems in which communities exist, and the interactions among these levels (Figure 2.1).

All levels of biological diversity are necessary for the continued survival of species and natural communities, and all are important for the well-being of humans. **Genetic diversity** is needed by any species in order to maintain reproductive vitality, resistance to disease, and the ability to adapt to changing conditions. Genetic diversity within domestic plants and animals is of particular value in the breeding programs necessary to sustain modern agricultural species. **Species diversity** represents the range of evolutionary and ecological adaptations of species to particular environments. The diversity of species provides people with resources and resource alternatives; for example, a tropical rain forest with many species produces a wide variety of plant and animal products that can be used as food, shelter, and medicine. **Community-level diversity** represents the collective response of species to different environmental conditions. Biological

Genetic diversity in a rabbit population

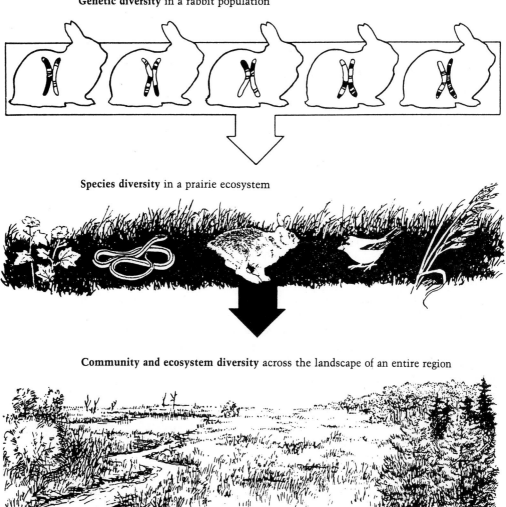

Species diversity in a prairie ecosystem

Community and ecosystem diversity across the landscape of an entire region

2.1 Biological diversity includes genetic diversity (the genetic variation found within each species), species diversity (the range of species in a given ecosystem), and community/ecosystem diversity (the variety of habitat types and ecosystem processes extending over a given region. (From Temple 1991; drawing by T. Sayre.)

communities such as deserts, grasslands, wetlands, and forests support the continuity of proper ecosystem functioning, providing beneficial services to people—services such as flood control, protection from soil erosion, and air and water filtering (Woodwell 1990; Odum 1993).

Biological diversity includes the entire range of species found on Earth. A **species** is generally defined in one of two ways:

1. A group of individuals that is morphologically, physiologically, or biochemically distinct from other groups in some characteristic (the morphological definition of species).
2. A group of individuals that can potentially breed among themselves and do not breed with individuals of other groups (the biological definition of species).

The morphological definition of species is the one most commonly used by **taxonomists**, biologists who specialize in the identification of unknown specimens and the classification of species (Box 1). The biological definition of a species is the one most commonly used by evolutionary biologists because it is based on measurable genetic relationships rather than on somewhat subjective physical features. In practice, however, the biological definition of a species is hard to use because it requires knowledge of which individuals are actually capable of breeding among themselves—information that is rarely available.

BOX 1 THE NAMES OF SPECIES AND THEIR CLASSIFICATION

Biologists throughout the world have agreed to use a standard set of names, often called scientific names, when discussing species. These names are often, but not always, derived from Latin or Greek. The use of scientific names avoids the confusion that can occur when using the common names found in everyday language. For example, several different plants are called bluebells in England and North America; if a biologist describes research on "bluebells," scientists elsewhere will not know which plant species was actually studied unless the biologist uses the scientific name. Conversely, the same species may be called by different names in different places; for example, the species *Daucus carota* is called wild carrot, Queen Anne's lace, bird's-nest plant, laceplant, parsnip, and rantipole in North America alone, and it has numerous other local names in Europe and Asia. Only the Latin name is standard across countries and languages.

Scientific species names essentially consist of two words. This system, known as **binomial nomenclature**, was developed in the eighteenth century by the Swedish biologist Carolus Linnaeus. In the scientific name for the Blackburnian warbler, *Dendroica fusca*, *Dendroica* is the genus name and *fusca* is the specific name. The genus name is somewhat like a person's family name in that many people can have the same family name (Sullivan), while the specific name is like a person's given name (Jonathan). The name *Dendroica fusca* refers to just one species of warbler; there are 27 other species of warbler that are similar enough to *Dendroica fusca*

Kingdom: Animalia

>1,000,000 species

Phylum: Chordata

±40,000 species

Class: Aves (birds)

8,600 species

Order: Passeriformes
        (songbirds)
5,160 species

Family: Parulidae
        (wood warblers)
125 species

Genus: *Dendroica*

28 species

Species: *Dendroica fusca*

Blackburnian warbler

Less specific

More specific

Blackburnian warblers (*Dendroica fusca*) are related to more and more other animals at successively higher levels of taxonomic organization.

to be included in the genus *Dendroica*.

Scientific names are written in a standard way to avoid confusion. The first letter of the genus name is always capitalized, whereas the species name is almost always lowercase. Scientific names are either italicized or underlined. Sometimes scientific names are followed by a person's name, as in *Homo sapiens* Linnaeus, indicating that Linnaeus was the person who first proposed the scientific name given to the human species. When many species in a single genus are being discusssed, or if the identity of a species within a genus is uncertain, the abbreviations spp. or sp., respectively, are sometimes used (e.g., *Dendroica* spp.). If a species has no close relatives, it may be the only species in its own genus. Similarly, a genus that is unrelated to any other genera may form its own family.

Taxonomy is the science of classifying living things. The goal of modern taxonomy is to create a system of classification that reflects the evolution of groups of species from their ancestors. The closely related field of **systematics** is the study of the diversity of the living world. Species that share the most recent common ancestor or are most similar to each other in their characteristics are grouped closest together. By grouping together similar species, identification of unknown species is usually made easier. In modern classification,

Similar *species* are grouped into a *genus*.

Similar *genera* are grouped into a *family*.

Similar *families* are grouped into an *order*.

Similar *orders* are grouped into a *class*.

Similar *classes* are grouped into a *phylum*.

Similar *phyla* are grouped into a *kingdom*.

Most modern biologists recognize five kingdoms in the living world: plants, animals, fungi, monerans (single-celled species without a nucleus and mitochondria, such as bacteria), and protists (more complex single-celled species with a nucleus and mitochondria).

Problems in distinguishing and identifying species are more common than many people realize (Rojas 1992; Standley 1992). Differences between species are not always as clear-cut as either of the definitions would imply. For example, a single species may have several varieties that have observable morphological differences, yet the varieties are similar enough that they are still considered to be a single biological species (Figure 2.2). Different breeds of dogs, such as German shepherds, collies, dachshunds, and beagles, all belong to one species and readily interbreed despite the conspicuous differences among them. Alternatively, there are closely related "sibling" species that appear very similar in morphology or physiology, yet are still considered biologically separate and do not interbreed. In practice, biologists often have difficulty distinguishing variation within a single species from variation between closely related species. Further complicating matters is the fact that what are otherwise distinct species may occasionally hybridize, producing intermediate forms that blur the distinction between species. Such hybridization is particularly common among plant species in disturbed habitats. Finally, for many groups of species, the taxonomic studies needed to determine species and identify specimens have not yet been done. The inability to clearly distinguish one species from another, whether due to similarities of characteristics or to confusion over the correct scientific name, often slows down efforts at species protection (Standley 1992).

The Origin of New Species

The biochemical similarity of all living species indicates that life on Earth probably originated only once, about 3.5 billion years ago. From one original species came the millions of species found on

2.2 The rat snake *Elaphe obsoleta* is a variable species that has many distinct subspecies. Each subspecies has its own geographical range and particular patterns of color and banding. The subspecies interbreed where their ranges overlap. (Redrawn from Conant 1958.)

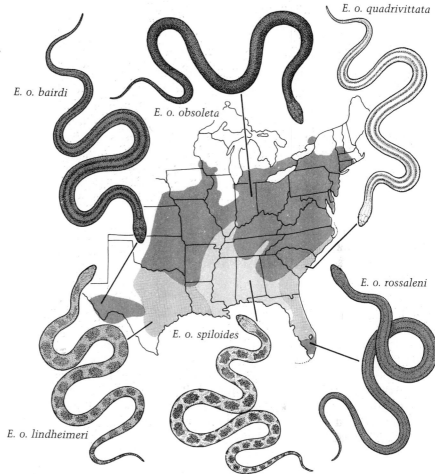

Earth today. The process of new species formation, known as **speciation**, is continuing today and will most likely continue in the future.

The process whereby one original species evolved into new, distinct species was first described by Charles Darwin and Alfred Russel Wallace more than 100 years ago. This theory of the origin of new species is widely accepted today in the scientific community. Darwin provided abundant and convincing evidence for his theory in his classic book, *On the Origin of Species* (1859). The wealth of new information provided by the fossil record and by modern molecular biology research has continued to support this theory.

The theory of evolution is both simple and elegant. Imagine a population of a species—mountain rabbits, for example—living in Canada. Individuals in the population tend to produce more offspring than can survive in that place. Most offspring will die before reaching maturity. In the population of rabbits, one pair of rabbits will produce numerous litters of six or more offspring, yet on average, in a stable population, only two of those offspring will survive. Individuals in the population show variations in certain characteristics, and some of these characteristics are inherited—they are passed genetically from parents to offspring. These genetic variations are caused by spontaneous changes in the chromosomes and by the rearrangement of chromosomes during sexual reproduction. Within the rabbit population, some individuals have thicker fur than others because of genetic differences. These differences in genetic characteristics will enable some individuals to grow, survive, and reproduce better than other individuals, an idea that is often referred to as "survival of the fittest." Our hypothetical thick-furred rabbits will be more likely to survive cold winters than rabbits with thinner fur. As a result of the improved survival ability provided by a certain genetic characteristic, the individuals with that characteristic will be more likely to produce offspring than the others; over time, the genetic composition of the population will change. After a series of cold winters, more thick-furred rabbits will have survived and produced thick-furred offspring, while more thin-furred rabbits will have died. More rabbits in the population now have thicker fur than in previous generations.

In the process of evolution, populations are constantly adapting genetically to changes in their environment. These changes may be biological (new food sources, new competitors, new prey) as well as environmental (changes in climate, water availability, soil characteristics). When a population has undergone so much genetic change that it is no longer able to interbreed with the original species from which it was derived, the population can be considered to be a new species; this process whereby one species is gradually transformed to another species is termed **phyletic evolution**.

In order for two or more new species to evolve from one original ancestor, there usually has to be a geographical barrier that prevents the movement of individuals between the various populations of a species. For terrestrial species, these barriers may be rivers, mountain ranges, or oceans that the species cannot readily cross. Speciation is particularly rapid on islands. Island groups such as the Galápagos and the Hawaiian Islands have many examples of insect and plant groups that were originally local populations of a single invading species. These local populations adapted genetically to the environments of particular unoccupied islands, mountains, and isolated valleys, and have diverged sufficiently from the original species to now be consid-

2.3 One of the most spectacular examples of adaptive radiation is displayed by the honey-creeper family. This family of birds, endemic to the Hawaiian islands, is thought to have arisen from a single pair of birds that arrived on the islands by chance. The shape and size of the bills are related to the foods eaten by each species: long bills are for feeding on nectar; short, thick bills are for cracking seeds; and short, sharp bills are for eating insects. (From Futuyma 1986.)

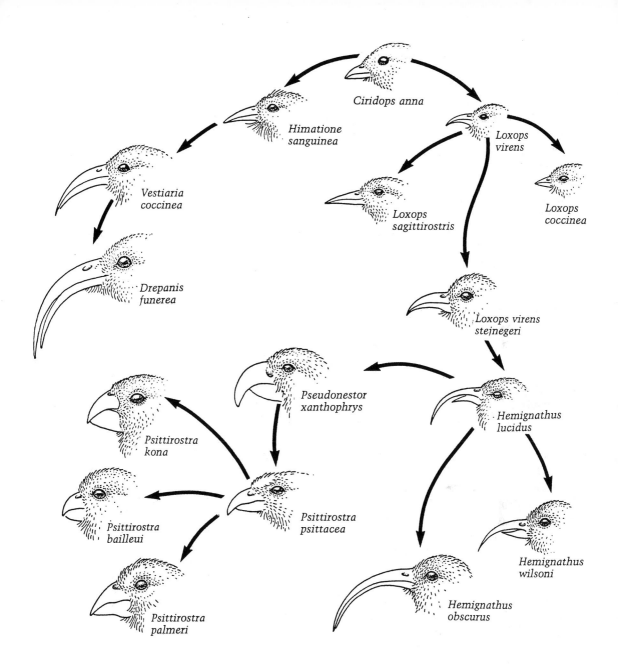

ered separate species. This process of local adaptation and subsequent speciation is known as **adaptive radiation**. One of the best-known examples of adaptive radiation is that of the Hawaiian honeycreepers, a group of specialized bird species that apparently derives from a single pair of birds that arrived by chance in the Hawaiian Islands tens of thousands of years ago (Figure 2.3; Scott et al. 1988).

29

The origin of new species is normally a slow process, taking place over hundreds, if not thousands, of generations. The evolution of new genera and families is an even slower process, lasting hundreds of thousands or even millions of years. However, there are mechanisms whereby new species can arise in just one generation. Particularly in plants, there may be unusual, unequal divisions of chromosome sets at the time of reproduction that result in offspring with extra sets of chromosomes; these offspring are known as **polyploids**. Polyploid individuals may be morphologically and physiologically different from their parents and, if they are well suited to the environment, may form a new species.

Even though new species are arising all the time, the present rate of species extinction is probably more than 1000 times faster than the rate of speciation. The situation is actually worse than this grim statistic suggests. First, the rate of speciation may actually be slowing down because so much of the Earth's surface has been taken over for human use and no longer supports evolving biological communities. There are now fewer populations of each species and thus fewer opportunities for evolution. Many of the existing protected areas and national parks may be too small in area to allow the process of speciation to occur (Figure 2.4). Second, many of the species threatened

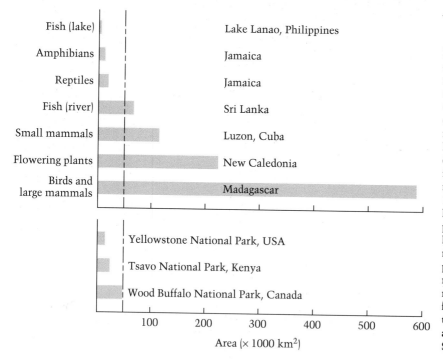

2.4   Certain groups of organisms apparently need a minimum area in order to undergo the process of speciation. For example, for small mammals, the smallest islands (Cuba and Luzon) on which a single species is known to have given rise to two species are 100,000 km². Even the largest national parks are probably too small to allow for the evolution of new species of flowering plants, birds, or mammals, although they might be large enough for the continued evolution of fishes, amphibians, and reptiles. (After Soulé 1980.)

with extinction are the sole remaining representatives of their genus or family; examples include the coelocanth fish of the Indian Ocean and the African cheetah. The extinction of these taxonomically unique species representing ancient lineages is not balanced by the origin of new species.

Genetic Diversity

At each level of biological diversity—genetic, species, and community—biologists study the mechanisms that alter or maintain diversity. Genetic diversity within a species is often affected by the reproductive behavior of individuals within populations. A **population** is a group of individuals that mate with one another and produce offspring; a species may include one or more separate populations. A population may consist of only a few individuals or millions of individuals, provided that the individuals actually produce offspring. A single individual of a sexual species would not constitute a population. Neither does a group of individuals that cannot reproduce constitute a population; for example, a group of 10 dusky seaside sparrows would not constitute a population if all of them were males.

Individuals within a population are often genetically different from one another. This genetic variation arises because individuals have slightly different forms of their **genes**, the units of the chromosomes that code for specific proteins (Tamarin 1993). These slightly different forms of a gene are known as **alleles**, and the differences can arise through **mutations**—changes that occur in the deoxyribonucleic acid (DNA) that constitute an individual's chromosomes. The various alleles of a gene may produce forms of a protein that differ in structure and function, in turn affecting the development and physiology of the individual organism.

Genetic variation increases when offspring receive unique combinations of genes and chromosomes from their parents via the **recombination** of genes that occurs during sexual reproduction. Genes are exchanged between chromosomes and rearranged during the crossing over that occurs during meiotic cell division, and new combinations are created when gametes from two parents unite to form a genetically unique offspring. Although mutations provide the basic material for genetic variation, the ability of sexually reproducing species to randomly rearrange alleles in different combinations dramatically increases the potential for genetic variation.

The total array of genes and alleles in a population is referred to as the **gene pool** of the population, while the particular combination of alleles that any individual possesses is referred to as its **genotype**. The **phenotype** of an individual represents the morphological, physio-

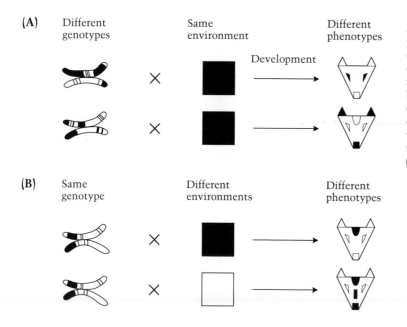

(A)

Different genotypes   Same environment   Different phenotypes

Development

×

×

2.5 The physical, physiological, and biochemical characteristics of an individual—its phenotype—are determined by its genotype and the environment (e.g., hot vs. cold climate; abundant vs. scarce food) in which the individual lives. (After Alcock 1993.)

(B)

Same genotype   Different environments   Different phenotypes

×

×

logical, anatomical, and biochemical characteristics of the individual that result from the expression of its genotype in a particular environment (Figure 2.5). Some characteristics of humans, such as the amount of body fat and tooth decay, are strikingly influenced by the environment, while other characteristics, such as eye color, blood type, and forms of certain enzymes are determined predominantly by an individual's genotype.

Sometimes individuals that differ genetically also differ in ways related to their survival or ability to reproduce, such as their ability to tolerate cold, their resistance to disease, or the speed at which they can run away from danger. If certain individuals are better able to survive and leave more offspring because of the alleles they possess, then **gene frequencies** in the population will change with subsequent generations. This phenomenon is called **natural selection**.

The amount of genetic variability in a population is determined by both the number of genes that have more than one allele (referred to as **polymorphic genes**) and the number of alleles for each polymorphic gene. The existence of a polymorphic gene allows individuals in the population to be **heterozygous** for the gene, that is, to receive two different alleles of the gene from their two parents. In a wide variety of plant and animal populations it has been demonstrated that heterozygous individuals have higher fitness than homozygous individuals: heterozygous individuals tend to grow, survive, and reproduce more successfully (Allendorf and Leary 1986; Ledig 1986). The mechanism for this appears to be that: (1) having two different forms of, say, an enzyme gives the individual greater flexibility in dealing with life's challenges; and (2) nonfunctional or harmful alleles received from one parent are masked by the functioning alleles received from the other parent. This phenomenon of increased fitness in highly heterozygous individuals, also referred to as **hybrid vigor**, is widely known in domestic animals (Falconer 1981).

Thus, genetic variability within a population can be measured as:

1. The number (and percentage) of genes in the population that are polymorphic (have more than one allele).
2. The number of alleles for each polymorphic gene.
3. The number (and percentage) of genes per individual that are polymorphic.

Genetic variability can affect diversity at the species level. A species consists of one or more populations, and although most mating occurs within populations, individuals occasionally move from one population to another, allowing the transfer of new alleles and genetic combinations between populations. This genetic transfer is referred to as **gene flow**. Variability and species diversity will be discussed at length in Chapters 11 and 12.

Genetic variation also occurs within domesticated plants and animals. In traditional societies, people preserved any new forms they found that were well-suited to their needs. Through generations of this process of **artificial selection** by humans, varieties of species were developed that were productive and useful to local conditions of soil, climate, and crop pests. This process has continued in modern agriculture with scientific breeding programs that manipulate genetic variation to meet human needs. Thousands of varieties of widespread crops, such as rice, potatoes, and corn, have been incorporated into the breeding programs of modern agriculture. Among animals, the huge number of breeds of domestic dogs, cats, rabbits, chickens, cattle, and pigs are evidence of the potential of artificial selection to alter gene pools; Figure 2.6 shows an example of some results of artificial selection. Genetic variation is also maintained in specialized collections of species used in scientific research, such as *Drosophila* fly stocks used in genetic research, tiny *Arabidopsis* plants used in plant science research, and mice used in physiological and medical research.

2.6   Pigeon hobbyists have bred and preserved unusual forms of domestic pigeons, all originally descended from the rock dove. (Illustration from Charles Darwin's *The Variation of Animals and Plants under Domestication*, 1866.)

Communities and Ecosystems

A **biological community** is defined by the species that occupy a particular locality and the interactions between those species. A biological community together with its associated physical environment is termed an **ecosystem**. Within an ecosystem, water evaporates from biological communities and the Earth's surface, to fall again as rain or snow and replenish terrestrial and aquatic environments. Soil is built up out of parent rock material and decaying organic matter. Photosynthetic plants absorb light energy, and this energy is released back into the atmosphere as heat during the lives of plants and animals, as well as after they die and decompose. Plants absorb carbon dioxide and release oxygen during respiration, while animals and fungi absorb oxygen and release carbon dioxide. Mineral nutrients, such as nitrogen and phosphorus, cycle between the living and the non-living compartments of the ecosystem.

The physical environment affects the structure and characteristics of a biological community; but the biological community can also have effects on the physical characteristics of the ecosystem. In a terrestrial ecosystem, for example, wind speed, humidity, and temperature in a given location can all be affected by the plants and animals present there. In aquatic ecosystems, such characteristics as water turbulence and clarity, water chemistry, and water depth affect the characteristics of the biological communities, but in turn biological communities such as kelp forests and coral reefs can affect the physical environment.

Within a biological community each species utilizes a unique set of resources that constitute its **niche**. The niche for a plant species might consist of the type of soil on which it is found, the amount of sunlight it receives, the amount of soil moisture it requires, the type of pollination system it has, and its mechanism of seed dispersal. The niche for an animal might include the type of habitat it occupies, the types of food it eats over the year, and its requirements for overwintering dens. The niche of each species is the space where that species exists in nature and is not outcompeted by other species.

The niche of a species often includes the stage in succession that the species occupies; **succession** is the gradual process of change in species composition, community structure, and physical characteristics that occur following natural and human-caused disturbance to a biological community. Particular species are often found at particular successional stages. For example, sun-loving butterflies and annual plants are found most commonly early in succession, in the months immediately after a hurricane or a logging operation has destroyed an old-growth forest. At this stage, the ground is receiving high levels of sunlight, with high temperatures and low humidity during the day. Over the course of years and decades, the forest is gradually reestablished. Different species, including shade-tolerant wildflowers and birds that nest in holes in dead trees, are found in these late-successional stages of mature trees and old-growth forest. Similar patterns of succession-stage specific organisms are found in other biological communities, such as grasslands, lakeshores, and the intertidal zones of oceans.

The composition of communities is often strongly affected by competition and predation (Price 1992; Ricklefs 1992; Terborgh 1992). Predators often dramatically reduce the numbers of populations of prey species and may eliminate some prey species from certain habitats. Predators may indirectly increase biological diversity in a community by keeping the densities of other species so low that competition for resources does not occur. A good example of this is the marine intertidal ecosystem in which a large sea star (starfish), *Pisaster*, feeds on 15 species of mollusks that cling to the rocks (Paine 1966). As long as the predatory sea star is present, the mollusk species do not compete with each other for space on the rocks, since they are eaten too fast to allow them to achieve high population densities. Under these circumstances, all 15 species are able to occupy the niche. If the sea star is removed, however, the mollusks increase in abundance and start competing for space on the rocks. In the absence of predation, competition between species reduces the number of species; eventually only a few of the original 15 species remain, with some rocks taken over by just one species. In plant communities as well, species diversity is often higher when intense grazing by animals prevents competition among plants than when grazers are absent (Harper 1977).

The number of individuals of a particular species that the re-

BOX 2 KELP FORESTS AND SEA OTTERS: SHAPING AN OCEAN ECOSYSTEM

Although the effects of human activities on the world's tropical and temperate forests have been given a lot of media attention in recent years, a third kind of forest has received very little notice in the popular press. Marine kelp forests, though unsung in magazines and newspapers, may be in the beginning stages of an equally alarming degradation. Kelp forests are communities that develop around any of a number of species of marine brown algae, such as the southern bull kelp (*Durvillaea antarctica*) and the giant kelp (*Macrocystis pyrifera*). These forests, found in the coastal waters of the world's oceans, supply food and shelter for enormous numbers of ocean fish, shellfish,

Forests of giant kelp provide the starting point and structure for a diverse biological community off the Pacific coast of North America. (Photograph by P. Cotter.)

and invertebrates (Estes et al. 1989). Like terrestrial forests, kelp and seaweed communities inhibit soil erosion: the presence of kelp reduces the impact of waves on the shoreline, preventing the destruction of coastal land (Lindstrom 1989).

In recent years, kelp deforestation has increased in many parts of the world, with a corresponding impact on commercial fish and shellfish. A number of different factors have contributed to the changes seen in communities of southern bull kelp, giant kelp, and other species of seaweed worldwide. One of these is the increased demand for kelp by human consumers and the food processing industry, which has resulted in the use of large-scale harvesting methods that limit or reduce the recovery of the kelp forest (Lindstrom 1989). However, an equally important but perhaps less obvious cause is the local extinction of predators, particularly the sea otter, that control the populations of animals that feed on kelp (Estes et al. 1989).

Sea otters, once widespread throughout the Pacific, were all but exterminated by fur traders. Confined mostly to the far northern Pacific islands for decades, the sea otter is now protected in the United States, and it has begun to re-colonize parts of its former range (Booth 1988; Estes et al. 1989). Sea otters prey on several species of sea urchins; when not controlled by predation, the sea urchins feed heavily on kelp and contribute to the decline of kelp forests. As the otters control the sea urchin populations, the kelp forests flourish, indirectly encouraging population increases among many commercial fish and shellfish. In some instances, fish species are 10 to 100 times more common in the presence of kelp than without it (Estes et al. 1989). Wherever otters have returned or have been reintroduced, significant changes have taken place in kelp communities: formerly deforested areas are again dominated by kelp within one to two years in the presence of otters.

It is ironic that many commercial fishermen resent sea otters—and indeed have sometimes even killed them—for eating commercial shellfish such as abalone, because otters are vital to the shellfish industry; without them, the kelp beds that support the shellfish and many other marine animals fall prey to sea urchins.

Sea otters are vital to the kelp community. They feed on invertebrates like sea urchins that in turn graze on kelp; without the otters to control the invertebrate populations, kelp forests decline because of overgrazing. (Photograph © NYZS/The Wildlife Conservation Society.)

sources of an environment can support is termed the **carrying capacity**. A population's numbers are often well below the carrying capacity when it is held in check by a predator population. If the predator is removed, the population may increase to a point where it reaches the carrying capacity, or may even increase beyond the carrying capacity to a point at which the environment is damaged (Box 2) and the population crashes.

The population size of a species may also be controlled by other species that compete with it for the same resources; for example, two heron species may eat the same type of fish. Population size may decline if a competing species becomes abundant, or may increase if the competitor becomes scarce or is eliminated from the community.

Community composition is also affected by **mutualistic relationships**, in which two species benefit each other. Mutualistic species reach higher densities when they occur together than when only one of the species is present. Common examples of such mutualisms are fruit-eating birds and plants with fleshy fruits; flower-pollinating insects and flowering plants; the fungi and algae that together form lichens; and corals and the algae that live inside them (Howe 1984; Bawa 1990). Another example involves plants that produce special chambers that are inhabited by ants. The ants protect the plants, and the ants' wastes are used by the plants as a source of nutrients; in turn, the ants have a secure home to occupy (Figure 2.7). At the extreme of mutualism, species may form a **symbiotic relationship**, in which two species are always found together and apparently cannot survive without each other. For example, the death of certain types of coral-inhabiting algae due to unusually high water temperatures in tropical areas may be followed by the weakening and subsequent death of their associated coral species.

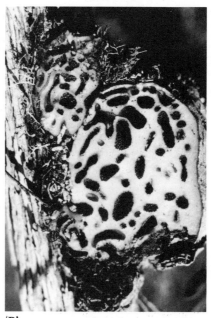

(A)　　　　　　(B)

2.7   A mutualistic relationship. (A) This epiphyte—a plant growing on the surface of another plant—in Borneo produces a tuber at its base that is filled with hollow chambers, as seen in (B). The chambers are occupied by ant colonies, which use some chambers as nesting sites and some as "dumps" for wastes and dead ants. The plant absorbs the mineral nutrients it needs for growth from these "dumps," while the ants obtain a safe nest. In the epiphyte–tree relationship shown in (A), the epiphyte benefits while the tree it grows on is neither benefitted nor harmed. (Photographs by R. Primack.)

Trophic Levels

Biological communities are organized into feeding levels called **trophic levels** that represent ways in which energy is obtained from the environment (Figure 2.8).

- **Photosynthetic species** (also known as **primary producers**) obtain their energy directly from the sun. In terrestrial environments, higher plants such as flowering plants, gymnosperms, and ferns are responsible for most photosynthesis, while in aquatic environments, seaweeds, single-celled algae, and cyanobacteria (blue-green algae) are the most important. All of these species use solar energy to build the organic molecules they need to live and grow. Only a tiny percentage of the light energy falling on the Earth's surface is actually captured by the photosynthetic process; the great majority of light energy is reflected back into space or absorbed by the air, ground, and water and radiated back as heat. And in fact most of the energy captured by each subsequent trophic level is used up in the basic metabolic processes of keeping the organisms alive, and is released into the environment as heat.

- **Herbivores** (also known as **primary consumers**) eat photosynthetic species. For example, in terrestrial environments gazelles and grasshoppers eat grass, while in aquatic environments crustaceans and fish eat algae. Because much plant material, such as cellulose and lignin, is indigestible to many animal species or is simply not eaten, only a small percentage of the energy captured by photosynthetic species is actually transferred to the herbivore level.

- **Carnivores** (also known as **secondary consumers** or **predators**) eat other animals. Primary carnivores (such as foxes) eat herbivores (such as rabbits), while secondary carnivores (such as bass) eat other carnivores (such as frogs). Since carnivores do not catch all of their potential prey, and since many body parts of the prey are indigestible, again only a small percentage of the energy of the herbivore trophic level is transferred to carnivores. Carnivores usually are predators, though some combine direct predation with scavenging behavior, and others, known as **omnivores**, have a substantial proportion of plant foods in their diets. In general, predators are larger and stronger than the species they prey on, but they usually occur in lower densities than their prey.

  Parasites form an important subclass of predators. Parasites are typically small in size and do not kill their prey immedi-

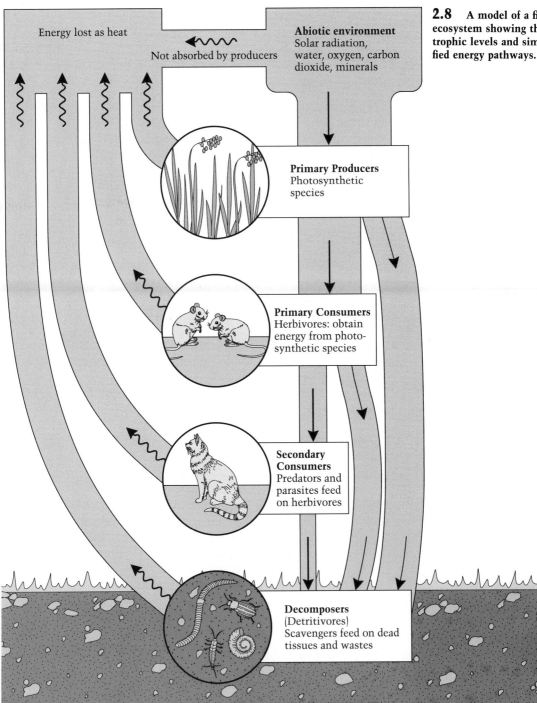

2.8 **A model of a field ecosystem showing the trophic levels and simplified energy pathways.**

Energy lost as heat

Not absorbed by producers

Abiotic environment
Solar radiation, water, oxygen, carbon dioxide, minerals

Primary Producers
Photosynthetic species

Primary Consumers
Herbivores: obtain energy from photosynthetic species

Secondary Consumers
Predators and parasites feed on herbivores

Decomposers
(Detritivores)
Scavengers feed on dead tissues and wastes

ately. The effects of parasites may range from imperceptibly weakening their prey to totally debilitating or even killing their prey over time. Parasites include a wide range of species, such as disease-causing bacteria and protists, intestinal worms, parasitic fungi, beetle larvae that feed on the inner bark of trees, and plants such as mistletoe and dodder that parasitize other plants. Parasites are often important in controlling the density of prey species. When their prey populations are at a high density, parasites can readily spread from one host individual to the next, causing an intense local infestation of the parasite and a subsequent decline in prey density. When prey densities are low, parasites are less able to move from one host to another, and the effects of parasites on the prey population are correspondingly weak.

- **Detritivores** (also known as **decomposers**) are species that feed on dead plant and animal tissues and wastes, breaking down complex tissues and organic molecules. Detritivores release minerals such as nitrogen and phosphorus back into the environment, where they can be taken up again by plants and algae. The most important detritivores are fungi and bacteria, but a wide range of other species play a role in breaking down organic materials; for example, vultures and other scavengers tear apart and feed on dead animals, dung beetles bury and feed on animal dung, and worms break down fallen leaves and other organic matter.

Principles of Community Organization

As a consequence of less energy being transferred to each successive trophic level in biological communities, there tends to be a lower **biomass** (total weight of living material) and fewer individuals at each level. The greatest biomass in an ecosystem will be that of the primary producers. And as a general rule in any community, there will be more individual herbivores than primary carnivores, and more primary carnivores than secondary carnivores (Price 1992). For example, a forest community generally contains more individual insects and insect biomass than insectivorous birds, and more insectivorous birds than raptorial birds (birds such as hawks that feed on other birds).

Although species can be organized into these general trophic levels, their actual requirements or feeding habitats within the trophic levels may be quite restricted (Freeland and Boulton 1992). For example, a certain aphid species may feed only on one type of plant, and a certain lady beetle species may feed on only one type of aphid. These specific feeding relationships have been termed **food chains**. The more common situation in many biological communities, however, is for a species to feed on several items at the lower trophic level, to compete for food with several species at its own trophic level, and in turn to be preyed upon by several species at the next higher trophic level. Consequently, a more accurate description of the organization of biological communities is a **food web** in which species are linked together through complex feeding relationships (Figure 2.9). Species at the same trophic level that use approximately the same resources in the environment are considered to be in a **guild** of competing species.

Species-specific ecological requirements are an important reason for the inability of many species to increase in abundance within a

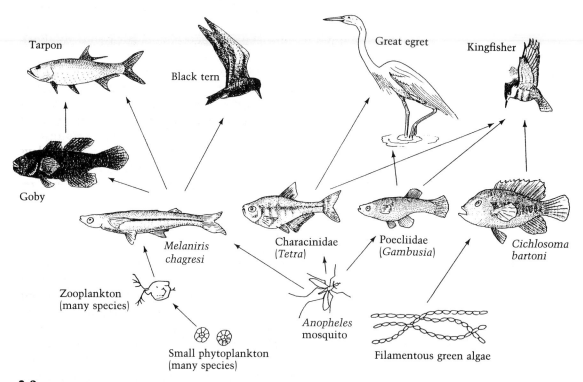

2.9   Diagram of an actual food web studied in Gatun Lake, Panama. Phytoplankton ("floating plants") such as green algae are the primary producers at the base of the web. Zooplankton are tiny, often microscopic, floating animals; they are primary consumers, not photosynthesizers, but they, along with insects and algae, are crucial food sources for fish in aquatic ecosystems. (Courtesy of G. H. Orians.)

community. For example, the diets of most herbivorous insects are restricted to just a few plant species, and many insects can only eat certain plant parts (Ehrlich and Raven 1964). The reason for this is partly that their mouthparts and body forms are only suited for certain feeding behaviors, but more important, their digestive systems are often surprisingly limited in their ability to extract nutrients and tolerate the toxic defensive chemicals produced by many plants. So even though a forest may be full of vigorously growing green plants, some insect species may be unable to complete their development and reproduce because they cannot get the specific food they require.

To survive, individuals of a species require a variety of resources, such as food, water, minerals, and a safe resting place; the continuous availability of these resources is an important component in the ability of a species to persist in an environment. A low supply of any one resource may limit the size of a population; the necessary resource that is in least supply may be the **limiting resource** on population size. In many communities, there are occasional episodes when one or several resources become limited and vulnerable species are eliminated from the community. For example, although the water is normally not a limiting resource to rain forest organisms, episodes of drought that last for weeks and even months do occasionally occur, even in the wettest rain forests. At such times, animal and plant species that need a constant supply of water may be eliminated from the community. In another example, bird species that are specialized to feed on insects may be unable to eat or feed their young during days or weeks when unusually cold, dry, or windy weather prevents insects from flying; in this situation, the insects are the limiting resource for the bird population.

Keystone Species

Within biological communities, certain species may be important in determining the ability of large numbers of other species to persist in the community. These crucial species have been termed **keystone species** (Paine 1966; Terborgh 1986; Howe 1984). Protecting keystone species is a priority for conservation efforts, because if a keystone species is lost from a conservation area, numerous other species might be lost as well (Box 3). Because many tropical insect species appear to be highly specialized in their feeding behavior, subsisting on just one or a few related plant species, Raven (1976) has argued that the extinction of each tropical plant species potentially results in an extinction cascade, with an additional loss of 10 to 30 insect species. With the recent evidence of a greater diversity of insect spe-

BOX 3 FLYING FOXES: THE DECLINE OF KEYSTONE SPECIES MAY LEAD TO MASSIVE EXTINCTIONS

The effects on an ecosystem of the extinction of a single species varies. In some cases, organisms previously dependent on an extinct species for food or shelter can adjust to the loss by using an alternative species. However, the loss of a species sometimes has a dramatic adverse effect upon other species, and can even lead to further extinctions. The severe decline and extinction of many species of pteropid bats, or "flying foxes," in the Old World tropics represents a potentially catastrophic example of the second type of effect, where "linked extinctions" follow in the wake of the loss of a keystone species (Cox et al. 1991). The relationship between these bats and many important plant species in the islands of the Pacific and Indian Oceans is so close that some biologists fear that the loss of flying fox species invites an ecological disaster that could profoundly affect human societies in these regions (Cox et al. 1991; Fujita and Tuttle 1991).

Flying foxes are widespread throughout the Old World tropics. The family Pteropodidae contains nearly 200 species, with approximately one-quarter of these in the genus *Pteropus*. Species of this genus are concentrated in the islands of the South Pacific, where they are the most important, and often the only, pollinators

Flying foxes—bats of the family Pteropodidae, such as these giant fruit bats—are vital pollinators and seed dispersers in Old World tropical forest communities. Many flying fox species are in danger of extinction. (Photograph © NYZS/The Wildlife Conservation Society.)

and seed dispersers for literally hundreds of species of tropical plants. The importance of these bats to maintaining plant diversity on Pacific islands cannot be overrated; one study in Samoa demonstrated that 80–100% of the seeds landing on the ground during the dry season were deposited by flying foxes (Cox et al. 1991). Many plant species are entirely dependent upon bats for pollination and seed dispersal; some have coevolved features, such as night-blooming flowers, that would prevent other potential pollinators from taking over the role in the bats' absence (Fujita and Tuttle 1991). Extinction of flying foxes is thus potentially devastating for these bat-dependent species. On Guam, where the two indigenous species of *Pteropus* are either extinct or virtually extinct, researchers have found indications that some plant species are not fruiting, that some of those that are fruiting have fewer fruits than normal, and that most fruits were not dispersed away from parent plants, so the offspring may be crowded out by the parent plants (Cox et al. 1991).

Many of the plants pollinated and dispersed by flying foxes throughout the region are of significant economic value to both local and international markets (Fujita and Tuttle 1991). At least 186 bat-dependent plant species produce a variety of products consumed by humans; these include important timber species such as ebony (*Diospyros melanoxylon*) and mahogany (*Calophyllum inophyllum*), medicinal species, and species that produce fibers, dyes, and other products. Fruits such as durian and wild bananas are also important bat-pollinated species. Flying foxes are also critical to the regeneration of large man-made clearings, since other forest animals are reluctant to venture into the open. Bats may drop 90–98% of the first seeds of woody plants that reach the clearings as they fly over (Thomas 1982). The implications for forest regeneration are obvious.

The main problem facing flying foxes is their vulnerability to human hunters. Flying foxes are hunted for food, for sport, or because they are considered pests by orchard farmers. Because the animals occur in large, conspicuous groups of up to a million animals and can be easily tracked to their home roost, hunters have little difficulty in bagging thousands of animals at a time. Moreover, the best hunting season for flying foxes coincides with the birthing season; many bats captured or killed at this time are pregnant or have newborn offspring. The high mortality rate of females and young from hunting, combined with a low reproductive rate and loss of habitat from deforestation, have caused a noticeable decline in many species of pteropid bats.

The conservation status of flying foxes is especially problematic because no baseline population or ecological data exist for most species (Fujita and Tuttle 1991). Despite recent action by international conservation organizations to protect flying foxes, few species are included as endangered or threatened species on CITES (Convention on International Trade in Endangered Species) lists, and legal protection of flying foxes is minimal throughout most of their ranges (Fujita and Tuttle 1991; Cox et al. 1991; Mickleburgh 1992). However, initial steps to protect these animals have been taken: commercial trade of all species of the genus *Pteropus* and five species from a second genus, *Acerodon*, has been restricted by CITES legislation, and seven of the most vulnerable species have been banned from international trade altogether (Fujita and Tuttle 1991). While these actions come too late to save many species, it is hoped that recognition of the importance of these keystone species, however delayed, will prevent further extinctions of flying foxes and the many plant species that depend on them.

One flying fox species that has been intensively investigated and protected is the Rodrigues fruit bat, which is endemic to Rodrigues Island in the Mascarene Islands (Cheke and Dahl 1981). The bat population dropped from about a thousand individuals in 1955 to less than a hundred by 1974, due to the destruction of its forest habitat by farming and fuelwood collection. The reduction in forest area reduced the bats' food supply and made them more susceptible to starvation and injury during the periodic cyclones that strike the island. Since 1974, the species has been legally protected both locally and under the CITES treaty. Its habitat has come under local protection, and the forest area of the island is being increased through a tree planting program. In response to these measures, the bat population has increased rapidly and appears to be stable. In addition, 11 successful captive breeding colonies have been established with a goal of eventually releasing captive-bred individuals into the wild (Carroll 1984). Hopefully, this program of legal protection, habitat protection and restoration, and captive breeding will provide a model for future flying fox protection efforts.

cies than previously suspected (see Chapter 3), it is even possible that the extinction of a single plant species could eliminate hundreds of insect species.

Among the most obvious keystone species are top predators, since these species are often important in controlling herbivore populations (Redford 1992). The elimination of even a small number of individual predators, constituting a minute amount of the community biomass, can result in dramatic changes in the vegetation and a great loss in biological diversity. Box 2 presented one such example in the sea otter; in another example, in many localities where gray wolves have been hunted to extinction by humans, deer populations have exploded. The deer have then severely overgrazed the habitat, eliminating many herbaceous species. The loss of these plants was detrimental to the deer and to other herbivores, including the insect community that fed on the plants. In addition, the loss of plant cover has led to soil erosion, contributing to losses of the species that inhabit the soil. In short, the elimination of a keystone species—in this case, the wolves, which kept deer populations in check—can create a series of linked extinction events that result in a degraded ecosystem with much lower biological diversity at all trophic levels.

The importance of grazing animals in patterning communities is also illustrated by the case of a black sea urchin of the genus *Diadema* (Kaufman 1986). This sea urchin was common in coral reefs throughout the tropical West Atlantic, where it fed on large, fleshy algae that were fairly uncommon. In 1981, following the unusually warm weather of an El Niño event, there was a massive die-off of *Diadema*, apparently caused by a viral epidemic. Without the *Diadema*, the fleshy algae increased dramatically in abundance, overgrowing and damaging coral reefs.

The importance of a keystone species may hinge on highly specialized relationships between the keystone species and other organisms. In many tropical forests, fig trees and fig vines (*Ficus* spp.) appear to be keystone species in the functioning of vertebrate communities (Lambert 1991). Fig flowers are pollinated by small, highly specialized fig wasps, which mature inside the developing fig fruit. Mature fig trees produce continuous fig crops, and generations of wasps are continually coming to maturity. As a consequence of this continuous fruit production, figs provide a reliable source of fruit to primates, birds and other fruit-eating vertebrates throughout the year, even during dry seasons (Terborgh 1986). While fig fruits do not have the high energy content of many preferred lipid-rich fruits or the high protein content of an insect diet, during drought periods the fig fruits serve as "famine food" that allows vertebrates to survive until their preferred foods are once more available. Even though fig trees and vines may be uncommon in the forest and the fruit may constitute only a small percentage of the total vertebrate diet, their persistence is necessary to the continued functioning of many species in the vertebrate community. In this case, the fig trees are a keystone species because so many other species rely on them for food, but the health of the *Ficus* population rests on the health of the wasp population; the relationship between the trees and the wasps forms the foundation of the entire community's health.

Many keystone species play less obvious roles that are nevertheless essential to maintaining biological diversity (Wilson 1987). Some inconspicuous detritivorous species play a significant role in the functioning of a community. For example, dung beetles exist at low density in tropical forests and constitute only a fraction of the biomass (Klein 1989). Yet the beetles are crucial to the community because they bury dung and carrion as a food source for their larvae. These buried materials break down rapidly, making nutrients available for plant growth. Seeds contained in the dung of fruit-eating animals are also buried, which may facilitate seed germination and the establishment of new plants. In addition, by burying and feeding on dung the beetles kill the parasites of vertebrates contained in the dung, helping to keep the vertebrate populations healthy.

The identification of keystone species has several important implications for conservation biology. First, as we have seen, the elimination of a keystone species from a community may precipitate the loss of many other species. Second, in order to protect a species of particular interest, such as a monkey, it may be necessary to protect the keystone species on which it depends either directly or indirectly, such as fig trees and dung beetles. Third, if the few keystone species of a community can be identified, these could be carefully protected or even encouraged if the area is being affected by human activity. For example, during selective logging, figs and other important fruit trees should be protected, while common trees that are not keystone species could possibly be reduced in abundance with little loss of biological diversity.

Keystone Resources

Nature reserves are typically compared and valued in terms of their size in hectares, because, in general, larger reserves contain more species than smaller reserves. However, area by itself may not be as significant as the range of habitats and resources that reserve contains. Particular habitats may contain critical limiting resources—"keystone resources"—that occupy only a small area of the habitat yet are crucial to many species in the community. For example:

- *Salt-licks and mineral pools* provide essential minerals for wildlife, particularly in inland areas with heavy rainfall. The distribution of salt licks can determine the abundance and distribution of vertebrates in an area.
- *Deep pools* in streams and springs may be the only refuge for fish and other aquatic species during the dry season when water levels drop. These water sources may be the only sources of drinking water for a considerable distance for terrestrial animals.
- *Elevational gradients* may be a crucial feature of tropical conservation areas. Many fruit- and nectar-feeding vertebrates and insects require a continuous supply of food, and one way they can accomplish this is by moving among communities searching for new food sources. A steep elevational gradient is typically occupied by a series of different plant communities, so migrating up or down a mountain slope is a common and efficient behavior for an animal in search of new food sources.
- *Mangroves* are communities of woody plants that occupy the narrow intertidal band in many tropical and subtropical regions of the world. Mangroves, complex communities themselves, are also important as nurseries for the juvenile stages of many marine species and as feeding grounds during high tide for many estuarine fishes and crustaceans.

Keystone resources may occupy only a small proportion of a conservation area yet be of crucial importance in maintaining many animal populations. The loss of a keystone resource could mean a rapid loss of animal species. For instance, removal of dead standing trees and fallen trees from forests may eliminate nesting and overwintering sites for many birds and other vertebrate species. When vertebrate species are lost, there may be linked extinctions of plant species that depend on the animals for pollination and seed dispersal.

Measuring Biological Diversity

In addition to the definition of biological diversity generally accepted by conservation biologists, there are many other definitions of biological diversity that have been developed in the ecological literature as a means of comparing the overall diversity of different communities (McMinn, n.d.). These concepts were used to test the theory that increasing levels of diversity leads to increasing levels of community stability (MacArthur 1955; McNaughton 1989; Pimm 1991). However, this line of research eventually led to the conclusion that there was no simple relationship between diversity and stability. The indices of biological diversity developed in the course of this research are primarily useful for comparing particular groups of species among communities. As Kimmins (1987) says: "Diversity can refer to all organisms in the community, but it is more frequently used to refer to one type or group of organisms. Thus we can talk about the diversity of vascular plants, of birds, of mammals, and of the soil fauna."

At its simplest level, diversity can be defined as the number of species found in a community, a measure known as **species richness**; but most definitions also include some measure of how evenly the total number (or abundance) of individuals is divided among species. For example, if there are 10 different bird species in a community of 60 individual birds, an even abundance would be 6 birds per species, while an uneven abundance would be 2 birds per species in 5 species and 50 birds in the sixth species. In the first case, no species would be considered to be dominant, while in the second case the community would be dominated by the sixth species.

In writing about diversity, Pielou (1969) maintained that diversity is "a single statistic in which the number of species and evenness are compounded." Many methods of calculating diversity have been proposed that combine these two types of information (Peet 1974), and while these methods generally give similar results, some are more strongly affected by sample size and species rarity than others. Also, it is problematic to evaluate communities in which individuals are of very different sizes, such as plant communities. If a community consists of a few large beech trees and many small painted trilliums, should the trilliums be considered dominant because there are more of them, or the beech trees because they contain most of the biomass of the community? There is no easy answer to these and many other questions about mathematical diversity indices, which makes such indices useful primarily for ecological comparisons and less useful for applied conservation biology. As the Committee of Scientists (1979) concluded, the U.S. Congress intended the term *diversity* "to refer to biological variety rather than any of the quantitative expressions now found in the biological literature." This is echoed in a later conference on the management implications of biological diversity: "Diversity indices must be used as an analytical tool and not used to define diversity" (Cooley and Cooley 1984). In their present form, diversity indices only capture part of the broad concept of biological diversity.

Mathematical indices of biodiversity have also been developed to connote species diversity at different geographical scales. The number of species in a single community is usually described as **alpha diversity**. Alpha diversity comes closest to the popular concept of species richness and can be used to compare the number of species in different ecosystem types. The term **beta diversity** refers to the degree to which species composition changes along an environmental gradient. Beta diversity is high, for example, if the species composition of moss communities changes at successively higher elevations on a mountain slope, but is low if the same species occupy the whole mountainside. **Gamma diversity** applies to larger geographical scales; it is defined as "the rate at which additional species are encountered as geographical replacements within a habitat type in different localities. Thus gamma diversity is a species turnover rate with distance between sites of similar habitat, or with expanding geographic areas" (Cody 1986). In practice, these indices will often be highly correlated. The plant communities of the Amazon, for instance, show high levels of diversity at the alpha, beta, and gamma scales (Gentry 1986).

Summary

1. The Earth's biological diversity includes the entire range of living species. Biological diversity also includes the genetic variation that occurs among individuals within a species. At a higher level, biological diversity includes the biological communities in which species live and their interactions with the physical and chemical environment.

2. Genetic variation within species arises through the mutation of genes and the recombination of genes during sexual reproduction. This genetic variation allows species to adapt to a changing environment through the process of natural selection. In some cases this process leads to the evolution of new species. Genetic variation within species also allows scientists to breed domestic plants and animals suitable for human use.

3. Species interact within biological communities through such processes as competition, predation, and mutualism. Species occupy distinct trophic, or feeding, levels within communities that represent the ways in which they obtain energy. Individual species often have specific feeding relationships with other species that can be represented as food chains and food webs.

4. Certain keystone species appear to be important in determining the ability of other species to persist in a community. These keystone species are often top carnivores but may also be inconspicuous species. The loss of a keystone species from a community can result in a cascade of extinctions of other species. Certain keystone resources, such as water holes and salt licks, may occupy only a small fraction of a habitat, but may be crucial to the persistence of many species in an area.

5. Mathematical indices of biological diversity have been developed to examine and compare patterns of species distribution at local and regional levels. Such mathematical indices are chiefly useful for examining particular groups of species rather than the full range of species and interactions found in nature.

Suggested Readings

Begon, M., J. L. Harper and C. Townsend. 1990. *Ecology*, 2nd Ed. Blackwell Scientific Publications, Oxford. A good introductory text on the principles of ecology.

Darwin, C. 1859. *On the Origin of Species*. John Murray, London. The classic work outlining a theory for the origin of new species that became the paradigm for modern biological thought.

Fujita, M. S. and M. D. Tuttle. 1991. Flying foxes (Chiroptera: Pteropodidae): Threatened animals of key ecological and economic importance. *Conservation Biology* 5(4): 455–463. A case study of keystone species.

Futuyma, D. 1986. *Evolutionary Biology*, 2nd Ed. Sinauer Associates, Sunderland, MA. A sophisticated textbook that provides background material on the processes of evolution and speciation.

Office of Technology Assessment of the U.S. Congress (OTA). 1987. *Technologies to Maintain Biological Diversity*. OTA-F-330. U.S. Government Printing Office, Washington, D.C. A clear statement of the definition of biological diversity.

Pimm, S. 1992. *The Balance of Nature*. University of Chicago Press, Chicago. An advanced text on the principles of community organization and their application.

Ricklefs, R. E. 1993. *The Economy of Nature*, 3rd Ed. W. H. Freeman, New York. A well-written textbook of the basic principles of ecology.

Standley, L. A. 1992. Taxonomic issues in rare species protection. *Rhodora* 94: 218–242. Taxonomic problems can complicate efforts at legal protection of rare species.

Tamarin, R. H. 1993. *Principles of Genetics*, 4th Ed. Wm. C. Brown, Dubuque, IA. A good background text for learning the principles of population and molecular genetics.

Terborgh, J. 1986. Keystone plant resources in the tropical forest. *In* M. Soulé (ed.), *Conservation Biology: The Science of Scarcity and Diversity*, pp. 330–344. Sinauer Associates, Sunderland, MA. A clear statement of the importance of keystone species.

CHAPTER 3

Where Is Biological Diversity Found?

The richest environments in terms of numbers of species appear to be tropical rain forests, coral reefs, large tropical lakes, and possibly the deep sea (Fischer 1960; Connell and Orias 1964; Pianka 1966; Grassle 1991). There is also an abundance of species in tropical dry habitats, such as shrublands, grasslands, and deserts (Mares 1966), and temperate shrublands with Mediterranean climates such as South Africa, southern California, and southwestern Australia. In tropical rain forests this diversity is primarily due to the great abundance of animal species in a single class—the insects; in coral reefs and the deep sea the diversity is spread over a much broader range of phyla and classes (Grassle et al. 1991; Ray and Grassle 1991). Diversity in the deep sea may be due to the great age and stability of the environment (Grassle and Maciolek 1992), as well as specialization on particular sediment types (Etter and Grassle 1992). The great diversity of fishes and other species in large tropical lakes is due to rapid evolutionary radiation in a series of isolated, productive habitats (Kaufman 1992). Marine systems contain representatives of 28 of the 33 animal phyla that exist today; 13 of these phyla are present only in the marine environment (Grassle et al. 1991). In contrast, only one phylum is found exclusively in the terrestrial environment, and no phylum is restricted to the freshwater environment. Four phyla are found only in symbiotic associations with other species.

Patterns of diversity are known primarily from the efforts of taxonomists who have methodically collected organisms from all areas of the world. Patterns of diversity are known only in broad outline for many groups of organisms because they have not been adequately collected. For example, 80% of the beetle species collected in a study in Panama were new to science, even though Panama is one of the best-known areas of the tropics (Erwin 1983; May 1992).

For almost all groups of organisms, species diversity increases toward the tropics. For example, Venezuela has 305 species of mammals, while France has only 113 species, despite the fact that the two countries have roughly the same land area (WRI/IIED 1988) (Table 3.1). The contrast is particularly striking for trees and other flowering plants; a hectare (100 m × 100 m) of forest in Amazonian Peru or lowland Malaysia might have 300 or more species growing as trees (diameter at breast height is greater than 10 cm), while an equivalent forest in temperate Europe or the United States would probably contain 30 species or less. In the case of tiger beetles, a particularly widespread and well-known insect group, major tropical regions of the world have over 300 species, while temperate regions have fewer than 150 species (Pearson and Cassola 1992). Within continents such as Australia and North America, the number of tiger beetle species increases toward the equator (Figure 3.1).

In general, there is a rough correspondence in the distribution of species richness among different groups of organisms. Continuing with this particularly well-documented example, it has been shown that tiger beetles, butterflies, and birds in North America and in Australia show patterns of species diversity that are similar to one another (Figure 3.2; Pearson and Cassola 1992). However, each group of living things may reach its greatest species richness in a different

TABLE 3.1
Number of mammal species in selected tropical and temperate countries

Tropical country	Number of species	Species per 10,000 km^2 [a]	Temperate country	Number of species	Species per 10,000 km^2 [a]
Angola	275	76	Argentina	255	57
Brazil	394	66	Australia	299	41
Colombia	358	102	Canada	163	26
Costa Rica	203	131	Egypt	105	31
Kenya	308	105	France	113	39
Mexico	439	108	Japan	186	71
Nigeria	274	82	Morocco	108	39
Peru	359	99	South Africa	279	79
Venezuela	305	92	United Kingdom	77	33
Zaire	409	96	United States	367	60
Average		96	Average		48

Source: After Reid and Miller 1989, using data from WRI/IIED 1988.

[a] To standardize for the size of the country and allow comparisons of species richness, the number of species found in 10,000 square kilometers (the approximate area of the island of Jamaica) is estimated from species–area formulas (see Chapter 4).

(A)

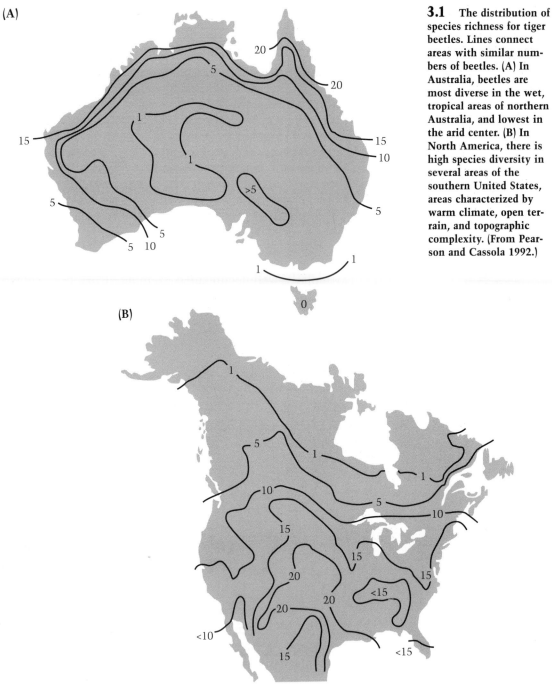

3.1 The distribution of species richness for tiger beetles. Lines connect areas with similar numbers of beetles. (A) In Australia, beetles are most diverse in the wet, tropical areas of northern Australia, and lowest in the arid center. (B) In North America, there is high species diversity in several areas of the southern United States, areas characterized by warm climate, open terrain, and topographic complexity. (From Pearson and Cassola 1992.)

(B)

3.2 Species diversities of butterflies and birds are highly correlated with tiger beetle diversity in North America and Australia. Each dot on the graph represents a square that is 275 km on a side. Overall, squares with many beetles also tend to have many bird species (A) and butterfly species (B). (From Pearson and Cassola 1992.)

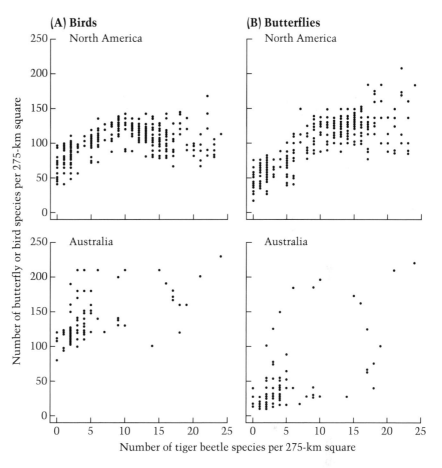

part of the world due to historical circumstances or the suitability of the site to its own needs. For example, in Africa, butterflies are most diverse in West Africa north of the equator and plants are most diverse in West Africa right along the equator, while birds, primates, and other mammals are most diverse in Central and East Africa (IUCN/UNEP 1986b). More work is needed on this topic of correlated patterns of species richness before definitive conclusions can be reached (Currie 1991).

Patterns of diversity in terrestrial species are paralleled by patterns in marine species, with an increase in species diversity toward the tropics. For example, the Great Barrier Reef off Australia has 50 genera of reef-building corals at its northern end where it approaches the tropics, but only 10 genera at its southern end farthest away from the tropics (Stehli and Wells 1971). In the case of sea squirts (tunicates), there are only 103 species in the Arctic but over 600 species in tropical waters (Fischer 1960). These increases in species diversity toward the tropics are paralleled by increases in planktonic species, such as foraminiferans, and increases in deep sea species (Buzas and Culver 1991).

Patterns of species richness are also affected by local variation in topography, climate, and environment (Diamond 1988a; Brown and Gibson 1983, Currie 1991). In terrestrial communities, species richness tends to decrease with increasing elevation, decreasing solar radiation, and decreasing precipitation. The lower richness of species in Africa than in South America or Asia may be due to Africa's lower rainfall and smaller total area of forest. Even within tropical Africa itself, areas of low rainfall in the Sahel have fewer species than forested areas with higher rainfall to the south. However, the extensive savannah areas of East and Central Africa have a richness and abundance of antelopes and other ungulate grazers not found on other continents. For mammals, the greatest abundance of species may occur at intermediate levels of precipitation rather than in the wettest or driest habitats (Western 1989; Mares 1992).

Species richness can also be greater where there is a complex topography that allows genetic isolation, local adaptation, and speciation to occur (Ray 1991; Pagel et al. 1991). For example, a sedentary species occupying a series of isolated mountain peaks may eventually evolve into several different species, each adapted to its local mountain environment. A similar process could happen in a fish that occupies a large drainage system that becomes divided into several smaller systems. Areas that are geologically complex produce a variety of soil conditions with very sharp boundaries between them, leading to a variety of communities and species adapted to one soil type or another. Among temperate communities, great plant species diversity is found in southwestern Australia, South Africa, and other areas with a Mediterranean climate of mild, moist winters and hot, dry summers (Myers 1991a). The shrub and herb communities in these areas are apparently rich in species due to their combination of considerable geological age and complexity of site conditions.

Historical factors are also important in defining patterns of species richness, with areas that are geologically older having more species than younger areas. For example, coral species richness is several times greater in the Indian Ocean and West Pacific than in the Atlantic Ocean, which is geologically younger (Figure 3.3). There are over 50 genera of coral in many of the Indo-Pacific areas, but only about 20 genera in the Caribbean Sea and adjacent Atlantic Ocean (Stehli and Wells 1971; Goldman and Talbot 1976). Areas that are geologi-

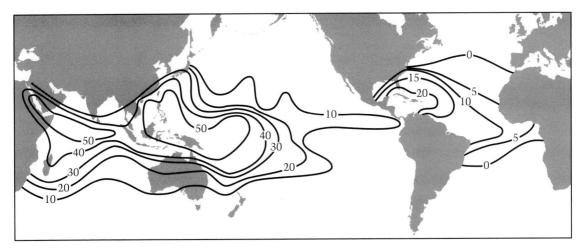

3.3 Global distribution of genera of reef-forming corals, showing major concentrations in the western Pacific and Indian Oceans, and a lesser concentration in the Caribbean. Numbers on the contour lines indicate the number of genera; the lines connect areas with similar numbers. (After Stehli and Wells 1971.)

cally older have had more time to receive species dispersing in from other parts of the world and more time for existing species to undergo adaptive speciation in response to local conditions.

Why Are There So Many Species in the Tropics?

There is ample evidence that tropical environments possess the greatest species diversity. Many theories have been advanced to explain this species richness (Fischer 1960; Connell and Orias 1964; Pianka 1966). The following are some of the most reasonable theories:

1. Over geological time the tropics have had a more stable climate than the temperate zones. In the tropics, locally adapted species could keep living in the same place while temperate species had to keep dispersing from north to south and back again in response to numerous episodes of glaciation. Only those temperate species that could migrate and be competitive within their new communities survived.

2. Because tropical communities are older than temperate communities, there has been more time for them to evolve. New species evolve in response to local conditions. A longer period of evolution has allowed a greater degree of specialization and local adaptation to occur in tropical areas. Many tropical species are quite specialized in their habitat requirements and reproductive behaviors.

3. The warm temperatures and high humidity in many tropical areas provide favorable conditions for many species that are unable to survive in the temperate areas. Species living in temperate zones must have physiological mechanisms to tolerate the cold and freezing conditions found there. These species may also have specialized behaviors, such as dormancy, hibernation, or migration, to help them survive the winter. The inability of many groups of plants and animals to live outside the tropics suggests that these adaptations are difficult to evolve.

4. In the tropics, there may be greater pressure from pests, parasites, and diseases, since there is no winter cold to reduce their populations. Ever-present populations of these pests prevent any single species or group from dominating communities, and there is opportunity for numerous species to coexist at low individual densities. In many ways, the biology of the tropics is the biology of rare species. In contrast, in the temperate zones there may be reduced pest pressure because of the winter cold, allowing one or a few competitively superior species to dominate the community and exclude many other species that are not as competitive.

5. Among plant species, rates of outcrossing (interbreeding with

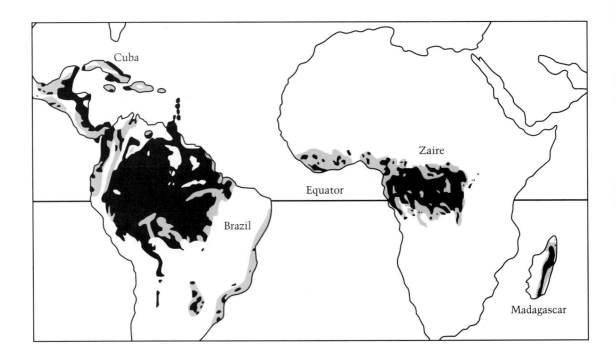

other individuals of the same species, as opposed to self-pollination) appear to be higher in tropical plant species than in temperate ones (Bawa 1992). Higher rates of outcrossing may lead to higher levels of genetic variability (see Chapter 11), local adaptation, and speciation.

6. Tropical regions receive more solar energy over the course of a year than temperate regions. As a result, many tropical communities have a higher productivity than temperate communities in terms of the number of kilograms of living material produced each year per hectare of habitat. This high productivity results in a greater resource base that can support a wider range of species. However, there is only a loose correlation between productivity and species diversity; some highly diverse communities occur in habitats with intermediate productivity.

Tropical Rain Forests

Tropical rain forests, ranking high among the world's most diverse ecosystems, are being lost at a rapid rate due to human disturbance (Figure 3.4; see Chapter 6). Even though the world's tropical forests occupy only 7% of the planet's land area, they contain over

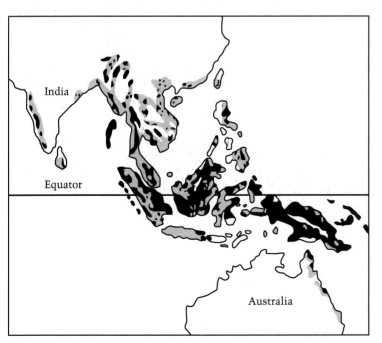

3.4 Tropical rain forests are found predominantly in wet, equatorial areas of America, Africa, and Asia. At one time tropical forests covered the entire shaded area, but in recent years human activities have resulted in the loss of a great deal of rain forest acreage, reducing the limits to the areas in black. (After Repetto 1990; maps based on data from the Smithsonian Institution.)

half of the world's species (Whitmore 1990; Conservation International 1990). This estimate is based to some degree on only limited sampling of insects and other arthropods, groups that are thought to contain the majority of the world's species (see Figure 3.7). Estimates of the number of undescribed insect species in tropical forests range from 5 million to 30 million (May 1992). If the 30 million figure is correct, it would mean that insects found in tropical forests may constitute over 90% of the world's species.

Information on other groups, such as plants and birds, is much more accurate. For flowering plants, gymnosperms, and ferns, about 86,000 species occur in tropical America; 38,000 species occur in tropical Africa and Madagascar; and 45,000 species occur in tropical Asia, including New Guinea and tropical Australia (Gentry 1982; Brenan 1978; Reid and Miller 1989). This total is about two-thirds of the estimated 250,000 plant species believed to occur worldwide. Over 100,000 of these plant species are found in tropical forests (Myers 1980).

About 30% of the world's bird species—1300 species in the American tropics, 400 species in tropical Africa, and 900 in tropical Asia— are dependent on tropical forests (A. Diamond 1985). This figure is probably an underestimate since it does not include species, such as migrant birds, that are only partially dependent on tropical forests. The figure also does not reflect the high concentrations of tropical forest birds living in restricted habitats, such as islands, that may be more vulnerable to habitat loss. In forested islands such as New Guinea, 78% of the non-marine birds are dependent on the forest for their survival (Beehler 1985).

Coral Reefs

Colonies of tiny coral animals (Figure 3.5) build the large coral reef ecosystems that are the marine equivalent of tropical rain forests in terms of their species richness and complexity (Wells 1989; Jackson 1991). One explanation for this richness is the high primary productivity of coral reefs, which produce 2500 grams of biomass per square meter per year, in comparison with 125 g/m^2 per year produced in the open ocean. The clarity of the water in the reef ecosystem allows sunlight to penetrate deep in the water and high levels of photosynthesis to occur (Whittaker 1975). Other possible factors are extensive niche specialization among coral species and adaptations to varying levels of disturbance (Jackson 1991; Ray 1991; Huston 1985).

The world's largest coral reef is the Great Barrier Reef off the east coast of Australia, with an area of 349,000 km^2. The Great Barrier

3.5 **Coral reefs in tropical waters are built up from the skeletons of billions of tiny individual animals. The intricate coral landscapes create a habitat for many other marine species. (Photograph by Les Kaufman, New England Aquarium.)**

Reef contains over 300 species of coral, 1500 species of fish, 4000 species of mollusks, and 5 species of turtles, as well as providing breeding sites for some 252 species of birds (IUCN/UNEP 1988). The Great Barrier Reef contains about 8% of the world's fish species even though it occupies only 0.1% of the ocean surface area (Goldman and Talbot 1976). The Great Barrier Reef is in turn part of the rich Indo-West Pacific region. The greater diversity of species in the Indo-West Pacific than in other marine areas is illustrated by the more than 2000 fish species found in the Philippine Islands, in contrast with 448 species in the mid-Pacific Hawaiian Islands and about 500 species in the islands of the Bahamas. By comparison, the number of marine fishes in temperate areas is low: the mid-Atlantic seaboard of North America has only 250 fish species, and the Mediterranean Sea has fewer than 400 species (Briggs 1974).

One notable difference between tropical forest species and coral reef species is that while many forest species are endemic (found only in one location and nowhere else) or have very restricted distributions, very few coral reef species show such patterns of narrow distribution (Vernon 1986). Many species of the coral reefs disperse widely in the ocean during their juvenile stages, so they are found over a wide area. Most Indo-Pacific coral species, for example, are widely distributed. It is only on isolated islands such as Hawaii that numerous endemic reef species are typically found; fully 20% of Hawaiian coral species are endemic to that area (Hourigan and Reese 1987). Because coral reef species are more widely distributed than tropical rain forest species, they are less prone to extinction by the destruction of a locality. However, this may be a taxonomic bias, since coral reef species are not as well known as terrestrial species; further research may reveal many locally distributed species, and as tropical reefs become damaged by human activity the possibility that such species might be lost is cause for serious concern.

How Many Species Exist Worldwide?

A strategy for conserving biological diversity must be based on a firm grasp of the numbers of species that exist in the world today and how those species are distributed. Certain groups of organisms are relatively well known, such as birds, mammals, and temperate flowering plants; but even in these groups a small number of new species are being discovered each year. Sometimes species are discovered when new research, often involving techniques of molecular systematics, reveals that what was originally thought to be a single species with a number of geographically distinct populations is really two or more species. Such a discovery occurred when scientists found that New Zealand's unique reptile, the tuatara, was in reality two separate species (Daugherty et al. 1990). In groups that are less well known, such as insects, spiders, mites, nematodes, and fungi, the number of described species is increasing at the rate of 1–2% per year (May 1992). Huge numbers of species in these groups have yet to be discovered and described, mostly in tropical areas but also in the temperate zones. At the present time, about 1.5 million species have been described in total. It is certain that at least twice this number of species remain undescribed, primarily in the tropics, leading to one estimate that there are about 5 million species in the world (Gaston 1991; May 1992).

New species are continually being discovered. Two of the most remarkable discoveries of new species in the present century involve "living fossils": species known from the fossil record, but thought to be extinct until living examples were found. In 1938 ichthyologists throughout the world were stunned by reports of a strange fish caught in the Indian Ocean. This fish, subsequently named *Latimeria chalumnae,* belonged to a group of marine fishes known as coelacanths that were common in ancient seas but were thought to have gone extinct 65 million years ago (Thomas 1991). Coelacanths are of particular interest to evolutionary biologists because they are considered to be possible ancestors of the first land animals (Fricke 1988; Gorr et al. 1991). Biologists had to search the Indian Ocean for 14 years until another coelacanth was found off Grande-Comore Island, between Madagascar and the African coast. Subsequent investigation has shown that there is a single population of about 200 individuals living approximately 200 meters offshore of Grande-Comore in underwater caves (Fricke and Hissman 1990). In recent years, the Republic of the Comoros has implemented a conservation plan to protect the coelacanths, including a ban on catching and selling the fish.

The situation of the dawn redwood, *Metasequoia glyptostroboides,* is a similarly remarkable discovery of a living fossil (Figure 3.6). In 1945 Chinese botanists working in northeastern Szechuan

3.6 The dawn redwood, *Metasequoia glyptostroboides,* is known in the wild from only one remote location in China. Following its scientific discovery in 1945, however, the species is now grown throughout the temperate world, prized for its unusual branching pattern, enlarged trunk, and delicate foliage. (Photograph by Peter Del Tredici, Arnold Arboretum.)

Province discovered several individuals of an unknown gymnosperm tree (Merrill 1948). A subsequent expedition and botanical investigations revealed that the tree belongs to a genus that was widely distributed across the Northern Hemisphere in the Mesozoic era, but had apparently gone extinct tens of millions of years ago. These unique trees are now protected by the local community and the Chinese government, and have been planted throughout the temperate world for their delicate foliage and pleasing branching pattern.

In both of these cases, the species discovered had been overlooked because their sole remaining habitats were located in out-of-the-way places. Both species came quickly to the attention of biologists because they are conspicuous—that is, they are easily visible to the naked eye and are clearly distinct from other species. Inconspicuous organisms, including most insects and microorganisms, are much less likely to be observed by chance outside their natural habitats, as the coelacanth was; even within their native environments, inconspicuous species are not as immediately obvious as the dawn redwood. Many inconspicuous species that live in remote habitats will not be found and catalogued unless biologists specifically go in search of them. This factor delays a thorough understanding of the full extent of biological diversity because inconspicuous species constitute the majority of species on Earth.

The most diverse group of organisms appears to be the insects, with about 750,000 species described already, or about half the world's known species (Figure 3.7). If the number of insect species can be estimated accurately, then it may be possible to determine the total number of species in the world. Since most insect species occur in tropical rain forests, recent studies in tropical America have attempted to sample entire insect communities using insecticidal fogging of whole trees (Erwin 1983). These studies have revealed an extremely rich and largely undescribed insect fauna in the canopies of these trees (Wilson 1991). Sampling of trees at localities across the Amazon further suggests that these insect communities are often very localized in distribution. On the basis of this evidence, Erwin (1982) has attempted to estimate the number of insect species in the world.

Erwin's reasoning goes as follows: In Panama, 1200 species of beetles were collected from the canopy of a single tree species, *Luehea seemannii*. About 800 of these species were herbivorous. He estimated that 20% of these herbivorous beetles (160 species) are specialist feeders on this particular tree species. Since beetles represent 40% of all insect species, there may be a total of 400 species of specialized insects that feed in the canopy of each tree species. Erwin estimated

3.7 A total of 1,413,000 species have been identified and described by scientists; the majority of these are insects and plants. Large numbers of insects, bacteria, and fungi are still undescribed, and the eventual number of identified species could reach 5 million or more. (Data from Wilson 1992.)

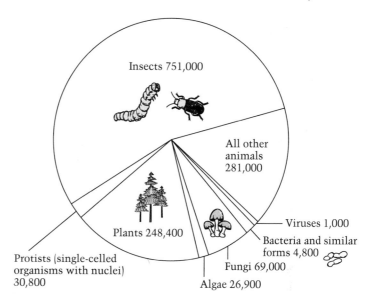

Insects 751,000

All other animals 281,000

Viruses 1,000

Bacteria and similar forms 4,800

Fungi 69,000

Algae 26,900

Plants 248,400

Protists (single-celled organisms with nuclei) 30,800

that canopy species represent only about two-thirds of the insect species on each tree species—other insects feed on roots, bark, and other plant parts—suggesting that there are 600 insect species specializing on each plant species. Since there are about 50,000 species of tropical trees, there may be as many as 30 million species of insects.

Each step in the calculation leading to the estimate of 30 million species is so speculative that many scientists do not accept this estimate, and keep to the earlier figure of about 5 million total species (Gaston 1991). In particular, the number of insect species specializing on each plant species may be as much as ten times too high, leading to an estimate of only 3 million species. However, Erwin's work is very significant in that it calls attention to the large numbers of undescribed species, develops a new approach to estimating insect numbers, and emphasizes the relationships among plant and animal species (Erwin 1991; May 1992).

Another approach to estimating biological diversity involves developing rules for determining how many species are involved in biological relationships (May 1988b, 1992). For example, in Britain and Europe there are about six times more fungus species than plant species. If this general ratio is applicable throughout the world, there may be as many as 1.6 million fungus species growing on the world's 270,000 plant species. Since only 69,000 species of fungi have been described so far, it is possible that there are over 1.5 million species of fungi waiting to be discovered, most of them in the tropics (Hawksworth 1991).

A third approach could be to assume that each species of plant or insect, which together form the majority of currently known species, has at least one species of specialized bacteria, protist, nematode, and virus, in which case the estimates of the number of species should be multiplied by five—bringing it to 25 million using traditional estimates, or to 150 million species if one accepts Erwin's estimates. Or, perhaps biological communities have a number of common species that can be readily assessed and used to estimate the number of rare species that are harder to find. Developing such approaches might allow ecological estimates to be made of the number of species in communities.

Inconspicuous species have not received their proper share of taxonomic attention. For example, mites and nematodes in the soil are small and hard to study. These groups could number in the hundreds of thousands of species if they were properly studied. Now that the role of nematode species (roundworms) as parasites of agricultural plants has been demonstrated, scientists have dramatically increased their efforts to collect and describe these minute animals. Consequently, the catalogue of this one group of organisms has grown from the 80 nematode species known in 1860 to about 15,000 species known today (May 1992).

Bacteria are also very poorly known (Hawksworth 1992). Only about 4000 species of bacteria are recognized by microbiologists because of the difficulty in growing and identifying specimens. However, recent work in Norway analyzing bacterial DNA hints that there may be more than 4000 species in a single gram of soil, and an equally large number of different species in marine sediments (Giovannoni et al. 1990; Ward et al. 1990). A lack of collecting has particularly hampered our knowledge of the number of species found in the marine environment (Grassle 1991). The marine environment appears to be a great frontier of biological diversity, with huge numbers of species and even entire communities still unknown. An entirely new animal phylum, the Loricifera, was described only in 1983 based on specimens from the deep seas (Kristensen 1983), and there are undoubtedly more species to be discovered.

Newly Discovered Communities

Entirely new biological communities are still being discovered, often in localities that are extremely remote and inaccessible. Often these communities consist of inconspicuous species, such as bacte-

3.8 Biologists are gaining access to the diverse world of the rain forest canopy by using techniques borrowed from technical rock climbing, along with a variety of specialized viewing platforms. (Photograph courtesy of Nalini Nadkarni.)

ria, protists, and small invertebrates, that have escaped the attention of earlier taxonomists. These communities may occur in highly specialized habitats that were not previously explored. As a result of specialized exploration techniques, particularly in the deep sea (Grassle 1991) and the forest canopy (Wilson 1991), these communities are being discovered and investigated, often with considerable public attention. Some recently discovered communities include:

- Diverse communities of animals, particularly insects, are adapted to living in the canopies of tropical trees (Figure 3.8), rarely if ever descending to the ground (Erwin 1983; Wilson 1991).

- A species-rich community of endemic bugs, crickets, and crane flies has been discovered occupying tube caves created by volcanic lava flows in Hawaii (Howarth 1973).
- Previously inaccessible sea caves in the Bahamas investigated by divers were found to have ancient crustacean species (Yager 1981).
- Calcium carbonate pebbles in the Persian Gulf were found to be inhabited by a rich assembly of cyanobacteria (blue-green algae) species (Figure 3.9), only one of which had been previously described by scientists (Al-Thukair and Golubic 1991).
- The floor of the deep sea, which remains almost entirely unexplored due to the technical difficulties of transporting equipment and people under high water pressure, has unique communities of bacteria and animals that grow around deep-sea geothermal vents (Box 4). Recent investigations have begun to catalogue some of the species in these communities, but scientists have been able to explore only a small portion of the vast ocean floor.

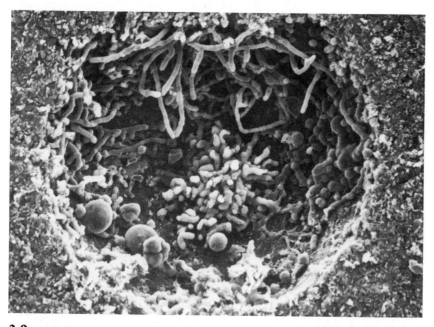

3.9　A recent survey in the Persian/Arabian Gulf region revealed a previously undescribed community of blue-green algae (cyanobacteria) that live inside calcium carbonate pebbles known as ooids. Each of the three cyanobacteria species growing inside this cavity has a distinctive appearance. (From Al-Thukair and Golubic 1991; courtesy of the authors.)

- A recent drilling project in Sweden turned up evidence of primitive anaerobic bacteria, known as archaebacteria, living within rocky fissures 5 kilometers beneath the Earth's surface (Gold 1992). If this evidence proves valid, it may be that large undescribed communities of bacteria live inside the Earth, existing on sulfur, methane, and other energy-rich gases.

BOX 4 CONSERVING A WORLD UNKNOWN: HYDROTHERMAL VENTS

Biologists are aware that many species have not been adequately studied and described, a fact that frequently hampers attempts at conservation. In recent years it has become apparent that there are also entire communities that remain undiscovered in the more remote parts of the Earth. The biota of deep-sea hydrothermal vents, discovered only recently with the invention of technology that enables scientists to explore great depths, demonstrates that species, genera, and even families of organisms exist about which scientists know nothing. Such organisms pose a significant problem for conservationists: How does one go about conserving species and communities that have not been discovered?

Hydrothermal vents are temporary underwater openings in the Earth's crust. The heat and minerals that escape from these vents support a profusion of species in the deepest parts of the ocean (Jannasch and Motti 1985; Grassle 1985; Gage and Tyler

Part of a hydrothermal vent community. Large tube worms (*Riftia pachyptila*) dominate the ecosystem. Crabs and mussels also make their home here. The energy and nutrients that support this community are derived from the hydrogen sulfide and minerals emitted by volcanic vents. (Photograph courtesy of V. Tunnicliffe.)

1991; Lutz 1991). Communities of large animals such as clams, crabs, fish, and tube worms (also known as pogonophorans or vestimentiferans) derive their energy from chemosynthetic microorganisms, which are the primary producers of the vent ecosystem (Fisher 1990). The vents themselves are short-lived, lasting at most a few decades and in some cases only a few years; however, the communities supported by these vents are thought to have evolved over the past 200 million years or more. Like many terrestrial communities, hydrothermal communities vary according to differences in their local environment. The species distribution of a hydrothermal community is dependent on the character of the vent, including temperature, chemical composition, and flow pattern of the hydrothermal fluid issuing from the vent (Grassle 1985).

Until deep-sea submersibles were developed in the 1970s, allowing scientists to take photographs and collect specimens at depths of over 2000 meters, biologists were completely unaware of the vent communities. However, since 1979, when the submersible *Alvin* was used to examine the vents around the Galápagos Rift in the Pacific Ocean, some 16 new families of animals—not including microorganisms—have been described (Grassle 1985; Tunnicliffe 1992). As investigation of deep-sea vents continues, more families will certainly be discovered, encompassing many new genera and species.

Scientists studying hydrothermal species may work for decades, yet only acquire minimal knowledge of the dynamics of vent communities because of the unique nature of the study sites: the vents are ephemeral, and they can be reached only by using expensive, specialized equipment. Hampered by these difficulties, biologists nevertheless need to think ahead to conservation problems that might face these species in the future. Industrial pollutants, for example, have damaged ocean species in shallower waters, and in theory could harm these communities should concentrations become high enough. How would conservationists respond to such a situation?

Though as yet this problem is strictly hypothetical, it illustrates a frustrating aspect of conservation biology: too little is known about too many species and ecosystems to undertake specific measures that might prevent future extinctions. New species, genera, and families are constantly being added to the list of known organisms, but a great many others are lost before they are even discovered. How does conservation biology deal with species and perhaps whole communities that are as yet unknown, but are in need of conservation nonetheless? Experience has shown that a specific conservation program created in ignorance of a species' behavior and biological needs can sometimes be worse than no program at all. Should we develop conservation programs despite our lack of information and hope for the best? Or do we continue our studies in the hope that the time lost will not prove fatal to the species? Since we know that restricting pollution has broadly positive effects on natural communities, pollution abatement programs may offer the best conservation strategy even when the biological communities are not thoroughly understood.

The Need for More Taxonomists

One problem the scientific community faces in describing and cataloguing the biological diversity of the world is the lack of trained taxonomists able to take on the job (Raven and Wilson 1992). At the present time there are only about 1500 taxonomists in the world competent to work with tropical species, with many of these based in temperate countries (NAS 1980). Unfortunately, this number is declining rather than increasing. When academic taxonomists retire, universities have tended to either eliminate the position or replace the retiring biologist with someone who is not a taxonomist. Many museums do not replace their taxonomists because of financial difficulties or changing priorities. Many members of the younger generation of taxonomists are preoccupied with mathematical theories of taxonomy and the new molecular approaches to taxonomy and do not have the interest in continuing the great tradition of cataloguing the world's biological treasures. At least a fivefold increase in the number of field taxonomists is needed to complete the task of describing the world's biological diversity before it is lost. One possible solution is for museums and projects worldwide to train talented local people in the basic elements of collecting and cataloguing in order to increase the amount of materials available to university- and museum-based taxonomists and to preserve a record of what species exist where.

Summary

1. The greatest biological diversity is in the tropical regions of the world, with large concentrations in tropical rain forests, coral reefs, tropical lakes, and the deep sea. In terrestrial habitats, species richness tends to be greatest at lower elevations and in areas with abundant rainfall. Areas that are geologically old and topographically complex also tend to have more species.

2. Tropical rain forests occupy only 7% of the Earth's land area yet are estimated to contain the majority of the Earth's species. The great majority of these species are insects as yet not described by scientists. Coral reef communities are also rich in species, with many of the species widely distributed. The deep sea also appears to be rich in species, but is not yet adequately explored.

3. About 1.5 million species have been described and at least twice that number of species remain to be described, leading to a conservative estimate that there are about 5 million species worldwide. Most of the undescribed species occur in the tropics.

4. While certain conspicuous groups, such as flowering plants, mammals, and birds, are reasonably well known to science, other groups, such as insects, bacteria, and fungi, have not been thoroughly studied. Recent attempts to collect all of the insects in tropical canopy trees have yielded mostly undescribed species, suggesting that far more species may exist than previously suspected.

5. New biological communities are still being discovered, especially in the deep sea and the forest canopy. For example, spectacular communities occupying deep-sea hydrothermal vents are a recent discovery. There is a vital need for more taxonomic scientists to study, classify, and help protect the world's biological diversity before it is lost.

Suggested Readings

Caulfield, C. 1985. *In the Rainforest.* Alfred A. Knopf, New York. A popular account of the issues in rain forest conservation.

Currie, D. J. 1991. Energy and large-scale patterns of animal- and plant-species richness. *American Naturalist* 137: 27–49. Detailed investigation of factors affecting species richness in North America.

Gage, J. D. and P. A. Tyler. 1991. *Deep-Sea Biology: A Natural History of Organisms at the Deep-Sea Floor.* Cambridge University Press, Cambridge. Covers a wide range of organisms and issues

Grassle, J. F. 1991. Deep-sea benthic diversity. *BioScience* 41: 464–469. Highlights the research problems and recent discoveries in the deep-sea environment.

Jackson, J. 1991. Adaptation and diversity of reef corals. *BioScience* 41: 475–482. A review article on the biology of the world's corals.

May, R. M. 1992. How many species inhabit the Earth? *Scientific American* 267 (October): 42–48. A review of the arguments for different estimates of the number of species.

Pearson, D. L. and F. Cassola. 1992. World-wide species richness patterns of tiger beetles (Coleoptera: Cicindelidae): Indicator taxon for biodiversity and conservation. *Conservation Biology* 6: 376–391. An excellent case study of the geographic patterns of species distribution, with applications for pinpointing species concentrations.

Raven, P. and E. O. Wilson 1992. A fifty-year plan for biodiversity surveys. *Science* 258: 1099–1100. A clear statement of the need for a comprehensive survey of the world's biodiversity.

Ray, G. C., J. F. Grassle and contributors. Marine biological diversity. *BioScience* 41: 453–465. An overview of marine biodiversity, with many other related articles in the same issue.

Whitmore, T. C. 1990. *An Introduction to Tropical Rain Forests.* Clarendon Press, Oxford. An excellent short presentation by a leading authority.

Wilson, E. O. 1991. Rain forest canopy: The high frontier. *National Geographic* 180 (December): 78–107. An authoritative and vivid account of diversity in the forest canopy.

Threats to Biological Diversity

Loss of Biological Diversity

We live at a historic moment in which the world's biological diversity is being rapidly destroyed. There are more species on Earth at the present geological time than in any other period, yet as a result of human activity, the current rate of extinction of species is greater now than at any time in the past. The loss of biological diversity is occurring at all levels; ecosystems and communities are being degraded and destroyed, and species are being driven to extinction. This loss is occurring in both the tropics and the temperate zones, and in both terrestrial and aquatic habitats (Williams and Nowak 1986). Unique populations and subspecies of remaining species are being eliminated. Even in the species that persist, a loss of genetic variation is taking place as the numbers of individuals in populations are reduced and populations are increasingly isolated from one another. The cause of this loss of biological diversity is the range of human activities that alter and destroy natural habitats to suit human needs. At the present time approximately 40% of the total net primary productivity of the terrestrial environment is used or wasted in some way by people; this represents about 25% of the total primary productivity of the world (Vitousek et al. 1986). Genetic variation is being lost even in domesticated species as farmers abandon traditional agriculture; in the United States, about 97% of the vegetable varieties that once were cultivated are now extinct (Cherfas 1993). In tropical countries, farmers are abandoning their local varieties in favor of high-yielding varieties for commercial sale (Bedigian 1991; Altieri and Anderson 1992). This loss of variability among food plants and its implications for world agriculture is discussed further in Chapter 17.

The most serious aspect of the loss of biological diversity is the extinction of species. Communities can be degraded and reduced in area, but as long as all of the original species survive, the communities still have the potential to recover. Similarly, genetic variation within a species will be reduced as population size is lowered; but species can potentially regain genetic variation through mutation, natural selection, and recombination. However, once a species is eliminated, the unique genetic information contained in its DNA and the special combination of characters that it possesses are unlikely ever to be repeated again. Once a species goes extinct, its chances for further evolution are lost, the communities that it inhabited are impoverished, and its potential value to humans will never be realized.

The meaning of the word *extinct* can vary somewhat depending on the context (Estes et al. 1989). A species is considered extinct when no member of the species remains alive anywhere in the world: "The dodo bird is extinct." If individuals of a species remain alive only in captivity or other human-controlled situations, the species is said to be *extinct in the wild* (Figure 4.1): "The Franklin tree is extinct in the wild but grows well under cultivation." In both of these situations the species would be considered *globally extinct.* A species

4.1 Although it can still be found in arboretums and other cultivated gardens, the Franklin tree, *Franklinia alatamaha*, is extinct in the wild. (Photograph by John A. Lynch, New England Wild Flower Society.)

is considered to be *locally extinct* when it is no longer found in an area it once inhabited but is still found elsewhere in the wild: "The gray wolf once occurred throughout North America; it is now locally extinct in Massachusetts." Some conservation biologists speak of a species being *ecologically extinct* if it persists at such reduced numbers that its effects on the other species in its community are negligible: "So few tigers remain in the wild that their impact on prey populations is insignificant."

Past Rates of Extinction

The diversity of species found on the Earth has been increasing since life first originated. This increase has not been steady, but rather has been characterized by periods of high rates of speciation, followed by periods of minimal change and episodes of mass extinction (Sepkoski and Raup 1986; Raup 1988; Wilson 1989). This pattern is visible in the fossil record, which has been examined by scientists interested in determining the number of species and families that existed in particular geological periods over time.

The best-studied fossils are marine animals, many of which have hard body parts that are preserved in rocks formed from marine sediments. The evolutionary history of a group of marine animals can therefore be readily traced. Marine animals first arose about 600 million years ago, in the Paleozoic. According to the fossil record, new families of marine animals appeared in a rapid and steady succession during the next 150 million years (Figure 4.2). For the next 200 mil-

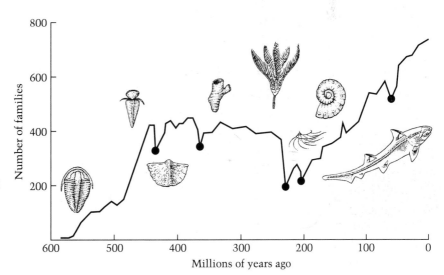

4.2 The number of families of marine organisms has been gradually increasing over geological time; this graph of their history clearly shows evidence of the five episodes of mass extinction discussed in Figure 4.3. (After Wilson 1989.)

lion years the number of families was more or less constant at around 400. Over the last 250 million years of the Mesozoic and Cenozoic eras, the diversity of families steadily increased to its present number of over 700. The fossil record of marine animals demonstrates the slow pace of evolutionary change, with new families appearing at a rate of roughly one per million years.

There have been nine episodes of high rates of extinction in the fossil record (Wilson 1989). Five of these episodes—during the Ordovician, Devonian, Permian, Triassic, and Cretaceous—could be called examples of natural mass extinction (Figure 4.3). Probably the most famous is the extinction of the late Cretaceous, 65 million years ago, during which dinosaurs became extinct and mammals achieved dominance in terrestrial communities. The most massive extinction took place at the end of the Permian, 250 million years ago, when 77–96% of all marine animal species are estimated to have gone extinct (Raup 1979). It is quite likely that some massive perturbation, such as wide-

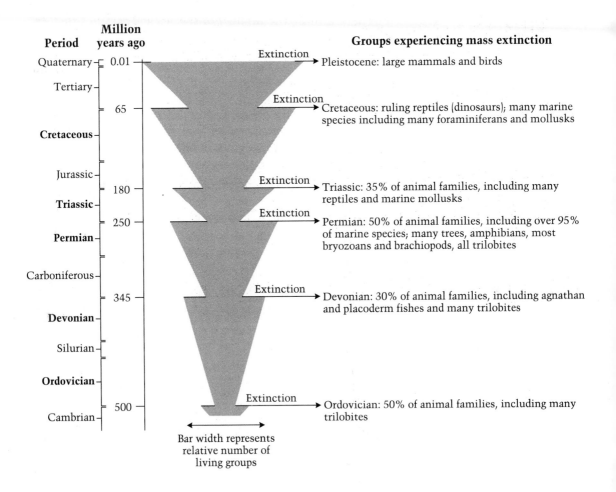

spread volcanic eruptions or a collision with an asteroid, caused such a dramatic change in the Earth's climate that many species were no longer able to exist. It took the process of evolution about 50 million years to regain the number of families lost during the Permian mass extinction.

Human-Caused Extinctions

The global diversity of species reached an all-time high in the present geological period. While species, genera, and families have been lost in the past, new ones have replaced them. The most advanced groups of organisms—insects, vertebrates, and flowering plants—reached their greatest diversity about 30,000 years ago. However, since that time humans have had a significant negative impact on the world's biota.

The first noticeable effects of human activity on extinction rates can be seen in the elimination of large mammals from Australia and North and South America at the time humans first colonized these continents. Shortly after humans arrived, 74–86% of the megafauna—mammals weighing more than 44 kg (100 lbs)—went extinct (Figure 4.4). These extinctions were arguably due to hunting by human bands (Martin 1973, 1986; Martin and Klein 1984), but other indirect anthropogenic effects also could have played a part: on all continents, there is an extensive record of prehistoric human alteration and destruction of habitat that has led to the reduction and extinction of species. For example, deliberate burning of savannahs, presumably to encourage plant growth for browsing by wildlife and so improve hunting, has been going on for 50,000 years in Africa (Murphy and Lugo 1986). For thousands of years, the total area of natural grassland and forest in North America, Central America, Europe and Asia has been steadily reduced to create pastures and farmlands to supply human needs.

How has human activity affected extinction rates? Extinction rates are best known for birds and mammals, since these species are relatively large, well-studied, and conspicuous. Extinction rates for other animals and for plant species, which account for 99.9% of the world's species, are just rough guesses at the present time. Yet extinction rates are uncertain even for birds and mammals, since species that are considered extinct have been rediscovered, and species that are presumed to be extant may actually be extinct. It has been argued that the number of extinct species is probably higher than is generally known because there are many remote areas scientists have not visited recently to determine the status of rare species there (Whitten et al. 1987; Diamond 1988b).

◄ 4.3 Although the total number of families and species of organisms has increased over the eons, during each of five episodes of natural mass extinction a large percent of these groups disappeared. The most dramatic period of loss occurred about 250 million years ago, at the end of the Permian period. A sixth episode, the Pleistocene extinction, incorporates the effects of hunting and habitat loss as human populations spread across the continents.

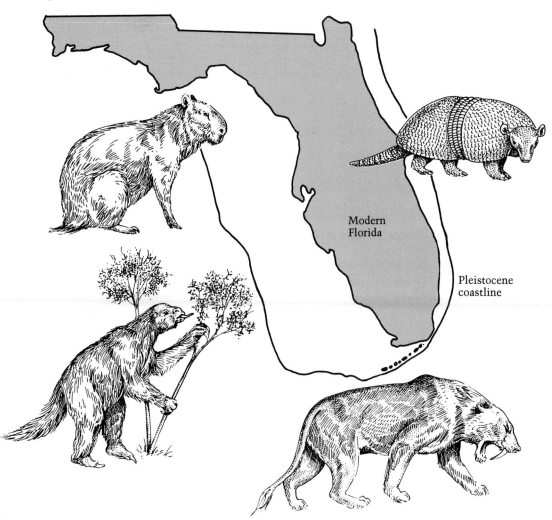

4.4   Many large mammals occupied Florida until the late Pleistocene (14,000–17,000 years ago), when they went extinct. Florida's land area then was greater than it is today; flooding caused by melting glaciers constricted the coastline to its present boundaries. Flooding does not completely explain the loss of these species, however, since a large land area remains. (From Eisenberg and Harris 1989.)

Based on the available evidence, 83 species of mammals and 113 species of birds are known to have gone extinct since the year 1600 (Table 4.1), representing 2.1% of mammal species and 1.3% of the birds (Nilsson 1983; IUCN 1988; Reid and Miller 1989). While this number initially may not seem alarming, the trend of these extinction rates is on the rise, with the majority of extinctions occurring in the last 150 years (Figure 4.5). The extinction rate for birds and mam-

TABLE 4.1
Recorded extinctions, 1600 to the present; numerous additional species have presumably gone extinct without being recorded by scientists

Taxon	Mainland[a]	Island[a]	Ocean	Total	Approximate number of species	Percent of taxon extinct since 1600
Mammals	30	51	2	83	4000	2.1
Birds	21	92	0	113	9000	1.3
Reptiles	1	20	0	21	6300	0.3
Amphibians	2	0	0	2	4200	0.05[b]
Fishes[c]	22	1	0	23	19,100	0.1
Invertebrates[c]	49	48	1	98	1,000,000+	0.01
Flowering plants[d]	245	139	0	384	250,000	0.2

Source: After Reid and Miller 1989; data from various sources.

[a] Mainland areas are those with land masses of 1 million km² or greater (the size of Greenland or larger); smaller land masses are considered to be islands.

[b] But see Box 9 in Chapter 6.

[c] The figures given are primarily representative of North America and Hawaii.

[d] The numbers for flowering plants include extinctions of subspecies and varieties as well as species.

4.5 Extinction rates for birds and mammals have been steadily increasing, with the most dramatic increase occurring within the last 150 years. (Data from Nilsson 1983; IUCN 1988.)

mals was about one species every 10 years during the period 1600–1700, but it rose to one species *every* year during the period 1850–1950. This increase in the rate of species extinction is an indication of the seriousness of the threat to biological diversity. Many species not yet technically extinct have been decimated by human activities and persist only in very low numbers. These species may be considered "ecologically extinct," in that they no longer play a role in community organization. About 2% of the world's remaining bird species and 5% of the mammals are in imminent danger of extinction if present threats to their existence are not halted (WRI/IIED 1987, 1988).

What is the natural rate of extinction in the absence of human influence? This rate can be estimated by looking at the fossil record, which documents the origin and exinction of species over time. From the fossil record, we can infer that an individual species lasts about 1 to 10 million years before it goes extinct or evolves into a new species (Raup 1978). Since there are perhaps 10 million species on the Earth today, it would be predicted that between 1 and 10 of the world's species would be lost each year as a result of natural extinction. These estimates are derived from studies of wide-ranging marine animals, and they may be lower than natural extinction rates for species with narrow distributions, which are more vulnerable to habitat disturbance. These estimates of natural "background extinction" rates appear to be applicable to terrestrial mammals and birds (Raup and Stanley 1978). The current observed rate of extinction of birds and mammals—1% per century, or 0.01% per year—is 100 to 1000 times greater than would be predicted based on the background rates of extinction. Putting it another way, about 100 species of birds and mammals were observed to go extinct between 1850 and 1950, but the natural rate of extinction would have predicted that at most only one of these species would have gone extinct. Therefore, the other 99 extinctions can be attributed to the effects of human activity.

The threat of extinction is greater for some groups of species than for others. Some groups are especially vulnerable for a combination of reasons, including high levels of human exploitation. For example, 17 of the world's 22 crocodile and alligator species are threatened with extinction not only because their habitat is disappearing but also because they are overhunted for their meat and skins. Throughout the world, large cat species (family Felidae) are hunted for sport, for their fur, and because they sometimes attack domestic animals and people. Slipper orchids have restrictive habitat requirements and are also overharvested by plant collectors. Within these and certain other groups, the majority of the species are in danger of extinction.

In most past geological periods, the extinction of existing species was balanced or even exceeded by the evolution of new species. However, the present rate of extinction far exceeds the known rate of evolution. Even presently known examples of rapid evolution, such as fruit flies that adapt to localized environments or polyploid plants that rapidly acquire new characteristics after their chromosomes double, have typically not produced unique new genera or families. Such unique evolutionary events require large numbers of generations and time measured in tens of thousands, if not millions, of years.

Extinction Rates on Islands

The highest species extinction rates during historic times have occurred on islands (IUCN 1988; Reid and Miller 1989). Most of the extinctions of birds during the last 350 years have occurred on islands (King 1985), and at least 90% of the endemic plants of oceanic islands are extinct or in danger of extinction (Dans et al. 1986). Island species are particularly vulnerable to extinction because many of them are endemic. A species is **endemic** to the location where it occurs naturally. A species might be endemic to a very large area; for example, the red maple tree is endemic to North America—it does not occur naturally elsewhere. However, the term is most often applied to species that are confined to a very small range; many island endemics are found on one particular island and in no other place. If the communities on that island are destroyed or damaged, or populations are intensively hunted or harvested, then the species will go extinct. In contrast, mainland species often have many populations over a wide area, so that the loss of one population is not catastrophic for the species.

Island species have usually evolved and undergone speciation with reduced levels of competition, predation, and threat of disease. In contrast, in mainland areas competition, predation, and disease have been powerful agents of natural selection. When predatory and competitive species from the mainland are introduced onto islands, they decimate the island species, which have not evolved any defenses against them (see Chapter 7).

The best-studied island group is the Hawaiian archipelago (Olson 1989; Olson and James 1982). There were 98 species of endemic birds in the Hawaiian islands before the arrival of the Polynesians in 400 A.D. The Polynesians introduced the Polynesian rat, the domestic dog, and the domestic pig; they also began clearing the forest for agriculture. As a result of increased predation and disturbance, about 50 of the bird species became extinct prior to the arrival of Europeans in 1778. The Europeans brought cats, new species of rats, the Indian mongoose, goats, cattle, and the barn owl; they also unwittingly brought bird diseases, and they cleared even more land for agriculture and human settlements. During the past 200 years, an additional 17 bird species have gone extinct. The extinctions over the past 1600 years account for roughly 70% of the original bird fauna. Many of the remaining species have low population numbers and are near extinction; ultimately, human disturbance and the introduction of exotic species could result in the extinction of all endemic bird species in the islands.

Island plant species are also threatened with extinction, mainly through habitat destruction (Table 4.2). In Hawaii, fully 91% of the naturally occurring plant species are endemic to the islands. About 10% of these endemic species have gone extinct, and 40% of the remaining endemics are threatened with extinction (Davis et al. 1986). In Madagascar, 80% of the plant species are endemic and threatened with extinction (Jenkins 1987; IUCN 1988). At this point about 80% of Madagascar's land area has been altered or destroyed by human activity, possibly leading to almost half of the endemic species being lost. To protect its rich natural heritage, Madagascar is moving toward setting up national parks, but only 1.5% of the island has been preserved so far (see Box 5 in Chapter 6).

A similar pattern can be found on other islands. The colonization of New Zealand by Polynesians in 1000 A.D. led to hunting, deforestation, and the introduction of the dog and the rat. Before the arrival

TABLE 4.2
Number of plant species and their status for various islands and island groups

Island(s)	Native species	Endemic species[a]	Percent endemic	Number of threatened endemics	Percent of endemics threatened
Ascension	25	11	44	9	82
Azores	600	55	9	23	42
Galápagos	543	229	42	135	59
Hawaiian	970	883	91	±353	40
Juan Fernandez	147	118	80	93	79
Madeira	760	131	17	86	66
New Caledonia	3250	2474	76	146	6
New Zealand	2000	±1620	81	±132	8
Norfolk	174	48	28	45	94
Rodrigues	145	40	28	36	90

Source: After Reid and Miller 1989; data from Davis et al. 1986 and Gentry 1986.

[a]Endemic species are found only on those islands. For example, of the 25 plant species living on Ascension Island, 11 are found nowhere else; these endemic species represent 44% of the Ascension flora.

of Europeans, all 13 species of the giant flightless moa and 16 other endemic bird species went extinct (Olson 1989).

European colonization of islands has generally been even more destructive than colonization by other peoples because of the greater amounts of clearing and the wholesale introduction of species characteristic of European occupation. As an example, between 1840 and 1880 more than 60 species of vertebrates, particularly grazing animals such as sheep, were deliberately introduced into Australia, where they displaced native species and altered many communities (K. Myers 1986). The first European visitors to the Mascarene Islands (Mauritius, Reunion, and Rodrigues) in the 1500s released monkeys and pigs (Nilsson 1983). These introduced animals and the subsequent Dutch colonization led to the extinction of the dodo, 19 other species of birds, and 8 species of reptiles. The impact of introduced predators on island species is highlighted by the example of the flightless Stephen Island wren, a bird that was endemic to a tiny island off New Zealand. Every bird on the island was killed by a single cat belonging to the lighthouse keeper (Diamond 1984); even one introduced predator can eliminate an entire species.

The vulnerability of island species to extinction is further illustrated by comparing the number of species that have gone extinct in mainland areas, on islands, and in the oceans from 1600 to the present (see Table 4.2). Of the 724 species of animals and plants known to have gone extinct, 351 species (about half the total) were island species, even though islands represent only a small fraction of the Earth's surface. In contrast, it has been estimated that only five species—four species of marine mammal and a limpet—have gone extinct in the world's vast oceans. This calculation is almost certainly an underestimate, because marine species are not as well known as terrestrial species, but it may also reflect a greater resiliency among marine species in response to disturbance. However, the significance of these marine losses may be greater than the numbers suggest, since many of the marine mammals are top predators that have a major impact on marine communities. Also, marine species have greater diversity at the phylum level than do terrestrial species, so the extinction of even a few marine species can represent a serious loss to global biological diversity.

The majority of the mammals, birds, and reptiles that have gone extinct have been island species, while the majority of fish and higher plant extinctions have been on mainland areas. In a recent survey of the rich freshwater fish fauna of the Malay Peninsula, the majority of the 266 species of fish known to exist on the basis of earlier collections could not be found (Mohsin and Ambak 1983). In

North America, over one-third of freshwater fish species are in danger of extinction (Moyle and Leidy, 1992).

Island Biogeography and Modern Extinction Rates

Studies of island communities have led to general rules on the distribution of biological diversity, synthesized as the **island biogeography model** of MacArthur and Wilson (1967). The central rule of this model is the **species–area relationship**, which states that islands with large areas have more species than islands with smaller areas (Figure 4.6). This rule makes intuitive sense, since large islands will tend to have a greater variety of local environments and community types than small islands. Also, larger islands allow greater geographic isolation and a larger number of populations per species, increasing the likelihood of speciation and decreasing the probability of local extinction of newly evolved as well as recently arrived species. The exact form of the species–area relationship can be accurately described by the empirical formula

$$S = CA^Z$$

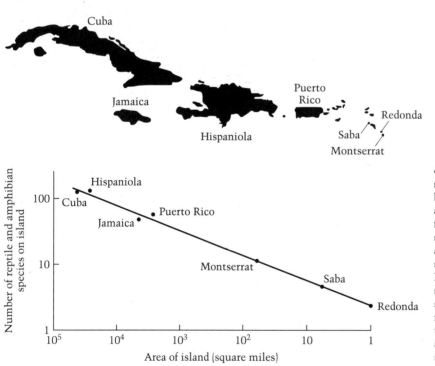

4.6 **The number of species on an island can be predicted from the area of an island. In this figure, the number of species of reptiles and amphibians is shown for seven islands in the West Indies. The number of species on large islands such as Cuba and Hispaniola far exceeds that on the tiny islands of Saba and Redonda. (From Wilson 1989.)**

where S is the number of species on an island, A is the area of the island, and C and Z are constants. The exponent Z determines the slope of the curve. The values for Z and C will depend on the type of island (tropical versus temperate, dry versus wet, etc.), and the type of species involved (birds versus fish, etc.) (Simberloff 1986a). Z values are typically around 0.25, with a range of values from 0.15 to 0.35 (Connor and McCoy 1979). Values of C will be high in groups that are high in species numbers, such as insects, and low in groups that are low in species numbers, such as birds.

Imagine the simplest situation, in which $C = 1$ and $Z = 0.25$, for raptorial birds on a hypothetical archipelago:

$$S = (1)A^{0.25}$$

Islands of 10, 100, 1000, and 10,000 km² in area would be predicted to have 2, 3, 6, and 10 species, respectively. It is important to note that a tenfold increase in island area does not result in a tenfold increase in the number of species; with this equation, each tenfold increase in island area increases the number of species by approximately a factor of 2.

The island biogeography model has been empirically validated to the point where it is now accepted by most biologists. However, the explanation for this relationship remains unresolved. As postulated by MacArthur and Wilson, the number of species occurring on an island represents a dynamic equilibrium between the arrival of new species (and also the evolution of new species) and the extinction rate of existing species (Figure 4.7). Starting with an unoccupied island,

4.7 The island biogeography model describes the relationship between the rates of colonization and extinction on islands. The immigration rate (black curves) on unoccupied islands is initially high, as species with good dispersal abilities rapidly take advantage of the available open habitats. The immigration rate slows as the number of species increases and sites become occupied. The extinction rate (gray curves) *increases* with the number of species on the island: the more species on an island, the greater the likelihood that a given species will go extinct at any time interval. Colonization rates will be highest for islands near a mainland population source, since species can disperse over shorter distances more easily than longer ones. Extinction rates are highest on small islands, where both population sizes and habitat diversity are low. The number of species present on an island reaches an equilibrium when the colonization rate equals the extinction rate (circles). The equilibrium number of species is greatest on large islands near the mainland, and lowest on small islands far from the mainland. (After MacArthur and Wilson 1967.)

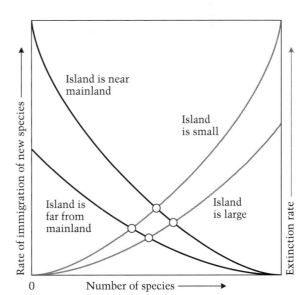

the number of species will increase over time as more species arrive than go extinct, until the rates of extinction and immigration are balanced. The extinction rate will be lower on large islands than on small ones because large islands have more habitat diversity, larger populations, and a greater number of populations. The rate of immigration of new species will be higher for islands near the mainland than for islands farther away, since mainland species are able to disperse to near islands more easily than to distant ones. The model predicts that for any group of organisms, such as birds or trees, the number of species found on large islands near a continent will be greater than on small islands far from a continent.

Species–area relationships have been used to predict the number and percentage of species that would go extinct if habitats were destroyed (Simberloff 1986a). The assumption is that if an island has a certain number of species, reducing the area of natural habitat on the island would result in the island being able to support only a number of species corresponding to that on a smaller island (Figure 4.8). This model has been extended from islands to national parks and nature reserves that are surrounded by damaged habitat (see Chapter 14); the reserves can be viewed as **habitat islands** in an inhospitable "sea" of unsuitable habitat. The models predict that when 50% of an island (or a habitat island) is destroyed, approximately 10% of the species occurring on the island will be eliminated. If these species are endemic to an area, they will go extinct. When 90% of the habitat is destroyed, 50% of the species will be lost; and when 99% of the habitat is gone, about 75% of the original species will be lost.

The habitat island approach has been applied to tropical rain forests in particular. Since the insects and plants in tropical forests account for the great majority of the world's species, estimating present and future rates of species extinction in rain forests gives an approximation of global rates of extinction. At present rates of deforestation, an estimated 15% of the plant species in the Neotropics will go extinct between 1986 and 2000, and 12% of Amazon bird species will go extinct (Simberloff 1986a). If deforestation continues until all of the forests except those in national parks and other protected areas were cut down, about two-thirds of all plant and bird species will be driven to extinction.

In the past, rates of deforestation have been more rapid in the Old World tropics—Africa, Madagascar, Asia and the Pacific region—than in the Neotropics, resulting in a loss of between 10% and 25% of the original species there. Losses of rain forest species in Asia and the Pacific are predicted to *increase* during the period 1990 to 2020 due to the combination of rapid population growth, increasing economic development, and a huge timber industry; 7–17% of all Asian and Pa-

4.8 The number of species present in an area increases asymptotically to a maximum value. As a result, if the area of habitat is reduced by 50%, the number of original endemic species going extinct may be 10%; if the habitat is reduced by 90%, the number of endemic species going extinct may be 50%. The shape of the curve is different for each region of the world and each group of species, but it gives a general indication of the impact of habitat destruction on species extinction and the persistence of species in the remaining habitat.

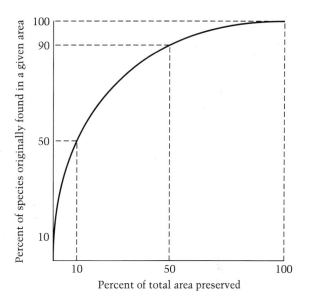

cific species may go extinct during this period (Reid and Miller 1989).

Using a conservative value of 1% of the world's rain forests being destroyed per year, Wilson (1989) estimated that 0.2–0.3% of all species—20,000 to 30,000 species if based on a total of 10 million species—would be lost per year. In more immediate units, 68 species would be lost per day, with 3 species lost each hour. Over the 10-year period from 1993 to 2003, approximately 250,000 species would go extinct. Other methods applied to the rates of extinction in tropical rain forests estimate a loss of between 2% and 11% per decade (Reid and Miller 1989). The variation in rates is caused by different estimates of the rate of deforestation and different values for the species–area curves. Regardless of which figure is the most accurate, all of these estimates indicate that hundreds of thousands of species are doomed to extinction within the next 50 years. These rates of species extinctions are without precedent since the great mass extinction of the Cretaceous period 65 million years ago.

These estimates make a number of assumptions and generalizations in calculating species extinction rates. First, all of them are based on typical values for the species–area curves. Groups of species with broad geographical ranges, such as marine animals and temperate tree species, tend to have lower rates of extinction than species with narrow distributions, such as island birds and freshwater fishes. Second, the models assume that all endemic species are eliminated from areas that have been cleared of forest. It is possible that many species can survive in isolated patches of forest and recolonize secondary forest that develops on abandoned land. A few primary forest species may also be able to adapt to plantations and managed forests. This adaptation to managed forests is particularly likely to be significant in Asian forests that are being selectively logged on a large scale. Third, the species–area models assume that areas of habitat are eliminated at random. In fact, species-rich areas are sometimes targeted for conservation efforts and national park status, as was the Hol Chan Marine Park in Belize (see Box 17 in Chapter 9). As a result, a greater percentage of species may be protected than is assumed in the species–area models. And fourth, the degree of habitat fragmentation may affect extinction rates. If remaining areas of habitat are divided into very small parcels or crossed by roads, then wide-ranging species as well as species requiring large population sizes may be unable to maintain themselves. Also, hunting by people, clearing land for agriculture, and the introduction of exotic species may be increased by forest fragmentation, leading to the further loss of species.

A vital question in conservation biology is, How long will it take for a given species to go extinct following a reduction in area or fragmentation of its habitat? When populations fall below a certain critical number of individuals, they are very likely to become extinct (see Chapter 11). In some species, a few individuals might persist for years or decades, and even reproduce, but their ultimate fate would be extinction. For woody plants in particular, individuals can persist for hundreds of years. Species that are doomed to extinction following habitat destruction have been called "the living dead": even though technically the species is not extinct while these individuals live, the population is no longer reproductively viable, so the species' future is limited to the life spans of the remaining individuals (Janzen 1986; Gentry 1986). Evidence from forest fragments and parks indicates that following the destruction of the surrounding habitat, species diversity of vertebrates may actually show a temporary increase as animals flee into the few remaining patches of forest (Bierregaard et al. 1992). Following this temporary increase, however, the number of species falls during the succeeding weeks, months, and years as species go extinct on a local scale and are not replaced by other species.

Local Extinctions

In addition to the global extinctions that are the primary focus of this chapter, many species are experiencing a series of local extinctions across their range. Formerly widespread species are now often restricted to a few small pockets of their former habitat. As an example, the American burying beetle (*Nicrophorus americanus*) was once found all across central and eastern North America, but is now found only in three isolated populations. Biological communities are impoverished by such local extinctions. The Middlesex Fells, a local conservation area in metropolitan Boston, contained 338 native plant species in 1894; only 227 native species remained when the area was surveyed 98 years later (Drayton 1993). Fourteen of the plant species that were lost had been listed as "common" in 1894. These large numbers of local extinctions serve as important biological warning signs that something is wrong with the environment. Action is needed to prevent further local extinctions as well as the global extinctions of species that are known to be experiencing local extinction on a massive and widespread scale.

Summary

1. The present geological period has more species than there have ever been in the past. However, the current rate of species extinctions is so rapid that it can be compared to the five episodes of natural mass extinction found at intervals in the geological record.

2. The effect of human activity has been to drive many species to extinction. Since 1600, 2.1% of the world's mammal species and 1.3% of the bird species have gone extinct. The rate of extinction is accelerating, and many extant species are teetering on the brink of extinction. The current observed rate of extinction for birds and mammals is between 100 and 1000 times greater than would be expected to occur naturally.

3. Species that occupy islands are most vulnerable to extinction, because these species occupy only a small area and they are often unable to defend themselves against humans and the exotic species and diseases humans bring to the islands. Among aquatic species, freshwater species appear to be more vulnerable to extinction than marine species.

4. An island biogeography model has been developed to predict the equilibrium number of species that might be found on islands of different areas and distances from the mainland. This model has been used to predict how many species will go extinct if human activity continues to destroy habitats at the present rate. The best evidence indicates that about 2.5% of the Earth's species will be lost over the next 10 years, with a loss of about 25,000 species per year.

5. Individuals of long-lived species that remain alive in severely disturbed and fragmented habitats can be considered "the living dead." The individuals may persist for many years, but the species will eventually die out due to lack of reproduction. Biological communities are gradually impoverished by the local extinction of species.

Suggested Readings

Adams, D. and M. Carwardine. 1990. *Last Chance to See.* Harmony Books, New York. A light but poignant account of the imminent threat of extinction facing many well-known species.

Janzen, D. H. 1986. The eternal external threat. *In* M. Soulé (ed.), *Conservation Biology: The Science of Scarcity and Diversity*, pp. 286–230. Sinauer Associates, Sunderland, MA. A superb essay on the causes of tropical extinctions, with vivid natural history examples.

Kaufman, L. and K. Mallory (eds.). 1986. *The Last Extinction*. MIT Press (in cooperation with the New England Aquarium), Cambridge, MA. Essays on the threats faced by groups of species, with a particularly strong treatment of marine examples.

MacArthur, R. H. and E. O. Wilson. 1967. *The Theory of Island Biogeography*. Princeton University Press, Princeton, NJ. This classic text outlining the island biogeography model has been highly influential in shaping modern conservation biology.

Myers, N. 1979. *The Sinking Ark: A New Look at the Problem of Disappearing Species.* Pergamon Press, New York. An important book that was an early warning of the biodiversity crisis.

Olson, S. L. 1989. Extinction on islands: Man as a catastrophe. *In* M. Pearl and D. Western (eds.). *Conservation Biology for the Twenty-First Century*, pp. 50–53. Oxford University Press, New York. Human impact on islands is demonstrated.

Sepkoski, J. and D. M. Raup. 1986. Periodicity in marine extinction events. *In* D. K. Elliot (ed.), *Dynamics of Extinction*, pp. 3–36. Wiley, New York. Past rates of extinction calculated using fossil evidence.

Simberloff, D. 1986. Are we on the verge of a mass extinction in tropical rain forests? *In* D. K. Elliott (ed.), *Dynamics of Extinction*, pp. 165–180. Wiley, New York. A demonstration of the application of the island biogeography model to calculate extinction rates.

Wilson, E. O. 1989. Threats to biodiversity. *Scientific American* 261(September): 108–116. How extinction rates are increasing due to human activities.

World Conservation Monitoring Centre. 1992. *Global Biodiversity 1992: Status of Earth's Living Resources*. World Resources Institute, Washington, D.C. The status of species and communities is shown.

CHAPTER 5

Vulnerability to Extinction

The protection of rare species is an important focus of conservation efforts. Rare species are considered to be especially vulnerable to extinction, while common species are considered less so. But the term *rare* has a variety of meanings in the biological literature, each of which has implications for conservation biology (Rabinowitz et al. 1986).

A species may be considered rare if it occupies a narrow geographical range. For example, the Venus's flytrap (*Dionaea muscipula*) occurs only in the savannahs of the coastal plain of the Carolinas in eastern North America. This contrasts with a species such as the red maple (*Acer rubrum*) that occurs in a variety of habitats throughout eastern North America. Many geographically rare species occupy islands, such as the Hawaiian Islands, that are surrounded by vast expanses of inhospitable ocean. Geographically rare species may also occupy isolated habitats such as high mountain peaks in the middle of lowlands, or lakes surrounded by a terrestrial landscape. Within their limited geographical range, however, a rare species may be locally abundant (Figure 5.1). Species may be geographically rare in only part of their range. For example, the sweet bay magnolia (*Magnolia virginiana*) is reasonably common throughout the southeastern United States, but in the New England region it occurs in only one population of 100 individuals in one particular swamp in Magnolia, Massachusetts (Primack et al. 1986). Species may have always had a narrow geographical range, or they may have formerly been widespread but are now restricted due to some human activity, such as habitat destruction.

A species may also be considered rare if it occupies only one or a few specialized habitats. Salt marsh cord grass (*Spartina patens*) is found only in salt marshes and not in other habitats; yet within this

5.1   An example of a rare wildflower, the Plymouth gentian (*Sabatia kennedyana*) is found only on the margins of coastal ponds in scattered locations in the eastern United States. At a few ponds, this species has large populations. (Photograph by Mark Primack.)

habitat, cord grass is quite common. (This example contrasts with common species that are found in many different habitats, such as the dandelion, which occupies a wide range of habitats, such as open meadows, roadsides, river edges, and mown lawns.)

Finally, a species may be considered rare if it is only found in small populations. A common species would at least sometimes have large populations.

These three components of rarity can be applied to the entire range of species, or to the distribution and abundance of species in a particular place. In a study of plants in the British Isles, one of the biologically best-known areas of the world, Rabinowitz et al. (1986) used these three components of rarity to designate eight categories of species (Table 5.1). Under this system, a species of wide geographical distribution, broad habitat specificity, and large local population size (such as lamb's quarters, *Chenopodium album*) would be considered a classic common species. In contrast, a species with a narrow geographical range, restricted habitat specificity, and always a small population size (such as the alpine lily, *Lloydia serotina*) would be considered a classic rare species. However, there are also species that combine elements of rarity and abundance, such as the Scottish bird's-eye primrose (*Primula scotica*), which has a narrow geographical range, a broad habitat specificity, and sometimes large popula-

TABLE 5.1
Categories of rarity for 160 plant species in the British Isles based on geographic distribution, habitat specificity, and local population size[a]

Local population size	Geographic distribution	
	Large	Small
Wide habitat specificity		
Somewhere large	58 spp.	6 spp.
Always small	2 spp.	0 spp.
Narrow habitat specificity		
Somewhere large	71 spp.	14 spp.
Always small	6 spp.	3 spp.

Source: After Rabinowitz 1981 and Rabinowitz et al. 1986.

[a] Based on the three criteria, 58 species are common; 3 species are rare by all criteria; and the remaining 99 species exhibit some traits of rarity.

tions; or the knotroot bristlegrass (*Setaria geniculata*), which has a broad geographical distribution and broad habitat specificity, but always occurs in small populations.

In analyzing these patterns of rarity and commonness, the most striking finding was that 149 out of 160 plant species in Britain have large populations at least somewhere in their ranges. This stands in apparent contradiction to the observations of community ecologists that at a local scale there tend to be only a few common species but many rare species. These contrasting observations can be reconciled because even though most species will have large local populations somewhere in their range, each species will be rare at the majority of sites that it occupies.

Most of the rare plant species in the British Isles (71 out of 102 species) have a wide range, are habitat specialists, and have large populations at least somewhere. The next-largest group of rare species (14 species) are geographically restricted habitat specialists that have large populations at least somewhere. Small numbers of species occupied the other categories of Table 5.1, with one exception: there appear to be no species with a narrow geographical range, broad habitat specificity, and always small populations.

This system of classification highlights priorities for conservation. Species with a narrow geographical range and specific habitat requirements that are always found in small populations require habitat protection and possibly habitat management to maintain their few, fragile populations. This also applies, to a somewhat lesser degree, to species with larger populations. However, where species have a narrow geographical distribution but a broad habitat specificity, experiments in which individuals are transported to unoccupied but apparently suitable locations to create new populations may be a strategy worth considering, since these species may have been unable to disperse outside of their narrow geographical area. Also, species with broad geographical ranges are less susceptible to extinction and less likely to need rescue efforts, since they tend to have more extant populations and more opportunities to colonize potentially suitable sites.

Endemic Species

A species that is found in only a single geographical area and nowhere else is said to be **endemic** to that area (see Chapter 4). Recent changes in species distributions caused deliberately or accidentally by people are not considered part of the species' natural distribution. For example, the panda is considered endemic to China, even though it now lives in zoos throughout the world; the English plantain is en-

TABLE 5.2
Total plant species and endemic plant species in some continental regions in temperate and tropical zones

Region	Area (km²)	Total number of species	Number of endemic species	Percent endemic species
Southern Africa	2,573,000	18,550	±14,800	80
Cape Region of South Africa	90,000	8,578	5,850	68
Southwest Australia	320,000	3,600	±2,450	68
Europe	10,000,000	10,500	±3,500	33
California, USA	411,000	5,046	1,517	30
Panama	75,000	6,800	±1,034	15
Northeastern North America	3,238,000	4,425	599	14
Texas, USA	751,000	4,196	379	9
Carolinas, USA	217,000	2,995	23	1
British Isles	308,000	1,443	17	1

Source: After Gentry 1986. Data compiled from various sources.

5.2 The number of plant species endemic to each state in the United States. For example, 379 plant species are found only in Texas and no-where else in the United States. California—with its large area and a vast array of habitats includ-ing deserts, mountains, seacoast, old-growth for-est, and myriad others—is home to more endemic species than any other state. The island archi-pelago of Hawaii, far from the mainland, hosts many endemic species despite its small area. (From Gentry 1986.)

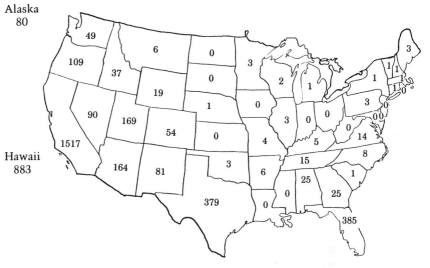

demic to Europe, but has been carried throughout the world in seed samples and soil, establishing itself as a common weed. Species may be endemic to a wide geographical area, such as the black cherry tree (*Prunus serotina*) found across North America, Central America, and South America; or species may be endemic to a small geographical area, such as an island. The giant Komodo dragon, for example, is known from only several small islands in the Indonesian archipelago; a more extreme instance is the Haleakala silversword plant, which is found in only one volcano crater on the island of Maui. Isolated geo-graphical units, such as remote islands, old lakes, and solitary moun-tain peaks, often have high percentages of endemic species. In con-trast, geographical units of equivalent area that are not isolated typically have much lower percentages of endemic species (Table 5.2).

One of the most noted examples in the world is the island of Madagascar. Here the tropical moist forests are spectacularly rich in endemic species: 93% of 28 primate species, 99% of 144 species of frogs, and over 80% of the plant species on the island are endemic to Madagascar (Box 5).

Among the United States, it is not surprising that Hawaii has a large number of endemic species (Figure 5.2); the large numbers of plant species endemic to California, Texas, and Florida reflect the fact that these states are large in area, were never covered by glaciers (and thus have had a longer time span for speciation to occur), con-tain a considerable variety of habitats, and have better growing condi-tions than many other U.S. states.

BOX 5 CONSERVING MADAGASCAR'S ENDEMIC FORESTS: A LAST-DITCH EFFORT?

The island of Madagascar in the Indian Ocean off the coast of East Africa has had a long and unique history. For at least 50 million years, the flora and fauna of this miniature continent have developed in isolation from the mainland species of Africa and Asia. An estimated 85% of all the plant and animal species found on Madagascar are endemic to the island (Cohn 1985; Fowler et al. 1989). Many genera and even some families of vascular plants on the island are found nowhere else (Koechlin et al. 1974). Furthermore, the irregular topography of the island contributes to abrupt changes in microclimatic conditions, creating different patterns of rainfall and seasonality in different regions of the islands. These conditions restrict the zones in which particular vegetation types may be found; consequently, some species endemic to Madagascar are found only in a small section of the island. Among the unique animals living in Madagascar's forests are 28 or more species of lemurs, thought to be the most ancient living representatives of the primate order (IUCN 1990).

Since the arrival of human beings on the island some 1500 years ago, much of the natural landscape of Madagascar has been eradicated (Rauh 1990). Almost 90% of the orig-

Endangered endemic plant species on Madagascar occur in all of the island's myriad microhabitats. In the arid areas in particular, vascular plants have speciated in unique patterns. The *Didierea* bush (left) and *Pachypodium lamerei* tree (right) are two examples. (Photographs courtesy of Dr. W. Rauh.)

inal forest cover is gone; only one-quarter of what remains is primary forest (Cohn 1985; Fowler et al. 1989). The need for agricultural land to support Madagascar's human population has destroyed most of the forests on the island, and the pressure on the remaining forests has not yet been halted. The island's population, estimated at 9 million in 1980, will exceed 12 million by the turn of the century; most of these people support themselves by farming using "slash-and-burn" methods, clearing pasture for zebu cattle, and cutting trees for fuel. As a result of these activities, the remaining forest is gradually be-

coming degraded, pushing already hard-pressed plant and animal species closer to extinction. In addition, competition from introduced species has further reduced the survival of many endemic plant species, often preventing them from recolonizing degraded areas altogether.

The prognosis for Madagascar's endemic species is desperately grim in light of these factors (Tattersall 1993). While some attempts at conservation have been made by the government of Madagascar, local officials are hampered by lack of funds, personnel, training, and popular awareness. For example, the Ankarana Special Reserve was created in 1956 to protect a tropical deciduous forest that contains a very high density of unique species (Hawkins et al. 1990). The reserve's forests are nevertheless subject to encroachment by farmers, loggers, and cattle herders, most of whom are completely unaware that the reserve even exists; there are no distinct boundaries denoting reserved land, no patrols to enforce the boundaries, and little public outreach to increase popular awareness and support of conservation issues. That the Ankarana Reserve has maintained much of its biological diversity is as much due to a geophysical accident as to the reserve's existence: the reserve encompasses rugged terrain that has many isolated pockets of primary forest hidden in canyons and sinkholes in the limestone massif, some of which are accessible only through a long series of underground caverns, and others which are not accessible at all by ground (Fowler et al. 1989; Hawkins et al. 1990). These canyons were recently surveyed by British and Malagasy scientists. All have populations of rare endemic plants and animals; many had never been visited by humans before,

and are relatively safe from the factors threatening other parts of Madagascar. These areas are very small, however, and are not sufficient to support viable populations of many species; the largest canyons in the reserve are accessible to human populations, and are becoming degraded as villagers begin to use them for agriculture.

Conservation on the island is not a hopeless cause, however. Intensive efforts by the International Union for the Conservation of Nature (IUCN), the World Wildlife Fund, and other international conservation organizations have promoted research, education, and fund-raising activities both in Madagascar and internationally (Fowler 1989). Improving surveillance of the existing reserves and the creation of new reserves are the current priorities. Maintaining threatened species' populations until the security of the remaining habitat can be established is another. Captive breeding programs sponsored by the Duke University Primate Center have successfully established populations of many of the island's 28 species of lemurs, providing the basis for a future reintroduction or release program once stable,

The aye-aye, *Daubentonia madagascariensis*, is one of the endangered lemur species of Madagascar. The aye-aye is the subject of conservation efforts in the field as well as captive breeding programs. All 28 lemur species—an entire order of primates found nowhere else in the world—are in danger of becoming extinct on Madagascar. (Photograph © David Haring, Duke University Primate Research Center.)

long-term habitat restoration and preservation programs become a reality (Cohn 1985). Additional proposals for sustainable development, particularly with regard to ecotourism, are currently under examination; the spectacular cavern system of the Ankarana Reserve, for example, has the potential to be of major economic importance as a tourist attraction. While the situation in Madagascar is alarming, enough of the island's intriguing natural diversity remains to present reason for hope.

Vulnerability to Extinction

When environments are damaged by human activity, the population sizes of many species will be reduced, and some species will go extinct. Ecologists have observed that not all species have an equal

5.3 Species of desert pupfish of the southwestern United States are highly endangered by the degradation and disappearance of their unique habitat—saline desert ponds. (Photo by Ken Kelley, San Diego Zoo.)

probability of going extinct; particular categories of species are most vulnerable to extinction (Ehrenfeld 1970; Terborgh 1974; Pimm et al. 1988). These species need to be carefully monitored and managed in conservation efforts.

Species with a very narrow geographical range. Such species may occur at only one or a few sites in a restricted geographical range, and if that whole range is affected by human activity, then the species may go extinct. Bird species on oceanic islands are good examples of species with restricted ranges that have gone extinct; many fish species confined to a single lake or a single watershed have gone extinct. Species of desert pupfish in the American Southwest, in which a species may be found only in one small pond, are extremely vulnerable to extinction (Figure 5.3)

Some species are confined to one unusual habitat type that is scattered and rare across the landscape. Examples of such habitats are serpentine outcrops in California, granite outcrops in the southeastern United States, and isolated high mountains in the northeastern United States. In commenting on the Sheffield flora in Britain, Hodgson (1986) said: "Most of the differences between common and rare species in the Sheffield region may be interpreted in terms of the availability of suitable habitats within the region, with common species occupying common, and rare species less common, habitats." Such examples illustrate the importance of habitat preservation in the conservation of species with a narrow range.

Species with only one or a few populations. Any one population of a species may go extinct as a result of chance factors, such as earthquakes, fire, an outbreak of disease, or human activity. Species with many populations are less vulnerable to extinction than are species that consist of only one or a few populations.

Species in which population size is small. Small populations are more likely to go locally extinct than large populations due to their greater vulnerability to demographic and environmental variation and loss of genetic variability (see Chapter 11). Species that characteristically have small population sizes, such as large predators or extreme specialists, are more likely to go extinct than species that sometimes have large populations. Also, when the population size of a species is reduced by habitat destruction and fragmentation, the loss of the species often follows.

Population size by itself seems to be one of the best predictors of the extinction rate of isolated populations (MacArthur and Wilson 1967; Richter-Dyn and Goel 1972; Pimm et al. 1988). An excellent example is provided by the survival of bird species at the Bogor Botanical Garden in Java, a woodland and arboretum that has been isolated for 50 years (Diamond et al. 1987). At this site, only 25% of the bird species that had small population sizes during the period between 1932 to 1952 survived in the 1980s, whereas all of the species that were initially common survived. The bird community of the Botanical Garden has also come to reflect the bird community of the surrounding disturbed countryside, indicating the significance of colonization in determining the species composition of isolated habitat fragments. These results were confirmed in studies of isolated forest fragments in Brazil; the persistence of individual forest species after several decades of isolation was related to the size of the forest fragment and the initial abundance of the species (Willis 1979; Bierregaard et al. 1992). Larger fragments had more forest species than smaller fragments, and species with high initial populations were far more likely to persist than species with low initial populations.

Species with low population density. A species with a low population density—with few individuals per unit of area—will tend to have only small populations remaining if its range is fragmented by human activities. Within each fragment the species may be unable to persist, and it will gradually die out across the landscape.

Species that need a large home range. Species in which individual animals or social groups need to forage over a wide area are prone to die off when part of their range is damaged or fragmented by human activity.

Species that have large body size. Large animals tend to have large individual ranges, require more food, and are more easily hunted out by humans. Top carnivores in particular are often killed by humans because they compete with humans for wild game, sometimes damage livestock, and are hunted for sport. Within guilds of species, often the largest species will be the most prone to extinction: i.e., the largest carnivore, the largest lemur, the largest whale. Sri Lanka can be used to illustrate this point, with the largest carnivores—leopards and eagles—and the largest herbivores—elephants and deer—having the greatest risk of extinction (Erdelon 1988). Countering this tendency to some degree is the tendency for species with large body size to live longer than smaller species. Also, among Neotropical forest mammals, large body size also tends to be correlated with a wider geographic distribution and a lower local density (Arita et al. 1990).

Species with low rates of population increase. Certain species (often called *K*-strategists) that live in stable habitats tend to delay reproduction to an advanced age and produce only a few, large young. Such species are more vulnerable to extinction than species (often called *r*-strategists) that produce many offspring at an earlier age and often occupy unstable, changing habitats. *K*-strategists are often unable to rebuild their populations fast enough to avoid extinction following habitat destruction.

Species that are not effective dispersers. Because the local environment is always changing, the ultimate fate of all populations is extinction. However, as a result of human activity, the pace of local extinction is increasing. The only long-term strategy for a species is to colonize new habitats when old ones are destroyed or become unsuitable. Species that are unable to cross the roads, farmlands, and disturbed habitats created by human activity are doomed to extinction as their original habitat becomes affected by pollution, exotic species, and global climate change. In particular, many animal species in isolated forest fragments are unable to cross pastures and colonize unoccupied areas of forest (Lovejoy et al. 1986).

The importance of dispersal in preventing extinction is illustrated by two studies from Australia. The first, a detailed analysis of the vertebrates of Western Australia, revealed that modern extinctions were almost exclusively confined to nonflying mammals, with few or no extinctions recorded among birds and bats (Burbidge and McKenzie 1989). Among the birds, species that are unable to fly or are poor fliers have shown the greatest tendency toward extinction. In the second study, which examined 16 nonflying mammal species in

Queensland rain forests, the most important characteristic determining the ability of species to survive in isolated forest fragments was their ability to use, feed on, and move through the intervening matrix of secondary vegetation (Laurance 1991a). While large-bodied, long-lived, low-fecundity species initially appeared to be more vulnerable to extinction, this effect disappeared when the abundance of each species in secondary vegetation was taken into account. This study highlights the importance of maintaining secondary vegetation to the survival of primary forest species.

Species that migrate. Species that migrate seasonally depend on two or more distinct habitat types. If either one of these habitat types is damaged, then the species may be unable to persist. Also, if barriers to dispersal are created between the needed habitats by roads, fences, or dams, the species may be unable to complete its life cycle. As an example, migratory bird species needing two widely spaced habitats have a greater risk of extinction than nonmigratory species (Pimm et al. 1988). Many animal species migrate among habitats in search of food, often along elevational and moisture gradients. Herds of wild pigs, grazing ungulates, frugivorous vertebrates and nectivorous birds are all examples of these. If these species are unable to migrate and are confined to one habitat type, they may not survive or, if they do survive, they may be unable to accumulate the nutritional reserves needed to reproduce.

Species with little genetic variability. Genetic variability within a population can sometimes allow a species to adapt to a changing environment (see Chapter 11). Species with little or no genetic variability will have a greater tendency to go extinct when a new disease, a new predator, or some other change occurs in the environment. For example, a lack of genetic variability is considered to be a contributing factor to the lack of disease resistance in the cheetah (O'Brien and Everman 1988).

Species with specialized niche requirements. Once a habitat is altered by human activity, the environment may no longer be suitable for specialized species. For example, wetland plants that require very specific and regular changes in water level may be rapidly eliminated when human activity affects the hydrology of an area. Soil arthropods and herbaceous plants may be eliminated when introduced livestock intensively graze native grasslands, compacting the soil. Species with specialized niche requirements often have small populations and so are more vulnerable to extinction. Specialized species may go extinct if species on which they depend go extinct (Gilbert 1980). For example, there are species of mites that feed only on the feathers of a single bird species. If the bird species goes extinct, so do its associated feather mite species.

Species that are characteristically found in stable environments. Many species are found in environments where disturbance is minimal, such as in old stands of equatorial rain forests and the interiors of rich temperate deciduous forests. Light-demanding plant species that thrive on disturbance may be relatively rare in these environments and confined to treefall gaps, landslips, ravines, and cliffs. When these forests are logged, grazed, burned and otherwise altered by human activity, many native species are unable to tolerate the changed microclimatic conditions (more light, less moisture, greater temperature variation) and influx of exotic species. Species able to tolerate these disturbed conditions come to predominate in the altered habitat at the expense of the original species.

Species that form permanent or temporary aggregations. Species that group together in specific places are highly vulnerable to local extinction. For example, bats may forage widely, but typically roost together in particular caves. Hunters entering these caves during the day can rapidly harvest every individual in the population. (Box 3 in Chapter 2 described this effect on pteropsid fruit bats, or "flying foxes.") Herds of bison, flocks of passenger pigeons, and schools of fish all represent aggregations of animals that have been exploited and completely harvested by people. Temporary aggregations include runs of salmon and alewives migrating up rivers; nets across rivers can catch virtually every fish and eliminate a species in a few days. Overly efficient harvesting of wild fruits for commercial markets can lead to a lack of new seedlings. Even though sea turtles may swim across vast stretches of ocean, egg collectors and hunters on a few narrow nesting beaches can threaten a species with extinction. Many species of social animals may be unable to persist when their population size falls below a certain number; they may be unable to forage, mate, and defend themselves. Such species may be more vulnerable to habitat destruction than species in which individuals are widely dispersed and are not social.

Species that are hunted or harvested by people. Overharvesting can rapidly reduce the population size of a species. If hunting and harvesting are not regulated either by law or by local customs, the species can be driven to extinction.

Combinations of Characteristics

The preceding discussion emphasized particular characteristics that make species vulnerable to extinction. However, these characteristics of extinction-prone species are not independent, but tend to group together into categories of characteristics. As an example, species with specialized diets tend to have low population densities—both characteristics of extinction-prone species. Often the characteristics of extinction-prone species vary among groups because of peculiarities of natural history; butterflies differ from jellyfish and cacti in characters associated with vulnerability to extinction. By identifying characteristics of extinction-prone species, conservation biologists can anticipate the need for managing populations of vulnerable species.

Ehrenfeld (1970) imagined that those species most vulnerable to extinction would have the full range of characteristics; for example,

> a large predator with a narrow habitat tolerance, long gestation period, and few young per litter. It is hunted for a natural product and/or for sport, but is not subject to efficient game management. It has a restricted distribution but travels across international boundaries. It is intolerant of man, reproduces in aggregates, and has nonadaptive behavioral idiosyncracies. Although there is probably no such animal, this model, with one or two exceptions, comes very close to being a description of a polar bear.

The Legal Status of Rare Species

To highlight the status of rare species for conservation purposes, the International Union for the Conservation of Nature (IUCN) has established five main conservation categories (IUCN 1984, 1988). These categories have proved to be useful at the national and international level by directing attention toward species of special concern and identifying species threatened with extinction for protection through international agreements such as the Convention on International Trade in Endangered Species (CITES):

1. *Extinct.* Species (and other taxa, such as subspecies and varieties) that are no longer known to exist in the wild. Searches of localities where they were once found and of other possible sites have failed to detect the species.
2. *Endangered.* Species that have a high likelihood of going extinct in the near future. Included are species whose numbers of individuals have been reduced to the point that the survival of the species is unlikely if present trends continue.

TABLE 5.3
Percent of species in some temperate countries that are threatened with global extinction

Country	Mammals		Birds		Reptiles		Amphibians		Plants	
	Number of species	Percent threat-ened[a]	Number of species	Percent threat-ened[a]	Number of species	Percent threat-ened[a]	Number of species	Percent threat-ened[a]	Number of species	Percent threat-ened[a]
Argentina	255	10.2	927	1.9	204	3.4	124	0.8	9,000	1.7
Canada	163	4.9	434	1.6	32	3.1	N/A	N/A	3,220	0.3
Japan	186	4.8	632	3.0	85	2.4	58	1.7	4,022	9.8
South Africa	279	7.2	N/A	N/A	N/A	N/A	N/A	N/A	23,000	5.0
United States[b]	367	10.3	1,090	6.1	368	4.6	222	6.3	20,000	8.5

Source: Data from WRI/IIED 1987, 1988.

[a] Threatened species include those in the IUCN categories "endangered," "vulnerable," and "rare."
[b] Includes Pacific and Caribbean islands.

3. *Vulnerable.* Species that may become endangered in the near future because populations of the species are decreasing in size throughout its range. The security of vulnerable species is not certain.
4. *Rare.* Species that have small total numbers of individuals, often due to limited geographical ranges or low population densities. Although these species may not face any immediate danger, their small numbers make them likely candidates to become endangered.
5. *Insufficiently known.* Species that probably belong in one of the conservation categories but are not sufficiently well known to be assigned to a specific category.

Using the IUCN categories, the World Conservation Monitoring Centre (WCMC) has evaluated and described the threats to about 60,000 plant and 2000 animal species in its series of Red Data Books (Table 5.3). The great majority of the species on these lists are plants, reflecting the recent trend of listing plant species in threatened habitats. However, there are also numerous listed species of fish (343), amphibians (50), reptiles (170), invertebrates (1355), birds (1037), and mammals (497). The IUCN system has been applied to specific geographical areas as a way of highlighting conservation priorities. Malaysia provides an example (Kiew 1991):

• Of 2830 tree species in peninsular Malaysia, 511 species are considered threatened.

- A large number of Malaysian herb species are endemic to single localities, such as mountaintops, streams, waterfalls, or limestone outcrops. These species are threatened with extinction if their habitat is destroyed.
- Of 2500 orchid species in Malaysian Borneo, 200 species are threatened.
- All 7 species of giant-flowered *Rafflesia* are threatened.
- Of peninsular Malaysia's rich palm flora, 47% of the species are threatened.
- Loss of rain forest habitat will lead to the extinction of an estimated 600 Malaysian butterfly species.
- All 5 species of sea turtles in Malaysia are considered endangered due to a combination of habitat loss, egg collecting, hunting, pollution of marine waters, unregulated tourism, and entanglement in fishing nets. In Sarawak, green turtle populations have been declining for 40 years and the species is on the verge of extinction.
- Thirty-five Malaysian bird species are listed as threatened, and many more species have experienced population declines associated with habitat destruction.
- Of 121 species of small mammals in peninsular Malaysia, 38 species are threatened with extinction.
- Over 80% of Malaysian Borneo's primate species are under some threat, due to a combination of habitat destruction and hunting pressure.

To help focus attention on the threatened species most in need of immediate conservation efforts, the IUCN has begun to issue lists of the world's "most threatened" plants and animals (Cahn and Cahn 1985). These lists include species of unique conservation value. Among the animals are the kago, a rare flightless bird that is the symbol of New Caledonia; the komprey, a primitive wild ox from Southeast Asia that has been hunted to near extinction; and the Orinoco River crocodile, which has been decimated by illegal trade in hides. The plant list includes the giant *Rafflesia* of Sumatran rain forests, which produces a flower about one meter across, and the African violet—the most common houseplant in the world, but only known in the wild from a few plants in fragments of mountain forest in central Tanzania.

The IUCN categories and WCMC Red Data Books are an excellent first step toward protecting the world's species; however, certain difficulties exist in using the category system (Fitter and Fitter 1987). First, each listed species must be studied to determine its population size and the trend in its numbers. Such studies can be difficult, expensive, and time-consuming. Second, a species must be studied over its whole range, which may present logistical difficulties. Third, the IUCN categories are not suitable for most tropical insect species, which are poorly known taxonomically and biologically, yet are threatened with extinction as rain forests are cut down. Fourth, species are often listed as endangered even if they have not been seen for many years, presumably on the assumption that they will be relocated if a thorough search is made. As an example, a survey of the natural history of the Indonesian island of Sulawesi showed that many endemic fish and bird species had not been seen for several decades; their status was unknown and not listed in the Red Data Books (Whitten et al. 1987). In such situations, species that have not been seen for many years and whose habitats have been heavily damaged by human activity should probably be listed as extinct or endangered until field studies had been undertaken to determine their true status (Diamond 1987). In other words, an argument could be made that the burden should be shifted from presuming a species to be safe until it is known to be threatened, to presuming a species to be threatened unless it is known to be secure and protected.

The most serious problem with the IUCN system is that the criteria for assigning species to particular categories are subjective. With greater numbers of people and organizations involved in assigning threat categories, there is potential for species to be arbitrarily assigned to particular categories. Given the legal restrictions that often accompany these assignments, with resulting financial implications to landowners, corporations, and governments, definitions need to be clarified to prevent arguments over the meaning of each category. To correct this situation, Mace and Lande (1991) have proposed a three-level system of classification based on the probability of extinction:

1. *Critical* species have a 50% or greater probability of extinction within 5 years or 2 generations, whichever is longer.
2. *Endangered* species have a 20% probability of extinction within 20 years or 10 generations.
3. *Vulnerable* species have a 10% probability of extinction within 100 years.

The criteria for assigning categories are based on the developing methods of population viability analysis (Chapter 12) and focus particularly on population trends and habitat condition. For example, a critical species has two of the following characteristics: total breeding population size less than 50 individuals, fewer than two populations containing more than 25 breeding individuals, more than 20%

decline in population numbers within 2 years or 50% within one generation, and population subject to catastrophic crashes every 5 to 10 years, in which half or more of the population dies. Species can also be assigned critical status as a result of observed or predicted habitat loss, ecological imbalance, or commercial exploitation. The advantage of this proposed system is that it provides a standard method of classification by which decisions can be reviewed and evaluated by other scientists according to accepted quantitative criteria.

Summary

1. A species may be considered rare if it has one of three characteristics: if it occupies a narrow geographical range; if it occupies only one or a few specialized habitats; or if it is always found in small populations. Isolated habitats such as islands, lakes, and mountaintops may have many endemic species that are found nowhere else.

2. Species most vulnerable to extinction have one or more of the following characteristics: very narrow range; one or only a few populations; and/or small population size. Additional risk factors include low population density; a large home range; large body size; low rate of population increase; poor dispersal ability; a need to migrate among different habitats; little genetic variability; specialized niche requirements; a need for a stable environment; and large aggregations. An extinction-prone species may display several of these characteristics.

3. To highlight the status of species for conservation purposes, the International Union for the Conservation of Nature (IUCN) has established five main conservation categories: extinct, endangered, vulnerable, rare, and insufficiently known. This system of classification is now widely used to evaluate the status of species and establish conservation priorities. Some scientists, however, recommend using a more quantitative three-category system that ranks species based on their statistical chances for extinction over a given period of time.

Suggested Readings

Diamond, J. M. 1987. Extant unless proven extinct? Or, extinct unless proven extant? *Conservation Biology* 1: 77–81. Determining the status of rare species in areas not regularly visited by biologists should be a priority.

Fitter, R. and M. Fitter. 1987. *The Road to Extinction.* IUCN, Gland, Switzerland. A critique of the IUCN species category system.

IUCN (International Union for the Conservation of Nature). 1988. *Red List of Threatened Animals.* IUCN, Gland, Switzerland. The authoritative source of information on the conservation status of animals.

Laurance, W. F. 1991. Ecological correlates of extinction proneness in Australian tropical rain forest mammals. *Conservation Biology* 5: 79–89. Excellent case study with statistical analysis.

Mace, G. M. and R. Lande. 1991. Assessing extinction threats: Towards a reevaluation of IUCN threatened species categories. *Conservation Biology* 5: 148–157. A proposal for a more objective and quantitative approach for assessing the vulnerability of species.

Pimm, S. L., H. L. Jones and J. Diamond. 1988. On the risk of extinction. *American Naturalist* 132: 757–785. A good summary of the factors that make species prone to extinction.

Rabinowitz, D., S. Cairnes and T. Dillon. 1986. Seven forms of rarity and their frequency in the flora of the British Isles. *In* M. E. Soulé (ed.), *Conservation Biology: The Science of Scarcity and Diversity*, pp. 182–204. Sinauer Associates, Sunderland, MA. An influential paper describing patterns of rarity and their significance to conservation.

Terborgh, J. 1974. Preservation of natural diversity: The problem of extinction-prone species. *BioScience* 24: 715–722. Why certain species are more vulnerable than others to extinction.

CHAPTER 6

Habitat Destruction, Fragmentation, and Degradation

Species and the communities to which they belong are adapted to local environmental conditions. As long as the conditions remain unchanged, species and communities tend to persist in the same place over time. The ranges of species will sometimes expand as a result of chance dispersal events, or contract as a result of increased competition and predation by other species. Species' ranges may also shift in response to changes in the landscape or climate. In the past such range changes have been very gradual. Even the major climate changes that took place in the Ice Ages, during repeated episodes of glaciation, took place over many thousands of years, so that species were able to adjust their ranges gradually in relation to the shifting climate (Gates 1993).

Human activity has disrupted this slow pattern of change in biological communities. Massive disturbances caused by people have altered, degraded, and destroyed the landscape on a vast scale, driving species and even communities to the point of extinction. Even when human activity does not directly eliminate a species, the population size of a species may become so small that the species is no longer viable and may eventually go extinct.

The major threats to biological diversity that result from human activity are habitat destruction, habitat fragmentation, habitat degradation (including pollution), the introduction of exotic species, the increased spread of disease, and the overexploitation of many species for human use.

Human Population and the Use of Resources

These six threats to biological diversity are all caused by an ever-increasing use of the world's natural resources by the expanding human population. Until the last few hundred years, the rate of human population growth was relatively slow, with the birth rate only slightly exceeding the mortality rate. The greatest destruction of biological communities has occurred during the last 150 years, during which the human population went from 1 billion in 1850, to 2 billion in 1930, to 5.3 billion in 1990, and will reach an estimated 6.5 billion by the year 2000 (Figure 6.1). Human numbers have increased because birth rates have remained high while mortality rates have declined as a result of both modern medical discoveries (specifically the control of disease) and the presence of more reliable food supplies (Keyfitz 1989). Population growth has slowed in the industrialized countries of the world but is still high in many areas of tropical Africa, Latin America, and Asia, where the greatest biological diversity is also found.

People use natural resources such as fuelwood, wild meat, and wild plants, and people convert vast amounts of natural habitat to agricultural and residential purposes, so population growth by itself is partially responsible for the loss of biological diversity. Some conservation biologists have argued strongly that controlling the size of the human population is the key to protecting biological diversity (Meffe et al. 1993). However, population growth is not the only cause of species extinction and habitat destruction. If one examines the situation throughout the world in both developing and industrial nations, species extinction and the destruction of ecosystems are not always caused by individual citizens obtaining their basic needs; inefficient and unequal usage of natural resources by people is also a major cause of the decline in biological diversity.

Some rural people destroy biological communities and hunt endangered species to extinction because they are poor and have no land of their own (Browder 1988; Downing et al. 1992). In many

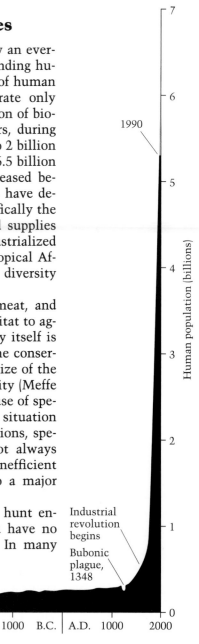

6.1 Human population has increased spectacularly since the seventeenth century. At current growth rates, the population will double in less than 40 years.

countries there is extreme inequality in the distribution of wealth, with the majority of the wealth (money, good farmland, timber resources, etc.) owned by a small percentage of the population. A common pattern in many countries of the developing world is that local farmers are forced off their land by large landowners and business interests, often backed up by the government, the police, and the army; the local farmers often have no choice except to move to remote, undeveloped areas and attempt to eke out a living through **shifting cultivation**. In this kind of subsistence farming, sometimes referred to as "slash-and-burn" agriculture, plots of natural vegetation are burned away and the cleared patches are farmed for two or three seasons, after which their fertility usually diminishes to a point where adequate crop production is no longer possible. The patch is abandoned and more natural vegetation must be cleared. Shifting cultivation is often practiced in such areas because the farmers are unwilling to spend the time and money necessary to develop more permanent forms of agriculture on land that they do not own and may not occupy for very long. Political instability, lawlessness, and war also force farmers off their land and into remote, undeveloped areas where they feel more safe. Rather than being called "shifting cultivators," however, these newly arrived people would be more accurately described as "shifted cultivators" in order to distinguish them from traditional farmers who have long inhabited rain forest areas.

In many cases, the factors causing habitat destruction are the large industrial and commercial activities associated with a global economy, such as mining, cattle ranching, commercial fishing, forestry, plantation agriculture, manufacturing, and dam construction, initiated with the goal of making a profit (Dasmann et al. 1973; Anderson 1992). Many of these projects are sanctioned by national governments and international development banks and are explained to the public as being developed to provide jobs, commodities, and tax revenues (Chapter 21). However, this exploitation of natural resources is often neither efficient nor cost-effective because the emphasis in these industries is on short-term gains, which frequently are made at the expense of the long-term viability of the natural resources, the environment, and, ultimately, the people and businesses that depend on the resources.

Governments in the developing world sometimes justify their development projects by the need to raise revenue to pay off their foreign debt. It is an unfortunate paradox of the international banking system that poor people in the developing world are driven deeper into poverty to pay the banks of wealthier countries, and that the biological riches of the tropical world are exhausted to pay national debts, rather than used to supply the needs of their inhabitants (Gillis 1991). The additional irony is that these developing countries often borrowed this money in the first place for large development projects designed to increase the national wealth; if these projects failed, the country was left not only poor but also in debt. Developing countries caught in this cycle are now asking for "debt forgiveness" and foreign aid in order to ward off social unrest, prevent epidemic disease, protect biological diversity, and address other problems stemming from their increasing poverty.

Once a mentality of short-term exploitation of resources is established in a community or a nation, it is difficult to restrain. Biological communities can often persist close to areas with high densities of people as long as the human activities in nature reserves are regulated by local custom and governments (Gomez-Pompa and Kaus 1988, 1992). Sometimes this regulation breaks down during times of war, political unrest, or social instability. When this happens there is often a scramble to use up and sell resources that have been sustainably used for generations. The higher the density of people, the more closely their activities must be regulated, and the greater the destruction that can result from a breakdown in authority (Homer-Dixon et al. 1993). The devastation that occurred in China's forests during the Cultural Revolution is a revealing example; strict regulations against cutting trees developed over centuries were no longer enforced, and farmers cut down trees at a tremendous rate, stockpiling wood for fuel, construction, and furniture-making (Primack 1988).

The responsibility for the destruction of biological diversity in species-rich tropical areas also lies with the unequal use of natural resources worldwide. People in the industrialized countries of the world (and the wealthy minority in the developing countries) consume a disproportionate share of the world's energy, minerals, wood products, and food (Figure 6.2). Overall, the lifestyle of the average American used 280 gigajoules of energy in 1987, in contrast to 110 gigajoules for the average Japanese citizen and only 1 gigajoule for the average citizen of Bangladesh (WRI/IUCN/UNEP 1992). This excessive consumption of resources in the developed world is not sustainable in the long run, and if this pattern is adopted by the expanding middle class in the developing world it will cause massive environmental disruption. The present unequal distribution of resources in the world results in the poverty, high mortality, and environmental damage experienced in much of the developing world, as well as being responsible for rampant pollution and habitat degradation in the industrialized world (Figure 6.3). The affluent citizens of the developed countries need to confront their excessive consumption of re-

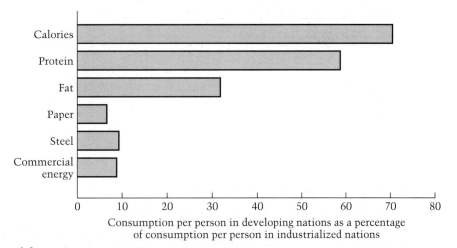

Consumption per person in developing nations as a percentage
of consumption per person in industrialized nations

6.2   **The percentage of resources that an average person in a developing country (75% of the world's population) uses in comparison to a person in a developed country (25% of the world's population). For example, the per capita consumption of paper, steel, and energy in the developing nations is less than 10% of that in developed countries. (From Ruckelshaus 1989, using data from the World Commission on Environment and Development.)**

sources and reevaluate their lifestyles as they make suggestions and offer aid to curb population growth and protect biological diversity in the developing world (Figure 6.4; Mies 1990; Parikh and Parikh 1991).

Habitat Destruction

> The primary cause of the decay of organic diversity is not direct human exploitation or malevolence, but the habitat destruction that inevitably results from the expansion of human populations and human activities. (Ehrlich 1988)

The major threat to biological diversity is loss of habitat, and the most important means of protecting biological diversity is habitat preservation. Habitat loss is the primary threat to the majority of vertebrate species currently facing extinction (Table 6.1; Prescott-Allen and Prescott-Allen 1978), and this is certainly true for invertebrates, plants, and fungi as well. In many countries of the world, particularly on islands and where human population density is high, most of the original habitat has been destroyed (Figure 6.5). More than 50% of the wildlife habitat has been destroyed in 49 out of 61 Old World

(A) Carbon dioxide emissions

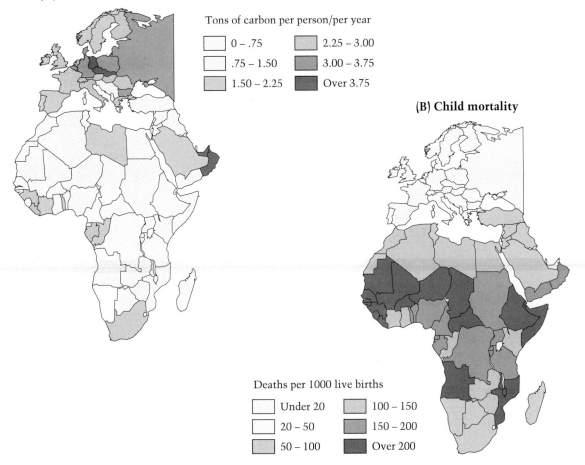

Tons of carbon per person/per year

0 – .75	2.25 – 3.00
.75 – 1.50	3.00 – 3.75
1.50 – 2.25	Over 3.75

(B) Child mortality

Deaths per 1000 live births

Under 20	100 – 150
20 – 50	150 – 200
50 – 100	Over 200

6.3   The distinction between industrialized/developed and developing nations can be seen in such aspects as the pollution residues from resource consumption, and mortality and poor health conditions in the less economically developed countries. These differences have a definite "northern hemisphere–southern hemisphere" dichotomy, illustrated by these maps of Europe, Africa, and the Middle East. Analogous maps of North and South America show the same dichotomy.

(A) Carbon dioxide emissions and the resulting carbon production can be used as a measure of resource use and impact on the environment. Worldwide the countries with the highest carbon emissions per capita are the United States and Germany; the lowest carbon emissions are found in Burundi (in Africa) and Bhutan (in Asia). (B) Child mortality is related to poverty, political instability, and lack of medical care. Rates of child mortality are much higher in poor countries than in rich ones. (Maps adapted from Ruckelshaus 1989.)

6.4 The wealthy developed countries of the world often criticize the poorer, developing nations for a lack of sound environmental policies, but seem unwilling to acknowledge that their own excessive consumption of resources is a major part of the problem. (Cartoon by Scott Willis, © *San Jose Mercury News*.)

TABLE 6.1

Factors responsible for some extinctions and threatened extinctions

Group	Habitat loss	Over-exploitation[b]	Species introductions	Predators	Other	Unknown
EXTINCTIONS						
Mammals	19	23	20	1	1	36
Birds	20	11	22	0	2	37
Reptiles	5	32	42	0	0	21
Fishes	35	4	30	0	4	48
THREATENED EXTINCTIONS[c]						
Mammals	68	54	6	8	12	—
Birds	58	30	28	1	1	—
Reptiles	53	63	17	3	6	—
Amphibians	77	29	14	—	3	—
Fishes	78	12	28	—	2	—

Column group header: *Percent due to each cause[a]*

Source: From Reid and Miller 1989, based on data from various sources.

[a] The values indicated represent the percentage of species that are influenced by the given factor. Some species may be influenced by more than one factor; thus some rows may exceed 100 percent.

[b] Overexploitation includes commercial, sport, and subsistence hunting and live animal capture for any purpose.

[c] Threatened species and subspecies include those in IUCN categories endangered, vulnerable, and rare.

6.5   Loss of wild habitat is a major cause of extinctions. In this scene from Rio de Janeiro, a small remaining patch of forest habitat is being squeezed by high-rise urban developments and expanding shantytowns. (Photograph by Paul Almasy.)

tropical countries (Table 6.2; IUCN/UNEP 1986b,c). In tropical Asia, fully 65% of the wildlife habitat has been lost, with particularly high rates of destruction reported for Bangladesh (94%), Hong Kong (97%), Sri Lanka (83%), Vietnam (80%), and India (80%). Perhaps the one bright note in Asia is that the two biologically rich countries of Malaysia and Indonesia still have about half of their wildlife habitats and are in the process of establishing extensive national parks and other protected areas. Sub-Saharan Africa has similarly lost a total of about 65% of its wildlife habitat with habitat loss being most severe in Rwanda (87%), Burundi (86%), Gambia (89%), Sierra Leone (85%),

TABLE 6.2
Wildlife habitat loss in some countries of the Old World tropics

Country	Original extent of habitat (× 1000 ha)	Habitat remaining (× 1000 ha)	Percent of habitat lost
AFRICA			
Angola	124,670	76,085	39
Burkina Faso	27,380	5,476	80
Burundi	2,570	359	86
Chad	72,080	17,299	76
Gambia	1,130	124	89
Ghana	23,000	4,600	80
Kenya	56,950	29,614	48
Madagascar	59,521	14,880	75
Rwanda	2,510	326	87
Senegal	19,620	3,532	82
Sierra Leone	7,170	1,076	85
South Africa	123,650	53,170	57
Zaire	233,590	105,116	55
Zambia	75,260	53,435	29
Zimbabwe	39,020	17,169	56
ASIA			
Bangladesh	14,278	857	94
Brunei	576	438	24
Burma	77,482	22,598	71
Hong Kong	107	3	97
India	301,701	61,509	80
Indonesia	144,643	74,686	49
Malaysia/Singapore	35,625	21,019	41
Philippines	30,821	6,472	79
Sri Lanka	6,470	1,100	83
Thailand	50,727	13,004	74
Vietnam	33,212	6,642	80

Source: IUCN/UNEP 1986.

Senegal (82%), Ghana (80%), Burkina Faso (80%), and Mauritania (81%). The biologically rich countries of Zaire and Kenya are relatively better off, still having about half of their wildlife habitat. In the Mediterranean region, which has been intensively occupied by people for thousands of years, only 10% of the original forest cover remains.

Comparison of habitat destruction across political boundaries gives an important perspective regarding the ways in which different nations use natural resources; for example, deforestation rates for various nations are shown in Figure 6.6. However, the crucial conservation factor for a species is the *total* amount of its original habitat remaining (IUCN/UNEP 1986d). Thus the fact that rhinos in Kenya have more suitable habitat than rhinos in Rwanda means little to the species except, perhaps, to indicate where the population has the best chance to persist in the wild, though possibly only for a limited period of time. For many important wildlife species, the majority of habitat in their original range has been destroyed, and very little of the remaining habitat is protected. Table 6.3 gives such statistics for some Asian primates. More than 95% of the original habitats of both the Javan gibbon and the Javan lutong has been destroyed, and they are protected on less than 2% of their original range. The orangutan, a great ape that lives in Sumatra and Borneo, has lost 63% of its range and is protected in only 2% of its range.

Threatened Rain Forests

The destruction of tropical rain forests has come to be synonymous with the loss of species (see Chapter 3). Tropical moist forests

TABLE 6.3
Range loss for some Southeast Asian primates and the percent of their original habitat now protected

Species	Original range (× 1000 ha)	Remaining range (× 1000 ha)	Percent of range lost	Percent now protected
Orangutan	55,300	20,700	63	2.1
Siamang	46,511	16,980	63	6.8
Bornean gibbon	39,500	25,300	36	5.1
Mentawai gibbon	650	450	31	22.9
Indochinese gibbon	34,933	8,753	75	3.1
Long-tailed macaque	38,318	12,332	68	3.4
Proboscis monkey	2,969	1,775	40	4.1
Douc langur	29,600	7,227	76	3.1
Javan lutong	4,327	161	96	1.6
Francois' leaf monkey	9,740	1,411	86	1.2

Source: IUCN 1986.

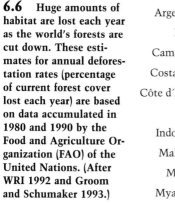

6.6 Huge amounts of habitat are lost each year as the world's forests are cut down. These estimates for annual deforestation rates (percentage of current forest cover lost each year) are based on data accumulated in 1980 and 1990 by the Food and Agriculture Organization (FAO) of the United Nations. (After WRI 1992 and Groom and Schumaker 1993.)

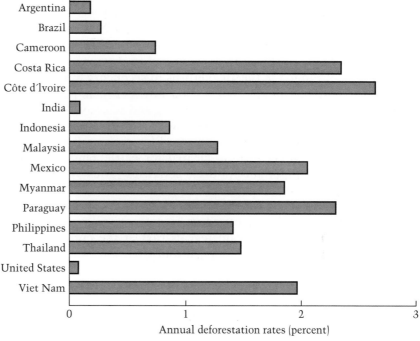

occupy 7% of the Earth's land surface, but are estimated to contain over 50% of its species (Myers 1986). These evergreen to partly evergreen forests occur in frost-free areas below about 1800 m in altitude and have at least 100 mm (4 inches) of rain per month in most years. Such forests are characterized by a great richness of species and a complexity of species interaction and specialization unparalleled in any other community.

The original extent of tropical rain forests has been estimated at 16 million km^2, based on current patterns of rainfall and temperature (Myers 1984, 1986, 1991b; Sayer and Whitmore 1991). A combination of ground surveys, airplane photos, and remote sensing data from satellites showed that in 1982 only 9.5 million km^2 remained, an area about equal in size to the continental United States. A re-census in 1985 showed a loss of almost another million km^2 during this three-year period (see Figure 3.4). At the present time, about 180,000 km^2 of rain forest are being lost per year—an area larger than the state of Florida—with 80,000 km^2 completely destroyed and 100,000 km^2 degraded to the point that the species composition and ecosystem processes of the community are greatly altered. Tropical forest ecosystems are readily degraded because the soils are often thin and nutrient-poor, and are readily eroded by heavy rainfall.

On a global scale, the primary cause of rain forest destruction is small-scale cultivation of crops by farmers (Figure 6.7A and B; Myers 1986, 1991b). Some of this land is converted to permanent farmland and pastures, but much of the area returns to secondary forest following shifting cultivation. Another 45,000 km² per year is destroyed through commercial logging in clear-cutting and selective logging operations. A further 25,000 km² is degraded for fuelwood production, mostly to supply local villagers with wood for cooking fires. The remaining 20,000 km² per year is cleared for cattle ranches (Figure 6.7C). The relative importance of these activities varies by geographical region, with logging being a more significant activity in tropical Asia, cattle ranching being more prominent in tropical America, and farming and fuelwood gathering more important in tropical Africa. Extending the projection forward reveals that at the current rate of loss, there will be virtually no tropical forest left after the year 2040 except in the relatively small areas under protection. The situation is actually even more grim than these projections indicate, since the world's population is still increasing and poverty is on the rise, putting ever greater demands on the dwindling supply of rain forest.

The destruction of tropical rain forests is frequently caused by the demand in industrialized countries for cheap agricultural products, such as rubber, palm oil, cocoa, and beef. During the 1980s Costa Rica had one of the world's highest rates of deforestation as a result of conversion of rain forests into cattle ranches (Downing et al. 1992). Much of the beef produced on these ranches was sold to the United States and other developed countries to produce inexpensive hamburgers. Adverse publicity resulting from this "hamburger connection," followed by consumer boycotts, led major restaurant chains in the United States to stop buying tropical beef from these ranches. Even though deforestation continued in Costa Rica after the boycott, the boycott was important in making people aware of the international connections promoting deforestation.

Another well-documented linkage involves the forests of Thailand and the livestock industry of the Netherlands. The Dutch import large quantities of tapioca (made from cassava), palm-kernel cake, and other tropical products to use as cattle feed. In order to supply this market, farmers in Thailand had increased their cultivation of cassava and other cattle-feed plants from 100,000 ha in 1965 to 1 million ha by the mid-1980s. Fully 25% of the deforestation in northeastern Thailand can be related to the clearing of land by the half-million families involved in cassava cultivation. What seems on

6.7 Rain forest displacement for agricultural purposes. (A) Swidden agriculture in northwest Amazonia; gardens are hewn from the forest with slash-and-burn techniques. Indigenous peoples have used such farming practices for centuries. However, when large numbers of people must eke out a living from the land, rain forest destruction is vast. (Photograph by Paul Patmore.) (B) Rice paddies take over rain forest in southwestern India. (Photograph by R. Primack.) (C) Again in Amazonia, forest land is burned to clear it for cattle pasture. Huge amounts of forest acreage are destroyed in this manner. (Photograph from The Woods Hole Research Center.)

(A)

(B)

(C)

the surface to be a minor Dutch import to supply a minor domestic industry in fact has major environmental effects (Netherlands National Committee for the IUCN 1988).

A priority for conservation biology is to help provide the information, programs, and public awareness that will allow the greatest amount of rain forest to persist once the present cycle of destruction ends. A number of places can be singled out as examples of how rapid and serious rain forest destruction can be.

Rondonia. This state in Amazonian Brazil was almost entirely covered by primary forest as late as 1975, with only 1200 km² cleared out of a total area of 243,000 km² (Myers 1986; Fearnside 1987, 1990). In the 1970s the Brazilian government built part of the Trans-Amazon Highway through Rondonia and built a network of lateral roads leading away from the highway into the forest (Figure 6.8A). The government also provided lucrative tax subsidies to allow corporations to establish cattle ranches in the region, and encouraged poor, landless people from coastal states to migrate to Rondonia with offers of free land. These incentives were necessary because the soils of the Amazon region are low in mineral nutrients, and the pastures and farmlands created by forest clearing are usually unproductive and unprofitable. During this rush for land amid government incentives, 10,000 km² of the forest was cleared by 1982 and an additional 6000 km² was cleared by 1985 (Figure 6.8B). By the late 1980s the population of Rondonia was growing at 15.8% per year, and deforestation was increasing by 37% per year. These rates of growth and deforestation are phenomenally high compared with most other parts of the world, whether industrial or developing. International protests against this environmental damage led the government to end its subsidies of the cattle industry. Without these subsidies, the forest clearing in Rondonia has slowed substantially. However, considerable environmental damage has already been done.

Madagascar. The moist forests of Madagascar, with their rich heritage of endemic species (see Box 5 in Chapter 5), originally covered 62,000 km²; they have been reduced to 26,000 km² by a combination of shifting cultivation, cattle grazing, and fire. The present rate of deforestation is about 3000 km² per year, which means that by the year 2000 there may be no moist forests left on Madagascar except in the 1.5% of the island under protection (Myers 1986).

The Atlantic coast of Brazil. Another area with high endemism is the Atlantic coast forest of northern Brazil. Fully half of its tree species are endemic to the area, and the region supports a number of rare

6.8 (A) A satellite photo of a new highway through the Amazon rain forest in Rondonia; the area shown covers about 24,500 km². Note the lateral roads that provide access into the forest. (B) With access to the interior and lucrative tax subsidies from the government, a Rondonian "land rush" resulted in massive deforestation as 16,000 km² were cleared over the space of a few years in the 1980s. (Photographs from The Woods Hole Research Center.)

and endangered animals, including such spectacular species as the golden lion tamarin. This forest originally occupied a long, continuous strip along the South American coast, covering a million square kilometers. The Atlantic coast of Brazil has been almost entirely cleared for sugar cane, coffee, and cocoa production; less than 5% of the original forest remains—a total area of 50,000 km^2 (Myers 1986). Although this total area may sound like a fairly large block of forest, the remnant forest is divided into numerous fragments that may be unable to support long-term viable populations of many wide-ranging species. The single largest patch of forest remaining is only 7000 km^2, and this patch is highly disturbed in places.

Coastal Ecuador. The coastal region of Ecuador contained a rich forest filled with endemic species that was minimally disturbed by human activity until 1960. At that time, roads were developed and forests cleared to establish human settlements and oil palm plantations. One of the only fragments to survive is the 1.7 km^2 Rio Palenque Science Reserve. This tiny conservation area has 1025 recorded plant species of which 25% are not known to occur anywhere else (Gentry 1986). Over 100 new plant species have been recorded at this site. Many of these species are known only from a single individual plant and are doomed to certain extinction.

Another depressing example is Centinella, an isolated ridge 8 kilometers east of Rio Palenque that was formerly covered by a cloud forest. Despite the small area of the ridge (20 km long × 1 km wide), possibly 90 species of flowering plants were endemic to Centinella. When the forests were destroyed, this local concentration of unique species was wiped out.

Other Threatened Habitats

Other habitat types besides the rain forest are equally threatened with destruction.

Tropical dry forests. The land occupied by tropical dry forests is more suitable for agriculture and cattle ranching than is land occupied by tropical rain forests. The moderate rainfall, in the range of 250 to 2000 mm per year, allows mineral nutrients to be retained in the soil where they can be taken up by plants. Because of their suitability for agriculture, human population density is five times greater in dry forest areas of Central America than in adjacent rain forests (Murphy and Lugo 1986). As a consequence, the Pacific Coast of Central America has less than 2% of its original extent of deciduous dry forest (Janzen 1988a).

Wetlands and aquatic habitats. Wetlands are of critical importance as habitats for fish, aquatic invertebrates, and birds, and are a resource for flood control, drinking water, and power production (Williams 1990; Mitchell 1992; Moyle and Leidy 1992). While many wetland species are widespread, certain wetland systems are known for their high levels of endemism. Isolated, specialized aquatic habitats, such as the saline ponds occupied by the desert pupfish in the southwestern United States and Mexico, support many endemic species. As another example, Lake Victoria has one of the richest endemic fish faunas in the world, but 250 of its species are in danger of extinction as a result of water pollution and the introduction of exotic species of fish (see Box 8, later in this chapter).

Wetlands are often filled in or drained for development, or they are altered by channelization of watercourses, dams and chemical pollution (Carrier 1991). In either case, the original communities are destroyed. Over the last 200 years, over half of the wetlands in the United States have been destroyed, resulting in 40–50% of the freshwater snail species in the southeastern United States becoming either extinct or endangered (Maltby 1988). Destruction of wetlands has been equally severe in other areas of the industrialized world, such as Europe and Japan.

In the last few decades, the major threat to wetlands in developing countries has been massive development projects involving drainage, irrigation, and dams organized by governments and often financed by international aid agencies (Maltby 1988).

Mangroves. One of the most important wetland communities in tropical areas is the mangrove forest. This special forest type occupies shallow, intertidal, coastal areas with saline or brackish water, typically where there are muddy bottoms, habitats similar to those occupied by salt marshes in the temperate zone. Mangroves are extremely important in preventing erosion and storm damage in coastal areas; as breeding grounds and feeding areas for shrimp and fish; and as a source of wood for poles, charcoal, and industrial production. In Australia, two-thirds of the species caught by commercial fishermen are dependent to some degree on the mangrove ecosystem (WRI/IIED 1986). Despite their great economic value, mangroves in Africa and other tropical areas are often destroyed to grow rice, which also grows in wet conditions. Mangroves are also cleared for commercial shrimp and prawn hatcheries, particularly in Southeast Asia, where as much as 15% of the mangrove area has been removed for aquaculture (ESCAP 1985). Mangroves have also been severely degraded by the overcollection of wood for fuel and construction in densely settled areas. The loss of mangroves is extensive in some parts of Southeast Asia; for example, the Philippines has lost over half of its mangroves in the last 100 years, and Thailand lost over 30% of its mangroves in the brief period from 1961 to 1979 (FAO 1982, 1985a).

6.9 Temperate grasslands are extremely valuable for protecting biological diversity and for agricultural purposes. (A) A natural grassland with numerous native species on the National Bison Range, federally protected land in the state of Montana. (B) Cattle graze on natural grassland. (C) Overgrazed grassland takes on the appearance of a desert and native species are eliminated. (Photographs courtesy of the U.S. Fish and Wildlife Service and U.S. Forest Service.)

Grasslands. Temperate grasslands are another habitat type that has been almost completely destroyed by human activity. It is relatively easy to convert large areas of grassland to farmland (Figure 6.9). Eastern Illinois and northwestern Indiana, for example, originally contained 13 million acres of tall-grass, black-soil prairie, but as of 1985 only 20 acres of this habitat remained undisturbed; the rest was converted to farmland (Sauders et al. 1987b). This remaining 20 acres is in turn divided up into many small fragments, widely scattered across the landscape.

Desertification

Many biological communities in seasonally dry climates are degraded into man-made deserts by human activities, a process known as **desertification** (Le Houérou and Gillet 1986; Breman 1992). Such communities include tropical grasslands, scrub, and deciduous forest, as well as temperate shrublands such as are found in the Mediterranean region, southwestern Australia, South Africa, Chile, and southern California. While these areas may be suitable for agriculture in some years, repeated cultivation, particularly during dry and windy years, leads to soil erosion and a loss of water-holding capacity in the soil (Figure 6.10). Communities may also be overgrazed by domestic livestock such as cattle, sheep, and goats, and woody plants may be

6.10 Soil erosion on cultivated and grazed land in Ethiopia. With general erosion of the soil and the loss of its ability to hold water, the vegetation has begun to die off and desertification is in progress. (Photograph courtesy WRI/ IIED.)

cut down for fuel. The result is a progressive degradation of the biological community and a loss of soil cover, to the point where the region takes on the appearance of a desert (Schlesinger et al. 1990). Worldwide, 9 million km² of arid lands have been converted to deserts by this process, with one-third of this area in Africa (Figure 6.11; Dregne 1983). The process of desertification is most severe in the Sahel region of Africa, in which most of the native large mammal species are threatened with extinction. The human dimensions of the problem are illustrated by the fact that the Sahel region is estimated to have 2.5 times more people than the land can sustainably support. Further desertification appears to be almost inevitable, except in the limited areas where intensification of agriculture is possible (Breman 1992).

Habitat Fragmentation

Habitats that formerly occupied wide areas are now often divided up into pieces by roads, fields, towns, and a wide range of other human activities. More than 30 years ago the effects of this fragmenta-

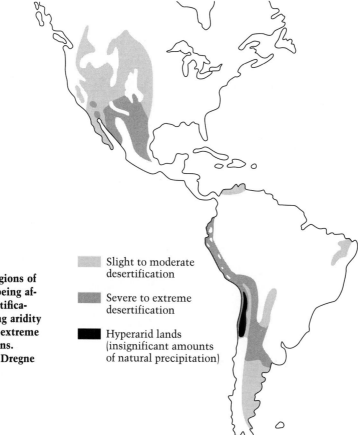

6.11 Arid regions of the world are being affected by desertification—increasing aridity and ever-more-extreme desert conditions. (Adapted from Dregne 1978.)

Slight to moderate desertification

Severe to extreme desertification

Hyperarid lands (insignificant amounts of natural precipitation)

tion on temperate forests were described by the plant ecologist Curtis (1956):

> Instead of an essentially continuous forest cover... the landscape now presents the aspect of a savannah, with isolated trees, small clumps or clusters of trees, or small groves scattered in a matrix of artificial grassland of grains and pasture grasses. Within the remnant forest stands, a number of changes of possible importance may take place. The small size and increased isolation of the stands tend to prevent the easy exchange of members from one stand to another. Various accidental happenings in any given stand over a period of years may eliminate one or more species from the community. Such a local catastrophe under natural conditions would be quickly healed by migration of new individuals from adjacent unaffected areas. In the isolated stands, however, opportunities for inward migration are small or nonexistent. As a result, the stands gradually lose some of their species, and those remaining achieve unusual positions of relative abundance.

Habitat fragmentation is the process whereby a large, continuous area of habitat is both reduced in area and divided into two or more

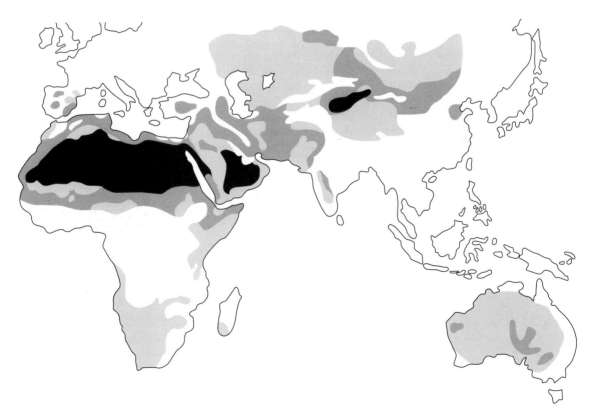

fragments (Wilcove et al. 1986; Lovejoy et al. 1986). When habitat is destroyed, there is often a patchwork of habitat fragments left behind (Figure 6.12). These fragments of the original habitat are often isolated from one another by a highly modified or degraded landscape. At the extreme, one can imagine a forested landscape in which most

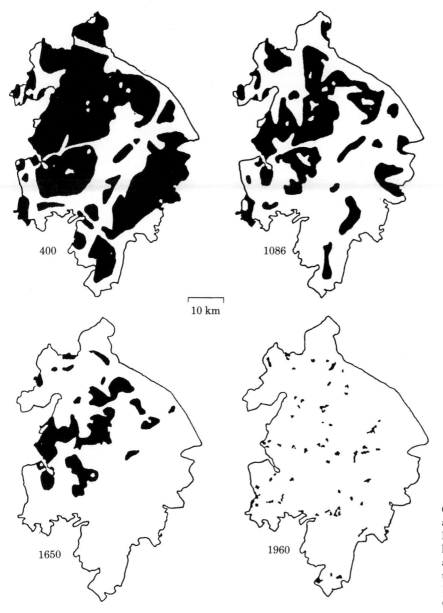

400

1086

10 km

1650

1960

6.12   The forested areas of Warwickshire, England (shown in black) have been fragmented and reduced in area over the centuries from 400 to 1960 A.D.. (From Wilcove et al. 1986.)

of the land is converted to agricultural fields, leaving small remnants of forest behind. This situation can be likened to the island model of biogeography (Chapter 4), with the forest fragments as habitat islands in an inhospitable agricultural sea. Fragmentation almost always occurs during a severe reduction in habitat area, but it can also occur even when area is reduced to only a minor degree if the original habitat is divided by roads, railroads, canals, power lines, fences, fire lanes, or other barriers to the free movement of species (Figure 6.13).

Habitat fragments differ from the original habitat in two important ways:

1. Fragments have a greater amount of edge for the area of habitat.
2. The center of each habitat fragment is closer to an edge.

A simple example will illustrate these characteristics and the problems that can occur because of them.

Consider a square conservation reserve 1000 m (1 km) on each side (Figure 6.14). The total area of the park is 1 km² (100 ha). The perimeter (or edge) of the park totals 4000 m. A point in the middle of the reserve is 500 m from the nearest perimeter. If domestic cats forage 100 m into the forest from the perimeter of the reserve and prevent forest birds from successfully raising their young, then only the 64 ha in the reserve's interior is available to the birds for breeding. Edge habitat, unsuitable for breeding, occupies 36 ha.

6.13   Even within national parks, fragmentation occurs when paved roads are built to allow access to visitors. Shown here are the park boundaries and paved roads (black) for three U.S. national parks; the parks are not shown to the same scale. (After Schonewald-Cox and Buechner 1992.)

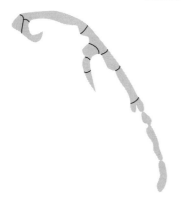

(A)   Cape Cod National Seashore, Massachusetts

(B)   Yellowstone National Park, Wyoming/Montana

(C)   Great Smoky Mountains National Park, North Carolina/Tennessee

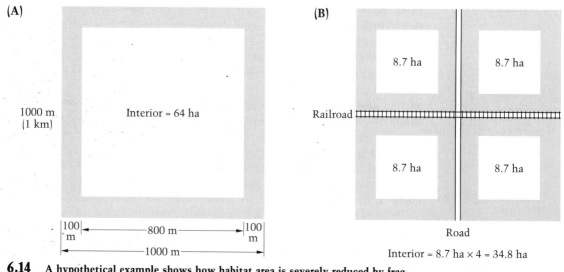

6.14   **A hypothetical example shows how habitat area is severely reduced by fragmentation and edge effects. (A) A 1-km² protected area. Assuming edge effects (gray) penetrate 100 m into the reserve, approximately 64 ha are available as usable habitat for nesting birds. (B) The bisection of the reserve by a road and a railway, although taking up little in actual area, extends the edge effects so that almost half the breeding habitat is destroyed.**

Now imagine the park being divided into four equal quarters by a north–south road 10 m wide and by an east–west railroad track that is also 10 m wide. The rights-of-way remove a total of 2 × 1000 m × 10 m of area (2 ha) from the park. Since only 2% of the park is being removed by the road and railroad, city planners argue that the effects on the park are negligible. However, the reserve has now been divided into four fragments, each of which is 495 m × 495 m in area. The distance from the center of each fragment to the nearest point on the perimeter has been reduced to 247 m, which is less than half of the former distance. Since cats can now forage into the forest from the road and railroad as well as the perimeter, birds can successfully raise young only in the most interior areas of each of the four fragments. Each of these interior areas is 8.7 ha, for a total of 34.8 ha. Even though the road and railroad removed only 2% of the reserve area, it reduced the habitat available to the birds by about half.

Habitat Fragmentation and Species Mobility

Habitat fragmentation results in a reduction of the area of the original habitat, a greater amount of edge habitat for a given area, and a reduced distance to the nearest edge. The effects of habitat loss on biological diversity are clear: the habitat being destroyed may contain the only site for a particular species. However, as the above example demonstrates, habitat fragmentation threatens the persistence of species in more subtle ways.

First, fragmentation may limit a species' potential for dispersal and colonization. Habitat fragmentation creates barriers to normal dispersal and colonization processes. In an undisturbed environment, seeds, spores, and animals move passively and actively across the landscape. When they arrive in a suitable but unoccupied area, new populations begin to develop at that site. Over time, populations of a species may build up and go extinct on a local scale as the species disperses from one suitable site to another. At a landscape level, a series of populations exhibiting this pattern of extinction and recolonization is sometimes referred to as a **metapopulation** (see Chapter 12).

When a habitat is fragmented, the potential for dispersal and colonization is often reduced. Many bird, mammal, and insect species of the forest interior will not cross even very short distances of open area (Lovejoy et al. 1986; Bierregaard et al. 1992). If they do venture into the open, they may find predators such as hawks, owls, flycatchers, and cats waiting at the forest edge to catch and eat them. Agricultural fields 100 m wide may represent an impassable barrier to the dispersal of many invertebrate species. When mammal dispersal is reduced by habitat fragmentation, then plants with fleshy fruits or sticky seeds that depend on the animals to carry their seeds will also be affected. The result is that isolated habitat fragments will not be colonized by many species that could potentially live there. As species go extinct within individual fragments through natural successional and metapopulation processes, new species will be unable to arrive due to barriers to dispersal, and the number of species present in the habitat fragment will decline over time. Most of the world's national parks and nature reserves are too small in area and too isolated to maintain populations of many species (Figure 6.15).

A second harmful aspect of habitat fragmentation is that it reduces the foraging ability of animals. Many animal species, either as individuals or as social groups, need to be able to move freely across the landscape to feed on widely scattered resources. A given resource may be needed for only a few weeks in a year, or even only once in a few years, but when a habitat is fragmented, species confined to a single habitat fragment may be unable to migrate over their normal home range in search of that scarce resource. For example, orangutans, gibbons, and other primates typically remain in forests and forage widely for fruits. Finding scattered trees with abundant fruit crops may be crucial during episodes of fruit scarcity. Clearings and

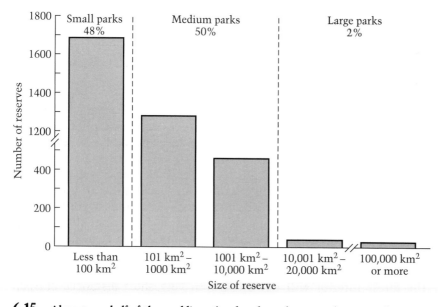

6.15   Almost one-half of the world's national parks and protected areas are less than 100 km², and 98% of the parks are less than 10,000 km². (Data from IUCN 1982.)

roads that break up the forest canopy may prevent these primates from reaching nearby fruiting trees because the primates are unable or unwilling to descend to the ground and cross the intervening open landscape. Fences may prevent the natural migration of large grazing animals such as wildebeest or bison, forcing them to overgraze an unsuitable habitat and eventually leading to starvation of the animals and degradation of the habitat. Barriers to dispersal can also restrict the ability of widely scattered animal species to find mates, leading to a loss of reproductive potential. Plants also may have reduced seed production if pollinators such as butterflies and bees are less able to migrate among habitat fragments to visit flowers (Bierregaard et al. 1992).

Habitat fragmentation may precipitate population decline and extinction by dividing an existing widespread population into two or more subpopulations, each in a restricted area. These smaller populations are then more vulnerable to inbreeding depression, genetic drift, and other problems associated with small population size (see Chapter 11). While a large area may have supported a single large population, it is possible that none of the fragments can support a subpopulation large enough to persist for a long period.

Edge Effects

Habitat fragmentation dramatically increases the amount of edge relative to the amount of interior habitat, as demonstrated above. The microenvironment at the fragment edge is different from that of the interior. Some of the more important **edge effects** include microclimatic changes in light, temperature, wind, and the incidence of fire (Lovejoy et al. 1986; Kapos 1989; Bierregaard et al. 1992). Each of these edge effects can have a significant impact on the vitality and composition of the species in the fragment.

Microclimatic changes in light and temperature. Sunlight is absorbed and reflected by layers of leaves, particularly in forest communities and other communities with a dense plant cover. In intact forests, less than 1% of the light energy may reach the forest floor. The forest canopy buffers the microclimate of the forest floor, keeping it relatively cool, moist, and shaded during the day, reducing air movement, and trapping heat during the night. When the forest is cleared, these effects are removed; as the forest floor is exposed to direct sunlight, the ground becomes much hotter during the day. Without the canopy to reduce heat and moisture loss, the ground is also much colder at night, and is generally less humid. These effects will be strongest at the edge of a habitat fragment and decrease toward the interior. In Amazonian forest fragments, such microclimatic effects were evident as far as 40 meters into the forest interior (Kapos 1989). Since plant and animal species are often precisely adapted to temperature, humidity, and light levels, these changes will eliminate many species from forest fragments. Shade-tolerant wildflower species of the temperate forest, late-successional tree species of the tropical forest, and humidity-sensitive animals such as amphibians are often rapidly eliminated by habitat fragmentation, leading to a shift in the species composition of the community.

The habitat edge is usually the most altered region of a fragment (Figure 6.16). Edges may have very high daytime temperatures when the angle of the sun is low, and very cold night temperatures due to the lack of buffering by other vegetation. However, a dense tangle of vines and fast-growing pioneer species often grows up at the forest edge in response to these altered conditions. These tangles of vegetation may create a barrier that reduces the effects of environmental disturbance on the interior of the fragment. In this sense, the forest edge plays an important role in preserving the composition of the forest fragment—but in the process, the species composition of the forest edge is dramatically altered and the area occupied by forest interior species is further reduced (Lovejoy et al. 1986). Over time the

6.16 Forest clearing for pasture in Brazil results in sharp edges that change the rain forest microclimate. (Photograph by R. Bierregaard.)

forest edge may be occupied by different species of plants and animals from those found in the forest interior.

Wind. Wind changes also have a significant effect in fragmented forest habitats. In an intact forest, wind velocity is substantially reduced by the tree canopies; the wind moves strongly over the treetops but is reduced to a gentle breeze within the forest. When a habitat is fragmented, wind is able to enter the habitat and move through the forest. The impact of wind will be greatest at the forest edge, which is subject to its full force, but the effects of air movement may be felt over a considerable distance, particularly in flat terrain. The increased wind and air turbulence directly damages vegetation, particularly at the forest edge. Trees that have grown up in the forest interior with minimal wind stress will have leaves and branches stripped off by the wind or even be blown down (Bierregaard et al. 1992; Laurance 1991b). Increased wind also leads to increased drying of the soil, lower air humidity, and higher water loss from leaf surfaces. The increased water stress that results may kill many forest interior plant species. Trees along the forest edge may die off and be replaced by a new suite of species better adapted to the new conditions. While such effects are most evident within 200 m of the forest edge, increased forest damage was noted up to 500 m inside the forest margins in Australian rain forests.

Increased incidence of fire. Increased wind, lower humidity, and higher temperatures make fires more likely. Fires may spread into habitat fragments from nearby agricultural fields that are burnt regularly, as in sugarcane harvesting, or from the irregular activities of farmers practicing slash-and-burn agriculture (Gomez-Pompa and Kaus 1992). Forest fragments may be particularly susceptible to fire damage when wood has accumulated on the edge of the forest as trees died or were blown down by the wind. In Indonesian Borneo, several million hectares of tropical moist forest burned during an unusual dry period in 1982 and 1983; a combination of forest fragmentation from farming and selective logging, the accumulation of brush after selective logging, and human-caused fires contributed to the extent of this environmental disaster (Leighton and Wirawan 1986).

Increased predation and competition from exotic and pest species. Habitat fragmentation increases the vulnerability of the fragments to invasion by exotic species and native pest species. The forest edge represents a disturbed environment in which many pest species of plants and animals can establish, increase in numbers, and then disperse into the interior of the fragment (Janzen 1983). For example, the seeds of wind-dispersed plants may be blown great distances into the interior of the fragment, where they colonize both natural disturbance sites and the gaps that occur because of increased tree mortality. Light-loving butterflies from open habitats may migrate up to 300 m into the forest interior (Lovejoy et al. 1986).

Omnivorous animals such as raccoons, skunks, and blue jays may increase in numbers along the forest edges, where they can eat foods of both undisturbed and disturbed habitats (Yahner 1988). These aggressive feeders will often seek out the nests of forest birds, often preventing successful reproduction for many bird species hundreds of meters from the nearest forest edge (Box 6). Nest-parasitizing cowbirds, which live in fields and edge habitats, use habitat edges as an invasion point into forest interiors where they take over the nests of forest songbirds. Similarly, predatory animals may decimate insect and amphibian populations in edge areas that were inaccessible to

BOX 6 SONGBIRD DECLINE IN NORTH AMERICA

The recent decline of songbirds throughout North America has led to a variety of theories to explain the phenomenon (Terborgh 1989, 1992a; Hagan 1992). Some researchers, observing that many of the birds missing from smaller, isolated forest fragments could still be found in larger tracts of forest, hypothesized that forest fragmentation was somehow responsible for the decreases in population of many species. Others, noting that the species that showed the most dramatic declines were those that migrated to the tropics during the winter (known as Neotropical migrants), suggested that the destruction of the tropical forest was affecting the songbirds' survival. Observations of songbird species, however, failed to adequately support either of these hypotheses. In the first instance, the sizes of forest tracts required by a species varied dramatically from one location to the next, so that the results of a study in one region of the United States were not applicable to populations in a different part of the country. In the second case, researchers found that while many Neotropical migrant birds are declining, not all of them winter in places that are undergoing deforestation. In contrast, a number of birds that winter in the areas of Latin America most profoundly affected by deforesta-

tion still have stable populations (Terborgh 1992a). Furthermore, the tropical deforestation hypothesis does not account for the declines in nonmigratory species, which, although not as severely hit as the migrants, have suffered significant losses. Habitat destruction, the most obvious explanation for songbird declines, does not seem to be the overriding factor; what, then, could be causing the problem?

The answer seems to lie in the way that forests and forest fragments were regarded by

scientists: if no net loss of area and no significant human activity took place within the forest, it was regarded as stable, suitable habitat for the songbird species. When many of these seemingly suitable forest tracts continued to be bypassed as breeding sites, some scientists began to wonder if perhaps another factor, not initially obvious, was present in the ostensibly stable forest tracts. Investigations in the early 1980s demonstrated that this was indeed the case: nest parasitism by cowbirds

Sightings of songbird species in Rock Creek Park, District of Columbia

Species	Mean number of pairs sighted		Percent change
	In 1940s	In 1980s	
MIGRANTS			
Red-eyed vireo	41.5	5.8	−86.0
Ovenbird	38.8	3.3	−91.5
Acadian flycatcher	21.5	0.1	−99.5
Wood thrush	16.3	3.9	−76.1
Yellow-throated vireo	6.0	0.0	−100.0
Hooded warbler	5.0	0.0	−100.0
Scarlet tanager	7.3	3.5	−52.1
Black-and-white warbler	3.0	0.0	−100.0
Eastern wood peewee	5.5	2.8	−49.1
NONMIGRANTS			
Carolina chickadee	5.0	4.3	−14.0
Tufted titmouse	5.0	4.5	−10.0
Downy woodpecker	3.5	3.0	−14.3
White-breasted nuthatch	3.5	3.1	−11.4

Source: Terborgh 1992a; National Audubon Society data.

and nest predation by raccoons, opossums, and other animals were very high in areas of songbird decline, particularly in the fragmented tracts (Wilcove 1985; Terborgh 1992a; Line 1993). All of these animals are attracted to areas occupied or disturbed by humans; cowbirds favor open agricultural land, while opossums and raccoons are common in suburbs, where lack of predators and the presence of garbage dumps create an ideal setting for scavengers.

Both nest parasitism and nest predation had a severe impact upon the songbird populations studied. A study in a large tract of forest near Shelbyville, Illinois, demonstrated that 80% of observed songbird pairs lost their eggs to nest predators. To compound the matter, more than two-thirds of the remaining songbird pairs ended up raising cowbird chicks rather than their own young (Terborgh 1992a). In other words, less than 7% of songbirds in this forest successfully reproduced. Given these statistics, the decline in songbirds is hardly surprising.

Furthermore, these observations partially explain the greater losses among Neotropical migrants. Migrant birds are more vulnerable to predation and parasitism because their breeding seasons are restricted by the time they require for migration. Many migrants also produce fewer offspring in a single breeding effort than do

One of the first Neotropical migrant songbirds to be lost as a result of tropical deforestation is Bachman's warbler, last seen in the 1960s. The Cuban forests in which this species overwintered were almost entirely cleared for sugarcane fields. The warbler is shown in this Audubon print with the flowering tree *Franklinia*, which is now extinct in the wild.

nonmigrants. Consequently, failure to produce young successfully is a more serious problem for migrant populations than for nonmigrant birds. The destruction of tropical forest habitat combined with this reproductive failure merely exacerbates the problem for migratory birds; thus, Neotropical migrant species have experienced a far sharper decline than have nonmigratory songbird species.

the predators until the fragmentation took place. Populations of deer and other herbivores can build up in edge areas, overgraze the vegetation, and eliminate certain plant species for distances of several kilometers into the forest interior (Alverson et al. 1988).

Habitat fragmentation also results in wild populations of animals being in closer proximity to domestic animals. Diseases of domestic animals can then spread more readily to the wild animals, which have little resistance to them. The effects of disease, and of exotic species in general, will be more thoroughly examined in the next chapter.

Four Studies of Habitat Fragmentation

Birds of eastern North America. Eastern North America is a region of high human density, wealth, and concern for conservation issues. Consequently, the effects of forest fragmentation on bird populations have been intensively studied in this area. Forest fragmentation has led to the disappearance of bird species, with the disappearance being most rapid in small patches of forest and among Neotropical migrant species that are typically found in the forest interior (Box 6). In one study, fifteen species of the forest interior, mostly Neotropical migrants, were eliminated from forest fragments of less than 0.7 km^2 (Whitcomb et al. 1981). If the forest fragment is near a large forested area and birds can fly between the two, then forest patches of 0.14 km^2 may allow some forest interior species to breed.

One extensive study surveyed bird populations in 270 upland forest patches in the coastal plain of Maryland (Lynch and Whigham 1984). Neotropical migrants, such as the wood thrush, the Eastern wood peewee, the Kentucky warbler, and the red-eyed vireo, were most abundant in large forest fragments composed of mature vegetation and a wide diversity of plant species, that were relatively near to other large forest fragments (Figure 6.17). The densities of resident nonmigratory species, such as the downy woodpecker, the Carolina chickadee, and the cardinal, were less affected or not affected at all by these site characteristics. Since each species has specific habitat requirements for breeding, each responds to forest fragmentation in different ways.

Dung and carrion beetles in central Amazonia. Dung beetles are keystone species in the forest ecosystems because of their function in burying dung and carrion as a food source for their larvae. As discussed in Chapter 2, nutrients from buried dung and carrion are more rapidly recycled in the ecosystem; the vertebrate parasite populations

6.17 The probability of sighting a wood thrush in a mature forest in Maryland is only about 20% in a forest fragment of 0.1 ha; it increases to about 80% in a forest fragment over 100 ha in area. (From Robbins 1991.)

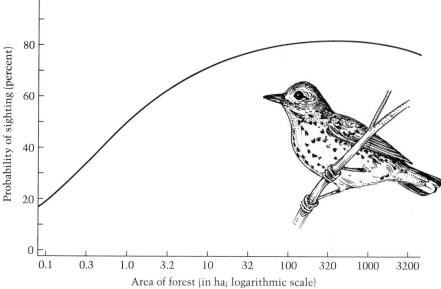

that reside in dung and carrion are killed, which reduces the level of disease in the vertebrate community; and seeds contained in the dung of fruit-eating animals are buried in a way which may facilitate seed germination and the establishment of new plants.

To investigate the effects of forest fragmentation on dung beetles, researchers studied forest fragments that had been established 2 to 6 years earlier in the Brazilian Amazon as part of the Minimum Critical Size of Ecosystems Project, also known as the Biological Dynamics of Forest Fragments Project (Figure 6.18; Klein 1989). The researchers trapped beetles in forest fragments of 1 ha, 10 ha, and continuous forest as well as clear cut areas. The study showed that the small forest fragments had fewer species of dung beetles, lower population densities for each species, and smaller-sized beetles than comparable undisturbed forest areas. In line with the lower beetle populations, dung decomposed at a lower rate in small forest fragments.

There are several reasons why beetle populations might be lower in small forest fragments. (1) The primates, birds, and other vertebrate species that the beetles depend on as sources of dung and carrion are at lower densities in small forest fragments, leading to a decreased food resource for the beetles. (2) The microclimate of small forest fragments is hotter and drier than that in continuous forest and may no longer be suitable for beetle larvae development. (3) Many

6.18   An innovative study in the Brazilian Amazon is experimentally investigating the effects of forest fragmentation on plant and animal communities. Forest fragments of specific sizes were preserved during the conversion of the site to commercial pastures. The forest square in the lower left is 10 ha in area (316 m × 316 m) and is surrounded on three sides by a 100-meter-wide strip of pasture. The small square fragment in the center is 1 ha. A large area of surrounding forest absorbed the birds displaced by the clearing. (Photograph by R. Bierregaard.)

beetle species are at low density and may go extinct in isolated forest fragments simply due to random fluctuations in population size. If clear-cut areas act as barriers to dispersal, then isolated forest fragments may be cut off from recolonization from nearby forested areas.

This study not only demonstrated that dung beetles are significantly affected by forest fragmentation, but also suggests that a reduction in beetle populations can have a widespread effect on community interactions and ecosystem processes.

Habitat fragmentation of the flattened musk turtle. The existence of numerous geographically separate river drainage systems in the southeastern United States has led to the evolution of a rich endemic fauna of mollusks, fish, and other aquatic species. Habitat modification caused by channelization, dam construction, chemical spills, siltation following construction and strip mining, and sewage discharge has made large areas of these stream systems uninhabitable by the native fauna (Moyle and Leidy 1992). The flattened musk turtle (*Sternotherus depressus*) is an endemic species of the Southeast, confined to river channels in the Warrior River Basin of central Alabama (Dodd 1990). This species is of special concern since it is protected by the U.S. Endangered Species Act. An extensive survey of the species revealed that it now occupies only 7% of its historic range. The species has been reduced to a series of isolated populations surrounded by habitat that has been rendered unsuitable for migration as a result of human activities and physical barriers. Many of the remaining populations appear to be vulnerable to local extinction due to disease, failure to breed (as shown by an absence of juveniles), the removal from rivers of old logs that the turtles need for basking, and collecting of turtles by people. This combination of threats to isolated populations and lack of opportunities for migration and recolonization suggests that the flattened musk turtle is in danger of extinction.

Raptorial birds on the island of Java. The island of Java has one of the densest agricultural populations in the world, with about 150 million people living in an area of 132,000 km^2. While the island was originally covered mainly by tropical moist forest, less than 10% of the original forest cover remains, with most of the land under intensive cultivation. Only a small number of forest reserves protect the remaining biological diversity of Java. A study was undertaken to determine the effects of forest fragmentation on the ten species of raptorial birds (birds of prey) that live in forest habitats, including the endangered Javan hawk eagle, *Spizaetus bartelsi* (Thiollay and Meyburg 1988). Biologists observed that the raptors used the reserves as nesting grounds and foraged from there out to a wide range of habitats, so that the reserves often had relatively high densities of birds. The number of bird species in a reserve increased with the size of the reserve. Many species, including the Javan hawk eagle, are probably not able to maintain populations in forest fragments of less than 20 to 100 km^2. Reserves of at least 300 km^2 are needed to preserve populations of all Javan forest raptorial birds. The study pointed out that non-forest raptorial species, such as the osprey, brahminy kite, and white-bellied sea eagle, do not have any of their nesting areas protected in reserves. As a result of habitat destruction, hunting and use of pesticides, these non-forest species are presently at very low populations and may soon vanish from Java.

Habitat Degradation and Pollution

Even when a habitat is unaffected by overt destruction or fragmentation, the communities and species in that habitat can be profoundly affected by human activities. Biological communities can be damaged and species driven to extinction by external factors that do not change the structure of dominant plants in the community, so that the damage is not immediately apparent. For example, in temperate deciduous forests, physical degradation of a habitat might be caused by frequent uncontrolled ground fires; these fires might not kill the mature trees, but the rich perennial wildflower community and insect fauna on the forest floor would be gradually eliminated. While occasional naturally occurring fires help to maintain natural conditions in isolated prairie fragments, frequent human-set fires in grasslands reduce species diversity, as only those species that are fire-tolerant can survive even though a grassland remains. Frequent boating and diving in coral reef areas typically degrade the community as fragile species are crushed by divers' flippers, boat hulls, and anchors. The most subtle form of environmental degradation is environmental pollution, the most common causes of which are pesticides, industrial chemicals and wastes, emissions from factories and automobiles, and sediment deposits from eroded hillsides. Pesticide pollution remains one of the most dangerous forms of environmental degradation, and the general effects of pollution on water quality, air quality, and even the global climate are cause for great concern. In some cases, such as the massive oil spills and 500 oil well fires that resulted from the Persian Gulf War, environmental pollution is highly visible and dramatic (Canby 1991).

Pesticide Pollution

The dangers of pesticides were brought to world attention in 1962 by Rachel Carson's influential book *Silent Spring*. The central thesis of this book was that DDT and other organochlorine pesticides, which are used on crop plants to kill insects and sprayed on water bodies to kill mosquito larvae, were finding their way up the food chain and harming wildlife populations. In particular, birds that ate large amounts of insects, fish, or other animals exposed to DDT frequently concentrated the pesticide in their tissues. Birds with high pesticide levels tended to lay eggs with abnormally thin shells, which cracked during incubation. As a result of failure to raise young, populations of these birds, particularly raptors such as hawks and eagles, were showing dramatic declines throughout the world (Box 7). In lakes and estuaries, DDT and other pesticides have become concentrated in predatory fish and in sea mammals such as dolphins. Recognition of this problem led to a ban on the use of DDT and other stable pesticides in many industrialized countries, including the United States, which outlawed them in 1972. The ban allowed the eventual recovery of many bird populations. But the continuing use of these classes of chemicals in other countries is still cause for concern.

Water Pollution

Water pollution has serious consequences for human populations; it can destroy important food sources, and it contaminates drinking water with chemicals that can cause immediate and long-term harm to human health. In the broader picture, water pollution often severely damages aquatic communities (Moyle and Leidy 1992). Rivers, lakes, and oceans are used as open sewers for industrial wastes and

BOX 7 PESTICIDES AND RAPTORS: HOW CHANGING HUMAN HABITS CAN HELP ENDANGERED SPECIES

Birds of prey such as the American bald eagle, the osprey, and the peregrine falcon are evocative symbols, representing power, grace, and nobility to people worldwide. Thus, when populations of these and other raptors began to decline in the 1950s and 1960s, concern for the birds prompted studies to determine the reasons for their decline. Although habitat degradation and hunting had affected some species, the most important factor in the decline of the raptors was reproductive failure due to pesticide pollution (Newton 1979). The primary culprits were the chemical pesticide DDT (dichlorodiphenyltrichloroethane), its breakdown products DDE and DDD, dieldrin, and related compounds. These pesticides cause thinning of the egg's shell, inhibit the proper development of the embryo, change the adult bird's behavior, and may even cause direct adult mortality.

Raptors are particularly vulnerable to these compounds because of their position at the top of the food chain. Toxic chemicals become concentrated at the top of the food chain through a process known as **biomagnification**. Pesticides are ingested and absorbed by insects and other invertebrates during pesticide application and remain in their tissues at fairly low concentrations. When fish, birds, or small mammals eat a diet of these insects, the pesticides become concentrated in the vertebrates' tissues; when raptors eat these animals, the pesticides are further concentrated to highly toxic levels. For example, DDT concentrations might be only 0.000003 parts per million (ppm) in lake water and 0.04 ppm in zooplankton, but up to 0.5 ppm in the minnows that eat the zooplankton, 2.0 ppm in the large fish that eat the minnows, and 25 ppm in the osprey that eats the large fish. Birds such as the osprey and the bald eagle are particularly susceptible because they rely heavily on fish, which absorb toxins draining into rivers and lakes from agricultural watersheds. The peregrine falcon, on the other hand, does not directly consume a large amount of fish, but it feeds on shorebirds and songbirds that eat fish, aquatic animals, and insects (Johnsgard 1990; Cade and Bird 1990). The peregrine's favored prey may concentrate the toxins further, so the peregrine is in more danger than the birds directly reliant upon fish.

Investigation of each of these species' reproductive success demonstrated that there was a noticeable relationship between the presence of DDT and other pesticides and reproductive failure. In particular, pesticide-induced thinning of eggshells among raptor populations on a global scale was correlated with raptor population decline and extirpation (Johnsgard 1990). More dramatic evidence of the damage done to raptors by DDT was the rapidity with which many populations recovered after DDT and other organochlorine pesticides were banned. The peregrine has made an astonishingly strong recovery in many parts of the world (Cade et al. 1988; Rowell 1991). Population increases in peregrines worldwide have occurred both with and without human assistance; in Great Britain, unexpected increases took place in the natural population, while other countries have established programs to release captive-bred falcons into their former ranges. Captive-bred peregrines reintroduced to eastern North America have become firmly established. It is an indication of the species' flexibility that many of the first successful nests established by these birds were located in urban areas (Johnsgard 1990; Cade and Bird 1990); the peregrine

A captive male peregrine falcon feeding young at the Cornell University "hawk barn" propagation facility. (Photograph courtesy of T. J. Cade, The Peregrine Fund.)

has apparently adapted well to nesting in high-rises and consuming pigeons in New York, Boston, and other eastern seaboard cities. Ospreys have made a similar comeback, as have bald eagles; the latter, however, have not recovered as thoroughly because of their greater decline in numbers, their lower reproductive rate, and continued illegal hunting (Brown and Amadon 1989; Porteous 1992). Bald eagles frequently scavenge the carcasses of birds or mammals shot by hunters; poisoning from lead pellets ingested by the eagles is a continuing threat to the species' recovery.

The decline and resurgence of the populations of raptors illustrates how severely wild species can be affected by pollution. Though the species are recovering from the damage inflicted by the use of DDT and other pesticides, the lessons learned from the decline of raptors will be important as decisions are made regarding the use of chemicals in the future. Because birds of prey are extremely sensitive to these and other pollutants, the responses of their populations to chemicals in their environment are strong indicators of the relative health of the environment. Such indicators could prove valuable to the welfare of human populations. For example, some of the fish species eaten by eagles and ospreys, such as salmon, are also consumed by humans; if the levels of chemicals in the fish are toxic to the raptors, there is a good chance that human consumers may also be poisoned. Though now banned in many industrialized countries, DDT is still used in many developing nations. Though the effects that this chemical may have on human populations are still unknown, scientists will be alert to its possible effects because of the damage it did to raptors.

The reestablishment of peregrines in the eastern United States has been a gradual process, costing around $3 million and involving over 300 people. (Graph courtesy of T. J. Cade, The Peregrine Fund.)

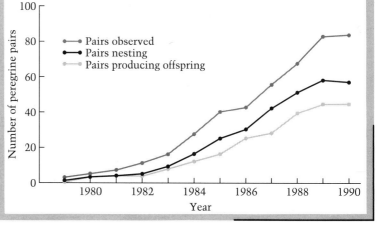

6.19 Chemicals and other waste products from a pulp mill gush from pipes into the Exploits River in Newfoundland. Industrial wastes are a major source of water pollution. (Photograph by J. N. A. Lott, McMaster U./Biological Photo Service.)

residential sewage (Figure 6.19). Pesticides, herbicides, petroleum products, heavy metals (such as mercury, lead, and zinc), detergents, and industrial wastes can kill organisms living in aquatic environments. Even if the organisms are not killed outright, these chemicals can make the aquatic environment so inhospitable that the species can no longer survive. And, in contrast to a dump in the terrestrial environment that has primarily local effects, toxic wastes in aquatic environments can diffuse over a wide area and be actively carried by currents. Toxic chemicals, even at very low levels, can be concentrated to lethal levels by aquatic organisms. Many aquatic environments are naturally low in essential minerals such as nitrogen and phosphorus, and aquatic species have adapted to the natural absence of these minerals by developing the ability to process large volumes of water and concentrate the minerals the water contains. When these species process polluted water, they concentrate toxic chemicals along with the essential minerals, eventually poisoning themselves. Furthermore, species that feed on aquatic species are exposed to the concentrated levels of toxic chemicals.

Even essential minerals that are beneficial to plant and animal life can become harmful pollutants at high levels. Human sewage, agricultural fertilizers, detergents, and industrial processes often release large amounts of nitrogen and phosphorus into aquatic systems, starting a process known as **cultural eutrophication**. Although small amounts of these nutrients can stimulate plant and animal growth, the quantities released by human activities are often detrimental (see Box 8). High concentrations of nutrients often result in thick "blooms" of algae at the water's surface. These algal blooms may be so dense that they outcompete other plankton species and shade out

BOX 8 CONSERVATION OF ENDEMIC FISH IN LAKE VICTORIA

The extinction of individual species usually does not take place in isolation; too frequently, a species is lost in conjunction with many other component species of a damaged ecosystem. Ecological changes affecting single species can have a domino effect upon other organisms, leading to catastrophic transformation of the entire ecosystem. This principle is illustrated by the recent, devastating changes in the ecology of Lake Victoria in East Africa. The lake, which is surrounded by Kenya, Tanzania, and Uganda, is one of the world's largest freshwater ecosystems. Until the early 1980s, it was also one of the most diverse in number of fish species. Prior to that time, Lake Victoria had over 350 endemic species of fish in its waters (Kaufman 1992). At present, however, only one native species and two introduced species inhabit the lake in significant numbers; all of the remaining species are endangered or extinct.

The rapid loss of endemic species has been correlated with an abrupt increase in the population of a single species of fish: the Nile perch, *Lates nilotica*, which was introduced into Lake Victoria in 1960 (Ogutu-Ohwayo 1990; Witte et al. 1992). Based solely on this information one might conclude that the Nile perch either consumed or outcompeted the native fishes, and was thus responsible for the recent losses. Such a conclusion, however, is only partially correct; while the Nile perch has played a role in the decline of endemic fishes in the lake, subtle ecological forces have also contributed to the losses.

The introduction of the Nile perch apparently did not have a significant impact on the Lake Victoria fish population until approximately 20 years after the fish was initially brought to the lake. In 1978, Nile perch constituted less than 2% of the lake's annual fishing harvest; by 1986, however, this species was nearly 80% of the total catch (Kaufman 1992). The endemic species were virtually gone from the lake, and the Nile perch had undergone an abrupt population explosion. While the perch was a prime consumer of many of the smaller native fish species, the ascendancy of the Nile perch was more than simply a case of an introduced species running amok.

One clue that other factors were contributing to native fish losses was the change in the occurrence of algal blooms in the lake's shallower waters. Algal blooms had been observed at intervals throughout the lake's history, but the frequency of these events increased noticeably in the early 1980s, at the same time that the perch population explosion took place (Ochumba and Kibara 1989). Increases in algal blooms are often associated with decreased oxygen levels in the lower strata of large bodies of water, which in turn makes the water less habitable for fish. Prior to 1978, Lake Victoria had fairly high oxygen levels at all depths; because of these aerobic conditions, fish were able to survive even in the deepest waters of the lake, which in some places exceeds 60 meters in depth. Studies done in 1987 revealed that Lake Victoria has severely depleted oxygen levels at depths below 25 meters, and is in the

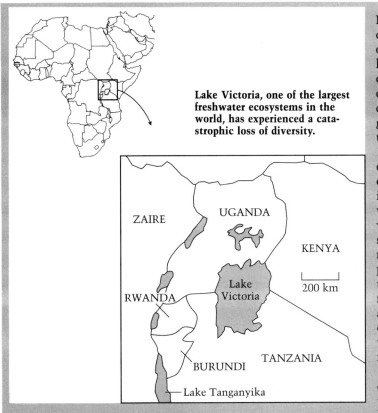

Lake Victoria, one of the largest freshwater ecosystems in the world, has experienced a catastrophic loss of diversity.

process of becoming anoxic—that is, completely lacking in oxygen below a certain depth (Ochumba and Kibara 1989; Kaufman 1992). The anoxic conditions effectively reduced the available habitat within the lake; fish species that preferred the deeper regions of the lake may have died out as a result, either because they could not adapt to the different conditions of shallower water or because they were unable to escape shallow-water predators, including the Nile perch. The mystery does not end there, however; algal blooms had occurred before, without

this devastating effect upon the native fauna. Why did the mass extinctions occur this time?

The answer is probably a combination of factors. The initial high population levels of the native species in the 1960s and 1970s probably were related to high inputs of nutrients from agricultural runoff and human settlements. The majority of these endemic species were haplochromine cichlids, which fed on the algae and other lake flora and fauna that increased because of the nutrient inputs. However, overfishing and predation by the

Nile perch contributed to a decline in native fish species over two decades; as the cichlids decreased in numbers, the excess nutrients and the lack of predation encouraged frequent algal blooms, which began the process of eutrophication (Ochumba and Kibara 1989; Kaufman 1992). The lack of oxygen in the lower depths of the lake drove the remaining native fishes to shallower waters, where they were more vulnerable to fishing nets and supported ever-increasing numbers of Nile perch. As the perch increased, the cycle continued in a downward spiral: fewer cichlids led to increased algal blooms, leading to decreased oxygen in deep water, which then further reduced the remaining cichlids.

Restoration of the once-diverse Lake Victoria ecosystem is one of the most challenging problems facing conservation biologists today. Captive breeding and restocking of the endangered cichlids will not work unless some method of restoring damaged ecosystem processes is developed. Transforming a eutrophic tropical lake of this size has never been attempted. If scientists are somehow successful in restoring oxygen levels, they must then deal with the factors that contributed to the problem in the first place: the excessive inputs of nutrients from human activity, overharvesting by fisheries, and the presence of the Nile perch.

bottom-dwelling plant species. As this process occurs, the biological diversity of the system begins to decline. When an algal mat becomes thick, its lower layers sink to the bottom and die. The bacteria and fungi that decompose the dying algae grow in response to this added sustenance and consequently absorb all of the oxygen in the water. Without oxygen, much of the remaining animal life dies off. The result is a greatly impoverished and simplified community, consisting only of species that tolerate polluted water and low oxygen levels.

Sediment eroding from logged or farmed hillsides can also harm aquatic ecosystems. The sediment covers plant leaves and other green surfaces with a muddy film that reduces light availability and so cuts down on the photosynthetic rate. Increasing water turbidity may prevent animal species from seeing in the water and reduce the depth at which photosynthesis can occur. Increased sediment loads are particularly harmful to coral communities, which often live in crystal-clear waters. Corals have delicate tentacles that strain food particles out of the clear water. When the water is filled with soil particles, the tentacles become clogged and the animal cannot feed. Many corals contain symbiotic algae species that provide carbohydrates to the corals. When soil particles cloud the water, the algae may not be able to photosynthesize, and the corals will not be able to use this source of energy.

Air Pollution

The air that surrounds the terrestrial environment has become contaminated and altered by human activities. In the past, people assumed the atmosphere was so vast that materials released into the air would be widely dispersed and their effects would be minimal. But today several types of air pollution are damaging ecosystems:

- *Acid rain.* Industries such as smelting operations and coal- and oil-fired power plants release large quantities of nitrates and sulfates, which combine with moisture in the atmosphere to produce nitric acid and sulfuric acid. These acids become part of cloud systems and dramatically lower the pH (the standard measure of acidity) of rain water (Figure 6.20). Acid rain in turn lowers the pH of soil moisture and water bodies such as ponds and lakes. By itself, the acidity is damaging to many plant and animal species. As the acidity of water bodies is increased by acid rain, many fish either fail to spawn or die outright (Figure 6.21). The acidity also inhibits the microbial process of decomposition, lowering the rate of mineral recycling and ecosystem productivity. Many ponds in industrialized areas of the world have

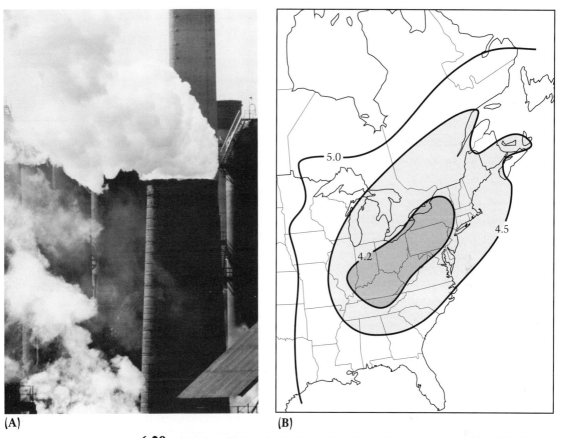

(A)

(B)

6.20 (A) Air pollution, in the form of emissions from factories and fossil-fuel burning power plants, is the major cause of acid rain. Acids in these emissions build up in the atmosphere and are precipitated in rain and snow that often falls many miles away from the source of the pollution. (Photograph by Lara Hartley/TERRAPHOTO-GRAPHICS.) (B) In North America, acid rain has its strongest effects in the eastern United States and Canada, because prevailing winds carry air pollution east from the industrial heartland. Lower pH values indicate greater acidity; normal rainfall has a pH of 5.0 or greater.

lost large portions of their animal communities as a result of acid rain (Box 9; France and Collins 1993).

- *Ozone*. Automobiles, power plants, and other industrial activities release hydrocarbons and nitrogen oxides as by-products of the burning of fuels in internal combustion engines. In the presence of sunlight, these chemicals react with the atmosphere to produce ozone and other secondary chemicals, producing **photochemical smog**. Although ozone in the upper atmosphere is im-

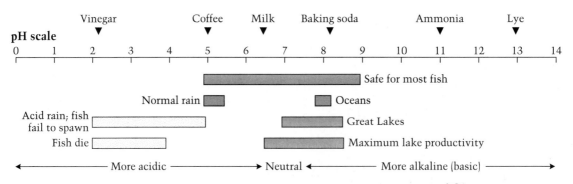

6.21 The pH scale, indicating ranges at which acidity becomes lethal to fish. Studies indicate that fish are indeed disappearing from heavily acidified lakes. (After Cox 1993, based on data from the U.S. Fish and Wildlife Service.)

portant in filtering out harmful ultraviolet radiation, high concentrations of ozone at ground level damage plant tissues and make them brittle.

• *Toxic metals.* Leaded gasoline, mining and smelting operations, and other industrial activities release large quantities of lead, zinc, and other toxic metals into the atmosphere. These compounds are directly poisonous to plant and animal life. The effects of these toxic metals are particularly evident in areas surrounding large smelting operations, where life can be destroyed for miles around.

The effects of air pollution on forest communities have been intensively studied because of the great economic value of forests in terms of wood production, watershed management, and recreation. It is widely accepted that acid rain damages and weakens many tree

BOX 9 WHY ARE FROGS AND TOADS CROAKING?

The realization that worldwide amphibian populations were in trouble came slowly to herpetologists. Biologists studying frogs, toads, salamanders, and other amphibians had been observing significant population declines in a variety of species for nearly a decade, but initially the problem seemed to be a localized phenomenon. However, at the First World Congress on Herpetology in 1989 in Canterbury, England, casual findings began to take on a disturbing significance: scientists from around the world were seeing a decline in amphibian populations (Phillips 1990; Wyman 1990). A workshop to address this possibility, hastily put together in 1990, confirmed that evidence existed for widespread population crashes among amphibian species. Species that were common less than two decades ago were becoming rare, some to the point of near extinction. Although many scientists had only anecdotal evidence for the population crisis, and although explanations for the phenomenon were speculative (Phillips 1990), reports from the United States, Central America, the

Amazon Basin, the Andes, Europe, and Australia repeated similar themes: habitat destruction and pollution, particularly acidic conditions, were contributing to the rapid decrease of many species.

Amphibians as a group are particularly sensitive to a number of global environmental problems, including temperature changes, increased ultraviolet radiation, habitat fragmentation, chemical pollution, and acid rain (Wyman 1990). The latter two factors are particularly dangerous to these animals, because the thin epidermis that characterizes amphibians leaves them vulnerable to chemical poisons, and because slight increases in acidity can destroy eggs and tadpoles. Increased acidity was identified as a probable contributor to the global decline, but given the variation among the habitats studied and the lack of direct evidence, the workshop participants could not come to any definitive conclusions until first-hand data became available. However, these initial speculations were not accepted by a small number of herpetologists, who felt that the declines in amphibian population could be related to some normal external event such as the El Niño–Southern Oscillation weather pattern, or that the decline simply represented typical extreme fluctuations of numbers in small amphibian populations (Pechmann et al. 1991).

One of the most thorough investigations of a declining amphibian species supported the acidification hypothesis. A study of the natterjack toad, *Bufo calamita*, in England specifically targeted acidity and pollution as factors to be tested (Beebee et al. 1990). The researchers took advantage of extensive documentation of both natterjack populations and land uses around sites to seek correlations between pollution and population decline; in addition, paleoecological reconstruction using sediment cores was also attempted. This study discovered that ponds that had formerly supported significant populations of the species had become gradually more acid through time, preventing the natterjacks from using the ponds as breeding grounds. An increase in heavy metals and soot particles was associated with the acidification, indicating that the process was most likely related to industrial activities. For the natterjack, even slight decreases in pH sharply increase the mortality of eggs and immature individuals; reconstructions of the acidity of one of the ponds suggested that pH had dropped from approximately 6.0 to between 4.0 and 4.5.

The specific factors contributing to the decline of the natterjack are certainly not valid for all amphibians, but the case of the natterjack is representative of conditions affecting many amphibian species. Acidification and chemical pollution of aquatic habitats are worldwide problems to which these animals are extremely vulnerable. These problems also may have subtle effects that are not immediately apparent, such as chemically mediated habitat fragmentation: even when a wetland appears relatively intact, different levels of acidity in the water and sediment may create a fragmentation effect, preventing amphibians from reproducing in certain areas of the wetland and inhibiting movement between aquatic habitats (Wyman 1990).

Biologists point out that the problems experienced by amphibians due to acidification and pollution are amplified by their relatively high position in the food chain and their hypersensitivity to their environment (Phillips 1990); amphibian species are thus some of the first to show the effects of pollution. The health of amphibian populations can serve as a yardstick for ecosystem vitality. The decline of species like the natterjack should serve as a warning that severe environmental degradation is taking place. Degradation is not irreversible—the natterjack study reported that removal of acidic sediments dramatically reduced the acidity in parts of one pond (Beebee et al. 1990)—but the ecosystem as a whole suffers less if the problem is addressed early.

6.22 Forests throughout the world are experiencing diebacks, thought to be caused in part by the effects of acid rain combined with insect attack and disease. These dead trees were photographed on Mt. Mitchell, North Carolina, in 1988. (Photograph by Jim MacKenzie, WRI.)

species and makes them more susceptible to attacks by insects, fungi, and disease (Figure 6.22). Widespread deaths of forest trees over large areas of Europe and North America have been linked to acid rain and other components of air pollution (Bormann 1982; Barrie and Hales 1984; Hinrichsen 1987; MacKenzie and El-Ashry 1988; Likens 1991). When the trees die, many of the other species in the forest will also go extinct on a local scale. Even when communities are not destroyed by air pollution, species composition may be altered as more susceptible species are eliminated. Lichens—symbiotic organisms composed of fungi and algae that can survive in some of the harshest natural environments—are particularly susceptible to air pollution. Since each lichen species has distinct levels of tolerance to air pollution, the composition of the lichen community can be used as a biological indicator of the level of air pollution.

Levels of air pollution continue to rise throughout the world. Increases in air pollution will be particularly severe in many Asian countries with their dense (and growing) human populations and increasing industrialization. The heavy reliance of China on high-sulfur coal and the rapid increase in automobile ownership in Southeast Asia are examples of potential threats to biological diversity in the region. The hope for controlling air pollution in the future depends on building motor vehicles with dramatically lower emissions of pollutants, increased development of mass transit systems, development of more efficient scrubbing processes for industrial smokestacks, and the reduction of overall energy use through conservation and efficiency measures.

Global Climate Change

Carbon dioxide, methane, and other trace gases in the atmosphere are transparent to sunshine, allowing light energy to pass through the atmosphere and warm the surface of the Earth. These gases and water vapor (in the form of clouds) trap the energy radiating from the Earth as heat, slowing the rate at which heat leaves the Earth's surface and radiates back into space. These gases have been called **greenhouse gases** because they function much like greenhouse glass, which is transparent to sunlight but traps the energy inside the greenhouse once it is transformed to heat (Figure 6.23). Another comparison would be to consider these gases as "blankets" on the Earth's surface. The denser the concentration of gases, the more heat is trapped near the Earth, and the higher the planet's surface temperature.

The greenhouse effect has been important in allowing life to flourish on Earth—without it the temperature on the Earth's surface would fall dramatically. The problem that exists today, however, is that concentrations of greenhouse gases are increasing so much as a result of human activity that they could affect the climate, creating an episode of global warming. During the past 100 years, global levels of carbon dioxide (CO_2) and other trace gases have been steadily increasing, primarily as a result of the burning of fossil fuels such as coal, oil, and natural gas (Graedel and Crutzen 1989). Burning forests to create farmland and burning firewood for heating and cooking also contribute to rising concentrations of CO_2. Carbon dioxide concentration in the atmosphere has increased from 290 parts per million (ppm) to 350 ppm over the last 100 years, and it is projected to reach 400 to 550 ppm by the year 2030 (Figure 6.24). Even if immediate massive efforts were made to reduce CO_2 production, there would be little immediate reduction in present atmospheric CO_2 levels, since each CO_2 molecule resides in the atmosphere for an average of 100

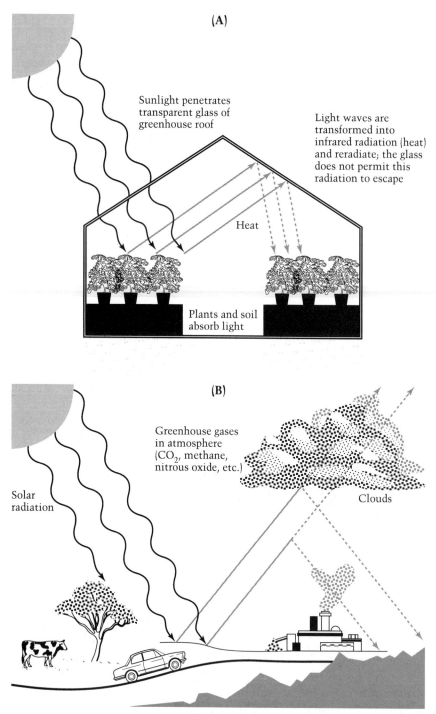

(A)

Sunlight penetrates transparent glass of greenhouse roof

Light waves are transformed into infrared radiation (heat) and reradiate; the glass does not permit this radiation to escape

Heat

Plants and soil absorb light

6.23 In the greenhouse effect, gases form a kind of blanket that acts like the glass roof of a greenhouse, trapping heat near the Earth's surface. (From Gates 1993.)

(B)

Greenhouse gases in atmosphere (CO_2, methane, nitrous oxide, etc.)

Clouds

Solar radiation

6.24   During the last 130 years, carbon dioxide concentrations (black curve) in the lower atmosphere have been steadily increasing. There is evidence of a global increase in surface air temperatures. Whether the observed temperature increases (gray "skyline") are being caused by increased concentrations of greenhouse gases is still being debated. (From Schneider 1989.)

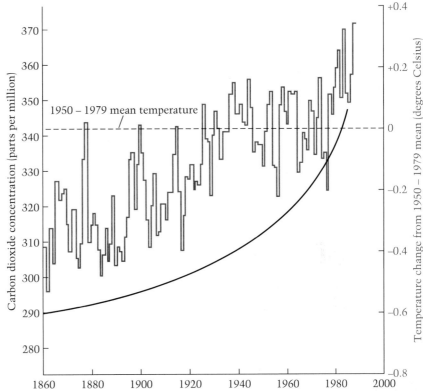

years before being removed by plants and natural geochemical processes. However, unless CO_2 production is reduced, CO_2 levels will continue to rise.

Another important greenhouse gas is methane, which has increased from 0.9 to 1.7 ppm in the last 100 years as a result of rice cultivation, cattle production, microbial activity in dumps, burning of tropical forests and grasslands, and methane release during fossil fuel production. Methane is far more efficient at absorbing heat than carbon dioxide, so methane is an important contributor to the greenhouse effect even at low concentrations. Reductions in methane levels may be difficult to achieve because methane production is closely associated with agricultural activity.

Many scientists believe that increased levels of greenhouse gases have affected the world's climate already, and that these effects will increase in the future (Peters and Lovejoy 1992; Gates 1993). The best evidence seems to show that world climate has warmed by 0.5° Celsius during the last century (Jones and Wigley 1990). Predicting

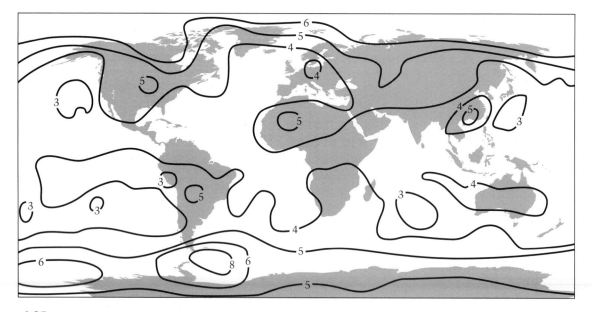

6.25   Computer-generated climate models incorporating increasing levels of greenhouse gases predict increases in surface air temperatures. The above model, one of
several generated by the Goddard Institute for Space Studies (GISS), shows predicted
temperature increases in °C. Many experts think a worldwide average temperature
rise of 3°C by the year 2100 is a reasonable scenario. All models agree that warming
will be greatest in the higher northern and southern latitudes and the polar regions.
(GISS model after Hansen 1985.)

future weather patterns is extremely complex and difficult, even with
all of the available weather data and using supercomputers. However,
there is a growing consensus among meteorologists that the world
climate will probably increase in temperature by an additional 2° to
6°C over the next century as a result of increased levels of carbon
dioxide and other gases (Figure 6.25). The increase could be even
greater if CO_2 levels continue to rise. The computer-generated models are still approximate because of uncertainty of the role of the
ocean in absorbing atmospheric carbon dioxide and uncertainty as to
how plant communities will respond to higher carbon dioxide levels
and temperatures.

Global warming is not a new phenomenon. During the past 2 million years, there have been at least 10 cycles of global warming and
cooling (Gates 1993). During warm periods, the polar ice caps have
melted, sea levels have risen to well above their present level, and
species have extended their ranges closer to the poles and migrated to

higher elevations. During cold periods, the ice caps enlarged, sea levels dropped below present levels, and species shifted their ranges closer to the equator and to lower elevations. While many species undoubtedly went extinct during these repeated episodes of range changes, the species we have today are survivors of global climate change. If species could adjust to changes in global climate in the past, will species be able to adjust to the predicted changes in global climate caused by human alteration of the atmosphere?

In fact, it seems likely that many species will be unable to adjust *quickly enough* to survive human-engendered global warming, which will occur far more rapidly than previous, natural, climate shifts. While the details of global climate change are being intensively investigated by scientists, the consequences of a rise in temperature are clear, and their impact on biological communities will be profound.

Changes in the temperate zones. Climatic regions in the northern and southern temperate zones will be shifted towards the poles. For the eastern deciduous forest of North America, species will have to migrate 500 to 1000 km northward over the next century to keep up with the expected temperature increase of 2°–6°C, or at rates of 5000 to 10,000 meters per year (Figure 6.26; Davis 1990; Davis and Zabinski 1992). Following the latest Pleistocene glaciation and subsequent increase in temperature, tree species migrated back into North America at a rate of 10 to 40 km per century, or 100 m to 400 m per year. Based on this observed rate of natural dispersal, it seems likely that many species will be unable to disperse northward rapidly enough to track the changing climate. Habitat fragmentation caused by human activities may further slow down or even prevent many species from migrating to sites where there is suitable habitat (Figure 6.27; Peters and Darling 1985; Peters 1988; Graham et al. 1990). Many species that have limited distribution and/or poor dispersal ability will undoubtedly go extinct, with widely distributed, easily dispersed species being favored in the newly establishing communities.

Changes in the tropics. The effect of global climate change on temperature and rainfall is expected to be less in the tropics than in the temperate zone. However, even small changes in the amount and timing of rainfall could have large effects on species composition, timing of plant reproduction, and susceptibility to fire. The impact of such changes on populations of migratory birds could be dramatic. Also, some predictions suggest that hurricanes could become more severe and more frequent in tropical areas, which would have major consequences for forest structure.

(A) Present   **(B) 2060 – 2100**

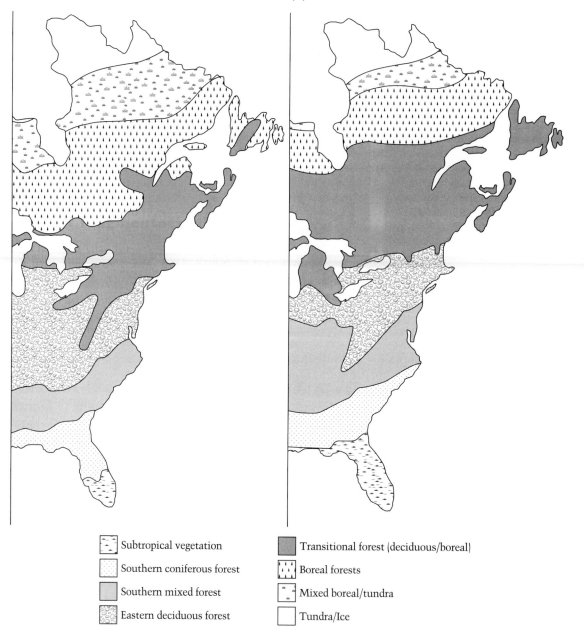

Subtropical vegetation

Southern coniferous forest

Southern mixed forest

Eastern deciduous forest

Transitional forest (deciduous/boreal)

Boreal forests

Mixed boreal/tundra

Tundra/Ice

◄ **6.26**   If predicted rises in temperature and changes in precipitation should occur, ranges of the various vegetation types would change. With their slow dispersal rates, trees could be hard-pressed to move quickly enough into new suitable habitat. (A) The current distribution of major forest types in eastern North America. (B) The projected distribution of these forest types under scenarios for a climate produced by a concentration of greenhouse gases double that of the present—an event that could occur within the next 70–100 years. The boreal forests (northern evergreen forests of spruce, fir, etc.) are expected to be especially affected. (From Gates 1993, after Solomon 1984.)

Rising sea level. Warming temperatures will cause melting of mountain glaciers and shrinkage of the Greenland ice sheet. As a result of this release of water over the next 50–100 years, sea levels could rise by 0.2 to 1.5 meters. This rise in sea level will flood low-lying coastal wetland communities. The rise could occur so rapidly that many species will be unable to migrate quickly enough to adjust to changing water levels. Particularly where human settlements, roads, and flood control barriers have been built adjacent to wetlands, the migration of wetland species will be blocked. It is possible that rising sea levels

6.27   If global climate increases as some models suggest, north-temperate tree species will have to disperse 250 km northward in order to find sites with a hospitable climate. Not only will these species encounter natural barriers such as mountains, oceans, rivers, and unsuitable terrain, but they face barriers created by people, such as agricultural fields, cities, suburbs, roads, and fences. (After Peters 1988.)

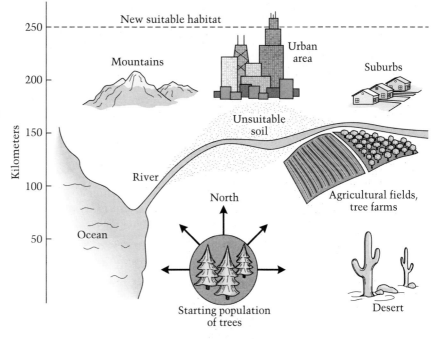

could destroy 25 to 80% of the coastal wetlands of the United States (Smith and Tirpak 1988). For low-lying countries such as Bangladesh, much of the land area could be under water within 100 years.

Rising sea levels could be detrimental to some coral reef species. Corals often grow at a precise depth in the water that has the right combination of light and water movement. If water levels rise at the expected rate of about 1 meter per century, this translates into a rise of 10 millimeters per year. A rate of 10 mm per year is the fastest that even fast-growing corals can grow (Grigg and Epp 1989). Many coral reefs will be unable to grow quickly enough to keep pace with the rise in sea level and will gradually be submerged and die. Even in those reefs that survive, the slower growing species will soon be eliminated. Damage to coral reefs could be compounded if ocean temperatures also rise (Brown and Ogden 1993). Abnormally high water temperatures in the Pacific Ocean in 1982 and 1983 led to the death of the symbiotic algae that live inside coral; the "bleached" coral then suffered a massive dieback, with an estimated death of coral over 70 to 95% of the area to depths of 18 meters (IUCN/UNEP 1988; Shinn 1989).

Changes in plant populations. Carbon dioxide is a necessary requirement for plant growth, and sometimes it occurs at concentrations too low to support maximum plant growth (Bazzaz and Fajer 1992). Some plant species will be able to utilize the increased carbon dioxide levels and will increase their growth rates, outcompeting other, less adaptable species that will not be able to grow as fast. Shifts in the abundance of herbivorous insect species may be pronounced as the abundance of their plant resources changes. Unpredictable fluctuations in the populations of plant and insect species could lead to the extinction of many rare species.

Global climate change has the potential to radically restructure biological communities and change the ranges of many species. The pace of this change could overwhelm the natural dispersal abilities of species. Because the implications of global climate change are so far-reaching, biological communities, ecosystem functions, and climate need to be carefully monitored in the future.

Summary

1. Massive disturbances to the environment caused by human activities are driving species and even communities to the point of extinction. These impacts will increase in the future as the human population reaches an expected 6.4 billion around the year 2000, with much of this population increase occurring in species-rich tropical countries. Slowing human population growth is part of the solution to the biological diversity crisis. In addition, industrial activities, logging, and agriculture are often unnecessarily destructive of the natural environment during the pursuit of short-term profit. The unequal divisions of wealth and natural resources among the countries of the world and citizens within individual countries are responsible for much of the poverty that leads to habitat loss.

2. The major threat to biological diversity is the loss of habitat, and the most important means of protecting biological diversity is preserving habitat. More than 50% of the wildlife habitat has been destroyed in 49 out of 61 Old World tropical countries. Many species have lost the great majority of their habitat and are protected in only a tiny percentage of their original range.

3. Tropical rain forests are being destroyed at a rapid rate. Out of an original extent of 16 million km^2, only 8.5 million km^2 of tropical rain forest was still present in 1985. By the year 2040, very little undisturbed rain forest will exist outside of the relatively small protected areas and national parks. Other habitats particularly threatened with destruction are tropical dry forests, wetlands in all climates, and temperate grasslands. Seasonally dry climates are often damaged by overgrazing, unsuitable agriculture, and burning, leading to soil erosion and desertification.

4. Habitat fragmentation is the process whereby a large continuous area of habitat is both reduced in area and divided into two or more fragments. These fragments are often isolated from one another by modified or degraded habitat. Habitat fragmentation can lead to the rapid loss of the species remaining, since it creates barriers to the normal processes of dispersal, colonization, and foraging. Particular fragments may lack the range of food types and other resources necessary to support permanent populations of certain species. Habitat fragments may experience altered environmental conditions and increased levels of pests, making them less suitable for the original inhabitants. In particular, the decline of numerous songbird species in North America has been linked to habitat fragmentation.

5. Environmental pollution eliminates many species from biological communities even where the structure of the community is not obviously disturbed. Pesticides used to control insects become concentrated in the bodies of birds of prey, leading to a decline in populations. Water pollution by petroleum products, sewage, and industrial wastes can kill species outright or eliminate them gradually. Increased sediment loads caused by soil erosion and excess nutrient inputs from sewage are particularly harmful to some aquatic communities. Acid rain, high ozone concentrations at the Earth's surface, and airborne toxic metals are aspects of air pollution that damage communities. Acid rain in particular is implicated in the observed worldwide decline in amphibian populations.

6. Global climate patterns may change within the coming century because of the large amounts of carbon dioxide entering the atmosphere, produced by the burning of fossil fuels. Predicted temperature increases could be so rapid that many species will be unable to adjust their ranges and will probably go extinct. Low-lying coastal communities may be submerged by seawater if polar ice caps start melting.

Suggested Reading

Bierregaard, R. O., T. E. Lovejoy, V. Kapos, A. A. Dos Santos and R. W. Hutchings. 1992. The biological dynamics of tropical rainforest fragments. *BioScience* 42: 859–866. A summary of the extensive long-term study of rain forest fragmentation in Brazil.

Carson, R. 1962. *Silent Spring.* Reprinted in 1982 by Penguin, Harmondsworth, England. This book describing the harmful effects of pesticides on birds created heightened public awareness when it was first published.

Ehrlich, P. R. and A. H. Ehrlich. 1981. *Extinction: The Causes and Consequences of the Disappearance of Species.* Random House, New York. A key work stimulating more attention on the loss of species.

Gates, D. M. 1993. *Climate Change and Its Biological Consequences.* Sinauer Associates, Sunderland, MA. A clear and thorough description of both past and predicted climate changes and their effects.

Harris, L. D. 1984. *The Fragmented Forest: Island Biogeographic Theory and the Preservation of Biotic Diversity.* University of Chicago Press, Chicago. Why habitat fragmentation leads to the loss of species.

Homer-Dixon, T., H. H. Boutwell and G. W. Rathjens. 1993. Environmental change and violent conflict. *Scientific American,* February, pp. 38–47. The relationships among population growth, conflict, and environmental degradation are outlined.

LeHouérou, H. N. and H. Gillet. 1986. Desertization in African arid lands. *In* M. Soulé (ed.), *Conservation Biology: The Science of Scarcity and Diversity,* pp. 444–462. Sinauer Associates, Sunderland, MA. The effects of encroaching desert on African species.

MacKenzie, J. and M. T. El-Ashry. 1988. *Ill Winds: Airborne Pollution's Toll on Trees and Crops.* World Resources Institute, Washington, D.C. The impact of air pollution on natural and human-dominated ecosystems.

Marsh, G. P. 1864. *Man and Nature; or, Physical Geography as Modified by Human Action.* Reprinted in 1965 (D. Lowenthal, ed.) by Harvard University Press, Cambridge, MA. An early attempt to describe how humans affect the natural environment.

Osborn, F. 1948. *Our Plundered Planet.* Little, Brown, Boston. An early work calling attention to habitat destruction.

Peters, R. L. and T. E. Lovejoy (eds.). 1992. *Global Warming and Biological Diversity.* Yale University Press, New Haven, CT. Species may have trouble keeping pace with global warming.

Terborgh, J. 1992. Why American songbirds are vanishing. *Scientific American,* May, pp. 98–104. A variety of theories are examined in light of recent evidence.

CHAPTER 7

Exotic Species Introductions, Disease, and Overexploitation

Habitat destruction, fragmentation, and degradation have obvious harmful effects on biological diversity. But even when biological communities are apparently intact, significant losses can be taking place as a result of changes caused by human activities. Three such changes are the introduction of exotic species, increased levels of disease, and excessive exploitation of particular species by people.

Exotic Species

The distributions of numerous species are restricted by their inability to disperse across major environmental barriers. Mammals of North America are unable to cross the Pacific to reach Hawaii, marine fishes in the Caribbean are unable to cross Central America to reach the Pacific, and freshwater fishes in one African lake have no way of crossing the land to reach other nearby, isolated lakes. Oceans, deserts, mountains, and rivers all restrict the movement of species. As a result of geographical isolation, patterns of evolution have proceeded in different ways in each major area of the world; for example, the biota of the Australia–New Guinea region is strikingly different from that of the adjacent region of Southeast Asia. Islands, the most isolated of habitats, have tended to evolve the most unique endemic biotas.

Humans have radically altered this pattern by transporting species throughout the world. In preindustrial times, people carried culti-vated plants and domestic animals from place to place as they set up new farming areas and colonies. Animals such as goats and pigs were set free on uninhabited islands by European seafarers to provide food on return visits. In modern times a vast array of species have been in-troduced, deliberately and accidentally, into areas where they are not native (Mooney and Drake 1986; Simberloff 1986b; Drake et al. 1989). Among the major ways this has occurred are:

- *European colonization.* Settlers arriving at new colonies re-leased hundreds of different species of European birds and mam-mals in places like New Zealand, Australia, and South Africa to make the countryside seem familiar and to provide hunting game.
- *Horticulture and agriculture.* Large numbers of plant species have been introduced and grown as ornamentals, as agricultural species, or as pasture grasses.
- *Accidental transport.* Large numbers of species have been intro-duced accidentally into new regions as a result of human activ-ity. Species are often transported unintentionally; common ex-amples include weed seeds that are accidently harvested with commercial seeds and sown in new localities, rats and insects that stow away aboard ships and airplanes, and disease and para-sitic organisms transported along with their host species. Ships frequently carry exotic species in their ballast. Soil ballast dumped in port areas brings in weed seeds and soil arthropods, and water ballast introduces algae, invertebrates, and small fish. One rice ship traveling from Trinidad to Manila had at least 41 species of exotic animals, mostly insects, hitching a ride (Myers 1934). This number is hardly remarkable: a recent survey of ships in an Oregon port showed that over 200 species of phyto-plankton and zooplankton were present in the water in the ships' ballast tanks (Carlton 1989).

The extent of this modern movement of human-transported spe-cies is unprecedented on a geological scale; it has been described by Elton (1958) as "one of the great historical convulsions of the world's flora and fauna." Many areas of the world are strongly affected by ex-otic species. Consider that approximately 4600 exotic species of plants have been recorded in the Hawaiian Islands, which is about three times the total number of native species there (St. John 1973). The Galápagos Islands have about as many exotic species of plants as they have native species. Many North American wetlands are completely dominated by exotic perennials: purple loosestrife from Europe dominates marshes in eastern North America, while Japanese honeysuckle forms dense tangles in bottomlands of the southeastern United States. More than half of the freshwater fish species in Massachusetts were introduced from elsewhere, and these exotic fishes constitute the majority of the fish biomass (Hartell 1992). Insects introduced deliberately, such as honeybees and bumblebees, and accidentally, such as fire ants and African honeybees, can build up huge populations. The effects of these exotic insects on the native insect fauna are undoubtedly large but mostly unknown.

The great majority of exotic species do not become established in the places to which they are introduced because the new environment is not suitable to their needs. However, a certain percentage of species do establish themselves in their new homes, and many of these increase in abundance at the expense of native species. These exotic species may displace native species through competition for limiting resources. Introduced animal species also may kill and eat native species to the point of extinction, or they may so alter the habitat that many natives are no longer able to persist.

Exotic Species on Islands

The effects of exotic species are generally greatest on islands and in continental areas that have experienced human disturbance (Coblentz 1990). The isolation of island habitats encourages the development of a unique assemblage of endemic species (Box 10; see also Chapter 2), but it also leaves these species particularly vulnerable to depredations by invading species (Gagné 1988). Only a limited number of organisms are capable of crossing large expanses of water without human assistance. Plants, birds, and invertebrates are among the types of organisms most commonly found on oceanic islands, with relatively few mammals present (Howarth et al. 1988). Thus island communities commonly include few large predators and grazers (Schofield 1989), and organisms representing the highest trophic levels, such as mammalian carnivores, may be absent altogether. Because they evolved in the absence of selective pressures from mammalian predators and grazers, many endemic island species have no defenses against them. Many endemic animals lack a fear of predators, and some birds have lost the power of flight and build their nests on the ground. Many island plants do not produce the bad-tasting, tough vegetative tissue that discourages herbivores, nor do they have the ability to resprout rapidly following damage.

Endemic species that thrive in the absence of these selective pres-

BOX 10 INTRODUCED SPECIES AND EXTINCTIONS IN ISLAND ECOSYSTEMS

The problem of introduced species is most pronounced in islands and archipelagos; the evolution of island species in isolation from the mainland makes them particularly vulnerable when pests, predators, and diseases are introduced by human colonists or visitors. The fragility of species endemic to islands and archipelagos has been dramatically illustrated by the recent history of multiple extinctions and 1989), and has greater humidity and topographical diversity; though the Hawaiian islands are five times as far from the mainland as the Galápagos, their age, climate, and topography combine to permit a higher level of biological diversity. Nevertheless, both archipelagos share a feature commonly found in island ecosystems: a high percentage of endemic species. Evolutionary radiation from a relatively species (Howarth et al. 1988; Howarth 1990). In addition to their unusual diversity, island ecosystems have particular value for evolutionary biologists as natural laboratories for the study of evolution (Scott et al. 1988).

But the same factors that make these island ecosystems so unique also leave them particularly vulnerable to damage caused by exotic species. Introductions of exotic species to

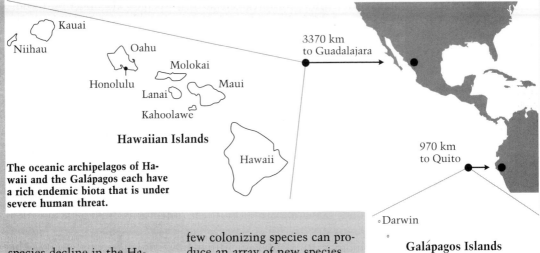

Hawaiian Islands

The oceanic archipelagos of Hawaii and the Galápagos each have a rich endemic biota that is under severe human threat.

species decline in the Hawaiian and Galápagos archipelagos.

The two archipelagos have several features in common. Both are volcanic in origin, and are a substantial distance from the nearest mainland coast. However, the Hawaiian chain is some 63 million years older than the Galápagos (Loope et al. 1988; Schofield few colonizing species can produce an array of new species (see Chapter 2). The finches that Darwin observed in the Galápagos and used to support his theory of the origin of new species are a well-known example of this principle. An extreme instance of this type of rapid evolution occurred in Hawaii, where one or two colonizing species of fruit fly evolved into over 800 different the Hawaiian and Galápagos archipelagos have had dramatic and devastating effects upon the endemic biota. In Hawaii, an initial wave of introductions accompanied the colonization of the islands by the Polynesians approximately 1000 years ago (Gagné 1988). Pigs and Polynesian rats are the most visible exotic species from the initial colonization, but invertebrates, earthworms, and a variety of plants may have accompanied the voyagers as well. The initial human colonization of the Hawaiian archipelago is thought to have resulted in a wave of extinctions. At present, paleontologists have documented at least 44 species of birds that became extinct after the arrival of the Polynesians; plant and invertebrate taxa have yet to be examined (Scott et al. 1988). Since the arrival of Europeans in 1778, many other exotic species have had a powerful impact upon native species. Black rats, feral dogs, cats, sheep, horses, cattle, goats, mongooses, and an estimated 3200 species of arthropods are only a few of the introduced species that have caused declines and extinctions among birds, insects, and plants in Hawaii in the past 200 years (Howarth et al. 1988; Loope et al. 1988; Howarth 1990). In addition, numerous plant species brought to the islands have become naturalized, often outcompeting endemic taxa. The number of naturalized exotic plants in Hawaii is now

Hawaiian bird extinctions, 1800 to the present

Common name	Species name	Became extinct in
Laysan rail	*Porzanula palmeri*	1944
Hawaiian brown rail	*Pennula millsi*	1964
Hawaiian spotted rail	*Pennula sandwichensis*	1893
Lanai thrush	*Phaeornis obscurus lanaiensis*	1931
Oahu thrush	*Phaeornis obscurus oahensis*	1825
Laysan millerbird	*Acrocephalus familiaris familiaris*	1923
Kioea	*Chaetoptila angistipluma*	1859
Oahu oo	*Moho apicalis*	1837
Molokai oo	*Moho bishopi*	1915
Hawaii oo	*Moho nobilis*	1934
Laysan apapane	*Himatione sanguinea freethi*	1923
Hawaiian mamo	*Drepanis pacifica*	1898
Black mamo	*Drepanis funerea*	1907
Lanai akialoa	*Hemignathus obscurus lanaiensis*	1894
Oahu akialoa	*Hemignathus obscurus ellisianus*	1837
Hawaii akialoa	*Hemignathus obscurus obscurus*	1895
Oahu nukupu'u	*Hemignathus lucidus lucidus*	1860
Oahu akepa	*Loxops coccinea rufa*	1893
Greater amakihi	*Viridonia sagittirostris*	1900
Lanai creeper	*Paroreomyza maculata montana*	1937
Ula-ai-hawane	*Ciridops anna*	1892
Greater Kona finch	*Psittirostra palmeri*	1896
Lesser Kona finch	*Psittirostra flaviceps*	1891
Kona finch	*Psittirostra kona*	1894

Source: From Williams and Nowak 1986.

greater than the number of endemic plant species. The impact of exotic species and habitat destruction has been so severe that Hawaii has the dubious distinction of having more recorded species extinctions than the entire rest of the United States (Gagné 1988).

The Galápagos archipelago has also experienced the effects of exotic species. The overall inhospitality of these arid, rocky islands has limited the amount of human colonization, and the extent of destruction of endemic species is less in the Galápagos than in Hawaii. Nonetheless, many species on these islands, particularly plants, are threatened by introduced species. Goats, cattle, and pigs are the primary culprits in the decline of many plant species (Schofield 1989); pigs also consume the eggs of iguanas and turtles, including those of the endangered Pacific green turtle, which nests on the islands. Introduced cultivated plants that have escaped, such as guava (*Psidium guajava*), quinine (*Cinchona succirubra*) and citrus trees, crowd out many native species (Schofield 1989).

Conservation biologists working in both archipelagos have sought to eradicate some of the more prominent and destructive introduced species. Hunting of feral goats, pigs, and other ungulates is encouraged, while domestic stock is kept closely penned. Introduced herbs and trees are eliminated by felling and burning. These measures are effective against larger species, but can do little to remove invertebrates and many herbaceous weeds. Occasionally, exotic species are introduced for the specific purpose of controlling previously introduced species; mongooses, for example, were brought to Hawaii to control exotic rodents (Gagné 1988). However, the strategy of bringing the exotics' predators can backfire: in the case of the mongooses, their effect upon rodent populations was minimal, but they destroyed the remnant population of flightless rails (Gagné 1988). While success in protecting the endemic biota of these islands has been limited, at least the problem has been identified, and the respective governments and conservation organizations have started to protect what remains.

sures often succumb rapidly when these pressures are introduced. Animals introduced onto islands have efficiently attacked, eaten, and outcompeted endemic animal species, and have grazed down native plant species to the point of extinction. Introduced plant species with tough, unpalatable foliage are better able to coexist with the introduced grazers than are the native plants, so the exotics begin to dominate the landscape as the native vegetation dwindles. Moreover, island species often have no natural immunities to mainland diseases; when exotic species are introduced to the island, they frequently carry pathogens or parasites which, though relatively harmless to the carrier, can devastate the native populations (Gagné 1988; Loope et al. 1988).

The introduction of one exotic species to an island may cause the local extinction of numerous native species. An equilibrium model in which the arrival of one exotic species on an island results in the loss of one native species would represent a great oversimplification of the real world. Several examples will illustrate this point.

- *Santa Catalina*. Forty-eight native plant species have been eliminated from Santa Catalina Island off the coast of California, primarily due to grazing by introduced goats and other alien mammals (Thorne 1967).

- *The brown tree snake.* The brown tree snake (*Boiga irregularis*; Figure 7.1) has been introduced onto a number of Pacific islands where it is devastating endemic bird populations. On Guam alone, the snake has reduced ten endemic bird species to the point of extinction (Savidge 1987).
- *Freshwater fishes.* The freshwater fish fauna of the island of Madagascar has extremely high levels of endemism, with 14 of its 23 genera found nowhere else. Recent surveys of freshwater habitats were able to locate only 5 of the known native freshwater fishes of the island (Reinthal and Stiassny 1991). Introduced fishes dominate all of the freshwater habitats. Most disturbing of all, only introduced fishes were found in the waters of the central plateau. The combination of habitat degradation and introduction of exotic fishes appears to be driving Madagascar's fish fauna to extinction.

7.1 The brown tree snake, *Boiga irregularis*, has been introduced onto many Pacific islands, where it devastates populations of endemic birds. (Photograph by Gordon Rodda, U.S. Fish and Wildlife Service.)

Exotic Species in Aquatic Habitats

Exotic species can have severe effects on lake communities and isolated stream systems. Freshwater communities are somewhat similar to oceanic islands in that they are isolated habitats surrounded by vast stretches of inhospitable and uninhabitable terrain. There has been a long history of introducing commercial and sport fish species into lakes where they do not naturally occur. Often these exotic fishes are larger and more aggressive than the native fish fauna, and

7.2 In the Flathead Lake and its tributaries in Montana, the food web was disrupted by the introduction of opossum shrimp (*Mysis relicta*). The natural food chain involved grizzly bear, bald eagles, and lake trout eating kokanee salmon; kokanee eating zooplankton (cladocerans and copepods); and zooplankton eating phytoplankton (algae). Opossum shrimp, introduced as a food source for the salmon, ate so much zooplankton that there was far *less* food available for the fish. Kokanee salmon numbers then declined radically, as did the eagle population that relied on the salmon. (After Spencer et al., 1991.)

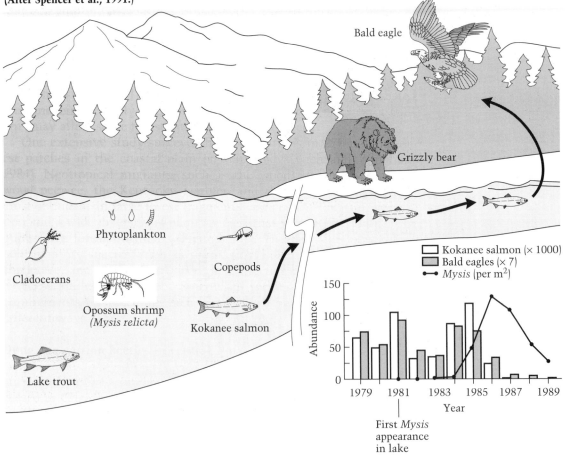

they eventually drive the local fishes to extinction. In a related case, the deliberate introduction of the freshwater opossum shrimp into the Flathead catchment of Montana was supposed to provide food for game fish (Figure 7.2); instead, the shrimp consumed the other invertebrate species the fish fed on, leading to a precipitous drop in salmon and bald eagle populations (Spencer et al. 1991).

The African Rift lakes, once among the most diverse aquatic communities in the world, have been affected by introduced species. After the Nile perch was introduced into Lake Victoria to start a commercial fishing industry, and the lake began to show signs of eutrophication, several hundred endemic cichlid fish species were apparently driven to extinction (Hughes 1986; Kaufman 1992). There are proposals by government agencies concerned with commercial fisheries to introduce the Nile perch to other Rift lakes as well (see Box 8 in Chapter 6).

Over 120 fish species have been introduced into marine and estuarine systems and inland seas; while some of the introductions have been deliberate attempts to increase fisheries, the majority of introductions were the unintentional result of canal building and the transport of ballast water in ships (Baltz 1991). Following many of these introductions, the native fish species disappeared or were greatly reduced in numbers as a result of competition with and predation by the exotic species.

Aggressive aquatic exotics are not confined to fish; they include plants and invertebrate animals as well. One of the most alarming recent invasions in North America was the arrival in 1988 of the zebra mussel (*Dreissena polymorpha*) in the Great Lakes. This small, striped native of the Caspian Sea was apparently a stowaway in the ballast tanks of a European tanker. Within two years zebra mussels had reached densities of 700,000 individuals per square meter in parts of Lake Erie, choking out native mussel species in the process (Stolzenburg 1992). One single individual of a native species was found with 10,000 tiny zebra mussels covering it in layers. Zebra mussels have been found in the Detroit, Cumberland, and Tennessee Rivers; as it spreads southward, this exotic species has the potential to cause enormous economic damage to fisheries, dams, and boats as well as to devastate the rich endemic aquatic communities it encounters.

Exotic Diseases

Disease-causing microorganisms are another important category of harmful exotics. Microorganisms can cause epidemics when they are introduced to a new locality where the native plants and animals have little resistance to them. In China, for example, most chestnut trees have resistance to chestnut blight, and only weak trees are killed by this fungal disease. However, when infected Chinese chestnut trees were brought to New York City early in this century, the chestnut blight spread to the American chestnut tree, which had no resistance to the disease. Within decades virtually all of the millions of chestnut trees in the United States had been killed by the disease. The list of species affected by exotic diseases is both diverse and extensive, and includes crayfish in Europe and black-footed ferrets in North America (see Box 11).

BOX 11 CRISIS AFTER CRISIS FOR THE BLACK-FOOTED FERRET

From time to time, even the most dedicated, energetic conservationists cannot prevent an endangered species from slipping into extinction. In the case of the black-footed ferret (*Mustela nigrepes*), however, after a difficult and controversial start, the ferret is making an impressive comeback with intensive human assistance. Several near-calamitous ferret population crises were brought about by the extreme sensitivity of this species to disease. In addition, conflict between local, federal, and private agencies, all working to "save" the ferret, created a highly emotional atmosphere that made implementing a recovery plan difficult. Ultimately, the fact that the species still exists, and apparently is beginning to thrive, is due to the persistence of wildlife biologists in overcoming the bad luck of the ferret.

Black-footed ferrets experienced a dramatic decline in their North American range during the first half of this century, presumably due to agricultural development of their prairie habitat and a deliberate, government-endorsed program by ranchers to kill off the prairie dog, the ferret's main prey species (Thorne and Williams 1988; Seal et al. 1989). In recent years, disease may have played a role in the further decline of the already fragmented black-footed ferret populations. The black-footed ferret was first thought to be extinct in the late 1970s, when the only known wild population—a small colony in South Dakota—died out (Williams 1986; Clark 1987).

Biologists with the U.S. Fish and Wildlife Service attempted a captive breeding program in the mid-1970s using animals captured from the South Dakota colony (Clark 1987, 1989). Although providing valuable information for future attempts, the experi-

ment was a failure. Four of the first six animals captured died almost immediately from vaccine-induced canine distemper after their human handlers inoculated them against this disease using a live-virus vaccine to which the black-footed ferrets were highly sensitive (Williams 1986; Thorne and Williams 1988). The vaccine had previously been demonstrated safe and effective in domestic ferrets. More ferrets were taken from the wild and added to the captive population, but this group was apparently too small to form a viable gene pool and compatible breeding pairs. Tumors and diabetes, possibly related to inbreeding, killed several of the captive ferrets, while others succumbed to disease, possibly transmitted through contact with humans (Williams 1986; Cohn 1991a). Only one of the captive females ever produced offspring, but none of her ten kits survived for more than a few days. With the loss of both the captive and wild populations by 1979, the species was thought to be completely extinct.

In the fall of 1981, however, a colony was found in Meeteetse, Wyoming. The Wyoming Game and Fish Department, designated in 1981 by the U.S. Fish and Wildlife Service to oversee ferret recovery, did not initially advocate another captive breeding program on the grounds that the wild population seemed to be surviving, no money or facilities had been committed for a captive breeding program, and the previous attempt had failed (Thorne and Williams 1988). Furthermore, administrative and policy disagreements among government departments and a private consulting firm stalled recovery initiatives. Critics charged that the state wanted to maintain control of the animals and for this reason was reluctant to hand over any ferrets to captive breeding facilities outside of Wyoming (Williams 1986; Cohn 1991a). Whatever the rationale, the delay proved costly to the ferrets. New census data in 1985 showed that the ferret population had dropped by 50 percent from its height of 129 animals in 1984 and was in immediate danger of extinction. A decision was made to capture some animals for breeding purposes (Williams 1986; Clark 1987). The first six black-footed ferrets captured in

October 1985 all died because two were infected in the wild with canine distemper. This led to the discovery that canine distemper was responsible for the rapid decline of the Meeteetse population.

At this point, biologists decided that drastic measures were required to save the ferret. All remaining black-footed ferrets were to be captured, vaccinated, quarantined, and sent to the Game and Fish Department's captive breeding center in Sybille, Wyoming. Six additional ferrets captured in October and November 1985 proved healthy, but a new obstacle to captive breeding was soon discovered. Only two males were initially present among the captive ferret population, both relatively young. Neither had ever mated before, and they showed no interest in the females during the first breeding season (Thorne and Williams 1988; Cohn 1991a). In the summer of 1986, the Wyoming Game and Fish Department made a concerted effort to capture the remaining ferrets. Through trial and error, and through the difficult capture of the last known free-ranging black-footed ferret, an experienced adult male, in 1987, the Sybille biologists were able to coax the males to mate. Once the remaining wild ferrets were added to the captive group, the captive population of black-footed ferrets stood at 18 individuals (Clark 1987;

A young black-footed ferret born at the captive colony in Sybille, Wyoming. (Photograph by LuRay Parker, Wyoming Fish & Game Department.)

Thorne and Oakleaf 1991).

The lessons of past failures, supplemented by advice on breeding, diet, and housing provided by the Captive Breeding Specialist Group of the IUCN (Clark 1989), finally paid off. Captive breeding has led to a virtual population explosion among the ferrets; between 1987 and 1991, the population jumped to 311 individuals. To prevent disease from wiping out the population, the ferrets were dispersed to several breeding centers around the country beginning in 1988 and are handled with extraordinarily strict hygienic precautions (Thorne and Williams 1988; Luoma 1992). Ferret numbers were sufficient in 1991 to permit the release of

Cages within exclosures allow ferrets to experience the range where they will eventually be released. The ferrets' caretaker is wearing a mask to reduce the chance of the ferrets' being exposed to human disease. (Photograph by LuRay Parker, Wyoming Fish & Game Department.)

mated to the Wyoming site for ten days in cages before being allowed to enter their new habitat through a tunnel. The program appears to be successful, with good survival of the animals. As of 1992, two litters had been born in the wild to released ferrets.

This particular happy ending, however beset by problems along the way, could not have occurred without human intervention. Failure to obtain a genetically diverse population of reasonable size may still threaten the viability of the captive-bred population (Thorne and Williams 1988). The combination of declining species numbers with extreme susceptibility to disease and squabbling among the government biologists and private conservation groups came within a whisker of destroying the black-footed ferret altogether.

49 animals into the wild, and an additional 90 were released in 1992 when the captive population reached 347 animals (Thorne, personal communication). The ferrets were accli-

The Ability of Exotic Species to Invade

Why are exotic species so easily able to invade and dominate new habitats and displace native species? One reason is the absence of their natural predators in the new habitat to control their population growth. For example, the prickly pear cactus (*Opuntia* sp.) was introduced into Australia from South America as an ornamental plant, but it subsequently spread over tens of thousands of hectares of grazing land. In its original habitat the species was partially controlled by a pyralid moth (*Cactoblastis cactorum*). When this moth was introduced into Australia to feed on the cactus, the cactus was eventually brought under control and now is found only in scattered small patches. The absence of its normal European predators may be one explanation for the ability of the introduced gypsy moth to build up to such enormous numbers in North America, where it damages huge tracts of deciduous forest.

Exotic species also may be better suited to take advantage of disturbed conditions than are native species. Human activity may create unusual environmental conditions, such as nutrient pulses, increased incidence of fire, or enhanced light availability, to which exotic species sometimes can adapt more readily than native species. The highest concentrations of exotics are often found in habitats that have been most altered by human activity. For example, in western North America, increased grazing by cattle and increased human-induced fires provided an opportunity for the establishment of exotic annual grasses in areas formerly dominated by native perennial grasses. In Southeast Asia, progressive degradation of forests results in a progressively smaller number of native species living in the habitat (Figure 7.3).

Exotic species are considered to be the most serious threat facing the biota of the United States park system. While the effects of habi-

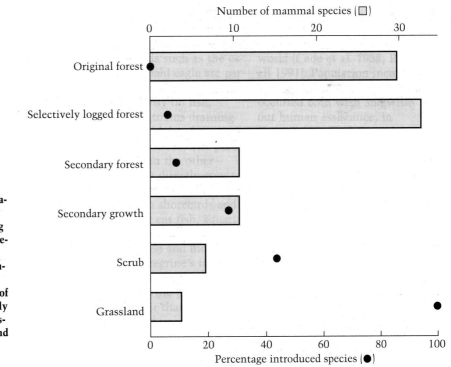

7.3 Progressive degradation of Southeast Asian forests by logging and farming not only decreases the number of species of non-flying native mammals, but increases the percentage of introduced species. Only introduced rats are present in the final grassland stage. (From Harrison 1968.)

tat degradation, fragmentation and pollution can potentially be corrected and reversed in a matter of years or decades as long as the original species are present, exotic species that are well established may be impossible to remove from communities (Coblentz 1990). Exotic species may have built up such large numbers, become so widely dispersed, and be so thoroughly entwined in the community that eliminating them may be extraordinarily difficult and expensive. An additional problem facing efforts to eliminate many exotic mammal species, such as feral goats and horses, from parks is that animal rights activists may try to block lethal control measures (Soulé 1990).

An additional class of exotics are species that have increased their ranges within continental areas because they are suited to the ways in which humans have altered the environment (Soulé 1990). Within North America, fragmentation of forests, suburban development, and easy access to garbage have allowed the numbers and ranges of coyotes, red foxes, and gulls to increase. As these aggressive species increase, they do so at the expense of other native species that are less competitive and less able to resist predation. The increase in these opportunistic species associated with human disturbance has been directly linked to the decline in numbers of many Neotropical migrant songbird species that nest in North American forests (see Box 6 in Chapter 6).

A special class of exotics are those introduced species that have close relatives in the native biota. When exotics hybridize with the native species and varieties, unique genotypes may be eliminated from local populations, and taxonomic boundaries become obscured (Cade 1983). Such appears to be the fate of native trout species when confronted by commercial species. In the American Southwest, the Apache trout (*Oncorhynchus apache*) has had its range reduced by habitat destruction and competition with introduced species. The species has also hybridized extensively with rainbow trout (*O. mykiss*), which was introduced as a sport fish (Dowling and Childs 1992).

Implications of Exotic Species for Human Health

The spread of exotic species has direct implications for the health of humans. The exotic killer bees and fire ants that are spreading in the New World not only displace the native insect species from their ecological niches, but can cause serious injuries to humans as well (Soulé 1990). As a result of the increasing movement of people, pets, wildlife, and materials from one part of the world to another, the potential for the spread of serious pests and disease-causing organisms has increased dramatically. Such organisms are already spreading: in

North America, the increased abundance of the intestinal parasite *Giardia lamblia* in beavers and elk requires that water be purified before drinking even in remote areas. The recent dramatic upsurge in Lyme disease and Rocky Mountain spotted fever, spread by infected ticks, has caused panic in some sectors of the general public in the United States. Yet these examples are minor in comparison to the potential problems that may result from human-induced changes to the environment. If world temperatures increase as predicted, the stage will be set for range expansions of mosquitoes, such as *Aedes albopictus* and *A. aegypti*, that can transmit dengue fever, yellow fever, equine encephalitis, filariasis, and hemorrhagic fever, and of disease-causing sandflies of the genus *Phlebotomus*. There is a serious potential for the environment of the temperate zone to become a more dangerous place as stinging and disease-causing species increase. In addition, if bird-watchers, hunters, hikers, and the general public thereby become frightened by and disenchanted with the outdoor experience, there is the potential of losing strong voices that support conservation efforts.

Increased Spread of Disease

Infection by disease organisms is a common feature of populations in the wild and in captivity (Scott 1988; May 1988a; Thorne and Williams 1988). These organisms include microparasites, such as bacteria, fungi, and protozoa, and macroparasites, such as helminth worms and parasitic arthropods. By living inside the tissues of a host, these parasites use and damage the host to the point of causing a weakened physiological condition, manifested by a lower probability of the host's surviving and reproducing. In many cases the infection will be undetectable by external examination of the host organism, while in other cases the host will show observable and measurable symptoms that indicate it is diseased.

The spread of disease in populations is facilitated by combinations of certain conditions. A high rate of contact between the host (such as a bighorn sheep) and the parasite (such as an intestinal worm) is one factor that encourages the spread of disease. In general, as a host population density increases, the load of parasites will also increase (Anderson 1982; Anderson and May 1980), as expressed by the percentage of hosts infected and the number of parasites per host. In addition, a high density of the infective stages of a parasite in the environment of the host population can also lead to increased incidence of disease. These parasites might come from the host population itself or from other species also infected by the parasite.

In some instances, human modifications of the environment have inadvertently increased the densities of disease-causing organisms (Cohen 1991b). For example, biologists in Texas were puzzled by the increased winter deaths of sandhill cranes—particularly the loss of 5000 birds over the winter of 1984–1985. Investigations revealed that the birds were eating unharvested, rotting peanuts on which a toxic fungus was growing. Thus the increased cultivation of peanuts could be directly linked to increased crane mortality. To halt the spread of the fungus and save the cranes, farmers now plow their fields before the winter to bury unharvested peanuts.

Another factor influencing the spread of disease is the susceptibility of the host to the infection. Individuals may fight off the infection, or the parasite may become established in the host's body. The susceptibility of a host is partially determined by the genotype of the individual, with some individuals having an immune system better able to resist the infection than others. The physiological condition of the host is also important: well-nourished individuals are better able to resist the infection (Isliker and Schurch 1981), and very old, young, or pregnant individuals may have less resistance to disease. Stress in individual animals has also been found to be a contributing factor to the susceptibility to disease (Vessey 1964).

Within populations, individuals vary in their susceptibility to particular diseases. Conservation biologists may now face a dilemma: whether to protect all individuals of a rare species from a potential disease in order to maintain population numbers and genetic variability, or whether to allow natural selection to take its course and allow the individuals that are genetically most susceptible to the disease to die off. If the disease only kills a few individuals and the population is still large, then probably the population has not been harmed and may even be better off for having weathered the disease. However, if the disease kills large numbers of individuals and the population becomes small, then many potentially valuable alleles will be lost from the population, and inbreeding depression may occur (see Chapter 11). But biologists may find it difficult to predict how virulent a disease will be in an isolated population of a rare species, particularly if the environmental conditions and population have been altered by human activity.

These basic principles of epidemiology have obvious practical implications for the captive breeding and management of rare species. First, both captive and wild populations of animals face increased direct pressure from parasites and diseases. In fragmented conservation areas, populations of animals may build up to unnaturally high densities and have high rates of disease transmission. In natural situations the level of infection is typically reduced when animals migrate away from their droppings, saliva, old skin and other bodily products.

However, in unnaturally confined situations such as habitat fragments, zoos, or even parks, the animals remain in contact with the
potential sources of infection, and disease transmission increases. In
zoos, colonies of animals are often caged together in a small area.
Consequently, if one animal becomes infected, the parasite can rapidly spread through the population.

Second, other effects of habitat destruction can indirectly increase
the animals' susceptibility to disease. When a host population is
crowded into a smaller area because of habitat destruction, there will
often be a deterioration in habitat quality and food availability, leading to a lowered nutritional status, weaker animals, and a greater susceptibility to infection. Crowding can also lead to social stress within
a species, which lowers the animals' resistance to disease. For example, territorial animals such as mountain lions will fight, sometimes
to the death, to prevent intrusion on their territories by other members of their species (Hornocker 1992); as the total area of habitat
available to such animals decreases, fights and injuries become more
frequent, lowering the animals' resistance to disease.

Third, in many conservation areas and zoos, species may come
into contact with other species that they would rarely or never encounter in the wild, so that infections may spread from one species
to another. A species that is common and fairly resistant to a parasite
can act as a reservoir for the disease, which can then infect a population of a highly susceptible species. In addition, as areas are fragmented and human activities affect the landscape, disease can spread
from domestic animals into wild populations. Even humans can infect some wild species.

A few examples illustrate these points.

- *Disease may be the single greatest threat to some rare species.*
  The last population of black-footed ferrets known to exist in the
  wild was destroyed by the canine distemper virus (Thorne and
  Williams 1988). One of the main goals of managing captive
  black-footed ferrets has been to protect the colony against canine distemper (Box 11).
- *Diseases may spread from one species to another.* The Mauritius pink pigeon (*Columba mayeri*; Figure 7.4), which is endangered in the wild, is being bred in captivity at the Rio Grande
  Zoo (Snyder et al. 1985; Vindevogel et al. 1985). Since this species often abandons its eggs, domestic pigeons were used as foster mothers. Unfortunately, the pink pigeon chicks died after
  about one week due to infection by the herpes virus. Subsequent
  examination showed that the apparently healthy domestic pigeons harbored this virus and spread it to the highly susceptible
  pink pigeon hatchlings.

7.4 Chicks of the endangered Mauritius pink pigeon are highly susceptible to viruses carried by apparently healthy domestic pigeons. (Photograph © NYZS/The Wildlife Conservation Society.)

- *Diseases that are not obvious in one species can kill related species.* Apparently healthy wildebeest, and possibly sheep, harbor the virus (*Cannochaetes* sp.) that causes malignant catarrhal fever (Thorne and Williams 1988). In zoos that maintain wildebeest, the virus can spread to other, highly susceptible endangered ruminants. Catarrhal fever has been responsible for deaths of the Pere David's deer (*Elaphurus davidianus*), a species that exists only in captivity, as well as the Indian gaur (*Bos gaurus*) and the Arabian oryx (*Oryx leucoryx*).
- *Captive species kept in crowded conditions can escalate the spread of diseases,* causing them to move very rapidly between species. An outbreak of herpes virus spread across the captive colony at the International Crane Federation, killing cranes belonging to several rare species. The outbreak was apparently related to the high density of birds in the colony (Docherty and Romaine 1983).
- *Diseases transmitted to new parts of the world have particularly powerful adverse effects on endemic species.* An important factor in the decline and extinction of many endemic Hawaiian birds is the introduction of the mosquito *Culex quinquefasciatus* and the malaria protozoan *Plasmodium relictum capistranode*. The malaria organism entered Hawaii in infected exotic birds. The present distribution and behavior of Hawaiian birds may still be modified by their susceptibility to malaria (Van Riper et al. 1986).

- *Humans can transmit diseases to captive animals.* Humans have been responsible for transmitting diseases such as tuberculosis, measles, and influenza to such animals as orangutans, colobus monkeys, and ferrets (Thorne and Williams 1988).
- *Once they are infected with exotic diseases, it may be impossible to return animals to the wild.* Captive Arabian oryx infected with the bluetongue virus of domestic livestock and orangutans affected by human tuberculosis could not be released into the wild as planned for fear of infecting free-ranging animals.

An outbreak of disease is often the final result of a complex series of factors that have affected a population. Forrester (1971) described the conditions that led to outbreaks of the lungworm-pneumonia disease in bighorn sheep: "poor range conditions, resulting in overcrowding and malnutrition, inclement weather, multiple parasitism, increase in the rate of transmission of protostrongylid lungworms due to conditions favoring the survival of the intermediate hosts and larvae of the lungworm, and secondary invasion by bacteria, resulting in pneumonia."

Overexploitation

People have traditionally harvested, hunted, and searched their immediate environment for the food and other resources that they need to survive. As long as human populations were low and the methods of exploitation were unsophisticated, people could often use the plants and animals of their environment without driving species to extinction. However, as human populations have increased, their use of the environment has escalated. Methods of harvesting have also become dramatically more efficient (Yost and Kellogg 1983; Redford 1992). Guns are used instead of blowpipes, spears, or arrows for hunting in the tropical rain forests and savannahs. Eskimo hunters in Alaska use snowmobiles and rifles instead of dogsleds and spears. Powerful motorized fishing boats harvest fish from all the world's oceans, while even small-scale local fishermen have outboard motors on their boats, allowing them to harvest more rapidly and over a wider area than previously possible. Overexploitation by humans threatens 37% of the endangered, vulnerable, and rare species of vertebrates (Prescott-Allen and Prescott-Allen 1978).

In traditional societies, controls often existed to prevent the overexploitation of natural resources. The rights to specific harvesting territories were rigidly controlled; hunting in certain areas was banned; there were often prohibitions against taking females, juveniles, and undersized individuals; certain seasons of the year and times of the day were closed for harvesting; and certain efficient types of harvesting were not allowed. These restrictions, which allowed traditional societies to harvest communal resources on a long-term, sustainable basis, are almost identical to the rigid fishing restrictions developed for many fisheries in industrialized nations (FAO 1988; McGoodwin 1990). Among the most highly developed restrictions were those of the traditional or "artisanal" societies of Micronesia and Polynesia (Johannes 1978). In these societies the resources of the reef and lagoon were clearly defined, and the possibility and consequences of overharvesting readily apparent. Even in traditional societies, intense exploitation has led to the decline and extinction of local species (Redford 1992). As an example, ceremonial cloaks worn by the Hawaiian kings were made from feathers of the mamo bird; a single cloak used the feathers of 70,000 birds of this now-extinct species.

In contrast with such traditional societies, in much of the world today resources are exploited as rapidly as possible. If a market exists for a product, local people will search their environment to find and sell it. Particularly when people are poor and hungry, they will use whatever methods are available to secure that product. In rural areas the traditional controls that regulated the extraction of natural products have generally weakened, and in many areas into which there has been substantial human migration, or where civil unrest and war have occurred, such controls may no longer exist at all (Homer-Dixon et al. 1993). In countries beset with civil conflict, such as Liberia, Somalia, Mozambique, Cambodia, Peru, the former Yugoslavia, and Afghanistan, there has been a proliferation of firearms among rural people and a breakdown of food distribution networks. In such situations the resources of the natural environment are available to whoever can exploit them. The most efficient hunter can kill the most animals, sell the most meat, and provide the most food for himself and his family. Animals are sometimes killed for target practice, or simply out of spite for the government.

Overexploitation of resources often occurs rapidly when a commercial market develops for a previously unexploited or locally used species. One of the most pervasive examples of this is the international trade in furs, in which species such as the chinchilla (*Chinchilla* spp.), vicuña (*Vicugna vicugna*), giant otter (*Pteronura brasiliensis*), numerous cat species, and some monkeys have been reduced to low numbers because of hunting for furs. Overharvesting of butterflies by insect collectors; orchids, cacti, and other plants by horticulturists; marine mollusks by shell collectors; and tropical fish for aquarium hobbyists are further examples where whole biological communities have been targeted to supply an enormous international demand. The legal and illegal trade in wildlife is responsible for the decline of many additional species (Poten 1991).

The pattern of many examples of overexploitation is distressingly similar. A resource is identified, a commercial market develops for that resource, the local human populace mobilizes to extract and sell the resource, the resource is extracted so thoroughly that it becomes rare or even extinct, and the market identifies another species to exploit. Commercial fishing fits this pattern well, with the industry working one species after another to the point of diminishing return. Commercial forestry is often similar, with species of decreasing desirability being extracted on successive cutting cycles, until there is little timber left in the forest. As a result of the extraction of timber with only a short-term perspective, almost all of the West Indies mahogany (*Swientenia mahoganii*) has been removed from the Caribbean islands, and the extensive cedar forests of Lebanon, which previously covered 500,000 ha, have now been reduced to a few isolated fragments (Chaney and Basbous 1978; Oldfield 1984). In Indonesian Borneo, ironwood has always been highly valued for its durable timber. But traditional sustainable patterns of extraction are giving way to a government-sanctioned industry that is rapidly removing all the available trees (Peluso 1992).

An extensive body of literature has developed in wildlife management, fisheries, and forestry to describe the **maximum sustainable yield** that can be obtained each year from a resource (Getz and Haight 1989). The maximum sustainable yield is the greatest amount of resource that can be harvested each year and replaced by population growth. Calculations using the population growth rate and the carrying capacity (the largest population that the environment can support) are used to estimate the maximum sustainable yield, which typically occurs when the population size is at half the carrying capacity. In highly controlled situations where the resource can be quantified, it is possible to achieve sustainable use. However, in many real-world situations, harvesting a species at the theoretical maximum sustainable yield is not possible, and attempts to do so lead to an abrupt species decline. In order to satisfy local business interests, harvesting levels are often set too high, resulting in damage to the resource base. Illegal harvesting may result in additional resource removal not seen in official records, and a considerable proportion of the remaining stock may be damaged during the harvesting operation. An additional difficulty presents itself if harvest levels are kept fairly constant even though the resource base is fluctuating; a normal harvest of a fish species during a year when fish stocks are low due to unusual weather conditions may severely injure or destroy the species.

Commercial exploitation of natural resources often follows a common pattern of increased dependency on technology, transportation, and new markets (Hart 1978; Wilkie et al. 1992; Ludwig et al. 1993). The pattern of wildlife harvesting usually occurs in the following sequence:

- The growth of towns, factories, logging camps, or mines creates a cash market for meat and other natural products.
- Traditional hunters and fishermen who formerly harvested primarily for their own needs begin to supply the cash market.
- The hunters and fishermen use the cash to buy guns and outboard motors for more efficient harvesting.
- Roads and motorboats allow middlemen to travel longer distances to bring the harvest to market, and provide access to more remote areas, particularly along logging roads. Often new marketable species are discovered and exploited as a result.
- Warehouses and refrigerators allow the harvest to be accumulated at distant collection points prior to shipping.
- The resources are depleted close to the markets, and hunters and fishermen have to search farther away and use less desirable species to supply the markets.
- International buyers and jet planes create a world market for wild species, further accelerating the overexploitation of natural resources.

Any number of examples could be given to illustrate this scenario: traditional hunters in remote areas of Borneo who supply the growing towns with wild boar meat (Caldecott 1988); Malay fishermen who use motorboats to collect edible jellyfish for sale in the high-priced Japanese market; the depletion of wild game for increasing distances around mining towns in Africa.

In the North Atlantic, one species after another has been overfished to the point of diminishing return. Current attention is focused on the Atlantic bluefin tuna, which has experienced a 90% population decline over the past ten years. Industry representatives claim that the species is doing fine, even as the catch continues to decline. A similar grim scenario can be recounted for other large fish prized for their flesh and for sport, such as the swordfish and the marlin. One of the most dramatic cases of overexploitation in recent years is that involving the overfishing of sharks. Over the last ten years, the fishing industry has been exploiting the shark fisheries of the North Atlantic at a rate approximately 60% higher than the sharks can sustain, in order to supply a growing international market. The boom in demand for shark meat and shark fins comes at a time when stocks of many commercial fish species are highly depleted, so that shark harvesting is a lucrative alternative for fishermen. As a result, the populations of many sharks are declining dramatically (Box 12). A similar case has been documented among small cetacean species (dolphins and porpoises) in the southeastern Pacific. In the same way that the depletion of commercial fish in the North Atlantic has led to an increase in the harvesting of sharks, fishermen from coastal Peru are increasingly harvesting dolphins and porpoises for the local market now that southeastern Pacific fish stocks have been depleted. While these small cetaceans were previously caught incidentally by fishermen, the exploitation of porpoises and dolphins is now becoming a common and deliberate practice (Van Waerbeck and Reyes 1990).

BOX 12 SHARKS: THE WORLD'S LEAST-FAVORITE ANIMAL IN DECLINE

Of the many plants and animals threatened by human exploitation, one of the least-loved is the shark. Public perception of these animals is based almost entirely upon news reports of attacks upon humans—which are actually much rarer than one might imagine (Manire and Gruber 1990)—and gruesome media images, such as the movie *Jaws*, which portray them as merciless, indiscriminate killers. For most people, a shark is little more than a triangular fin, an appetite, and a mouthful of very sharp teeth. The shark's appearance does little to dispel its fearsome reputation. For conservationists concerned with rapidly dwindling shark populations worldwide, the shark's bad reputation is a public relations nightmare.

The single quality that redeems these animals in the public eye is not one that encourages conservation: sharks are a popular item on menus in Chinese restaurants. Shark fishing has become a booming business in the past decade (Waters 1992). In Asia, shark-fin soup is a delicacy that has created high demand for several species of shark; shark

Shark fishing in Florida. The sharks are caught by vacationers on a pleasure cruise, displayed for photographs, and then discarded. (Photograph © Paige Chichester.)

fins may bring up to $44 per kg (Manire and Gruber 1990; Waters 1992). The repugnant and wasteful practice called "finning," in which a captured shark is flung back into the water to die after its fins are amputated, has spurred some public sympathy for sharks and has led to calls for banning the practice. A more serious problem, however, is the tendency for sharks to become "by-catches" of commercial fishing done with drift gill nets. More than half of the estimated 200 million annual shark kills are related to accidental gill net catches; sharks caught in this manner are usually simply discarded.

High shark mortality has conservationists concerned for several reasons. Sharks mature very slowly, have long reproductive cycles, and produce only a few young at a time (Manire and Gruber 1990; Waters 1992). Fishes such as salmon, which have also been overharvested, can recover rapidly because of the large numbers of offspring they produce annually; sharks do not have this capability. A second problem is that harvesting of sharks by commercial and private fishing concerns is largely unregulated. A few countries, notably Australia and South Africa, have enacted legislation to stem shark losses, but major contributors to commercial shark fishing either see no need for action or are delaying proposed regulations. Recent government-imposed quotas on the total catch of coastal and deep-sea sharks in United States waters are a step in the right direction, but may still be too high to let many vulnerable species recover to their original numbers. Finally, the decimation of shark populations is occurring at a time when very little is known about more than a handful of individual species. Though more than 350 species of sharks exist, management proposals treat all sharks as a single entity because, lacking specific information, management by species is not possible. Species that have been studied, including the lemon shark (*Negaprion brevirostris*), have demonstrated a precipitous decrease in the number of juveniles observed in the past five years.

The decline of shark populations is a matter for concern in and of itself, but it is also an important factor in a larger problem. Sharks are among the most important predators in marine ecosystems; they feed upon a variety of organisms and are distributed throughout oceans, seas, and lakes worldwide (Manire and Gruber 1990). Terrestrial ecologists have already observed the benefits of predation for prey populations and the problems that occur when predators are removed from an ecosystem. The decline of sharks could have a significant, and possibly catastrophic, effect upon marine ecosystems. Ironically, sharks have fulfilled their role for some 400 million years, making them one of the longest-lived groups of organisms on the planet; yet their future in the next two decades or so depends upon a change in human attitudes and perceptions. Conservationists have their work cut out for them: they must persuade world governments to look beyond the shark's terrifying aspect and act to preserve an animal vital to the health of the world's oceans.

The hope for many overexploited species is that as they become rare it will no longer be commercially viable to harvest them and their numbers will have a chance to recover (Figure 7.5). Unfortunately, populations of many species, such as the rhinoceros and certain wild cats, may already have been so severely reduced that they will be unable to recover. In some cases rarity can even increase demand: as the rhinoceros becomes more rare, the price of its horn goes

7.5   Industrialized countries can develop large markets for products gathered from the wild, often leading to overexploitation of resources. Here a buyer of raccoon skins lies on top of his stock. Although the raccoon is not currently threatened, such overharvesting by trappers imperils many other fur-bearing species. (Photograph courtesy of In Defense of Animals.)

up, making it even more valuable as a commodity on the black market.

One of the most emotional debates on the harvesting of wild species has involved the commercial whaling industry (Nagasaki 1990). After recognition that many whale species had been hunted to dangerously low levels, the International Whaling Commission eventually banned all commercial whaling in 1986, but still allowed limited hunting of whales for research. Despite that ban, certain species, such as the blue whale and the right whale, remain at densities far below their original numbers (Best 1988; Kraus 1990), although other species appear to be making a comeback (Box 13).

BOX 13 ENDANGERED WHALES: MAKING A COMEBACK?

Whales are among the most intriguing of mammals. They are among the largest and possibly most intelligent animals on earth, with complicated social organization and communication systems. The discovery of the whale's complex, unique songs captured the public imagination, resulting in strong public support for research on whales and for legal measures to protect them.

Scientists have only recently begun to comprehend the complexity of whale behavior, since studies of many whale species are difficult for several reasons. First, whales are difficult to track in the open ocean. Radio tracking devices commonly used for land animals are difficult to use in water. Second, whales are often very far-ranging, traveling in the course of a year from tropical to polar seas. Finally, many whale species have become so rare that finding the animals in the open ocean is truly a needle-in-a-haystack search; severe overhunting, ended only within the past decade, has decimated populations of such species as blue, bowhead, humpback, and right whales. All of these species together probably have fewer than 35,000 individuals remaining (Darling 1988).

Whales come in two varieties: the baleen whales, which include some of the larger species such as the humpback, gray, and right whales, and the toothed whales, notably such common species as the pilot whale, porpoises and dolphins, and the well-known orca or "killer" whale (Darling 1988). Baleen whales feed upon schools of small fish or tiny marine shrimp, while toothed whales commonly feed upon larger fish. Due to their large size, whales have little to fear from predators. The orca is the only predator, other than humans, that attacks the larger whales; smaller whales, such as porpoises, are sometimes killed by sharks as well as by orcas.

For nearly four centuries, however, the greatest threat to all whale species has been human activity (Klinowska 1991). Particularly in the nineteenth century, when whale bone and oil made from whale blubber became important commercial products on an international scale, extensive hunting of whales was the primary cause of declines in whale populations (Best 1988). Several species that were hunted preferentially, particularly the right (in the sixteenth through eighteenth centuries), the bowhead (in the eighteenth and nineteenth centuries), and the blue (in the early twentieth century), were pushed to the brink of extinction (Best 1988; Kraus, personal communication). Right whales—so named because nineteenth-century whalers considered them the "right" whale to hunt—were never a very populous species, originally numbering around 50,000 before overhunting brought the population to its present size of roughly 3000 animals. Hunting of right whales was made illegal in 1935, by which time the species had already become severely depleted, but the ban was often violated (Best 1988).

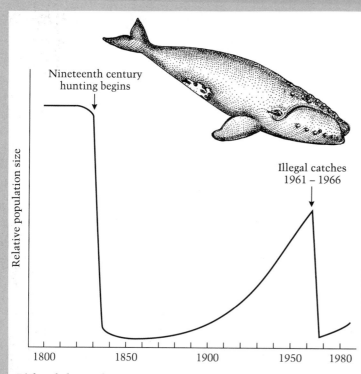

Right whale populations around the island of Tristan da Cunha in the South Atlantic were greatly diminished by whaling in the mid-nineteenth century and in the 1960s, as shown in this graph. (From Best 1988.)

Such extensive exploitation created an imbalance between the birth rate and the mortality rate of most whale species; right whales, for example, produce only a single offspring every three to four years, and thus could not replace the adults lost to hunting each year. Despite the declines in many species, hunting continued; an estimated half-million whales were killed by commercial whalers between 1940 and 1986.

The International Whaling Commission has regulated whale hunting since its creation in 1946, in an attempt to prevent species extinctions. Bans upon whaling have been in effect in parts of the world where whale populations have been most severely affected by hunting. In 1986 the IWC instituted a moratorium on all commercial killing of whales worldwide, against the protests of nations such as Japan, Norway, Russia, and Iceland (Donovan 1986; Darling 1988). Some of these nations still hunt one common species of whale, the minke whale, by claiming that the whales are required for scientific studies, thus employing a loophole in the IWC agreement (Darling 1988; Harwood 1990; Nagasaki 1990). Such efforts to continue whaling have caused conflicts between nations; Japanese ships, for instance, are barred from fishing in United States waters partly as a result of their defiance of the ban. Nevertheless, annual kills of

Whale populations

Species	Numbers prior to whaling[a]	Present numbers	Main diet items
BALEEN WHALES			
Blue	>200,000	5,000–10,000	Plankton
Bowhead	65,000	8,000	Plankton
Fin	>500,000	120,000	Plankton, fish
Gray	24,000	21,000	Crustaceans
Humpback	>125,000	12,000	Plankton, fish
Minke	>350,000	350,000	Plankton, fish
Right	50,000	3,000	Plankton
Sei	>250,000	>50,000	Plankton, fish, squid
TOOTHED WHALES			
Beluga	>100,000	>50,000	Fish, crustaceans
Narwhal	Unknown	30,000	Fish, squid, crustaceans
Sperm	>2,000,000	>1,000,000	Fish, squid

Source: After *National Geographic*, December 1988.

[a] Pre-exploitation population numbers are highly speculative.

whales have dropped dramatically since 1980, and no endangered species are hunted by whaling ships. Since then, different species have had variable recovery rates. Right whales, though protected since 1936, have not recovered in the North Atlantic and North Pacific, due to several factors that continue to inhibit population growth, including illegal hunting, environmental pollution, and possibly inbreeding (Best 1988). Humpback numbers, on the other hand, have more than doubled in some areas since the early 1960s, an increase of nearly 10% an-nually (Paterson and Paterson 1989). Gray whales appear to have recovered to their previous levels of about 20,000 animals after being hunted to less than 1000 whales. The southern minke appears to be quite abundant, with an estimated 350,000 individuals.

However, direct hunting is not the only cause of whale mortality. Right whales frequently are killed when they collide with ships, and this may be occurring in other less studied species as well (Kraus 1990). Furthermore, each year thousands of dolphins and an unknown number of whales suffocate when they become entangled in drift nets, gill nets, and other deep-sea fishing equipment intended for tuna, cod, and other commercial fish. Large whales, primarily humpbacks, are released from such entanglements at regular intervals off the coast of Massachusetts by a team of specialists from the Center for Coastal Studies in Provincetown, which responds to citizen reports of distressed or beached whales in much the same way a paramedic squad rescues accident victims.

Many smaller whales, such as dolphins and porpoises, have shown substantial population declines as a result of being accidentally caught in nets, hunted directly in increasing numbers, and harmed by pollution in bays and rivers (Van Waerbeck and Reyes 1990; Norris 1992). In enclosed, shallow waters, noise pollution caused by ships, mining, and military activities may interfere with the social communication and foraging abilities of these whales. Dolphins in tropical waters of the eastern Pacific Ocean are particularly vulnerable to fishing-related fatalities because groups of them often travel with schools of tuna; an estimated 100,000 dolphins die in seine nets each year (Darling 1988). Such anthropogenic sources of mortality apparently account for up to one-third of annual deaths in right whales (Kraus 1990), and may be as significant in other whale species. Efforts to limit or ban drift-net fishing have been only partly effective, and have caused still further acrimony between the conservation groups and nations trying to protect whales and the nations dependent upon fishing. One approach has been to establish international certification that tuna have been caught using "dolphin-friendly" methods. Only continued innovative conservation efforts will prevent the extinction of many vulnerable species.

In the coming years, whales and people will come into increasing conflict over marine resources. Fin, humpback, minke, and sperm whales eat the same fish and squid that the commercial fishing fleet is harvesting intensively in the North Atlantic Ocean. In Japan, fishermen have organized killings of dolphins that they see as competitors for fish. As harvesting of marine resources becomes ever more efficient and marine habitats are damaged and destroyed, it grows increasingly difficult to avoid driving whales and other marine species to extinction and disrupting the balance of ocean ecosystems.

Conclusion

Often a combination of factors acting simultaneously or sequentially can overwhelm a species, as illustrated by the case of the large freshwater mussel *Margaritifera auricularia*. This species was formerly known from Western Europe to Morocco, but now only occurs in one river and its adjoining canals in Catalonia, Spain (Altaba 1990). The main reason for the decline of the species has been overcollecting for its attractive shell and pearls, which were used as ornaments by humans as far back as the Neolithic. Overcollecting led to the mussel's disappearance from rivers in Central Europe in the 15th and 16th centuries, while pollution, destruction of freshwater habitats, and overcollecting continued to reduce its range in recent times. The mussel is also affected by the loss of other species, since its larval stage needs to attach to certain species of fish to complete its life cycle. Unless strict conservation measures are implemented to prevent overcollecting, control water quality, maintain fish stocks, and protect the habitat, this important species will soon become extinct.

Threats to biological diversity come from a number of different directions, but their underlying cause is the same: increasing levels of destructive human activity. It is often easy to blame a group of poor, rural people or a certain industry for the destruction of biological diversity, but the real challenge is to understand the national and international linkages that promote the destruction and to find viable alternatives. These alternatives must involve stabilizing the size of the human population, finding a livelihood for rural people in developing countries that does not damage the environment, providing incentives and penalties that will convince industries to value the environment, restricting international trade in products that are obtained by damaging the environment, and a willingness on the part of people in developed countries to reduce their consumption of the world's resources and pay fair prices for products that are produced in a sustainable, nondestructive manner.

Summary

1. Humans have deliberately and accidentally moved thousands of species to new regions of the world. On some islands, there are now more exotic plant species than there are native species. Some of these exotic species grow aggressively and eliminate native species. Island species are most vulnerable to exotic species; introduced grazers overgraze the native plants, while introduced predators eat defenseless island animals. Aquatic communities throughout the world are often dramatically altered by the introduction of fishes and other exotic species, done with the intention of enhancing commercial and sport fishing.

2. Human activities may increase the incidence of disease in wild species. The levels of disease and parasites often increase when animals are confined to a nature reserve rather than being able to disperse over a wide area. Also, animals are more susceptible to infection when they are under stress. Animals held in captivity are particularly prone to higher levels of disease, which sometimes spreads between related species of animals.

3. Overexploitation threatens about one-third of the endangered vertebrates in the world, as well as other groups of species. Growing rural poverty, increasingly efficient methods of harvesting, and the globalization of the economy combine to exploit species to the point of extinction. Traditional societies had customs to prevent overharvesting of resources, but these customs are breaking down. Even when a species is not completely eliminated by overexploitation, the population size may become so low that the species is unable to recover.

Suggested Readings

Baltz, D. M. 1990. Introduced fishes in marine systems and inland seas. *Biological Conservation* 56: 151–177. Summary of fish introductions throughout the world and their effects on native fishes.

Coblenz, B. E. 1990. Exotic organisms: A dilemma for conservation biology. *Conservation Biology* 4: 261–265. The elimination of exotic species poses a number of complex problems.

Darling, J. D. 1988. Working with whales. *National Geographic* 174(6): 886–908. An excellent popular account of the issues and problems involved in protecting whale populations.

Drake, J. A. et al. (eds.). 1989. *Biological Invasions: A Global Perspective.* John Wiley & Sons, Chichester. The impact of invasions on a wide range of ecosystems.

Getz, W. M., and R. G. Haight. 1989. *Population Harvesting: Demographic Models of Fish, Forest, and Animal Resources.* Princeton University Press, Princeton, NJ. The theory behind calculating harvest levels.

Loope, L. L., O. Hamann and C. P. Stone. 1988. Comparative conservation biology of oceanic archipelagos: Hawaii and the Galápagos. *BioScience* 38: 272–282. This article and others in the same issue give an outstanding overview of these two archipelagos.

Ludwig, D., R. Hilborn and C. Walters. 1993. Uncertainty, resource exploitation, and conservation: Lessons from history. *Science* 260: 17, 36. An excellent short statement about why commercial exploitation so often destroys its resource base.

McGoodwin, J. R. 1990. *Crises in the World's Fisheries: People, Problems and Politics.* Stanford University Press, Stanford, CA. Describes the factors leading to the overexploitation of one species after another.

Mooney, H. A. and J. A. Drake (eds.). 1986. *Ecology of Biological Invasions of North America and Hawaii.* Ecological Studies, Vol. 58. Springer-Verlag, New York. A strong collection of papers.

Safina, C. 1993. Bluefin tuna in the West Atlantic: Negligent management and the making of an endangered species. *Conservation Biology* 7: 229–234. The difficulties of implementing an effective international policy on tuna.

Scott, M. E. 1988. The impact of infection and disease on animal populations: Implications for conservation biology. *Conservation Biology* 2: 40–56. An excellent review article.

Soulé, M. 1990. The onslaught of alien species and other challenges in the coming decades. *Conservation Biology* 4: 233–239. Exotic species not only pose a threat to native species but also to human health.

Thorne, E. T. and E. S. Williams. 1988. Disease and endangered species: The black-footed ferret as a recent example. *Conservation Biology* 2: 66–74. This species is emerging as a classic conservation subject.

The Value of Biological Diversity

CHAPTER 8

Direct Economic Values of Diversity

Decisions on protecting species, communities, and genetic variation often come down to arguments over money: How much will it cost? And how much is it worth? Standard economics provides one method of assigning a value to anything, even biological diversity. The economic value of something is most often considered to be the amount of money people are willing to pay for it. The economic perspective is only one possible way of assigning value; other ethical, aesthetic, scientific, and educational methods are available as well. At this point in history, however, economic valuation is one of the principal methods used by government and corporate leaders in making major policy decisions. A major problem with standard economics is that it has tended not to assign value to natural resources; thus the costs of environmental damage have been ignored, the depletion of natural resource stocks disregarded, and the future value of resources discounted in economic calculations.

To remedy this short-sighted perspective, a major new discipline has been developing that integrates economics, environmental science, and public policy, and includes valuations of biological diversity in economic analyses (McNeely 1988; Costanza 1991). This discipline is known as **environmental economics** or **ecological economics**. Conservation biologists are increasingly using the methodology and vocabulary of environmental economics, since government officials, bankers, and corporate leaders can be more readily convinced of the need to protect biological diversity if there is an economic justification for doing so (Schumacher 1973). Governments need to allocate their resources in the most efficient manner possible, and a well-considered argument for conservation on economic grounds will often effectively support arguments based on biological, ethical, and emotional grounds.

199

The newly emerging field of environmental economics has begun to strengthen the conservation movement by assigning monetary values to species, communities, and ecosystems (McNeely et al. 1990; McNeely 1988; Peterson and Randall 1984; Brown and Goldstein 1984; Costanza 1991; Gillis 1991). These initial efforts have been simplistic in many cases due to the difficulties of assigning economic benefits to such variables as the amelioration of the climate and the potential future use of presently unused species (Green and Tunstall 1991; Daly and Cobb 1989). The hidden costs of environmental degradation that occurs during income-producing activities such as logging, agriculture, and fishing also make it hard to determine the real value of natural resources (Repetto 1992). In addition, economists are still developing appropriate methods for determining the long-term costs and benefits associated with the disruption of a biological community during economic activity. For example, during feasibility studies for a large logging operation that will remove a forest, monetary values could be assigned to the cost of replacing naturally occurring resources (such as game meat, medicinal plants, and wild foods) with equivalent ones produced by human activity (in this case by agriculture), or alternatively, to the cost of restoring the forest community or resource to its original condition after it is destroyed. These different strategies are likely to have very different costs and produce very different results. Despite these difficulties, environmental economists are continuing their evaluation of the economic benefits of biological diversity.

Many conservationists would argue that any attempt to place a strictly monetary value on biological diversity is inappropriate and potentially corrupting, since many aspects of the natural world are unique and thus truly priceless (Ehrenfeld 1988). Supporters of this stance correctly point out that there is no way to put a value on the wonder people experience when they see an elephant in the wild or a truly beautiful natural landscape; nor can an economic value be realistically assigned to the human lives that have been—and will be—saved through discoveries of medicinal compounds in wild species. Nevertheless, economic models have much to contribute to the debate over the protection of biological diversity. For example, the U.S. Department of the Interior estimated the value of potential oil production in the Arctic National Wildlife Refuge at $79.4 billion, but did not make any comparable calculation of the biological and recreational value of the polar bears, caribou, birds, wolves, and other wildlife in what could be viewed as "America's Serengeti" (Nash 1991). In general, it is to the advantage of conservationists to participate in the development of economic models to improve their accuracy and to appreciate their limitations. In addition, estimates of the monetary value of biological diversity often provide surprisingly strong evidence for the crucial role of biological diversity in local economies and support for the need to protect natural communities.

Valuation of Natural Resources in National Economies

Recent attempts have been made to include the loss of natural resources in calculations of the widely used Gross Domestic Product (GDP) and other indices of national productivity (Gillis 1991; Repetto 1992). The problem with the GDP as a measure of national well-being is that it is really a measure of all economic activity in a country, not just beneficial activity. As a result, non-sustainable activities (including overfishing of coastal waters and destructive strip-mining) and unproductive activities (unnecessarily heating unoccupied residential buildings in winter and cooling them in summer) cause the GDP to increase, even though these activities are actually irrelevant or destructive to the long-term well-being of the country. The economic costs associated with environmental damage can be considerable, and often offset the gains apparently attained through agricultural and industrial development (Repetto 1990a, 1992). In Costa Rica, for example, the value of the forests destroyed during the 1980s greatly exceeded the income produced from forest products, so that the forestry sector actually represented a drain on the wealth of the country. The cost of this destruction of timber was so great that it even exceeded the value of the interest paid on foreign debt. Similarly, the costs associated with soil erosion actually decreased the value of agriculture by 17%. When a country's GDP does not include environmental costs and the loss of natural resources, the country may appear to be achieving impressive economic gains even though it is actually on the verge of economic collapse. Unregulated national fisheries are classic examples of the need to monitor assets: large catches and high profitability may drive the industry into increased investment and excessive fishing levels that initially seem impressively successful, but gradually lead to the destruction of one commercial species after another, and eventually to the collapse of the entire industry (McGoodwin 1990). The hidden costs that can be associated with superficial economic gains are effectively demonstrated by the case of the *Exxon Valdez* oil spill in Alaska in 1989. The spill cost billions of dollars to clean up, damaged the environment, and wasted 11 million gallons of oil, yet the event was recorded as a net economic gain because expenditures associated with the cleanup increased the United States' GDP and provided employment for cleanup crews hired from throughout the United States. Without consideration of the hidden environmental costs and long-term damage to natural resources, a disaster like the *Valdez* spill can easily be misrepresented as economically beneficial.

One attempt to include natural resource depletion, pollution, and income distribution in measures of national productivity has been the development of the Index of Sustainable Economic Welfare (ISEW) (Daly and Cobb 1989). This index includes such factors as the loss of farmlands, the filling in of wetlands, the effects of acid rain, and the effects of pollution on human health. Using the ISEW, the U.S. economy apparently has not improved during the period 1956 to 1986, even though the standard GDP shows a dramatic improvement. While such measures as the ISEW are still at a preliminary stage of development, they suggest that many modern economies are achieving their apparent growth only through a non-sustainable consumption of natural resources. As these resources run out, the economies on which they are based may be seriously disrupted.

The environmental effects of large projects are increasingly being evaluated by national governments, international banks, and corporations before the projects are approved. These evaluations often take the form of **environmental impact assessments** that consider the present and future effects of the project on the environment. The environment is often broadly defined to include not just immediate natural resources, but also air and water quality, the lives of indigenous people, and endangered species. In its most comprehensive form, such an evaluation would be a cost–benefit analysis that compares the values gained by a project against the costs of the project and the values lost (Randall 1987). In theory, if the analysis shows that a project is profitable, it should go forward, while if the project is unprofitable, it should be stopped. In practice, cost–benefit analysis is notoriously difficult to carry out, because the valuations of the benefits and costs may be difficult to assign and may change over time. For example, when a new paper mill is being constructed in a forested area, it is difficult to estimate the future price of paper, the future wage scale, the future need for clean water, the value of other plant and animal species in the forests being harvested, and the extent of damage to nearby aquatic systems. In the past, natural resources used or damaged by large projects were either ignored in environmental impact assessments or grossly undervalued (Norton 1986a,b).

Assigning Values to Biological Diversity

A major focus of environmental economics has been finding methods for valuing the components of biological diversity. A number of approaches have been developed to assign economic values to genetic variation, species, communities, and ecosystems. These values include the marketplace (or harvest) value of resources, the value provided by unharvested resources in their natural state, and the future value of resources. For example, the Southeast Asian wild gaur could be valued for the meat that could be harvested from its current populations, its value for ecotourism, or its future potential in domestic cattle breeding programs. There is as yet no universally accepted framework for assigning values to biological diversity, but a variety of outlines have been proposed (Bishop 1987; Myers 1983; Oldfield 1984). One of the most useful is that used by McNeely (1988) and McNeely et al. (1990). In this framework, values are divided between **direct values**, which are assigned to those products harvested by people, and **indirect values**, which are assigned to benefits provided by biological diversity that do not involve harvesting or destroying the resource.

Direct values can be further divided into **consumptive use value** for goods which are consumed locally, and **productive use value**, for products that are sold in markets.

Indirect values can be assigned to components of biological diversity that provide economic benefits to people, but do not involve the consumption of the resource (see Chapter 9). These benefits include water quality, soil protection, recreation, education, scientific research, regulation of climate, and providing future options for human society. Indirect value can also be assigned to the **existence value**—how much people are willing to pay to protect a species from going extinct.

Direct Values

Direct values (also known in other frameworks as **use values** and **commodity values**) are assigned to those products that are directly harvested and used by people. These values can often be readily calculated by observing the activities of representative groups of people, by monitoring collection points for natural products, and by examining import and export statistics.

Consumptive Use Value

One kind of direct value is **consumptive use value**, which can be assigned to goods such as fuelwood and game that are consumed locally and do not appear in the national and international marketplaces. People living close to the land often derive a considerable proportion of the goods they require for their livelihood from the environment around them. These goods do not typically appear in the GDP of countries since they are neither bought nor sold (Repetto et al. 1989). However, if rural people are unable to obtain these products, as might occur following environmental degradation, overexploitation of natural resources, or creation of a protected reserve, then their standard of living will decline, possibly to the point where they are unable to survive.

Studies of traditional societies in the developing world show how extensively these people use their natural environment to supply their needs for fuelwood, vegetables, fruit, meat, medicine, cordage, and building materials (Lewis and Elvin-Lewis 1977; Myers 1979; Simpson and Connor-Ogorzaly 1986). In one study of Amazonian Indians it was found that about half of the species of rain forest trees in the area were used for some specific product other than fuelwood (Prance et al. 1987). About 80% of the world's population still relies principally on traditional medicines derived from plants and animals as their primary source of treatment (Farnsworth 1988). Over 5000 species are used for medicinal purposes in China (Figure 8.1A), while 2000 species are used in the Amazon basin (Schultes and Raffauf 1990; WRI/IUCN/UNEP 1992).

One of the most crucial requirements of rural people is protein, which they obtain by hunting wild animals for meat (Figure 8.1B). In many areas of Africa, wild meat constitutes a significant portion of the protein in the average person's diet: in Botswana about 40%, in Nigeria about 20%, and in Zaire 75% (Sale 1981; Myers 1988b). In Nigeria, over 100,000 tons of giant rats are consumed each year, while in Botswana, over 3 million kg of springhare are eaten per year. This wild meat includes not only birds, mammals, and fish, but also adult insects, snails, caterpillars, and grubs. In certain areas of Africa, insects may constitute the majority of the dietary protein as well as supplying critical vitamins. In areas along coasts, rivers, and lakes, wild fish represent an important source of protein. Throughout the world, 100 million tons of fish, mainly wild species, are harvested each year (FAO 1988). Much of this catch is consumed locally.

Consumptive values can be assigned to wild meat by considering how much people would have to pay to buy equivalent domestic meat if wild meat were no longer available. One detailed example of this approach was an attempt to estimate the number of wild pigs harvested in Sarawak, East Malaysia, by native hunters, in part by counting the number of shotgun shells used in rural areas and interviewing hunters. This pioneering and somewhat controversial study estimated that the consumptive value of the wild pig meat was approximately $40 million per year (Caldecott 1988).

Consumptive value can also be assigned to fuelwood used for heating and cooking, which is gathered from forests and shrublands

(A) (B)

8.1 (A) A wide variety of plants and other natural products are used in Chinese medicine. (Photograph by Catherine Pringle/Biological Photo Service.) (B) River fish are the main source of protein for Tukanoan people living in the Amazon basin. (Photograph © Paul Patmore.)

(Figure 8.2). In countries such as Nepal, Tanzania, and Malawi, the great majority of primary energy comes from fuelwood and animal dung (Pearce 1987). The value of these fuels can be assigned by considering how much people would have to pay for kerosene or other fuels if they were unable to obtain fuel from their environment. In many areas of the world, rural people have used up all the local sources of fuelwood but do not have the money to buy fuel. This situation, which has been termed the "poor man's energy crisis," forces people to walk great distances to obtain fuelwood and leads to ever-widening circles of deforestation.

In many areas of the world people developed systems for extracting resources from the natural environment in ways that prevented overuse of renewable resources. In the past, certain species of wild fruit trees could never be cut down; the breeding season of the year was taboo for hunting; families owned hunting territories that other families were not allowed to enter. These systems were organized at the village and tribal level, and were enforced through strong social pressures. For example, traditional Sherpa villages in Nepal had the custom of *shingo nava*, in which men were elected to be forest

8.2 One of the most important natural products required by local people is fuelwood, particularly in Africa and southern Asia. Here a woman in Upper Volta gathers kindling. (World Bank Photo by Yosef Hadar, © IBRD.)

guards. These men determined how much fuelwood people could collect and what trees could be cut, and so protected the common resource. People violating the village rules were made to pay fines, which were used for village activities (Furer-Haimendorf 1964).

In many parts of the world, however, these systems have broken down as a cash economy develops and people are exposed to outside influences. People now frequently sell natural resources in town markets to obtain cash. As social controls break down at the village level, the villagers themselves as well as outsiders begin to extract the resources in a destructive and non-sustainable manner, in what has been termed "the tragedy of the commons" (Hardin 1968, 1985). When an individual exploits the common resources in a non-sustainable way, that individual obtains the benefits of the resources, but the costs of this type of use are spread throughout the village. As the resources become depleted, many villagers are unable to obtain the natural products that they need for their own consumption, and often have to pay high prices in the towns for many of the products that they formerly obtained from their natural environment without cash.

While dependency on local natural products is primarily associated with the developing world, there are rural areas of the northern United States and Canada where hundreds of thousands of people are dependent on fuelwood for heating and on wild game for meat. Many of these people would be unable to survive in such remote areas if they had to buy fuel and meat.

Productive Use Value

Productive use value is a direct value that is assigned to products that are harvested from the wild and sold in commercial markets, at both the national and international levels. These products are typically valued at the price that is paid at the first point of sale minus the costs incurred up to that point, rather than the final retail cost of the products; as a result, what may appear to be minor natural products may actually be the starting points of major manufactured products. For example, wild cascara bark is gathered in the western United States and is the major ingredient in certain brands of laxatives; the purchase price of the bark is about $1 million, but the final retail price of the medicine is $75 million (Prescott-Allen and Prescott-Allen 1986). The range of products obtained from the natural environment and then sold in the marketplace is enormous, but the major ones are: fuelwood, construction timber, fish and shellfish, medicinal plants, wild fruits and vegetables, wild meat and skins, fibers, rattan, honey, beeswax, natural dyes, seaweed, animal fodder, natural perfumes, ivory, and plant gums and resins (Myers 1983, 1984). A special category includes wild species that are periodically collected for use in scientific and medical research and agricultural breeding.

At present, timber is among the most significant products obtained from natural environments, with a value of over $75 billion per year (Reid and Miller 1989). Timber products are being exported at a rapid level from many tropical countries to earn foreign currency, to provide capital for industrialization, and to pay foreign debt. In tropical countries such as Indonesia and Malaysia, timber products are among the top export earners, accounting for billions of dollars per year (Figure 8.3A). Non-wood products from forests, including game, fruits, gums and resins, rattan, and medicinal plants, also have a great productive use value (Figure 8.3B). For example, the value of non-timber forest products exported by Indonesia is about $200 mil-

(A)   (B)

8.3 (A) The timber industry is a major source of revenue in many tropical countries. Here "monkey puzzle" trees—a species with a very narrow distribution pattern—are harvested in Chile. (Photograph by Alejandro Frid/Biological Photo Service.) (B) Non-timber products are often important in local and national economies. Many rural people supplement their incomes by gathering natural forest products to sell in local markets. Here a Land Dayak family in Sarawak (Malaysia) sells wild honey and edible wild fruits. (Photograph by R. Primack.)

lion per year (Gillis 1986). In India, non-timber products account for 63% of the total foreign exchange earned from the export of forest products (Gupta and Guleria 1982). So these non-timber products, which are sometimes erroneously called "minor forest products," are in reality very important economically.

Although many countries focus on logging and timber exports as the key to economic development, other forest products may be more valuable in the long run. Researchers have attempted to compare the productive value of Amazonian tropical rain forests that are being logged with those that are sustainably harvested for fruits and latex (Peters et al. 1989). For a species-rich rain forest in the Peruvian Amazon 30 km from Iquitos, the greatest value in a single year can be realized by harvesting and selling all of the timber, but then no more timber can be harvested for decades. The greatest long-term value comes from gathering fruits, latex, and other non-timber products, which can be brought to market and sold every year. The economic value of the land for cattle ranching is less than half that of collecting rain forest products. In this analysis, the net benefit of selective logging is also surprisingly low because of the damage it does to latex- and fruit-producing trees.

This method of valuation assumes that there are stable markets for these forest products; admittedly, the profitability of harvesting fruits will decline dramatically with increasing distances from large towns due to transportation costs and spoilage, and the prices of these products may fall if more people bring them to market (Tremaine 1993; Salafsky et al. 1993). However, when the added values of wild game and medicinal plants are included, maintaining and utilizing the rain forest in this way may sometimes be more productive than the alternatives of intensive logging, converting the forest into commercial plantations, or establishing cattle ranches (Grigg 1989; Noonan and Zagata 1982). Careful harvesting of trees in ways that minimize damage to the surrounding biological community and soil, combined with the gathering of non-timber products, will also be a profitable approach that justifies maintaining the land in forest (Poore et al. 1989, 1991).

The incidental preservation of species diversity that occurs when forests and other communities are sustainably harvested adds further value to activities that are not destructive. Many biological species have great productive use value in their ability to provide new founder stock for industry and for the genetic improvement of agricultural crops (National Academy of Science 1972; Prescott-Allen 1986; Prescott-Allen and Prescott-Allen 1986). These uses differ from the traditional productive uses that involve continuous harvesting from the wild. In the case of crop plants, a wild species or variety might have a special gene that confers pest resistance or increased yield. This gene needs to be obtained from the wild only once; the gene then can be incorporated into the breeding stock of the crop species and stored in gene banks. The continued genetic improvement of cultivated plants is necessary not only for increased yield, but also to guard against pesticide-resistant insects and more virulent strains of fungi, viruses, and bacteria (Hoyt 1988; Frankel and Soulé 1981). Catastrophic failure of crop plants can often be directly linked to low genetic variability: the 1846 potato blight in Ireland, the 1922 wheat failure in the former Soviet Union, and the 1984 outbreak of citrus canker in Florida were all related to low genetic variability among crop plants (Pluncknett et al. 1987). To overcome this problem, new, resistant varieties of agricultural species are constantly being substituted for susceptible varieties. The source of resistance often comes from genes obtained from wild relatives of crop plants, and also from local varieties of the domestic species grown by traditional farmers.

Development of new varieties can have a noticeable economic impact; for example, genetic improvements in United States crops were responsible for increasing the value of the harvest by an average of $1 billion per year from 1930 to 1980 (OTA 1987). Genes for high sugar content and large fruit size from a single collection of wild tomatoes from Peru have been transferred into domestic varieties of tomatoes, resulting in an enhanced value of $80 million to the industry (Iltis 1988). The discovery of a wild perennial relative of corn in the Mexican state of Jalisco is potentially worth billions of dollars to modern agriculture because it could allow the development of a high-yielding perennial corn crop without the need for annual plowing (Iltis 1988; Norton 1988).

A special category of species collected in the wild and then used extensively are biological control agents (Julien 1987). Often a plant or animal species is introduced deliberately or accidentally into a new locality far from its original range. Such species sometimes increase in abundance in an explosive fashion, becoming troublesome pests and noxious weeds. Such introduced species may drive native species to the brink of extinction and damage ecosystems (see Chapter 7). Biologists have sometimes controlled these introduced species by searching the pest species' original habitat for a control species that limits its population. This control species can be brought to the new locality, where it can be released to act as a biological control agent. One classic example is the case of the prickly pear cactus (*Opuntia* sp.), which was introduced into Australia from South America as a hedgerow plant. The cactus spread out of control and took over millions of hectares of rangeland. However, in its native habitat, the larvae of a *Cactoblastis* moth feeds on this cactus. The moth was successfully introduced into Australia, where it has reduced the cactus to comparative rarity. Thus pristine habitats can be of great value as reservoirs of natural pest control agents.

Wild species of plants and animals that are found to have economic value to people on a local scale can be grown on plantations and ranches, and some are cultured in laboratories. The wild populations of the species provide the initial breeding stock for these colonies, and are a source of material for genetic improvement of the species. In the case of some species with medicinal uses, the useful chemical can be synthesized more cheaply than it can be grown once the chemical structure has been identified; aspirin, one of the most valuable and widely used medicines, was originally developed in this manner. In these cases the wild stock acts as the crucial blueprint for the efforts at synthesis. The structure of natural chemicals can be slightly altered by processing to produce a new structure with more desirable effects; classic medical examples of these are the pain-killing drugs: morphine, which is modified from opium, derived from the poppy plant, and novocaine and procaine, which are synthetic forms derived from the structure of cocaine, obtained from the coca plant.

Twenty-five percent of prescriptions used in the United States contain active ingredients derived from plants, and many of the most important antibiotics, such as penicillin and tetracycline, are derived from fungi and other microorganisms (Farnsworth 1988; Eisner 1991). Most recently, the fungus-derived drug cyclosporine has proved to be a crucial element in the success of heart and kidney transplants. Cyclosporine suppresses the body's immune response, preventing the rejection of the foreign tissue. Many other important new medicines have been first identified in animals; poisonous animals such as snakes, arthropods, and marine species have been a rich source of chemicals with valuable medical applications (Wachtel 1991). The 20 pharmaceuticals most used in the United States are all based on chemicals first identified in natural products; these drugs have a combined sales value of $6 billion per year. One outstanding example of such a species is the rose periwinkle (*Catharanthus roseus*) from Madagascar; two potent drugs derived from this plant are effective at treating Hodgkin's disease, leukemia, and other blood cancers (Eisner 1991). Treatment using these drugs has increased the survival rate of childhood leukemia from 10% to 90%. How many more such valuable plants will be discovered in the years ahead—and how many will go extinct before they are discovered?

The biological communities of the world are being actively searched for new plants, animals, fungi, and microorganisms that can be used to fight human diseases that are difficult to treat, such as cancer and AIDS (Box 14; see also Chapter 9). These searches are being carried out by government research institutes and pharmaceutical companies. As one example, the U.S. National Cancer Institute initiated an $8 million program in 1987 to test extracts of thousands of wild species for their effectiveness in controlling cancer cells and the AIDS virus (Booth 1987). To facilitate the search for new medicines and to profit financially from new products, the Costa Rican government established the National Biodiversity Institute (INBio) to collect biological products and supply samples to drug companies (Figure 8.4). The Merck Company has signed an agreement to pay INBio $1 million for the right to screen samples and will pay royalties to INBio on any commercial products that result from the research.

The large ecosystems of the world often have enormous productive value. Forests provide timber and other natural products. The extensive rangelands of the world provide fodder for sheep, cattle, and other domestic grazers. Coastal areas, the open ocean, rivers, and

8.4   Taxonomists at INBio are sorting and classifying Costa Rica's rich array of species. In the offices shown here many species of plants and insects are cataloged. (Photograph by Steve Winter.)

BOX 14   TRADITIONAL MEDICINES AND RAIN FOREST CONSERVATION

An argument sometimes presented for the conservation of tropical rain forests is the potential economic value of their diverse plant, animal, and insect species for commercial use, particularly as medicines (Eisner 1991). This theme was highlighted in the 1992 movie *Medicine Man*. While this line of reasoning correctly points out that many organisms in these forests have not even been scientifically described, let alone tested in laboratories for useful chemical properties, it generally overlooks the fact that many organisms, particularly plants, have been used for centuries by traditional healers, shamans, and herbalists.

Anthropologists and ethnobotanists who study traditional healing increasingly are finding that healers in nonindustrial societies have a significant body of knowledge regarding the medicinal properties of plants, and that this knowledge could be valuable to Western medical practices (Schultes and Raffauf 1990; Goleman 1991). For example, in Central America, *Dioscorea* tubers, inedible relatives of yams, are harvested from the forest and used by traditional healers to treat urinary tract ailments; *Dioscorea* is known to contain steroids and has been used in the past by major pharmaceutical companies to

make an early form of birth control pills and other Western medicines.

This knowledge has led some conservationists to transform their estimates of the rain forests' *potential* economic value to an evaluation of the *actual* monetary worth of the forest based upon the local uses of different plant species (Balick and Mendelsohn 1992). By taking an inventory of the plant species in specific plots of forest in Belize, determining the quantities of raw plant matter that can be used by traditional healers, and calculating the prices that healers pay for these raw materials from local collectors, ethnobotanists discovered that a 1-ha forest plot could bring a farmer greater income than an agricultural field of the same size, even after labor costs are worked into the equation (Balick and Mendelsohn 1992). Though the value of a plot varies according to the specific plants available in a given location, even the least lucrative plot produced significantly more than the estimated income from agriculture. These calculations are encouraging to conservationists because they provide concrete support for the potential economic value of sustainable harvests of rain forest species. To a conservationist, the obvious importance of this type of land use is that, unlike agriculture or cattle ranching, collecting plants does relatively little damage to the tropical forest, yet provides the landowner with a reasonable income. However, it remains to be seen how many people can actually make a living collecting medicinal plants; it may be that this is a specialized activity that will only support a small number of individuals. Also, increased harvesting may depress the price to the point where collecting the plants does not provide enough money to the collectors. For these reasons and others, the optimistic economic calculations of this study have been sharply questioned (Tremaine 1993).

In purely economic terms, the value of the tropical forest takes on global significance

when traditional medicine is taken into consideration. For example, a significant portion of health care at the study site in Belize is provided by traditional practitioners (Balick and Mendelsohn 1992). Billions of people in developing countries use plants for primary health care, although the traditional healers are slowly disappearing (Goleman 1991). Many rural people simply cannot afford Western-style doctors and medicines; for them, traditional medicines are the only reasonable option. Many others actually prefer traditional medicines to commercially available products, even when the latter are inexpensive. Conservation of the rain forests may therefore represent a means of preventing a severe medical crisis in tropical countries—a crisis that would certainly affect the economies of the industrialized nations, since the affected nations would probably need increased medical and humanitarian aid to cope with the problem.

Yet the information provided by the studies is incomplete. For instance, researchers have not yet learned how much harvesting a plot of forest can withstand before extraction of plant resources harms either the individual plant species or the forest as a whole (Balick and Mendelsohn 1992). In long-settled parts of the world such as India, rain forests have been gradually overharvested of their medicinal plants. This question might become vital in the future

Ethnobotanists work with local people to collect medicinal plants and gather information on their use. Here a native of Suriname is interviewed. (Courtesy of Mark Plotkin and Conservation International.)

should particular plants demonstrate their value in medical or commercial laboratories; if, for instance, a plant in the Belizean rain forest were found to cure or prevent AIDS, the international demand for that plant could skyrocket beyond the sustainable level of harvesting. In such a case, the plants might have to be grown in cultivation or the active ingredient synthesized. At the moment, the pharmaceutical plants of the rain forest supply mainly local health needs, but conservation measures need to be firmly in place before the hypothetical discovery is made in the laboratory. Otherwise, the impact upon the forest and the rural people who depend upon it could be catastrophic.

large lakes produce vast quantities of seaweed, shellfish and fish that are harvested for human use. In an innovative study, Prescott-Allen and Prescott-Allen (1986) evaluated the productive value of all wild species to the United States economy. They concluded that 4.5% of the U.S. GDP depends in some way on wild species, for an amount averaging about $87 billion per year. The percentage would be far higher for developing countries that have less industry and a higher percentage of the population living in rural areas.

Summary

1. Standard economics has tended to ignore the costs of environmental damage and the depletion of natural resources. The new field of environmental economics is developing methods for valuing biological diversity and in the process is providing arguments for its protection. While some conservation biologists would argue that biological diversity is priceless and should not be assigned economic values, economic justification for biological diversity will play an increasingly important role in debates on the use of natural resources.

2. Many countries that show annual increases in their Gross Domestic Product may actually have stagnant or even declining economies when depletion of natural resources and damage to the environment are included in the calculations. Large development projects are increasingly being analyzed by environmental impact assessments and cost–benefit analyses before being approved.

3. A number of methods have been developed to assign economic value to biological diversity. In one method, values are divided between direct values, which are assigned to products harvested by people, and indirect values, which are assigned to benefits provided by biological diversity that do not involve harvesting or destroying the resource.

4. Direct values can be further divided into consumptive use value and productive use value. Consumptive use value is assigned to products that are consumed locally, such as firewood, local medicines, and building materials. These goods can be valued by determining how much money people would have to pay for them if they were unavailable in the wild. If overexploitation makes these wild products unavailable, then the living standard of people that depend on them will decline. Productive use value can be assigned to products harvested in the wild and sold in markets, such as commercial timber, fish and shellfish, and wild meat. Species collected in the wild have great productive use value in their ability to provide new founder stock for domestic species and for the genetic improvement of agricultural crops. Wild species have also been a major source of new medicines.

Suggested Readings

Bishop, R. C. 1987. Economic values defined. *In* D. J. Decker and G. R. Goff (eds.), *Valuing Wildlife: Economic and Social Perspectives*, pp. 24–33. Westview Press, Boulder, CO. A clear system for valuing wildlife.

Costanza, R. (ed.). 1991. *Environmental Economics: The Science and Management of Sustainability*. Columbia University Press, New York. Leading economists grapple with the need to include environmental factors in their analyses.

Daly, H. E. and J. B. Cobb, Jr. 1989. *For the Common Good: Redirecting the Economy toward Community, the Environment, and a Sustainable Future*. Beacon Press, Boston. New methods for evaluating economic wealth are demonstrated.

Gore, A. 1992. *Earth in the Balance: Ecology and the Human Spirit*. Houghton Mifflin, Boston. A popular account of the need to establish a better balance between development and the environment, written by the U.S. vice president.

McNeely, J. A. 1988. *Economics and Biological Diversity: Developing and Using Economic Incentives to Conserve Biological Resources*. IUCN, Gland, Switzerland. Case studies are used to demonstrate the economic justifications for preserving biodiversity.

Myers, N. 1983. *A Wealth of Wild Species*. Westview Press, Boulder, CO. The range of species currently and potentially used by people.

Oldfield, M. L. 1984. *The Value of Conserving Genetic Resources*. Sinauer Associates, Sunderland, MA. The arguments for protecting the genetic variation of domestic species.

Panayotou, T. and P. S. Ashton. 1992. *Not by Timber Alone: Economics and Ecology for Sustaining Tropical Forests*. Island Press, Washington, D.C. Arguments for maintaining forests.

Peters, C. M., A. H. Gentry and R. Mendelsohn. 1989. Valuation of a tropical forest in Peruvian Amazonia. *Nature* 339: 655–656. This paper drew attention to the great economic potential of managed rain forests.

Prescott-Allen, C., and R. Prescott-Allen. 1986. *The First Resource: Wild Species in the North American Economy*. Yale University Press, New Haven, CT. Innovative examination of the economic importance of wild species to a modern economy.

Reid, Walter V., and Kenton R. Miller. 1989. *Keeping Options Alive: The Scientific Basis for Conserving Biodiversity*. World Resources Institute, Washington, D.C. An excellent summary of conservation biology with a strong emphasis on environmental economics.

Schumacher, E. F. 1973. *Small is Beautiful: Economics as if People Mattered*. Harper & Row, New York. An influential early attempt to develop alternative economic analyses.

CHAPTER **9**

Indirect Economic Values

Indirect values can be assigned to aspects of biological diversity, such as environmental processes and ecosystem services, that provide economic benefits to people without being harvested and destroyed during use. Because these benefits are not goods or services in the usual economic sense, they do not typically appear in the statistics of national economies, such as the Gross Domestic Product (GDP) (Repetto 1992). However, they may be crucial to the continued availability of the natural products on which the economies depend. For example, mountain forests prevent soil erosion and flooding, which could damage human settlements and farmlands in nearby lowland areas. Likewise, coastal estuaries are areas of rapid plant and algal growth that provide the starting point for food chains leading to commercial stocks of fish and shellfish. The protected waters of coastal estuaries serve as nurseries for the juvenile stages of many marine fish. In the post-World War II era these estuaries have been severely damaged by industrial pollution, dumping of untreated residential sewage, and coastal development. The U.S. National Marine Fisheries Service has estimated that damage to these estuaries has cost the United States over $200 million per year in lost productive value of commercial fish and shellfish and in lost consumptive value of fish caught for sport (McNeely et al. 1990).

In thinking about how we might value the indirect use values of ecosystems, consider this summary of the consequences of deforestation:

> We must find replacements for wood products, build erosion control works, enlarge reservoirs, upgrade air pollution control technology, install flood control works, improve water purification plants, increase air conditioning, and provide new recreational facilities. These substitutes represent an enormous tax burden, a drain on the world's supply of natural resources, and an increased stress on the natural system that remains. (F. H. Bormann 1976)

Nonconsumptive Use Value

Biological communities provide a great variety of environmental services that are not consumed through use. This **nonconsumptive use value** is sometimes relatively easy to calculate, as in the case of the value of wild insects that pollinate crop plants. About 100 species of crop plants in the United States require insect pollination of their flowers (USDA 1977). The value of these pollinators could be assigned either by calculating how much the crop increases in value through the actions of the wild insects, or by how much the farmer would have to pay to hire honeybee hives from a commercial beekeeper.

Determining the value of other ecosystem services may be very difficult, particularly at the global level. Economists are just beginning to think about the value of plant communities in absorbing carbon dioxide, thereby minimizing the expected trend toward global warming. A major attempt is being made to determine the value of these resources (Oldfield 1984; Peterson and Randall 1984; Sinden and Worrell 1979; Green and Tunstall 1991; Costanza 1991). In addition, biologists are trying to understand how many species can be lost from a community before the community structure is altered and the ecosystem functions are impaired. The following is a partial listing of the general benefits of conserving biological diversity that typically do not appear on the balance sheets of environmental impact assessments or in GDPs.

Ecosystem Productivity

The photosynthetic capacity of plants and algae allows the energy of the sun to be captured in living tissue. The energy stored in plants is sometimes harvested by humans directly as fuelwood, fodder, and wild foods. This plant material is also the starting point for innumerable food chains leading to all of the animal products that are harvested by people. The destruction of the vegetation in an area through overgrazing by domestic animals, overharvesting of timber, or frequent fires will destroy the system's ability to make use of solar energy, eventually leading to a loss of production of plant biomass and the deterioration of the animal community (including humans) that lives at that site (Likens et al. 1977).

Protecting Water Resources

Biological communities are of vital importance in protecting watersheds, buffering ecosystems against extremes of flood and drought, and maintaining water quality (Ehrlich and Mooney 1983; Likens 1991). Plant foliage and dead leaves intercept the rain and reduce its impact on the soil, and plant roots and soil organisms open up and aerate the soil, increasing its capacity to absorb water. This increased water-holding capacity reduces the flooding that occurs after heavy rains and allows a slow release of water for days and weeks after the rains have ceased.

These ecosystem functions are not completely maintained when the land is converted to farmland or tree plantations. For example, forested areas in peninsular Malaysia have only half as much runoff of water per hectare following heavy rains as do nearby rubber tree and oil palm plantations, and the forests release stored water during the dry periods at twice the rate of the plantations (Daniel and Kulasingham 1974). Unprecedented catastrophic floods in Bangladesh, India, the Philippines, and Thailand have been associated with recent extensive logging in watershed areas, and have led to calls by local people for bans on logging. Flood damage to India's agricultural areas has led to massive government and private tree planting programs in the Himalayas. In the industrial nations of the world wetlands protection has become a priority to prevent flooding of developed areas. In the region surrounding Boston, Massachusetts, the value of marshland has been estimated at $72,000 per hectare per year solely on the basis of its role in reducing flood damage (Hair 1988).

There is an increasing recognition of the fact that when dams, reservoirs and new croplands are developed, natural communities on the highlands above these projects must be protected to ensure a steady supply of high-quality water. For example, in Sulawesi, the Indonesian government borrowed $1.2 million from the World Bank to establish the Dumoga-Bone National Park to protect the watershed above a major agricultural project in the adjacent lowlands (Hufschmidt and Srivardhana 1986; McNeely 1987). This combining of development projects with watershed protection can be a major factor in the conservation of biological communities, and acknowledges the nonconsumptive use value of uncultivated land.

In many rural areas of the developing world, people settle near natural water sources to obtain water for drinking, washing, and irrigation. As hydrological cycles become disrupted by deforestation, soil erosion, and dam projects, and as water quality deteriorates due to pollution, people are increasingly unable to obtain their water needs from natural systems. The cost of boiling water, buying bottled water, or building new wells, rain catchment systems, pipes, and water pumps gives some measure of the consumptive value of water from surface sources. The increases in waterborne disease and intestinal ailments, and subsequent lost days of work, that occur as water quality declines add to estimates of the economic value of water.

Protecting Soils

Biological communities protect and maintain the soil, which in turn is vital to the productivity of the community. Soil of the appropriate structure and composition has enormous value to agriculture, ranching, and forestry. Soil can take hundreds or possibly thousands of years to form from the parent rock material. Plant roots and fungal hyphae bind soil particles together so that the soil is not readily washed away. When the vegetation is disturbed by logging, farming, and other human activities, the rates of soil erosion and even landslides increase rapidly, decreasing the value of the land for human activities (Likens et al. 1977; Ehrlich and Mooney 1983). Damage to the soil limits the ability of the plant life to recover following disturbance and can render the land useless for agriculture. In addition, the soil particles suspended in water from runoff can kill freshwater animals, coral reef organisms, and the marine life in coastal estuaries. The silt also makes the water undrinkable for human communities along the rivers, leading to a decline in human health. Increased soil erosion can lead to premature filling of the reservoirs behind dams, leading to a loss of electrical output, and may create sandbars and islands, reducing the navigability of rivers and ports.

Regulation of Climate

Plant communities are important in moderating local, regional, and probably even global climate conditions. At the local level, trees provide shade and transpire water, which reduces the local temperature in hot weather. This cooling effect reduces the need for fans and air conditioners, and increases the comfort and work efficiency of people. Trees are also locally important as windbreaks and in reducing heat loss from buildings in cold weather.

At the regional level, transpiration from plants recycles rainwater back into the atmosphere so that it can return again as rain. Loss of vegetation from regions of the world such as the Amazon Basin and West Africa may result in a reduction of average annual rainfall (Fearnside 1990). At the global level, plant growth is tied into carbon cycles; a loss in vegetation cover results in reduced uptake of carbon dioxide by plants, contributing to the rising carbon dioxide levels that are leading to global warming. Plants are also the source of oxygen on which all animals, including people, depend for respiration.

Waste Disposal

Biological communities are capable of breaking down and immobilizing pollutants such as heavy metals, pesticides, and sewage that have been released into the environment by human activities (Odum 1989; Greeson et al. 1979). Fungi and bacteria are particularly important in this role. When such ecosystems are damaged and degraded, expensive pollution controls have to be installed and operated to take over these functions.

An excellent example of this ecosystem function is provided by the New York Bight, a 2000-square-mile (5200 km²) embayment at the mouth of the Hudson River. The New York Bight acts as a sewage disposal system into which is dumped the waste produced by 20 million people in the New York metropolitan area (Young et al. 1985). The New York Bight provides these sewage treatment services for free. Until recently, the Bight was able to break down and absorb this onslaught of sewage because of the high degree of bacterial activity and water mixing in the area. However, the Bight is now showing signs of stress, such as fish die-offs and beach contamination, that suggest that the system is overloaded. The system is being further strained by the progressive filling in and development of coastal estuaries and marshes, which assist in the breakdown and assimilation process. If the New York Bight becomes overwhelmed and damaged by this combination of sewage overload and development, an alternative waste disposal system will have to be found, involving massive waste treatment facilities and giant landfills and costing tens of billions of dollars.

Species Relationships

Many of the species harvested by people for their productive use value depend on other wild species for their continued existence. Thus, a decline in a wild species of little immediate value to humans may result in a corresponding decline in a harvested species that is economically important. For example, the wild game and fish harvested by people are dependent on wild insects and plants for their food. A decline in insect and plant populations will result in a decline in animal harvests. Similarly, plants often require wild insect species to act as pollinators. Crop plants also benefit from birds and predatory insects, such as praying mantises, which feed on pest insect species that attack the crops. Many useful wild plant species depend on fruit-eating animals, such as bats and birds, to act as seed dispersers (Fujita and Tuttle 1991).

One of the most economically significant relationships in biological communities is that between many forest trees and crop plants and the soil organisms that provide them with essential nutrients. Fungi and bacteria break down dead plant and animal matter, which they use as their energy source; in the process, they release mineral nutrients such as nitrogen into the soil where they can be used by plants for further growth. Mycorrhizal fungi greatly increase the ability of plant roots to absorb water and minerals, and certain mutualistic bacteria convert nitrogen into a form that can be taken up by plants. In return, the plants provide the mutualists with photosynthetic products that help them grow. The poor growth and massive dieback of many trees throughout North America and Europe may be attributable in part to the deleterious effects of acid rain and air pollution on mycorrhizal fungi (Box 15).

BOX 15 THE DECLINE OF FUNGI IN FORESTS: A PREMONITION OF DISASTER

With the exception of gourmet mushrooms, fungi are generally not highly regarded by humans. Fungi evoke images of rotting wood, spoiled food, mildewed rugs, and other damp, smelly, unpleasant things. Yet fungi may be vital to the continued existence of both temperate and tropical forests. Many fungi exist in symbiotic relationships with forest trees that are essential to the trees' health; thus, the decline of many species of fungus in recent decades is cause for great concern among biologists.

The best body of evidence for a decline among fungus species comes from Europe, where records of mushroom collecting date back to 1912. Scientists combining these rec-

The presence of mycorrhizal fungi greatly enhances the growth and health of most tree species. The spindly young pine seedlings on the left have no mycorrhizae; those on the right have large developments of fungi on their roots.

ords with long-term observations of different mushroom species have noticed a distinct decline in both the overall number of species and the population sizes of each species. One 20-year study in the Netherlands, for example, observed a significant drop in the average number of mushroom species found in 1000-m^2 plots; roughly 65% of the species originally counted in the plots vanished during the study period (Cherfas 1991). In addition, the population size of many species has declined, as has the size of individual mushrooms. Records for annual crops of edible chanterelle and bolete mushrooms in Germany demonstrate that both the average size of the crop and the average size of the individual mushrooms have declined since 1950. The decline in mushrooms is not limited to edible mushrooms, however, ruling out overexploitation as a prime factor in the reduction of the population. In addition, similar declines among North American species are suspected, but since no data comparable to the European records exist, any estimates of declines and extinctions of North American fungi are purely guesswork.

Fungus extinctions have cataclysmic implications for forests worldwide. Many soil fungi live in a close symbiotic relationship with the roots of tree species; experiments on tropical tree species have shown that some species cannot grow at all in the absence of fungi, and many others grow faster and are generally healthier with fungi than without (Janos 1980; Medina and Huber 1992). Soil fungi improve the plant's resistance to herbivory and extremes of temperature, and apparently increase the plant's ability to take up nutrients from the soil, which is vitally important in the nutrient-poor soils of the tropics. Mass extinctions of fungi thus could conceivably precede large-scale losses of trees in both tropical and temperate forests.

These data present several questions for conservation biologists: first, how widespread is this problem? So far, the only long-term and relatively complete data set available is for European temperate forest fungi. Almost nothing is known of the condition of fungus species in the Americas or elsewhere. Second, what is causing the decline in fungus species? Initial investigations suggest air pollution, particularly the presence of excess nitrogen and sulfur in rain or soil (Cherfas 1991). The exact problem is unclear; biologists are not certain whether pollution damages the fungi directly, or whether pollution-related damage to the trees does indirect damage to the fungus by inhibiting the symbiosis between the two organisms. In either case, more specific information is needed before a third question may be addressed: how can the trend be reversed?

Recreation and Ecotourism

People use natural environments for recreation in a variety of ways (Figure 9.1). A major focus of recreational activity is on the nonconsumptive enjoyment of nature through activities such as hiking and bird-watching (Rolston 1989a,b; Duffus and Dearden 1990). The monetary value of these activities, sometimes called the **amenity value**, can be considerable. For example, 84% of Canadians participate in nature-related recreational activities that have an estimated value of $800 million per year (Fillon et al. 1985). In the United States, almost 100 million adults and comparable numbers of children are involved each year in some form of nondestructive nature

9.1 Wildlife is used in a variety of consumptive and nonconsumptive ways by both traditional and modern societies. The range of this use and the value of wildlife to people is increasing all the time. (After Duffus and Dearden 1990.)

Consumptive Uses

Commercial hunting • Sport hunting • Subsistence hunting • Commercial fishing • Sport fishing • Subsistence fishing • Fur trapping • Hunting for animal parts and pet trade • Indirect kills through other activities (pollution by-catch, road kills) • Eradication programs for animals posing real or perceived threats

Low-Consumptive Uses

Zoos and animal parks • Aquariums • Scientific research

Nonconsumptive Uses

Birdwatching • Whale-watching • Photography trips • Nature walks • Commercial photography and cinematography • Wildlife viewing in parks, reserves and recreational areas

---

recreation, spending $4 billion in the process on fees, travel, lodging, food, and equipment (Shaw and Mangun 1984a,b). The value of these recreational activities may be even greater than these numbers suggest, since many visitors to parks indicate they would be willing to pay even higher admission fees (Hvengaard et al. 1989).

In places of national and international significance for conservation or exceptional scenic beauty, the nonconsumptive recreational value often dwarfs that of other local industries. The value of biological diversity looms particularly large when the money spent off-site by visitors on food, lodging, equipment, and other goods and services purchased in the local area is included. As an example, mineral and timber harvesting and agriculture have been traditionally regarded as the key to the economic vitality in the area of Yellowstone National Park. (Power 1991). However, a careful analysis of the local economy reveals that the recreational activities and scenic beauty that attract new business, temporary visitors, and retirees, now account for the majority of income and jobs in the area (Box 16).

Ecotourism is a dramatically growing industry in many developing countries, earning approximately $12 billion per year worldwide (Box 17). Ecotourists visit a country and spend money wholly or in part to experience its biological diversity and to view particular flagship species (Lindberg 1991; Ceballos-Lascurain 1993). By charging high visitor fees, Rwanda has developed a gorilla-tourism industry that was the country's third largest foreign currency earner until recent civil disturbances (Vedder 1989). Ecotourism has traditionally been a key industry in East African countries such as Kenya and Tanzania (Western and Henry 1979) and is increasingly part of the tourist picture in many American and Asian countries. In the early 1970s, an

BOX 16 INDUSTRY, ECOLOGY, AND ECOTOURISM IN YELLOWSTONE PARK

Yellowstone National Park is the oldest and most famous of the protected areas of the U.S. National Park System. Such is its place in the American consciousness that federal policies affecting the natural landscape of the park often attract intense public scrutiny; a storm of public criticism and counterreaction followed the dramatic forest fires of 1988, for example. Nevertheless, public policies that support the extraction of timber, oil, and other natural resources within the park and in nearby parts of Wyoming, Montana, and Idaho generally go unchallenged; the industries that benefit from these policies argue that such activities are necessary for the economic health of the local communities surrounding the park. Recent studies of the regional economy of the Yellowstone area indicate that this argument is increasingly less valid; the economic health of the communities surrounding Yellowstone has come to rely primarily on the tourism industry and on the new businesses and retired people that move to the area because of its perceived higher quality of life. (Power 1991). Though extractive industry was a significant force in the regional economy two decades ago, it may now be detrimental to the economic well-being of local residents because it harms what

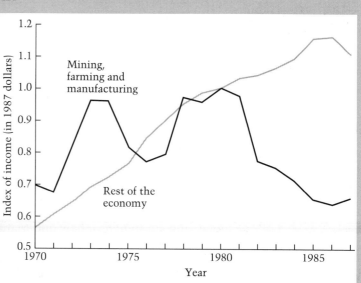

In the Greater Yellowstone region, income from the extractive industries has fluctuated over the last 20 years, while income from the rest of the economy—including recreation industries, tourism, service industries, and retirement communities—has grown steadily. The region's economy has become increasingly independent of the extractive industries. The two income lines are standardized for a value of one in 1980, when extractive industries provided about 20% of the region's total income. (From Power 1991.)

has become the major resource for these communities: the wildlife and natural landscape of Yellowstone National Park.

One of the industries in Yellowstone that presently provides the largest boost to the economy of the region—ecotourism—is also the one that does the least damage to the ecosystem of the park. Ecotourism does have its drawbacks: the noise and pollution brought by the passage of millions of tourists an-

nually, the disruption and alteration of animal behavior when the animals are constantly exposed to human presence, and the threat of human-caused soil erosion and fire. The latter is perhaps the most visible and fearsome form of disturbance related to ecotourism; nevertheless, even the damage caused by anthropogenic fires pales in comparison with the damage done by logging and mining activities. The reason that extractive industry is so much more damaging than ecotourism is simple: ecotourism, while it can pollute and alter habitats, does not actively destroy them.

Even the wildfires that seem to cause monumental destruction in the park are in fact part of the natural regime of the Yellowstone ecosystem (Jeffery 1989). In 1988, fires raged out of control for nearly two months, despite the best efforts of 10,000 firefighters that eventually included the U.S. Army. In the aftermath of the fires, ecologists pointed out that such fires provide necessary services to the Yellowstone ecosystem. By opening up old-growth forests and removing deadwood, fires create areas that can be colonized by sun-loving grasses and wildflowers. These plant species attract a variety of insects and mammals that are not found in mature forests. Fires also increase the rate of decomposition of dead trees and plants, providing nutrient-rich ash that encourages rapid recolonization of the burned areas. Fires simultaneously increase the biological diversity of the park and augment the available food supply for the herds of deer, bison, and elk. Moreover, tourism increased in the months following the 1988 fires (Jeffery 1989), and the Park Service videotape of the fires is a popular item in Yellowstone's souvenir shops.

In contrast, logging and mining do not encourage regeneration of the natural environment, and have many detrimental effects. Clear-cutting, a common logging practice in which forested slopes are simply cleared of trees, can induce massive sheet erosion, particularly if steps are not immediately taken to replace the vegetation removed during logging. Siltation of streams from this erosion can kill fish and other aquatic species, and the loss of soil nutrients retards regrowth of vegetation. Mining practices often introduce harmful chemical by-products, including cyanide, into the environment. These practices are ultimately not cost-effective for several reasons: first, they lower the potential for future extraction by damaging the soil and water resources needed to regenerate timber; second, they lower the potential of the region for tourism, retirement communities, and new businesses by damaging the natural beauty of the area; and third, they create hidden costs by lowering water quality for residents of the area, who must then pay more to have clean drinking water.

BOX 17 CONSERVATION AND ECOTOURISM ON THE BELIZEAN REEF

In May of 1991, the First World Congress on Tourism and the Environment was held in Belize City, Belize. The decision to act as host to the conference reflected the growing determination of the Belizean government to safeguard the country's natural resources. Though Belize is one of the smallest and least populous of the Central American nations, it has an advantage not shared by many of its neighbors: most of the land in Belize has been only minimally damaged by human activities. Belize is therefore in the unique position of having the opportunity to plan its economic development with an eye to preserving a large segment of its natural resources. Ecotourism has been lauded as one means of encouraging economic growth without destroying irreplaceable ecosystems; however, as Belize and other Central American nations have discovered recently, even this seemingly benign industry has drawbacks for conservation.

Like its northern neighbor, Mexico, Belize has a number of coastal islands in the Caribbean that are perfect spots for tourists seeking sea and sun. While Cancun and Cozumel are well-known tourist haunts, the 250-kilometer-long barrier reef and associated cays that fringe the Belizean coast have been a well-kept secret among diving enthusiasts until fairly recently. Within the last decade, however, a brisk tourist trade has become established on several of the coastal islands. One of the larger islands on the northern coast, Ambergris Cay, boasts an airport, numerous resort hotels, gift shops, restaurants, an equestrian club, and many diving and snorkeling tours of the outer reef and islands. The increased influx of visitors has benefited Belize economically, but there are signs that tourism is starting to have negative effects on the reef ecosystem (Valdez 1992). For example, a popular spot for diving and snorkeling is the Hol Chan Marine Reserve, located off the southern tip of Ambergris Cay. The 45-acre reserve has been successful in sustaining populations of reef fauna in its five years of operation; local fishermen and diving instructors unanimously agree that the reserve has more fish than anywhere else in the vicinity, and that the fish are significantly larger than outside the reserve. However, the high volume of tourist traffic

in the area has the potential to damage that section of the reef. On any given morning, the reserve may host between 30 and 50 scuba divers and snorkelers. Despite repeated warnings from guides, visitors frequently touch or stand on delicate coral, which can result in irreparable damage to these organisms. Fish in the reserves are virtually tame, having learned that tourists often give handouts of bread crumbs and crackers. Shy creatures such as stingrays and octopuses get no respite even after dark; night diving is a popular activity at Hol Chan, and it is not unusual to see 10 or 20 divers' lights bobbing in the depths after sunset.

Hol Chan's popularity is partly due to its accessibility; while Belize has a number of other spectacular marine preserves, most of them are a significant distance away from the mainland, have few facilities, and often require a visitor to be an experienced scuba diver. One response to the pressure on Hol Chan is to disperse the impact of tourism by creating more reserves that are accessible to average tourists. Plans for marine reserves at Glover's Reef, South Water Cay, Sapodilla Cay, and other sites are presently in various stages of development (Phillips 1992; Valdez 1992). However, coastal ecosystems other than the reef have also suffered indirectly from the increase in tourism. The popularity of Hol

Chan and other locales near Ambergris has stimulated development of nearby islands; deforestation and erosion related to construction of tourist facilities has damaged or destroyed nearby mangroves, which are vital natural hatcheries for many species of fish. Future development plans will presumably take this problem into account; the South Water Cay site, currently being surveyed by Coral Cay Conservation, a British-based nonprofit conservation group, will have up to 300 kilometers of habitat set aside in the central portion of the barrier reef. Included in the plan are mangrove and sand cays, in addition to the marine and reef habitats that are the primary tourist attraction.

Nevertheless, the government of Belize is struggling to keep control of the threats posed by too-rapid development of the reef ecosystem. An incident reported on Belize's national radio service in August 1992 highlighted the government's problem: an ambitious foreign property owner, anxious to stake a claim in the expanding tourist market, used dynamite to blast several holes in the reef to create a boat passage, with the goal of increasing his guests' access to the outer islands. While this type of wanton and illegal destruction represents an extreme example, it serves as a warning that even a seemingly innocuous industry like ecotourism can have unintended, catastrophic consequences to the environment if not carefully monitored. There is little doubt that ecotourism is a profitable venture that can provide developing nations like Belize with income without severely damaging the environment; however, like any other industry, proper safeguards are necessary if the industry is to be environmentally sustainable.

9.2 Facilities for ecotourists sometimes create a tropical fantasy that disguises and ignores the realities of Third World problems. (Cartoon from E.G. Magazin, Germany.)

estimate was made that each lion at Amboseli Park in Kenya could be valued at $27,000 per year in tourist revenue, while the elephant herd was worth $610,000 per year (Western and Henry 1979); those values are certainly much higher today. Ecotourism is estimated to bring in ten times more revenue than either hunting or plantation agriculture in some African countries. Ecotourism has the potential to provide one of the most immediate justifications for protecting biological diversity. However, the danger of this industry is that tourist facilities will provide a sanitized fantasy experience, rather than allowing visitors to be aware of or even see the serious social and environmental problems that endanger biological diversity (Figure 9.2).

Sport hunting and fishing are also major recreational activities generating hundreds of millions of dollars, particularly in rural economies. In Britain, the National Federation of Anglers had 446,000 members in 1977. The value of these sport activities can be calculated on the basis of how much money is actually spent during hunting and fishing trips. However, their value to sportsmen and women is often far greater since these people would often be willing to spend even more money if necessary to continue their activities (Hammack and Brown 1974). These activities could possibly be listed in the section on consumptive use value, but the food value of the animals caught by sport fishermen and hunters is probably so tiny in comparison with the time and money spent that these activities are arguably better categorized as nonconsumptive recreation. The increasingly common practice of sport fishermen releasing fish rather than keeping them emphasizes the recreational aspect of the activity.

Educational and Scientific Value

Many books, television programs, and movies produced for educational and entertainment purposes are based on nature themes. Increasingly, natural history materials are being incorporated into school curricula (Hair and Pomerantz 1987). Again, the nonconsumptive value of these educational programs is enormous. A considerable number of professional scientists as well as highly motivated amateurs are engaged in ecological observations that have nonconsumptive value (Figure 9.3). While these scientific activities provide economic benefits to the areas surrounding field stations, their real value lies in their ability to increase human knowledge, enhance education, provide better knowledge about the environment, and enrich the human experience.

Environmental Monitoring

Species that are particularly sensitive to chemical toxins can serve as an "early warning system" for monitoring the health of the envi-

9.3   Most people find interacting with other species to be an educational and uplifting experience. Here people greet a minke whale that is being rescued after it became entangled in a trawler's gill net; the float behind the whale was attached to the net to keep the whale at the surface so it could breathe. Later, rescuers were able to release the whale from the netting. Such meetings—usually taking place at greater distances, as in a more traditional "whale watch" setting, or on "photo safaris" in Africa—can enrich human lives. (Photograph by Scott Kraus, New England Aquarium.)

ronment (Hellawell 1986). Some species can even serve as substitutes for expensive detection equipment. Among the best-known indicator species are lichens, which grow on rocks and strongly absorb chemicals in rainwater and airborne pollution (Hawksworth 1990). Lichens will even absorb toxic materials to the point of dying off, so that the distribution and abundance of lichens can identify areas of contamination around sources of air pollution, such as smelters. Aquatic filter feeders, such as mollusks, have also proved to be effective monitoring species, since they process large volumes of water and concentrate toxic chemicals, such as poisonous metals, PCBs, and pesticides, in their tissues. The California Mussel Watch Program, started in 1977, has expanded to include 135 coastal and freshwater sites at which mussel (*Mytilus* spp.) and clam (*Corbicula fluminea*) tissues are sampled and analyzed for toxic compounds, highlighting areas of serious water pollution (Stevens 1988).

Option Value

The **option value** of a species is its potential to provide an economic benefit to human society at some point in the future. As the needs of society change, so must the methods of satisfying those needs (Myers 1984). Often, the solution lies in previously untapped animal or plant species. Health agencies and pharmaceutical companies are making a major effort to collect and screen species for compounds that have the ability to fight cancer, AIDS, and other human diseases (Farnsworth 1988; Plotkin 1988). The recent discovery of a potent anticancer chemical in the Pacific yew is only the most recent discovery in this search (Box 18). Another example is the ginkgo tree, *Ginkgo biloba*, a species that occurs in the wild only in a few isolated localities in China (Figure 9.4A). During the last 20 years a $500 million a year industry has developed around the cultivation of the ginkgo tree (Figure 9.4B) and the manufacture of medicines made from its leaves, widely used in Europe and Asia to treat problems of blood circulation and stroke (Del Tredici 1991; Del Tredici et al. 1992). The extent of this search for valuable natural products is wide-ranging. Entomologists search for beneficial insects that can be used as biological control agents, microbiologists search for bacteria that can assist in biochemical manufacturing processes, and wildlife biologists are identifying wild species that can potentially produce animal protein more efficiently and with less environmental damage than existing domestic species.

The growing biotechnology industry is finding new uses of species to reduce pollution, to develop new industrial processes, and to fight against new diseases and drug-resistant forms of old diseases that are

BOX 18 THE PACIFIC YEW: FROM WORTHLESS TO INVALUABLE

The case of the Pacific yew, *Taxus brevifolia*, is a remarkable illustration of how unsustainable use of an ecosystem can backfire by threatening a species with great potential value to humans. Though this tree is presently common in the Pacific Northwest, the conifer forests in which it grows are threatened by extensive logging. These old-growth forests are of very high value to timber companies (Simberloff 1987). Until recently, most loggers considered the Pacific yew to be worthless in comparison with many of the other species in the forests, such as the Douglas fir. A common practice of timber companies was to cut and discard any Pacific yew trees encountered as they extracted the more valuable species.

The discovery that a highly complex molecule called taxol found in the needles and bark of the Pacific yew can effectively treat many forms of cancer (Daly 1992; Joyce 1993) abruptly changed the attitudes of many people toward the species. Cancer is one of the leading causes of death throughout the world, afflicting millions of people each year; the prospect of a cure derived from the Pacific yew is therefore of concern to millions of doctors and patients. In particular, taxol has proved to be far more effective than standard drugs in treating advanced cases of breast and ovarian cancer, two leading causes of mortality in women. Because the tree is not endangered, it is not protected by the Endangered Species Act. Nevertheless, its medical value is so great that the U.S. Forest Service has developed regulations prohibiting the "poaching" of yew bark on federal lands.

Yet the attempts of the federal government to limit use of the tree may not actually help to conserve the species. Ironically, the importance of the tree for cancer treatment may accelerate the decline of the species. As demand for the compound taxol increases, harvesting of yew bark, both legal and illegal, has skyrocketed. Removal of bark generally kills the tree if a large amount is taken; if harvest of yew bark is to be sustainable, each tree can only contribute a very small amount before being left to heal. However, the extraction of taxol from yew bark is not a highly productive process. Each tree contains only a small amount of the compound, so that the bark of 3 to 12 trees is needed to treat a single patient (Daly 1992; Joyce 1993). Despite federal limitations, the demand for taxol may soon lead to overharvesting of the Pacific yew that outstrips the ability of the population to recover. In 1992, hundreds of thousands of trees were stripped of their bark to meet the insatiable demand for taxol. Anticipating an even larger demand, scientists are actively trying to synthesize the molecule and to locate alternate sources in other yew species, and forestry companies are establishing yew plantations. When this occurs, there will be less pressure to cut wild Pacific yews.

The story of the Pacific yew demonstrates what many conservation biologists have argued all along: that conservation of species is important for the future benefits they could bring to mankind. The millions of Pacific yew trees that were wasted in the past would have been available to researchers, doctors, and thousands of cancer patients had sustainable practices been in place before the old-growth forests of the Pacific Northwest were stripped. While we cannot change the past, we can learn from it: the few remaining old-growth forests may hold the key to curing other diseases, and should be kept as intact as possible. The present economic benefits of clear-cutting the old-growth forests are negligible when compared with the future cost of treating cancer, heart disease, AIDS, and other diseases that affect large numbers of people.

9.4 (A) Ginkgo trees are preserved in the wild in the Tian Mu Shan forest reserve in China; no other wild populations exist. This species is the basis of a pharmaceutical business worth hundreds of millions of dollars each year. (B) Because of the valuable medicines made from their leaves, ginkgo trees are now cultivated as a crop. Each year the woody stems sprout new shoots and branches, which are harvested. (Photographs by Peter Del Tredici, Arnold Arboretum.)

(A)

(B)

threatening human health. In some cases, newly discovered species or well-known species have been found to have exactly those properties needed to deal with a significant human problem. New techniques of molecular biology are allowing unique, valuable genes found in one species to be transferred to another species. If biological diversity is reduced in the future, the ability of scientists to locate and utilize a broad range of species for the benefit of people will be reduced.

Some of the most promising new species being investigated by industrial scientists are the bacteria that live in extreme environments, such as deep-sea thermal vents (see Box 4 in Chapter 3). Bacteria that thrive in unusual chemical and physical environments can often be adapted to special industrial applications of considerable economic value. One of the most important techniques of the multibillion dollar biotechnology industry, the polymerase chain reaction (PCR) for multiplying copies of DNA, depends on an enzyme that is stable at high temperatures; this enzyme was originally derived from a bacterium endemic to natural hot springs in Yellowstone National Park. An exciting development in the search for valuable bacteria species is the $43 million DEEPSTAR (Deep-Sea Environment Exploration Program: Suboceanic Terrane Animalcule Retrieval) project in Japan (Myers and Anderson 1992). The project includes plans to build vessels in which scientists can explore the oceans for new bacteria to a depth of 6500 meters, and laboratories in which the bacteria can be cultured at pressures of up to 1000 atmospheres. The leader of the project, Koki Horikoshi, has had a long career of searching for unusual bacteria. In his previous "Superbugs" project, Horikoshi investigated bacteria in high-pH (alkaline) environments and isolated an enzyme that could digest cellulose. It turned out that this enzyme is effective in removing dirt from cotton clothing, and it is now part of "Attack" laundry detergent, which is the leading Japanese brand. The race is on to discover other microbial systems of economic value.

The option value of species could be estimated by examining the effect on the world economy of wild species only recently utilized by humans. Consider a hypothetical example: if, during the last 20 years, newly discovered uses of 100 previously unused plant species accounted for $100 billion of new economic activity in terms of increased agriculture, new industry, and improved medicines, and there are 250,000 species of presently unused plant species, then a rough calculation might be that each unused plant species has the potential to provide an average of $400,000 worth of benefits to the world economy in the next 20 years. These types of calculations are at a very preliminary stage at the present time. The calculations also assume for convenience that the average value of a species can be determined. However, there may be just one species that has an enormous potential value because of its ability to cure disease, to be the foundation of a new industry, or to prevent the collapse of a major agricultural crop. If this species became extinct before it was discovered, it would be tremendous loss to the global economy, even if the majority of the world's species were preserved. This represents a powerful argument for the preservation of all species. As Aldo Leopold (1953) commented, "If the biota, in the course of aeons, has built something we like but do not understand, then who but a fool would discard seemingly useless parts? To keep every cog and wheel is the first precaution of intelligent tinkering." Stating this in another way, the diversity of the world's species can be compared to a manual on how to keep the Earth running effectively. The loss of a species is like tearing a page out of the manual. If we ever need the information from that page in the Earth manual to save ourselves and the Earth's other species, we will be out of luck.

Existence Value

Many people throughout the world care about wildlife and plants and are concerned with their protection (Walsh et al. 1984; Randall 1986). This concern may be associated with a desire to someday visit the habitat of a unique species and see it in the wild, or it may be a fairly abstract identification. Particular species, so-called "charismatic animals" such as pandas, lions, elephants, and many birds, elicit strong responses in people. People value these emotions in a direct way by joining and contributing millions of dollars each year to conservation organizations that protect species. Citizens also show their concerns by directing their governments to spend money on conservation programs and to purchase land for habitat protection. For example, the government of the United States has already spent more than $20 million to protect a single rare species, the California condor. The citizens of the state of Wisconsin have indicated in surveys that they would be willing to spend about $26 million to protect the bald eagle (Figure 9.5), a species that is endangered in the state and is a national symbol (Boyle and Bishop 1987).

Such **existence value** can also be attached to biological communities, such as tropical rain forests and coral reefs, and to areas of scenic beauty. People and organizations are increasingly contributing money to ensure the continuing existence of these habitats. In a survey relating to the spectacularly scenic Grand Canyon, which has been marred in recent years by sulphur dioxide air pollution from a nearby power plant, U.S. citizens indicated that they would be willing to pay $1.30 to $2.50 per household per year to have improved

9.5   The bald eagle, symbolic bird of the United States. Many people in the United States have indicated a willingness to pay to protect its continued existence. (Photograph by Jessie Cohen, National Zoological Park.)

pollution control equipment installed at the power plant, an amount well over $100 million per year (Nash 1991). Another survey taken over several years showed a steadily increasing number of Americans indicating their approval of the statement that "environmental improvements must be made, regardless of the cost" (Figure 9.6; Rucklehaus 1989). The money spent to protect biological diversity, particularly in the developed countries of the world, is on the order of hundreds of millions of dollars per year, and represents the existence value of species and communities—the amount that people are willing to pay to prevent species from going extinct.

Common Property Resources and Environmental Economics

Many natural resources, such as clean air, clean water, soil quality, rare species, and even scenic beauty are considered to be **common property resources** that are owned by society at large. These resources are often not assigned a monetary value. People, industries, and governments use and even damage these resources without paying more than a minimal cost, and sometimes pay nothing at all (Hardin 1968, 1985). For example, in many countries power plants

9.6 Increasingly, people throughout the world are unwilling to tolerate damage to the natural environment. In the United States, Times/CBS polls have been tracking public attitudes since 1981. People were asked to respond to the statement, "Protecting the environment is so important that requirements and standards cannot be too high, and continuing environmental improvements must be made regardless of cost." The final polls shown here were taken just after the tanker *Exxon Valdez* spilled 11 million gallons of oil onto Alaska's coastline. (From Ruckleshaus 1989.)

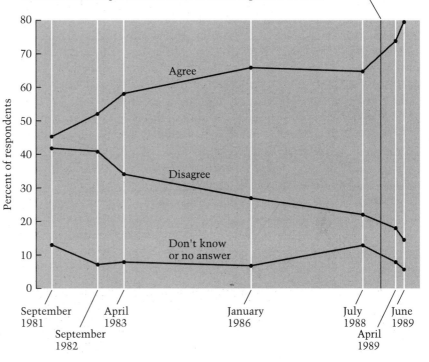

Assertion: "Protecting the environment is so important that requirements and standards cannot be too high, and continuing environmental improvements must be made regardless of cost."

and factories burn fossil fuels, giving off massive clouds of air pollutants. These air pollutants cause acid rain downwind, damage terrestrial and aquatic ecosystems, and may even drive species to extinction (Likens 1991; Odum 1993). Typically the industries do not pay for their use of the atmosphere—a common property resource—as a waste disposal system. As another example, municipalities and industries release sewage and wastes into water bodies, often without considering or paying for the impact of these pollutants on the environment. Other human users of these resources are not consulted, nor are they compensated when their ability to use the common resource is eliminated. In the more complete systems of accounting being developed by environmental economics, the use of such common property resources will be made part of the internal cost of doing business, instead of being regarded as external to the accounting process. When people and organizations have to pay for their actions, they will be likely either to stop damaging the environment or become much more careful (Pearce et al. 1989).

Concluding Remarks: The Economic Value of Biological Diversity

Chapters 8 and 9 have attempted to describe how environmental considerations are being incorporated into more inclusive economic models. While these models are a positive development, they can also be viewed as a willingness to accept the present world economic system as it is with only minor changes. Some environmental philosophers feel that the present system, which is characterized by pollution, environmental degradation, and extinctions at an unprecedented rate, needs more substantial changes. Perhaps the most damning aspect of the present system is the overconsumption of resources by a minority of the world's citizens and their consumer culture, while the majority of people worldwide live in poverty. Given a world economic system in which millions of children die each year from disease, malnutrition, crime, and war, and in which thousands of unique species go extinct each year due to habitat destruction, do we need to make minor adjustments or major structural changes? An alternative approach for protecting biological diversity and improving the human situation would be to dramatically lower the consumption of resources in the developed world, halt population growth, reduce the need to exploit natural resources, and greatly increase the value that is placed on the natural environment and biological diversity. Some suggestions for bringing this about include higher taxes on fossil fuels, penalties for inefficient energy use and pollution, and mandatory recycling programs. Third World debt could be reduced or eliminated, and investment redirected to activities that provide the most benefits to the greatest number of people in poverty. And finally, financial penalties for damaging biological diversity could be developed and made so severe that industries would be more careful of the natural world.

Summary

1. Indirect values can be assigned to aspects of biological diversity that provide economic benefits to people but are not harvested or damaged during use. Nonconsumptive use values of ecosystems include ecosystem productivity, important as the starting point for all food chains; protection of soil and water resources; the interactions of wild species with commercial crops; and regulation of climate.

2. Biological diversity is the foundation of a growing recreation and ecotourism industry. The numbers of people involved and the amount of money spent on such activities are surprisingly large. In many countries, particularly in the developing world, ecotourism represents one of the major sources of foreign income. Even in industrialized countries, the economy in areas around national parks is increasingly dominated by the recreation industry. Educational materials and the mass media draw heavily on themes of biological diversity, creating materials of considerable value.

3. Biological diversity also has an option value in terms of its potential to provide future benefits to human society, such as new medicines, biological control agents, and new crops. The biotechnology industry is developing innovative techniques to take advantage of new products and biochemical processes found in the living world.

4. People are often willing to pay money in the form of taxes and voluntary contributions to ensure the continued existence of unique species, biological communities, and landscapes; this amount represents the existence value of biological diversity.

5. The economic valuation of biological diversity provides one possible method for justifying increased protection of species and communities. However, assigning economic value assumes a willingness to accept the present economic system with minor changes. Given a world in which there is an unequal distribution of resources among people, and in which thousands of species go extinct each year, we may need to ask: Do we need minor adjustments to the present economic system, or major structural changes?

Suggested Readings

Duffus, D. A. and P. Dearden. 1990. Non-consumptive wildlife-oriented recreation: A conceptual framework. *Biological Conservation* 53: 213–231. Wildlife use is viewed within a very broad framework, with an emphasis on nonconsumptive value.

Ehrlich, P. R. and H. A. Mooney. 1983. Extinction, substitution and ecosystem services. *BioScience* 33: 248–254. Outstanding brief review of the value of ecosystem functions.

Green, C. H. and S. M. Tunstall. 1991. Is the economic evaluation of environmental resources possible? *Journal of Environmental Management* 33: 123–141. A critique of some of the methods currently used.

McNeely, J. A., Kenton R. Miller, Walter V. Reid et al. 1990. *Conserving the World's Biological Diversity.* IUCN, WRI, CI, WWF-US, World Bank, Gland, Switzerland. Outstanding summary of the value of biodiversity and strategies for preservation.

Myers, F. W. and A. Anderson. 1992. Microbes from 20,000 feet under the sea. *Science* 255: 28–29. Exciting account of the search for useful bacteria in the deep ocean.

Plotkin, Mark J. 1988. The outlook for new agricultural and industrial products from the tropics. *In* E. O. Wilson and F. M. Peter (eds.), *Biodiversity*, pp. 106–116. National Academy Press, Washington DC. Researchers are actively searching for valuable new species.

Power, T. M. 1991. Ecosystem preservation and the economy in the Greater Yellowstone area. *Conservation Biology* 5: 395–404. Ecosystem preservation and recreational benefits are surprisingly important.

Orians, G. H., G. M. Brown, W. E. Kunin, and J. E. Sweirzbinski (eds.). 1990. *The Preservation and Valuation of Biological Resources.* University of Washington Press, Seattle. A collection of papers by leading authorities.

Repetto, Robert. 1992. Accounting for environmental assets. *Scientific American,* June. A popular article describing how environmental degradation decreases national wealth.

Whelan, T. (ed.) 1991. *Nature Tourism: Managing for the Environment.* Island Press, Washington, D.C. Case histories of successful ecotourism, and overviews of important questions.

World Resources Institute. 1993. *Biodiversity Prospecting: Using Genetic Resources for Sustainable Development.* WRI, Washington, D.C. Designing successful programs for using biodiversity, including discussions of policy and legislation.

CHAPTER 10

The Ethical Value of Biological Diversity

While economic arguments are often advanced to justify the protection of biological diversity, there are also strong ethical arguments for doing so (IUCN 1980; Naess 1989; Rolston 1988a, 1989a,b). These arguments have foundations in the value systems of most religions, philosophies, and cultures, and thus can be readily understood by the general public. They appeal to a respect for life, a reverence for the living world, a sense of intrinsic value in nature, and a concept of divine creation. In the United States the rights of species are protected under the Endangered Species Act, which asserts "that Congress intended endangered species to be afforded the highest of priorities." The basis of this protection was the "esthetic, ecological, educational, historical, recreational and scientific value" of species. Economic value was not included in this official rationale.

Ethical arguments for preserving biological diversity appeal to the nobler instincts of people, and are based on widely held truths. People will accept these arguments on the basis of their belief systems (Hargrove 1986a, 1989). In contrast, arguments based on economic grounds are still being developed, and may eventually prove to be inadequate, highly inaccurate, or unconvincing (Warford 1987; Norton 1988). As Ehrenfeld (1988) has stated,

> It is certain that if we persist in this crusade to determine value where value ought to be evident, we will be left with nothing but our greed when the dust finally settles. I should make it clear that I am referring not just to the effort to put an actual price on biological diversity but also to the attempt to rephrase the price in terms of nebulous survival value. . . . . Economic criteria of value are shifting, fluid and utterly opportunistic in their practical application. This is the opposite of the value system needed to conserve biological diversity over the course of decades and centuries.

Economic arguments by themselves might provide a basis for valuing species, but they might also be used (and misused) to decide that we ought not to save a species, or that we ought to save one species and not another. In economic terms, a species that has a small physical size, low population numbers, a limited geographical range, an unattractive appearance, no immediate use to people, and no relationship to any species of economic importance will be given a low value; such qualities may characterize a substantial proportion of the world's species, particularly insects, other invertebrates, fungi, non-flowering plants, bacteria, and protists. Costly attempts to preserve these species may not have any short-term economic justification.

Key Ethical Arguments

Several ethical arguments can be made for preserving all species, regardless of their economic value. The following assertions, based on the intrinsic value of species, are important to conservation biology because they provide the rationale for protecting rare species and species of no obvious economic value.

Each species has a right to exist. All species represent unique biological solutions to the problem of survival. On this basis, the survival of each species needs to be guaranteed, regardless of its abundance or its importance to humans. This is true whether the species is large or small, simple or complex, ancient or recently evolved, economically important or of little immediate value. All species are part of the community of living beings and have just as much right to exist as humans do. Each species has value for its own sake, an **intrinsic value** unrelated to human needs (Naess 1986; Norton 1988). Not only do humans not have the right to destroy any species, but people have the responsibility of taking action to protect species from going extinct as the result of human activities (Table 10.1). This argument envisions humans as part of a larger biotic community in which we respect and revere all species.

One incident emphasizing this broader appreciation of life is a change in the sign at a scenic trailside in the Rocky Mountains (Rolston 1987). For many years, the sign read, "Please leave the flowers for others to enjoy," but then the old sign was replaced by a new one saying, "Let the flowers live!" This change in signs suggests a recognition that plants have a right to live for their own sake, not just for their ability to give pleasure to people.

Many writers, particularly those writing from the perspective of the animal rights movement, have had difficulty with assigning

10.1 Government agencies judged the continued existence of the endangered plant Santa Barbara live-forever (*Dudleya traskiae*) to be more valuable than the common rabbits on its island home. The rabbits, which fed on the plant's leaves, were killed to stop their destruction of this fragile plant species. (Photograph courtesy of Center for Plant Conservation, St. Louis.)

rights to species. Singer (1979), for one, argues that "species as such are not conscious entities and so do not have interests above and beyond the interest of individual animals that are members of a species." However, Rolston (1985b) counters that on both biological and ethical grounds, species rather than individual organisms are the appropriate target of conservation efforts. All individuals eventually die, but it is the species that continues, evolves, and sometimes forms new species. In a sense, individuals are just the temporary representatives of the species. Individuals require special protection from human activity when their loss threatens the continued existence of the species. As an example of the primacy of the rights of species, the U.S. National Park Service killed hundreds of rabbits on Santa Barbara Island to protect a few plants of the endangered species *Dudleya traskiae*, Santa Barbara live-forever (Figure 10.1). In this case, one endangered species, with a curiously optimistic name, was judged to be more valuable than hundreds of individual animals of a common species.

How can we assign rights of existence and legal protection to non-human species when they lack the self-awareness that is usually associated with the morality of rights and duties? It could be further argued that non-animal species, such as mosses and fungi, even lack a nervous system to sense their environment. However, species do assert their will to live through their production of offspring and their continuous evolutionary adaptation to a changing environment. The premature extinction of a species due to people destroys this natural process, and could be regarded as something more than a single killing: a "superkilling" (Rolston 1985), since it kills future generations of the species and eliminates the processes of evolution and speciation.

TABLE 10.1
Some notable vertebrate species and subspecies that have gone
extinct in North America since 1492 as a result of human activity

Common name	Scientific name	Region	Date of extinction
FISHES			
Blackfin cisco	*Coregonus nigrapinnis*	Great Lakes	1960s
Yellowfin cutthroat trout	*Salmo clarki macdonaldi*	Colorado	1910
Silver trout	*Salvelinus agassizi*	New Hampshire	1930s
Tecopa pupfish[a]	*Cyprinodon nevadensis calidae*	California	1974
AMPHIBIAN			
Relict leopard frog	*Rana onca*	Utah, Arizona, Nevada	1960
REPTILES			
Iguana	*Leiocephalus eremitus*	Navassa Island, West Indies	1800s
St. Croix racer	*Alsophis sancticrucis*	St. Croix, U.S. Virgin Islands	1900s
BIRDS			
Labrador duck	*Camptorhynchus labradorium*	Northeastern North America	1878
Heath hen	*Tympanuchus cupido cupido*	Eastern U.S.	1932
Great auk	*Pinguinus impennis*	North Atlantic	1844
Passenger pigeon	*Ectopistes migratorius*	Central and eastern North America	1914
Carolina parakeet	*Conuropsis carolinensis carolinensis*	Southeastern U.S.	1914
Hawaii akialoa	*Hemignathus obscurus obscurus*	Hawaii[b]	1895

[a] This fish became extinct even though it was protected by the U.S. Endangered Species Act.
[b] A more complete list of Hawaiian bird extinctions appears in Box 10 in Chapter 7.

TABLE 10.1 *(continued)*

Common name	Scientific name	Region	Date of extinction
MAMMALS			
Puerto Rican ground sloth	*Acratocnus odontrigonus*	Puerto Rico	1500
Giant deer mouse	*Peromyscus nesodytes*	Channel Islands of California	1870
Puerto Rican paca	*Elasmodontomys obliquus*	Puerto Rico	1500
Atlantic gray whale	*Eschrichtius gibbosus gibbosus*	Atlantic Coast	1750
Florida red wolf	*Canis rufus floridianus*	Southeastern U.S	1925
Texas gray wolf	*Canis lupus monstrabilis*	Texas, New Mexico	1942
Sea mink	*Mustela macrodon*	New Brunswick, New England	1890
Caribbean monk seal	*Monachus tropicalis*	Florida, West Indies	1960
Stellar's sea cow	*Hydrodamalis stellari*	Alaska	1768

Source: From Williams and Nowak 1986.

All species are interdependent. Species interact in complex ways as part of natural communities. The loss of one species may have far-reaching consequences for other members of the community; other species may go extinct as well, or the entire community may become destabilized as the result of cascades of species extinction (see Boxes 2 and 3). As we learn more about global processes, we are also finding out that many chemical and physical characteristics of the atmosphere, the climate, and the ocean are linked to biological processes in a self-regulating manner. The idea that the Earth is a superecosystem, in which the biotic community has a role in creating and maintaining conditions suitable for life, is sometimes called the Gaia hypothesis (Lovelock 1988). If this is the case, our instincts toward self-preservation may impel us to preserve biodiversity. When the natural world prospers, we prosper. We are obligated to conserve the system as a whole, because that is the appropriate survival unit.

In a colorful metaphor, Ehrlich and Ehrlich (1981) imagine that species are rivets holding together the Earthship carrying all species, including humans, in its travel through time. Species going extinct are like rivets popping out of the structure. Many species can go extinct and the Earthship still holds together, because it is well built. However, when enough species go extinct the Earthship will fall apart, and humans will go extinct as well. Myers (1979) develops a similar metaphor of a sinking ark. These two metaphors, in which the species (as rivets) prevent the Earthship-ark from crashing (or sinking) represent a reversal of the original Biblical story in which Noah built an ark at God's instruction to preserve each species (Rolston 1985); rather than people saving biodiversity, biodiversity is seen as the salvation of people.

Humans must live within the same ecological limitations as other species do. All species in the world are constrained by their biological carrying capacity in the environment. Each species utilizes the resources of its environment to survive, and the numbers of a species decline when its resources are damaged. Humans also must be careful to minimize damage to their natural environment during their activities, since such damage not only harms other species but harms people as well. Our duties to other humans, as well as to others in the biotic community, require us to live within sustainable limits. People in the industrialized countries can put this belief into practice by taking strong actions to reduce their excessive and disproportionate consumption of natural resources. Many religions advocate a lifestyle that minimizes human impact on other species and on the environment. A primary element of the Indian religions of Hinduism, Jainism, and Buddhism is the ethical concept of *ahimsa*, avoiding unnecessary harm to life. In attempting to live out this concept, many religious people become vegetarians and live as simply as possible.

People must take responsibility for their actions. In the rush to generate profits and satisfy their immediate concerns, people often ignore the effects of their actions on the environment and on other species. Much of the pollution and environmental degradation that occurs is unnecessary and could be minimized with better planning. When people's basic needs are met, they will have more time and energy to devote to larger issues. People will be more likely to take responsibility for protecting biological diversity when citizens of all countries have full political rights, a secure livelihood, and an awareness of environmental issues.

People have a responsibility to future generations. If we degrade the natural resources of the Earth and cause species to become extinct, future generations will have to pay the price in terms of a lower standard of living and quality of life. As Rolston (in press) predicts, "it is safe to say that in the decades ahead, the quality of life will decline in proportion to the loss of biotic diversity, though it is often thought that one must sacrifice biotic diversity to improve human life." Therefore, people of today should try to use resources in a sustainable manner, such that the amounts used do not damage species and communities. We might imagine that we are borrowing the Earth from future generations, and when they receive it back from us they will expect to get it in good condition. A special example of this is the wonder that children experience on seeing new animals and plants in the wild. As species are lost, children are deprived of one of their most exciting experiences in growing up.

Resources should not be wasted. Technology and policy should be directed toward using natural resources in the most efficient manner possible, so that human demands on the environment can be minimized. For example, if water is recycled or used more efficiently, then we will not have to disrupt as many hydrological cycles. Or, if selective logging is practiced carefully, there will be far less soil erosion and unnecessary destruction of forests. This is commonly called the **stewardship argument**, and sees humans as caretakers. Human responsibility for protecting species is explicitly described in the Judeo-Christian Bible as well. In the Book of Genesis, the creation of the Earth's biological diversity was a divine act, after which "God saw that it was good" and "blessed them." In saving Noah from the great flood, God also provides instructions for building the ark, an early species rescue project, saying "Keep them alive with you."

A respect for human life and human diversity is compatible with a respect for biological diversity. An appreciation of the complexity of human culture and the natural world leads people to respect life in its diverse forms. Attempts to bring peace among the nations of the world and an end to poverty, crime, and racism will benefit people and biological diversity simultaneously, since violence within and among human societies is one of the principal destroyers of biological diversity. Human maturity leads naturally to an "identification with all life forms" and "the acknowledgement of the intrinsic value of these forms" (Naess 1986). This view sees an expanding circle of moral obligations, moving outward from oneself to include duties to relatives, the social group, all humanity, animals, all species, the ecosystem, and ultimately the whole Earth (Figure 10.2; Noss 1992).

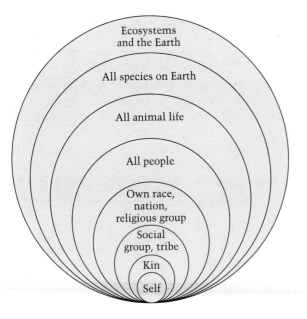

10.2   An ethical sequence in which the individual extends concern outward beyond the self to progressively more inclusive levels. (From Noss 1992.)

Nature has spiritual and aesthetic values that transcend economic value. Throughout history, religious thinkers, poets, writers, artists, and musicians of all varieties have used themes of nature in positive ways. A loss of biological diversity diminishes the ability of people to tap into this wellspring of inspiration. For many people, an essential quality of this inspiration involves experiencing nature in an undisturbed setting. Simply reading about species or seeing them in museums, gardens, and zoos will not suffice. Nearly everyone enjoys wildlife and landscapes aesthetically, and many persons regard Earth as a divine creation which has goodness and value that ought to be respected.

Biological diversity is needed to determine the origin of life. Two of the central mysteries in the world of philosophy and science are how life originated and how the diversity of life found on Earth today came about. Thousands of biologists are working on these problems and are coming ever closer to the answers. However, when species go extinct, important clues are lost, and the mystery then becomes harder to solve.

Deep Ecology

During the twentieth century, ecologists and philosophers have been increasingly articulate about the need for changes in the human lifestyle. As stated by Aldo Leopold (1949), "A thing is right when it tends to preserve the integrity, stability, and beauty of the biotic community. It is wrong when it tends otherwise." This recognition that a true belief in the value of nature would lead to a questioning of the destructive practices that are common in modern society and are often taken for granted, led Paul Sears to call ecology a "subversive science." In the 1960s and 1970s, Paul Ehrlich and Barry Commoner, among others, demonstrated that professional biologists and academics can use their knowledge of environmental issues to create and lead political movements to protect species and ecosystems, with Commoner even running for President in 1980. Such political movements are now widespread across the world, and include activist conservation organizations such as Greenpeace and EarthFirst! as well as "green" political parties in Europe. During the last five years, over 200 action groups dedicated to halting rain forest destruction have formed in the United States alone.

One of the most well-developed environmental philosophies is described in *Deep Ecology: Living As If Nature Mattered* (Devall and Sessions 1985; Sessions 1987). **Deep ecology** begins with the premise that all species have value in themselves, and that humans have no right to reduce this richness. Since present human activities are destroying the Earth's biological diversity, existing political, economic, technological, and ideological structures must be changed. One of the most important changes involves enhancing the **life quality** of people, emphasizing improvements in environmental quality, aesthetics, culture, and religion, rather than higher levels of material consumption. The philosophy of deep ecology includes an obligation to work to implement the needed changes through a commitment to personal lifestyle changes and political activism (Naess 1989). Deep ecology urges professional biologists, philosophers, and all concerned people to escape from their narrow, everyday concerns and to act and live "as if nature mattered."

Summary

1. Protecting biological diversity can be justified on ethical grounds as well as on economic grounds. The value systems of most religions, philosophies, and cultures provide justification for preserving species that are readily understood by people. These justifications support protecting even species that have no obvious economic value to people.

2. The most central ethical argument is that species have a right to exist based on an intrinsic value unrelated to human needs. People do not have the right to destroy species and must take action to prevent the extinction of species. Species, rather than individual organisms, are the appropriate target unit for conservation efforts; it is the species that evolves and undergoes speciation, whereas individuals are just the temporary representatives of the species.

3. Species interact in complex ways in biological communities. The loss of one species may have far-reaching consequences, with other members of the biological community going extinct as well.

4. People must learn to live within the ecological constraints of the planet. People must learn to minimize environmental damage since it hurts humans as well as other species. People must learn to take responsibility for the effects their actions have on the environment and other species. People also have a responsibility to future generations of humans to keep the Earth in good condition.

5. An appreciation of human diversity and the living world leads to an expanding circle of moral obligations, leading from duties to oneself, to include relatives, the social group, all humanity, animals, all species, the ecosystem, and eventually, the whole Earth. Biological diversity has provided generations of writers, artists, musicians, and religious thinkers with inspiration. A loss of species in the wild cuts people off from this wellspring of creative experience.

6. Deep ecology is a philosophy that advocates major changes in the way society functions in order to protect biological diversity. Advocates of this philosophy are committed to personal lifestyle changes and political activism in the environmental movement.

Suggested Readings

Armstrong, S. J. and R. G. Botzler (eds.). 1993. *Environmental Ethics: Divergence and Convergence.* McGraw-Hill, New York. Short essays that stimulate discussion.

Bormann, F. H. and S. R. Kellert (eds.). 1991. *Ecology, Economics and Ethics: The Broken Circle.* Yale University Press, New Haven. Leading authorities present the urgent need for an interdisciplinary approach to solve global problems.

Devall, B. and G. Sessions. 1985. *Deep Ecology.* Gibbs Smith Publisher, Salt Lake City, Utah. An environmental philosophy and a call for action.

Hargrove, E. C. (ed.). 1986. *Religion and the Environmental Crisis.* University of Georgia Press, Athens. What religious traditions have to say about conservation and environmental issues.

Hutchins, M. and C. Wemmer. 1986. Wildlife conservation and animal rights: Are they compatible? *In* M. W. Fox and L. D. Mickley (eds.), *Advances in Animal Welfare Science 1986/87*, pp. 111–137. Humane Society of the United States, Washington, D.C. A good discussion of the overlap and differences between conservation biology and the animal rights movement.

Johns, D. M. 1990. The relevance of deep ecology to the Third World: Some preliminary comments. *Environmental Ethics* 12: 233–252. Important discussion of differing cultural perspectives on environmental issues, published in a highly relevant journal.

Leopold, A. 1949. *A Sand County Almanac: and sketches here and there.* Oxford University Press, New York. Classic work on the need for establishing a better balance between people and the natural world.

List, P. C. (ed.). 1992. *Radical Environmentalism: Philosophy and Tactics.* Wadsworth Publishing, Belmont, CA. Activist organizations and extreme philosophical positions challenge mainstream environmentalists and provoke debate.

McPhee, J. 1971. *Encounters with the Archdruid.* Farrar, Straus, and Giroux, New York. Unique book describing an exchange of ideas among environmentalists and developers while on wilderness trips.

Naess, A. 1989. *Ecology, Community, and Lifestyle.* Cambridge University Press, Cambridge. Reasons for the deep ecology movement by a leading proponent.

Norton, B. G. 1988. Commodity, amenity and morality: The limits of quantification in valuing biodiversity. *In* E. O. Wilson and F. M. Peter (eds.), *Biodiversity*, pp. 200–205. National Academy Press, Washington, D.C. Environmental economics is only one way of valuing nature.

Rolston III, H. 1988. *Environmental Ethics: Values In and Duties To the Natural World.* Temple University Press, Philadelphia. In this and other references (listed in the bibliography), a leading environmental philosopher lays out the ethical arguments for preserving biological diversity.

PART **IV**

Conservation at the Population Level

CHAPTER 11

The Problems of Small Populations

No population truly lasts forever. Due to changing climate, succession, disease, and a range of rare events, the ultimate fate of every population is extinction. But extinction of species caused by human activities is now occurring at a rate roughly 1000 times faster than the natural rate of extinction, and is also occurring far more rapidly than new species can evolve. So the real question is whether a population goes extinct sooner rather than later, and what factors cause the extinction. Will a population of African lions last for over 1000 years and go extinct only after a change in climate, or will the population go extinct after 10 years due to hunting by humans and introduced disease? Because endangered species are made up of one or more populations, protecting populations is the key to preserving species.

Individuals from large, long-lived natural populations have the greatest probability of colonizing unoccupied habitat and forming new populations. This process of new population formation, described by metapopulation models (see Chapter 12), is being curtailed by human activities to the point where some species can no longer survive. A species is particularly vulnerable to extinction when it has only a few small populations. Once a population is reduced below a certain number of individuals by habitat loss, habitat degradation, habitat fragmentation, or overharvesting by humans, it tends to dwindle rapidly toward extinction (Franklin 1980; Gilpin and Soulé 1986; Soulé 1987). Field biologists have provided abundant empirical evidence that small populations are more likely to go extinct than large populations, and this observation is supported by several lines of ecological and genetic argument. This observation of rapid extinction in small populations has led to the concept of **minimum viable population size**, defined generally as the smallest number of individuals necessary to give a population a high probability of surviving over a specified period of time (see Chapter 12).

253

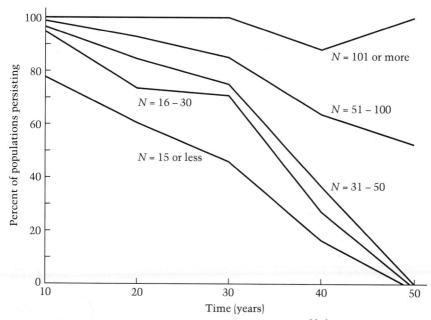

11.1   The relationship between the size of a population of bighorn sheep and the percentage of populations that persist over time. The numbers on the graph indicate population size (*N*); populations with more than 100 sheep almost all persisted beyond 50 years, while populations with fewer than 50 individuals died out within 50 years. (After Berger 1990; photograph by Mark Primack.)

One of the best-documented cases of minimum viable population size comes from a study of the persistence of 120 bighorn sheep (*Ovis canadensis*) populations in the deserts of the southwestern United States (Berger 1990). Some of these populations have been followed for up to 70 years. The striking observation was that 100% of the populations with fewer than 50 individuals went extinct within 50 years, while virtually all of the populations with more than 100 individuals persisted for this time period (Figure 11.1). No single cause was evident for the extinction of most of the populations that died out; rather, a wide variety of factors appeared to cause the population extinctions. For bighorn sheep the minimum population size is at least 100 individuals, and populations below 50 cannot maintain their numbers even in the short term.

Small populations are subject to rapid decline in numbers and local extinction for three main reasons:

- Genetic problems due to loss of genetic variability, inbreeding, loss of heterozygosity, and genetic drift
- Demographic fluctuations due to random variations in birth and death rates
- Environmental fluctuations due to variation in predation, competition, disease, and food supply; and natural catastrophes resulting from single events that occur at irregular intervals, such as fires, floods, volcanic eruptions, storms, or droughts

Loss of Genetic Variability

Genetic variability is important in allowing populations to adapt to a changing environment. **Genetic variability** occurs as a result of individuals having different forms of a gene, known as alleles. Within a population, these alleles may vary in frequency from common to very rare. New alleles arise in a population through random mutations, as described in Chapter 2.

In small populations, allele frequencies may change from one generation to the next simply due to chance, a process known as **genetic drift**. When an allele is at a low frequency in a small population, it has a significant probability of being lost in each generation. For example, if a rare allele constitutes 5% of the gene pool in a population of 1000 individuals, then 100 copies of the allele are present (1000 individuals × 2 copies/individual × 0.05 gene frequency), and the allele will not quickly be lost from the population. However, with a population size of 10 individuals, only one copy of the allele is present (10 individuals × 2 copies/individual × 0.05 gene frequency) and there is a good chance that the rare allele will be lost from the population in the next generation.

Considering the general case of an isolated population in which there are two alleles per gene, Wright (1931) proposed a formula to express the expected drop in heterozygosity per generation (ΔF) for a population of breeding adults (N_e):

$$\Delta F = \frac{1}{2N_e}$$

According to this equation, a population of 50 individuals would have a decline in heterozygosity of 1% (1/100) per generation due to

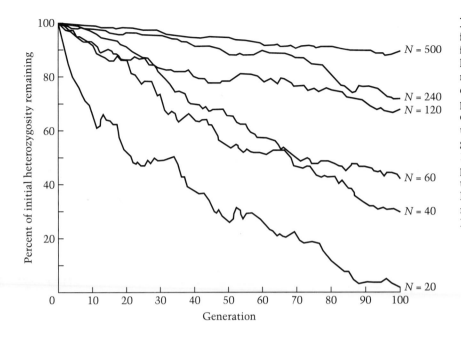

11.2   Genetic drift affects small populations far more strongly than large populations. In this representation of the average of 25 simulated populations, a population of 500 individuals (top trace) still has 90% of its genetic variability after 100 generations, while a population of 20 individuals (bottom trace) has less than 5% of its original variability. (From Lacy 1987.)

the loss of rare alleles; a population of 25 individuals would show a decline of 2% (1/50, or 2/100) per generation (Figure 11.2).

This formulation demonstrates that significant losses of genetic variability that can occur in isolated small populations. However, migration of individuals among populations and the regular mutation of genes tends to increase the amount of genetic variability within the population and balance the effects of genetic drift. Even a low frequency of movement of individuals between populations will tend to minimize the loss in genetic variability associated with small population size (Figure 11.3; Lacey 1987). If only one new immigrant arrives every generation in an isolated population of 120 individuals, the impact of genetic drift will be negligible; gene flow appears to be the major factor preventing the loss of genetic variability in Galápagos finches (Grant and Grant 1992). On the other hand, the mutation rates found in nature—around 1 in 1000 to 1 in 10,000 per gene per generation—are ineffective at countering genetic drift in small populations of 120 individuals or less. Mutation rates would have to be about 1 in 100 or greater to maintain the level of heterozygosity.

How many individuals are needed to maintain genetic variability? Conservation biologists have attempted to determine how large populations must be to prevent the loss of genetic variability. In an early attempt, Franklin (1980) suggested that 50 individuals might be the

11.3 The effects of immigration and mutation on genetic variability in 25 simulated populations of size $N = 120$ individuals. (A) In an isolated population of 120 individuals, even low rates of immigration from a larger source population prevent the loss of heterozygosity from genetic drift. In the model, an immigration rate as low as 0.1 (1 immigrant per 10 generations) increases the level of heterozygosity, while genetic drift is negligible with an immigration rate of 1. (B) It is more difficult for mutation to counteract genetic drift. In the model, the mutation rate m must be 1% per gene per generation ($m = 0.01$) or greater to affect the level of heterozygosity. Since this mutation rate is far higher than what is observed in natural populations, mutation appears to play a minimal role in maintaining genetic variability in small populations. (From Lacy 1987.)

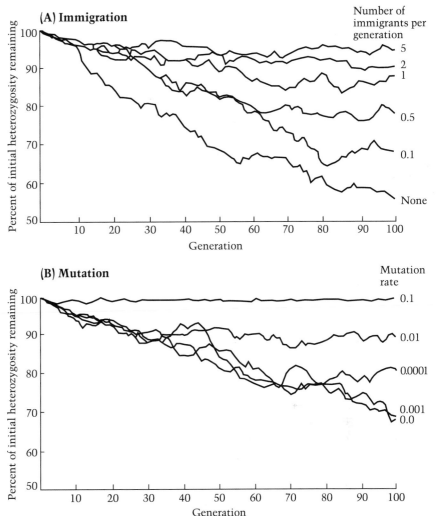

minimum number necessary to maintain genetic variability. This figure is based on the practical experience of animal breeders, which indicates that animal stocks can be maintained with a loss of 2–3% of the variability per generation; using Wright's formula, with a population of 50 individuals, only 1% of the variability will be lost per generation, which would be erring on the safe side. However, since this figure of at least 50 individuals is based on work with domestic animals, its applicability to the wide range of wild species is uncertain. Using data on mutation rates in *Drosophila* fruit flies, Franklin (1980) suggested that in populations of 500 individuals, the rate of new genetic variability arising through mutation might balance the variability being lost due to small population size. This range of values has been referred to as the 50/500 rule: isolated populations need to have at least 50 individuals and preferably 500 individuals to maintain genetic variability.

Effective Population Size

The 50/500 rule is difficult to apply in practice since it assumes that a population is composed of N individuals that all have an equal probability of mating and having offspring. However, many individuals in a population do not produce offspring due to factors such as age, poor health, sterility, malnutrition, small body size, and social structures that prevent some animals from finding mates. As a result of these factors, the **effective population size** (N_e) of breeding individuals is often substantially smaller than the actual population size. Since the rate of loss of genetic variability is based on the effective population size, the loss of genetic variability might be quite severe even when population size is high (Kimura and Crow 1963; Franklin 1980). An effective population size that is smaller than expected can exist under any of the following circumstances:

Unequal sex ratio. By random chance, the population may consist of unequal numbers of males and females. If, for example, a population of a monogamous (one male and one female form a long-lasting pair bond) goose species consists of 20 males and 6 females, only 12 individuals will be involved in mating activity. In this case, the effective population size is 12, not 26. In other animal species, social systems may prevent many individuals from mating even though they are physiologically capable of doing so: in elephant seals, the dominant male may control a large group of females and prevent other males from mating with them, whereas among African wild dogs the dominant female in the pack may have all of the pups.

The effect of unequal numbers of breeding males and females on N_e can be described by the general formula

$$N_e = \frac{4N_m N_f}{N_m + N_f}$$

where N_m and N_f are the numbers of breeding males and breeding females, respectively, in the population. In general, as the sex ratio of breeding individuals becomes increasingly unequal, the ratio of the effective population size to the number of breeding individuals (N_e/N) also goes down (Figure 11.4). Only a few individuals of one sex are making a disproportionately large contribution to the genetic

11.4 The effective population size (N_e) declines when the number of males and females in a breeding population of 100 individuals is increasingly unequal.

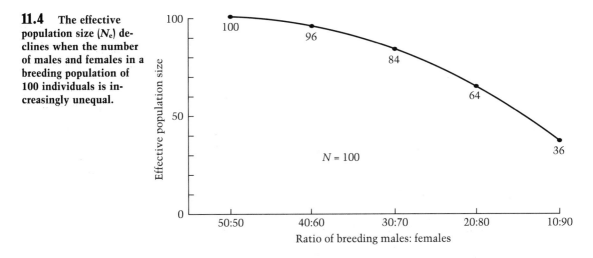

makeup of the next generation, rather than the equal contribution found in monogamous mating systems. For example, consider a population of seals that contains 6 breeding males and 60 breeding females, with each male mating with 10 females. Calculations demonstrate that the effective population size is actually 22, not 66, because of the relatively few males involved in the matings.

Variation in reproductive output. In many species the number of offspring varies substantially among individuals. This phenomenon is particularly true in plants, where some individuals may produce a few seeds while others produce thousands of seeds. Unequal production of seeds leads to a reduction in N_e since a few individuals in the present generation will be disproportionately represented in the gene pool of the next generation. In general, when the variation in reproductive output is high, then the effective population size is lower than the actual population size. For a variety of species, Crow and Morton (1955) estimated that variation in offspring number reduces effective population size by a factor of 60–85%. In many annual plant populations that consist of large numbers of tiny plants producing one or a few seeds and a few gigantic individuals producing thousands of seeds, N_e could be reduced even more.

Population fluctuations. In some species, population size varies dramatically from generation to generation. Particularly good examples of populations with this variation are checkerspot butterflies in California (Murphy et al. 1990), annual plants, and amphibians (Pechmann et al. 1991). In populations that show such extreme fluctuations, the effective population size is somewhere between the lowest and the highest number of individuals. The effective population size over a time period of t years for such a population can be calculated using the number of individuals breeding in any one year:

$$1/N_e = 1/t(1/N_1 + 1/N_2 + \cdots + 1/N_t)$$

The effective population size tends to be determined by the years with the smallest numbers. A single year of drastically reduced population numbers will substantially lower the value of N_e; this principle is involved in a phenomenon known as a population bottleneck, discussed later in this chapter.

As an example, consider a butterfly population that is monitored for 5 years and has 10, 20, 100, 20, and 10 breeding individuals in the 5 successive years. In this case,

$$1/N_e = 1/5(1/10 + 1/20 + 1/100 + 1/20 + 1/10) = 31/500$$

N_e is the reciprocal of 31/500; thus

$$N_e = 500/31 = 16.1$$

In this example, it can be seen that effective population size over the course of these 5 years is above the lowest population level of 10 but well below the maximum number of 100 and the average population size of 32.

These examples demonstrate that the effective population size is often substantially less than the total number of individuals in the population. Particularly where there is a combination of these factors, such as fluctuating population size, numerous nonreproductive individuals, and an unequal sex ratio, the effective population size may be far lower than the number of individuals alive in a good year. As described below, lower effective population size leads to a more rapid loss of alleles from the population.

Estimating Effective Population Size Using Allele Frequency

Another means of estimating effective population size is to measure the loss in genetic variability over time in repeatedly censused populations (Box 19). The rationale for this approach is that the rate of loss of genetic variability over time due to genetic drift is directly correlated to N_e. This approach was used to examine eight large captive populations of *Drosophila melanogaster* fruit flies using a technique for evaluating the number of variable alleles (Briscoe et al. 1992). All eight populations had been maintained separately with a population size of about 5000 individuals each for between 8 and 365 generations. Despite the large population size, N_e varied between 185

BOX 19 RHINO SPECIES IN ASIA AND AFRICA: GENETIC DIVERSITY AND HABITAT LOSS

In recent decades, conservationists have focused an extraordinary effort on restoring the numbers of rhinoceros throughout their original ranges. The task is monumental: the five species of rhinoceros that inhabit Asia and Africa are all critically endangered. Habitat destruction is a special threat to the three species of the Asian forests, and the illegal killing of rhinos for their horns is a particular problem for the two African species. Rhino losses are so severe that it is estimated that only 10,000 individuals of all five species survive today (Cohn 1988a; Ashley et al. 1990). The most numerous of the five is the white rhinoceros, *Ceratotherium simum*; this species numbers approximately 5700 wild animals, while the rarest species—the elusive Javan rhinoceros, *Rhinoceros sondaicus*—is thought to number perhaps 50 in Indonesia and 5 to 15 in Viet Nam (IUCN/SSC African Rhino Specialist Group, Nov. 1992 meeting). The population decline of each species is alarming enough, but the problem is apparently exacerbated by the fact that many of the remaining animals live in very small, isolated populations. The black rhino, *Diceros bicornis*, for example, numbers about 2400 in all, but these individuals are in approximately 75 small,

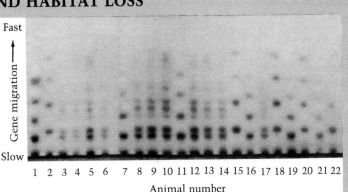

Starch gel electrophoresis reveals that the Chitwan population of the greater one-horned rhinoceros *(R. unicornis)* shows high levels of genetic variation. In this technique, proteins (in this case, an enzyme called LDH) produced by different allele forms of a gene migrate at varying rates across a starch gel plate, appearing as bands at different distances from the starting point at the bottom of the gel. Each column represents an individual animal. Note that animals 10 and 11 have bands at different positions, indicating that they are genetically different for the enzyme LDH. (From Dinerstein and McCracken 1990.)

widely separated subgroups (Ashley et al. 1990). Some biologists fear that many of these small populations may not be viable as a result of inbreeding depression, genetic deficiencies resulting from matings among closely related individuals.

The question of genetic viability in rhino populations is not as simple as it appears at first. Genetic diversity varies greatly among rhino species. Recent studies of the greater one-horned, or Indian, rhinoceros (*Rhinoceros unicornis*) in Nepal indicate that despite its small total population—an estimated 1200 animals—the genetic diversity of this species is extremely high (Cohn 1988; Dinerstein and McCracken

1990), contradicting the common assumption that small populations automatically have low heterozygosity. The combination of long generation times and high individual mobility may have allowed the Indian rhino to maintain its genetic variability despite passing through a population bottleneck. In contrast, six living subspecies of the black rhinoceros are known to exist, representing adaptations to local environmental conditions throughout the species' range. Each of these subspecies appears to have very low genetic variability; in three cases, *D. bicornis brucii*, *D. bicornis longipes*, and *D. bicornis chobiensis*, only a few dozen animals remain to represent these variants (Ashley et al. 1990).

On the basis of genetic viability alone, one might assume that the Indian rhino has the advantage over the black rhino and is more likely to survive. However, the Indian rhino faces a different and possibly more deadly threat: habitat loss. Though no immediate threat of inbreeding exists for the Indian rhino, the critical pressure upon this species' range since the nineteenth century has reduced its numbers dramatically, from possibly tens of thousands of animals to less than 1000 by the 1960s (Dinerstein and McCracken 1990). The geographical range of this animal, which originally covered northern India, Pakistan, southern Nepal, Burma, and Bangladesh, has been almost completely taken over by human settlement. Indian rhino populations in parks and sanctuaries have increased dramatically, and are genetically healthy (Cohn 1988; Dinerstein and McCracken 1990); however, the species will always be limited to these small remnant habitats, with no opportunity to return to its former range or numbers.

In contrast, much of the range of the black rhino in Africa is still open, and is not likely to be subject to human encroachment at any time in the near future. A degree of heterozygosity could be restored to the remaining population by bringing all black rhinos together in a single breeding population. If the black rhino population is managed for genetic diversity in this manner, it is conceivable that this species can be fully restored to its original numbers and range. Yet the problem of microenvironmental adaptations remains: if conservationists placed all black rhinos together in a sanctuary to increase genetic diversity in the species, would they risk losing small adaptive differences that might prove crucial to the survival of local populations? Analysis of mitochondrial DNA in populations of black rhinos (Ashley et al. 1990) indicates that the relationship between most individuals is so close that small adaptive changes are probably not relevant; the need to diversify the black rhino population outweighs any possible costs. Maintaining genetic diversity is contingent upon controlling outside threats to the breeding population, including illegal poaching. Optimal park conditions must also be maintained to ensure that all adult individuals reproduce.

Conservation of rhinos is therefore a task that must be tailored to the specific circumstances of particular species and populations. For species threatened with habitat loss, such as the Indian rhino, sanctuaries and habitat preservation may be the most important methods of preserving the species. Others, such as the black rhino, may require management to increase genetic diversity, including breeding programs and protection of the remnant population. The rarest species, the Sumatran and Javan rhinos, may require a combination of approaches. They need habitat protection, since both of these Asian species are under severe pressure from logging and conversion of forest to agricultural land; and breeding programs to increase and maintain genetic diversity. For each of these rhinos, there is no single, all-encompassing answer; the problems and circumstances of conserving the species must be evaluated individually.

and 253, or only about 4% of the population size. These results suggest that merely maintaining large population sizes may be insufficient to prevent the loss of genetic variability in captive populations. In such cases, genetic variability can be more effectively maintained by controlling breeding, possibly by subdividing the population and allowing limited migration.

Another example involves the threatened winter run of Chinook salmon in the Sacramento River (Bartley et al. 1992). Despite there being 2000 adults in the mating population, the effective population size is only 85. This number may not be sufficient to maintain population variation. A similarly low effective population size (of only 132) was noted in a commercial salmon hatchery, even when up to 10,000 fish were present in the spawning population. Both wild and hatchery-raised fish populations apparently have effective population sizes well below the adult population size and are in danger of losing significant amounts of genetic variability. In the case of fish hatcheries, managers have recognized that more adults have to be used in breeding programs, more tanks have to be used to subdivide the population, and progeny from different parents have to be maintained.

Field Evidence for Lowered Genetic Variability in Small Populations

Correlation of population size with genetic variability. In New Zealand, the wind-pollinated dioecious conifer *Halocarpus bidwillii* occurs naturally in discrete populations in subalpine habitats, with population sizes ranging from 10 to 400,000 individuals. Studies of this species showed a strong correlation of population size with genetic variability: large populations had the greatest levels of heterozygosity, the highest percentage of polymorphic genes, and the largest mean number of alleles (Figure 11.5; Billington 1991). Populations smaller than 8000 individuals appeared to suffer a loss of genetic variability, with the lowest variability in the smallest populations.

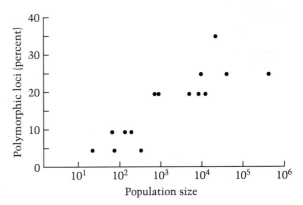

11.5   The level of genetic variability is directly correlated with population size in populations of a New Zealand coniferous shrub, *Halocarpus bidwillii*. This pattern holds true for the percentage of loci that are polymorphic as well as for the mean number of alleles per gene and the level of heterozygosity. (From Billington 1991.)

Rare species with little or no genetic variability. A number of rare species have been found to lack genetic variability. The rare aquatic annual plant *Howellia aquatilis* (Campanulaceae) was found to lack genetic variability both among individuals within populations and among populations (Lesica et al. 1988). The lack of variability was postulated to be due to fluctuations in population size and a tendency for flowers to self-fertilize. A similar lack of genetic variability is apparent in four populations of the rare Furbish's lousewort, *Pedicularis furbishiae*, an endemic plant of Maine (Waller et al. 1988). In this case, the lack of genetic variability is attributed to genetic drift in a series of temporary populations established on riverbanks after colonization by a few seeds. In a classic study, Babbel and Selander (1974) showed that a lupine species (*Lupinus*) with a restricted range had less genetic variability than a more widespread species. Similarly, when 11 pairs of species were compared, rare plant species were consistently found to have lower genetic variability than common species in the same genus (Karron 1987). An extensive review of studies of genetic variability in plants showed that only 8 of 113 plant species had no measurable genetic variability and that these often had very limited ranges (Hamrick et al. 1979).

Genetic Bottlenecks

A population may occasionally be severely reduced in size due to some environmental or demographic event that kills all but a few individuals (Carson and Templeton 1984; Barrett and Kohn 1991). This phenomenon is referred to as a **genetic bottleneck**. When a population is greatly reduced in size, rare alleles in the population will be lost if no individuals possessing those alleles survive (Carson 1983). With fewer alleles present and a decline in heterozygosity, the overall fitness of the individuals in the population declines. A special category of bottleneck, known as the **founder effect**, occurs when a few individuals leave a large population to establish a new population (Figure 11.6).

The lions of Ngorongoro Crater in Tanzania provide a well-studied example of a genetic bottleneck (Packer et al. 1991; Packer 1992). The lion population consisted of 60 to 75 individuals, until an outbreak of biting flies reduced the population to 9 females and 1 male in 1962 (Figure 11.7). Two years later, 7 additional males immigrated into the crater; there has been no further immigration since that time. The small numbers of founders, the isolation of the population, and the variation in reproductive success among individuals has apparently created a genetic bottleneck even though the population has

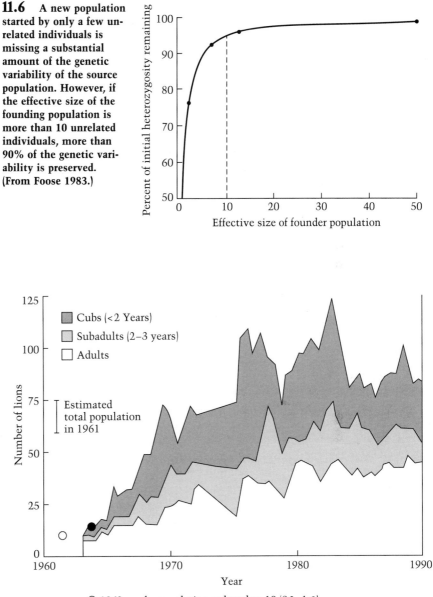

11.6 A new population started by only a few unrelated individuals is missing a substantial amount of the genetic variability of the source population. However, if the effective size of the founding population is more than 10 unrelated individuals, more than 90% of the genetic variability is preserved. (From Foose 1983.)

○ 1962 crash: population reduced to 10 (9♀, 1♂)

● 1964: 7 males immigrate

11.7 The Ngorongoro Crater lion population was about 61 individuals in 1961 before crashing in 1962. Since that time the population has returned to its original numbers, but the isolated location and lack of immigration since 1964 have apparently created a genetic bottleneck. (From Packer et al. 1991.)

subsequently increased to 75–125 animals. In comparison with the large Serengeti lion population nearby, the Crater lions show reduced genetic variability, high levels of sperm abnormalities (Figure 11.8), and reduced reproductive rates.

The cheetah is a species that has undergone a severe reduction in numbers in recent times as a result of habitat alteration and hunting by humans. While the species occurred widely throughout Africa, Asia, Europe, and North America 20,000 years ago, today only about 1500 to 5000 cheetahs remain, confined exclusively to Africa (O'Brien et al. 1985a,b; O'Brien and Evermann 1988). The vulnerability of cheetahs to extinction is illustrated by the extinction of four other species in its genus, *Acinonyx*, in the late Pleistocene. Evidence for the genetic effects of a population bottleneck in the cheetah comes from studies showing exceptionally low levels of genetic variability in two isolated populations. These low levels are what would be expected after 10–20 generations of brother–sister matings in an isolated population. As a possible consequence of the loss of heterozygosity, cheetahs are known to have only 10% of the sperm count of related felid species, and over 70% of their sperm are aberrant in some way. Their lack of genetic variability may be one reason why cheetahs are unusually vulnerable to disease.

Population bottlenecks need not always lead to reduced heterozygosity. If the population expands rapidly in size after a temporary bottleneck, average heterozygosity in the population may be restored even though the number of alleles present is severely reduced (Nei et al. 1975; Allendorf 1986) (Figure 11.9). An example of this phenomenon is the high level of heterozygosity found in the greater one-horned rhinoceros in Nepal (see Box 19). At one point, the population was less than 30 breeding individuals, but by 1988 it had recovered to almost 400 individuals (Dinerstein and McCracken 1990).

Genetic Consequences of Small Populations

Small populations have greater susceptibility to a number of deleterious genetic effects, such as inbreeding depression, loss of evolutionary flexibility, and outbreeding depression, leading to a decline in population size and a greater probability of extinction.

Inbreeding depression. In large populations of most animal species, individuals do not normally mate with close relatives (Ralls et al. 1986). Individuals often disperse away from their place of birth, or are inhibited from mating with relatives through unique individual odors or other sensory cues. In many plants, a variety of morphological and physiological mechanisms encourage cross-pollination and prevent

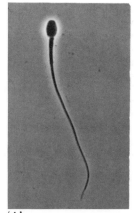

(A)

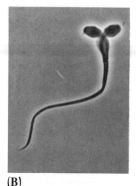

(B)

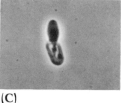

(C)

11.8   Males of the isolated and inbred population of lions at Ngorongoro Crater in Tanzania exhibit a high level of sperm abnormalities. (A) Normal lion sperm. (B) Bicephalic ("two-headed") sperm. (C) Nonfunctional sperm with a coiled flagellum. (Photographs by D. Wildt.)

11.9 Effects of a population bottleneck on genetic variation for a population that is reduced to 2 individuals (solid lines) and one that is reduced to 10 individuals (dashed lines). Loss of heterozygosity is also affected by the population's growth rate (r); a population with a slow growth rate per generation ($r = 0.1$) will experience more severe loss of heterozygosity and take longer to "rebound" than a population with a more rapid rate of growth ($r = 1.0$). Thus, for example, elephant populations, which have a very low growth rate, will be more severely affected by bottlenecks than are fast-growing populations such as those of mice or rabbits. (After Nei et al. 1975.)

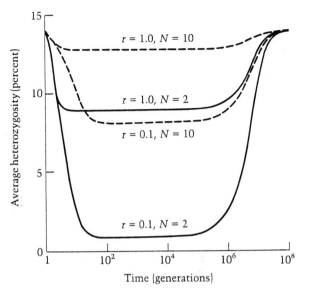

self-pollination. In some cases, particularly when no other mates are available, these mechanisms fail to prevent inbreeding. Mating among close relatives such as parents and their offspring, siblings, and cousins results in **inbreeding depression**, characterized by fewer offspring, or offspring that are weak or sterile.

The most plausible explanation for inbreeding depression is that it allows the expression of harmful alleles (Selander 1983; Barrett and Kohn 1991; Charlesworth and Charlesworth 1987). Outbreeding not only increases the level of heterozygosity in the population, but also allows many rare but harmful recessive alleles that arise by mutation to accumulate unexpressed in populations; as long as these alleles are rare, the function of the genes will be performed by the more common dominant allele of the gene. When the population size of an outcrossing species declines, close relatives may be forced to mate with one another because no other mates are available. This inbreeding allows the rare, harmful recessive alleles to become expressed in the homozygous form, with resulting harmful effects on the offspring.

Evidence for the existence of inbreeding depression comes from population genetics theory and from studies of human populations, captive animal populations, and cultivated plants (Darwin 1876; Charlesworth and Charlesworth 1987). In a wide range of captive mammal populations, matings among close relatives, such as parent–offspring matings and sibling matings, resulted on average in offspring with a 33% higher mortality rate than in outbred animals (Figure 11.10; Ralls et al. 1986; Ralls et al. 1988). Inbreeding depression

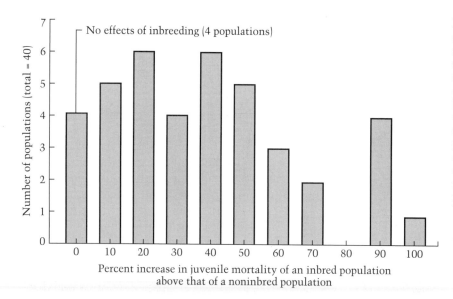

Percent increase in juvenile mortality of an inbred population
above that of a noninbred population

11.10   A high degree of inbreeding (such as matings between mother and son, father and daughter, or brother and sister) results in a "cost of inbreeding." In the above data, based on a survey of 40 inbred mammal populations, the cost is expressed as a percentage of increase in juvenile mortality above the juvenile mortality of outbreeding animals of the same species. (From Ralls et al. 1988.)

can be a severe problem in small captive populations in zoos and domestic breeding programs. The effects of inbreeding in the wild are unknown, but would be expected to be very significant (Barrett and Kohn 1991). In writing about the tendency of species to avoid breeding with close relatives, Charles Darwin (1868) commented: "That any evil directly follows from the closest inbreeding has been denied by many persons; but rarely by any practical breeder; and never, as far as I know, by anyone who has largely bred animals which propagate their kind quickly."

The variety of genetic factors that affect small populations is illustrated by studies of gray wolves on Isle Royale, Canada (Wayne et al. 1991). A pair of wolves established a population on Isle Royale in Lake Superior around 1949. The population had increased to about 50 individuals in 1980 but had dropped to only about 14 by 1990, with many females not breeding and few pups sighted. Decreased availability of food, the effects of canine provirus disease, and loss of genetic variability have been put forth as possible explanations for the population decline. Loss of genetic variability and heterozygosity seems to be the most likely explanation, since there have been only 2 to 3 breeding pairs of wolves for 5 to 7 generations, with a predicted 39–65% loss of genetic variability. Genetic studies demonstrated that the island's wolf population has lost 50% of its genetic variability compared with mainland populations. The wolves of Isle Royale are as genetically similar to each other as siblings and are probably descended from a single mother. These genetic results suggest that inbreeding depression and an unwillingness of close relatives to mate could explain the decline of population size.

Outbreeding depression. When individuals from two separate species mate, their offspring are often weak or sterile due to the lack of compatibility of the chromosomes and enzyme systems inherited from their different parents (Templeton 1986). The hybrid offspring may also no longer have the precise mixture of genes that allowed individuals to survive in a particular set of local conditions. **Outbreeding depression** can also result from matings between different subspecies, or even matings between divergent genotypes or populations of the same species (Waser et al. 1987; Waser and Price 1989). Excessive outbreeding is rare in nature since individuals have strong behavioral, morphological, and physiological mechanisms to ensure that mating occurs only within a species. Related species may also live in different geographical areas and in different types of habitats. In captivity, however, different species may breed together deliberately or accidentally because no suitable mate is available. Domestic horses and donkeys, for example, are commonly bred to produce mules; while these animals are not weak—on the contrary, they are quite strong—they are almost always sterile. Such outbreeding depression must be guarded against in captive breeding programs.

Outbreeding depression may be particularly important in plants, where mate selection is to some degree a matter of the chance movement of pollen (Waser and Price 1989). A rare plant species growing near a closely related common species may become overwhelmed by the pollen of the common species (Ellstrand 1992). The offspring of such hybridization events are often sterile. Even when the hybrids are not sterile, the genetic identity of the rare species becomes lost as its small gene pool is mixed into the much larger gene pool of the common species. The seriousness of this threat is illustrated by the fact that more than 90% of California's threatened and endangered plants occur in close proximity to another species in the same genus, with which the rare plants could possibly hybridize (Anonymous 1989).

Loss of evolutionary flexibility. Loss of genetic variability may limit the ability of a population to respond to long-term changes in the environment (Allendorf and Leary 1986). Rare alleles and unusual combinations of alleles that confer no immediate advantages may be uniquely suited for a future set of environmental conditions. When rare alleles are lost in small populations and heterozygosity declines, the population has fewer genetic options available. A small population has a lower probability than a large population of having the necessary genetic variability to adapt to long-term environmental changes, and so will be more likely to go extinct. For example, in many plant populations, a small percentage of the individuals have alleles to tolerate toxic metals, such as zinc and lead, even when these metals are not present (Antonovics et al. 1971; Antonovics 1976). If toxic metals are present, whether as a result of natural processes, mining activities, or environmental pollution, these alleles increase dramatically in frequency in the population. However, if the population is small and the genotypes for metal tolerance are not present, the population could go extinct.

Demographic Variation

In an idealized stable environment, a population would increase until it reached the carrying capacity (K) of the environment. At that point the average birth rate (b) per individual would equal the average death rate (d), and there would be no net change in population size. In any real population, individuals do not usually produce the average number of offspring, but rather may have no offspring, somewhat fewer than the average, or more than the average. For example, in an ideal stable giant panda population, each female would produce an average of two surviving offspring in her lifetime, but field studies show that individual females vary widely around this number of two offspring. However, as long as population size is large, the average provides an accurate description of the population. Similarly, the average death rate in a population can only be determined by examining large numbers of individuals.

Once population size drops below about 50 individuals, individual variation in birth rates and death rates begins to cause the population size to fluctuate randomly up or down (Gilpin and Soulé 1986; Menges 1992). If population size fluctuates downward, then the resulting smaller population will be even more susceptible to demographic fluctuations in the next generation, possibly resulting in extinction. Random fluctuations upward in population size are eventually bounded by the carrying capacity of the environment, and the population again may begin to fluctuate downward. Consequently, once a population becomes small because of habitat destruction and fragmentation, demographic variation begins to become important and the population has a higher probability of going extinct. Based on population theory, Richter-Dyn and Goel (1972) and MacArthur and Wilson (1967) argued that random demographic variation, also known as **demographic stochasticity**, will be greater as population size gets smaller, resulting in a greater probability of extinction due to chance alone. The chance of extinction is also greater in species that have low birth rates, such as elephants, since these species take longer to recover from a chance reduction in population size.

As a simple example, imagine a population of three hermaphroditic individuals that live for one year, reproduce, and then die. Each individual has a 33% probability of producing 0, 1, or 2 offspring, resulting in an average birth rate of 1 per individual; in this instance, there is theoretically a stable population. However, when these individuals mate, there is a 1-in-27 chance (0.33 × 0.33 × 0.33) that no offspring will be produced in the next generation and that the population will go extinct. Consider also that there is a 1-in-9 chance that only one offspring will be produced in the next generation (0.33 × 0.33 × 0.33 × 3); since this individual will not be able to find a mate, the population will be doomed to extinction in the next generation. There is also a 22% chance that the population will decline to 2 individuals in the next generation. This simple example illustrates that random variation in birth rates can lead to demographic stochasticity and extinction in small populations. Similarly, random fluctuations in the death rate can lead to fluctuations in the population size. When populations are small, a random high mortality in one year might eliminate the population altogether.

When populations drop below a critical number, there is also the possibility of deviations from an equal sex ratio and a resulting decline in birth rate. For example, imagine a population of 4 birds that includes 2 mating pairs of males and females, in which each female produces an average of 2 surviving offspring in her lifetime. In the next generation, there is a 1-in-16 chance that only male birds will be produced (as well as a 1-in-16 chance that only female birds will be produced), in which case no eggs will be laid to produce the following generation. There is a 50% (8-in-16) chance that there will be either 3 males and 1 female or 3 females and 1 male in the next generation, in which case only one pair of birds will mate and the population will decline. Though these are hypothetical situations, two examples from nature illustrate this point. First, the last five surviving individuals of the extinct dusky sparrow were all males, so there was no opportunity to establish a captive breeding program. And second, the last 30 individuals left in Illinois of the rare plant *Hymenoxys acaulis* var. *glabra* are unable to produce viable seeds when cross-pollinating among themselves because they belong to the same self-infertile mating type (De Mauro 1989). Pollen has to be brought in from Ohio plants in order for the Illinois plants to produce seeds.

In many animal species, small populations may be unstable due to the inability of the social structure to function once the population falls below a certain number. Herds of grazing mammals and flocks of birds may be unable to find food and defend themselves against attack when numbers fall below a certain level. Animals that hunt in packs, such as wild dogs and lions, may need a certain number of individuals to hunt effectively. Many animal species that live in widely dispersed populations, as do bears or whales, may be unable to find mates once the population density drops below a certain point; this is known as the **Allee effect** (Dennis 1981). In this last case, the average birth rate will decline, making the population size even smaller and worsening the problem. In plant species, as population size decreases, the distance between plants increases; pollinating animals may not visit more than one of the isolated, scattered plants, with a resulting loss of seed production (Bawa 1990). This combination of random fluctuations in demographic characteristics, unequal sex ratios, and disruption of social behavior contributes to instabilities in population size, often leading to local extinction.

Environmental Variation and Catastrophes

Random variation in biological communities and in the natural environment can also cause variation in the population size of a species. For example, the population of an endangered rabbit species might be affected by fluctuations in the population of a deer species that eats the same types of plants as the rabbits, in the population of a fox that feeds on the rabbits, and in the presence of parasites and diseases affecting the rabbits. Fluctuations in the physical environment might also strongly influence the rabbit populations; rainfall during an average year might encourage plant growth and allow the population to increase, while dry years might limit plant growth and cause rabbits to starve. Random environmental variation, also known as **environmental stochasticity**, affects all individuals in the population, in contrast to demographic variation, which causes variation among individuals within the population.

Natural catastrophes at unpredictable intervals, such as droughts, storms, earthquakes, fires, and cyclical die-offs of the surrounding biological community, can also cause large variation in populations. Natural catastrophes can kill a certain percentage of a population or even eliminate the population from an area. Even though the probability of a natural catastrophe in any one year is low, over the course of decades and centuries, natural catastrophes will have a higher likelihood of occurring.

As an example of environmental variation, imagine a rabbit population of 100 individuals in which the average birth rate each spring is 0.2 and an average of 20 rabbits are eaten each summer by foxes. On average, the population will maintain its numbers at exactly 100 individuals, with 20 rabbits born each year and 20 rabbits eaten each year. However, if there are three successive years in which the foxes eat 40 rabbits per year, the population size will decline to 80 rabbits, 56 rabbits, and 27 rabbits in years 1, 2, and 3 respectively. If there are then three years of no fox predation, the rabbit population will increase to 32, 38, and 46 individuals in years 4, 5, and 6. Even though the same average rate of predation occurred over this six-year period, the variation in year-to-year predation rates would cause the rabbit population size to decline by 50%. At a population size of 46 individuals, the rabbit population will rapidly go extinct when subjected to the average rate of 20 rabbits eaten per year by foxes.

Extensive modeling efforts by Menges (1992) and others have shown that random environmental variation is generally more important than random demographic variation in increasing the probability of extinction in populations of moderate size. Environmental variation can increase the risk of extinction even in populations showing positive population growth under the assumption of a stable environment. In general, introducing environmental variation into population models, in effect making them more realistic, results in populations having lower growth rates, lower population sizes, and higher probabilities of extinction. Menges (1992) introduced environmental variation into models of plant populations that had been developed by field ecologists. These models, before the inclusion of environmental variation, suggested that the minimum viable population size (the smallest number of individuals that would allow the population to persist; see Chapter 12) for plants was about 140 mature individuals (Figure 11.11). When moderate environmental variation was included, however, the minimum viable population size increased to 380 individuals.

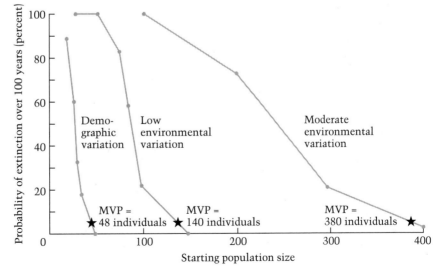

11.11 The effects of demographic variation, low environmental variation, and moderate environmental variation on the probability of extinction of a population of the Mexican palm, *Astrocaryum mexicanum*. (In this study, the minimum viable population size, shown as stars, was defined as the population size at which there is a less than 5% chance of the population going extinct within 100 years.) (From Menges 1991; data from Pinero et al. 1984.)

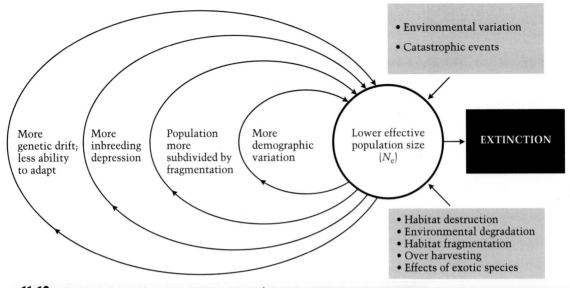

11.12   Extinction vortices progressively lower population sizes, leading to local extinctions of species. Once a species enters a vortex, its population size becomes progressively lower, which in turn enhances the negative effects of the vortex. (Adapted from Guerrant 1992 and Soulé and Gilpin 1986.)

Extinction Vortices

The smaller a population becomes, the more vulnerable it is to demographic variation, environmental variation, and genetic factors that tend to reduce population size even more and drive the population to extinction. This tendency of small populations to decline toward extinction has been likened to a **vortex effect** (Gilpin and Soulé 1986). For example, a natural catastrophe, environmental variation, or human disturbance could reduce a large population to a small size. This small population could then suffer from inbreeding depression, with an associated lowered juvenile survival rate. This increased death rate could result in an even lower population size and even more inbreeding. Similarly, demographic variation will often reduce population size, resulting in even greater demographic fluctuations and a greater probability of extinction. These three factors—environmental variation, demographic variation, and loss of genetic variability—act together so that a decline in population size caused by one factor will increase the vulnerability of the population to the other factors (Figure 11.12). For example, a decrease in orangutan population size caused by forest fragmentation may result in inbreeding depression, further lowering population size; the lowered population size may disrupt the social structure and the ability to find mates, leading to an even smaller population that is more vulnerable to an unusual environmental event. This combination of factors may drive the population into a rapid decline in numbers and eventually lead to extinction. Once a population has declined to a small size, it will often go extinct unless highly favorable conditions allow the population to increase to its minimum viable population size. Another alternative is a careful program of population and habitat management to minimize the effects of small population size (Schonewald-Cox et al. 1983), as described in later chapters.

Summary

1. Biologists have observed that small populations have a greater tendency to go extinct than large populations. Small populations are subject to rapid extinction for three main reasons: loss of genetic variability, demographic fluctuations, and environmental variation or natural catastrophes.

2. As a population becomes smaller, it tends to lose genetic variability just by chance, a process known as genetic drift, leading to inbreeding depression and a lack of evolutionary flexibility. Experience with captive animals suggests that isolated populations should have at least 50 breeding individuals and preferably 500 individuals to maintain genetic variability. The key component in considering population size is the effective population size, based on the number of individuals that are actually producing offspring. The calculated effective population size is often much lower than simply the number of individuals alive because many individuals are not reproducing, there may be an unequal sex ratio, there may be variation among individuals in number of offspring produced, and populations may show large fluctuations over time.

3. In small populations, variation in reproductive rates and mortality rates can cause populations to fluctuate randomly in size, leading to population extinction. Environmental variation can also cause random fluctuations in population size, with infrequent natural catastrophes sometimes causing major reductions in population size. Once a population's size has been reduced, it is even more vulnerable to random fluctuations in size and eventual extinction. The combined effects of demographic variation, environmental variation, and genetic factors have been compared to a vortex that tends to drive small populations to extinction.

Suggested Readings

Barrett, S. C. and J. R. Kohn. 1991. Genetic and evolutionary consequences of small population size in plants: Implications for conservation. *In* D. Falk and K. Holsinger (eds.), *Genetics and Conservation of Rare Plants*, pp. 3–10. Oxford University Press, New York. Excellent review of the genetics of small populations with numerous references to plant studies.

Berger, J. 1990. Persistence of different-sized populations: An empirical assessment of rapid extinctions in bighorn sheep. *Conservation Biology* 4: 91–98. An excellent case study.

Carson, H. L. and A. R. Templeton. 1984. Genetic revolutions in relation to speciation phenomena: The founding of new populations. *Annual Review of Ecology and Systematics* 15: 97–131. New populations are often genetically unique.

Charlesworth, D. and B. Charlesworth. 1987. Inbreeding depression and its evolutionary consequences. *Annual Review of Ecology and Systematics* 18: 237–268. The implications of inbreeding are discussed by two leading scientists.

Falk, D. A. and K. E. Holsinger (eds.). 1991. *Genetics and Conservation of Rare Plants*. Oxford University Press, New York. Conservation efforts involving plants require some special considerations.

Franklin, I. R. 1980. Evolutionary change in small populations. *In* M. E. Soulé and B. A. Wilcox (eds.), *Conservation Biology: An Evolutionary–Ecological Perspective*, pp. 135–149. Sinauer Associates, Sunderland, MA. Seminal paper outlining the problems of small populations.

Gilpin, M. E. and M. E. Soulé. 1986. Minimum viable populations: Processes of species extinction. *In* M. E. Soulé (ed.), *Conservation Biology: The Science of Scarcity and Diversity*, pp. 19–34. Sinauer Associates, Sunderland, MA. Excellent summary of the extinction vortex facing small populations.

Lacy, R. C. 1987. Loss of genetic diversity from managed populations: Interacting effects of drift, mutation, immigration, selection and population subdivision. *Conservation Biology* 1: 143–158. Clearly presented simulations of various realistic scenarios.

Menges, E. S. 1992. Stochastic modeling of extinction in plant populations. *In* P. L. Fiedler and S. K. Jain (eds.), *Conservation Biology: The Theory and Practice of Nature Conservation, Preservation and Management*, pp. 253–275. Chapman and Hall, New York. Clear presentation of extinction models.

Ralls, K., P. H. Harvey and A. M. Lyles. 1986. Inbreeding in natural populations of birds and mammals. *In* M. E. Soulé (ed.), *Conservation Biology: The Science of Scarcity and Diversity*, pp. 35–56. Sinauer Associates, Sunderland, MA. The evidence for inbreeding depression in animals.

Schonewald-Cox, C. M., S. M. Chambers, B. MacBryde and L. Thomas (eds.). 1983. *Genetics and Conservation: A Reference for Managing Wild Animal and Plant Populations*. Benjamin/Cummings, Menlo Park, CA. Excellent set of early papers on the genetic problems of small populations.

Soulé, M. E. (ed.). 1990. *Viable Populations for Conservation*. Cambridge University Press, Cambridge. Leading authorities discuss the problems of small populations.

Templeton, A. R. 1986. Coadaptation and outbreeding depression. *In* M. E. Soulé (ed.), *Conservation Biology: The Science of Scarcity and Diversity*, pp. 105–116. Sinauer Associates, Sunderland, MA. Reasons why hybridizations between species are often unsuccessful.

CHAPTER 12

Population Biology of Endangered Species

In order to successfully maintain species under the restricted conditions imposed by human activities, conservation biologists must determine the stability of populations under certain circumstances. Will a population of an endangered species persist or even increase in a nature reserve? Alternatively, is the species declining, and does it require special attention to prevent it from going extinct? The ability of a species to persist in a protected area can often be predicted using the methods of population biology and the mathematics of population viability analysis. These techniques can be used to estimate the minimum viable population size (MVP), the smallest number of individuals needed to maintain a long-term population. Even without human disturbance, a population of any species may be stable, increasing, or decreasing. Populations can also fluctuate in size, sometimes increasing and sometimes decreasing. In general, the effect of widespread human disturbance is to destabilize populations of native species, often sending them into sharp decline.

The Metapopulation

Over the course of time, populations of a species may go extinct on a local scale, and new populations may form on other nearby, suitable sites. Many species in ephemeral habitats, such as streamside herbs, and species with widely fluctuating population sizes are characterized by a shifting mosaic of temporary populations known as a **metapopulation** (Menges 1990; Murphy et al. 1990). A metapopulation may be characterized by one or more **core** populations, with fairly stable numbers, and several **satellite** areas with fluctuating populations (Bleich et al. 1990). Populations in the satellite areas may go extinct in unfavorable years, but the areas are recolonized by migrants from the core population when conditions become more favorable (Figure 12.1). In some species, every population is short-lived, and the distribution of the species changes dramatically with each generation.

The target of a population study is typically one or several populations, but an entire metapopulation may need to be studied if this would result in a more accurate portrayal of the species. Metapopulation models have the advantage of recognizing that local populations are dynamic and that there is movement of organisms from one local population to another (Harrison and Quinn 1989; Silvertown 1991; Hanski 1989; Olivieri et al. 1990). The recognition that infrequent colonization events and migration are occurring also allows biologists to consider the impact of founder effects and genetic drift on the species. The following are several examples in which the metapopulation approach has proved to be more useful than a single-population description in understanding and managing species:

- The European nuthatch, *Sitta europaea*, occupies fragments of forest ranging in area from 0.3 to 30 ha within the agricultural landscapes of Western Europe (Verboom et al. 1991). The populations in any one fragment are dynamic, but populations in small fragments are more prone to extinction than populations in larger and better-quality fragments. The colonization rate of unoccupied forest fragments depends on the density of the nuthatch population in surrounding forest patches.
- The endemic Furbish's lousewort, *Pedicularis furbishiae*, occurs along a river in Maine subject to periodic flooding (Menges 1990). Flooding often destroys some existing populations, but creates conditions suitable for new populations. Studies of any single population would give an incomplete picture of the species, since the populations are short-lived. The metapopulation is really the appropriate unit of study, and the watershed is the appropriate unit of management.
- The checkerspot butterfly (*Euphydryas* spp.) has been extensively studied in California (Ehrlich and Murphy 1987; Murphy et al. 1990). Individual butterfly populations often go extinct, but dispersal and colonization of unoccupied habitat allow the species to survive. Environmental stochasticity and a lack of habitat variation at a particular site often cause extinction in local populations, with the largest and most persistent populations found in large areas that have both moist, north-facing slopes and warmer, south-facing slopes. Butterflies migrating

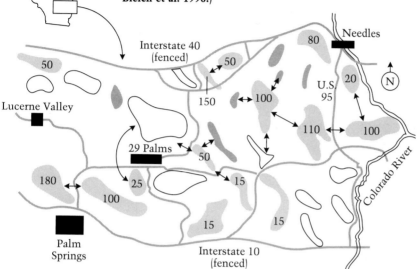

12.1   Mountain sheep in the southeastern California desert have a shifting mosaic of populations best described as a metapopulation. Mountain ranges shown in light gray had populations of the sizes indicated in 1990; open mountain ranges were unpopulated in 1990 but had resident populations in the past. Dark gray mountains have never had resident populations. Arrows indicate observed migrations of sheep. (After Bleich et al. 1990.)

out from these core populations often colonize the unoccupied satellite areas.

In metapopulation situations, destruction of the habitat of one central, core population might result in the extinction of numerous smaller populations that depend on the core population for periodic colonization. Also, human disturbances that inhibit migration, such as fences, roads, and dams, might reduce the rate of migration among habitat patches, and so reduce the probability of recolonization after local extinction (Lamberson et al. 1992). Metapopulation models highlight the dynamic nature of population processes and show how the elimination of a few populations could lead to the local extinction of a species over a much wider area. Effective management of a species often requires an understanding of these metapopulation dynamics.

Natural History and Autecology

The key to protecting and managing a rare or endangered species is to have a firm grasp of its biological relationship to its environment and the status of its populations. In the past, this information was called the **natural history** of the species, while as a scientific discipline it was known as **autecology**. With more information concerning a rare species' natural history, land managers are able to make more effective efforts to maintain the species and identify factors that place it at risk of extinction (Gilpin and Soulé 1986; Simberloff 1988).

The following are categories of autecological questions that are important to answer in order to implement effective population-level conservation efforts. For most species, only a few of these questions can be answered without further investigation. Yet decisions on management may have to be made before this information is available or while it is being gathered. The exact types of information gathered will obviously depend on the characteristics of the species (single-celled species vs. fungi vs. plants vs. animals).

- *Morphology*. What does the species look like? What is the shape, size, color, surface texture, and function of its parts? How does the morphology of the species change over its geographical range? Do all of the individuals in the population look the same? How does the shape of its body parts relate to their function and help the species to survive in its environment? How large are the new offspring, and are they different in appearance from adults?
- *Physiology*. How much food, water, minerals, and other necessities does an individual need to survive, grow, and reproduce? How efficient is an individual at using its resources? How vulnerable is the species to extremes of climate, such as heat, cold, wind, and rain? When does the species reproduce, and what are its special requirements during reproduction? How susceptible is the species to disease and parasitism? Are the juveniles particularly vulnerable to disease, a harsh climate, and predation?
- *Behavior*. How do the actions of an individual allow it to survive in its environment? How do individuals in a population mate and produce offspring? In what ways do individuals of a species interact among themselves, either cooperatively or competitively? In what ways do the individuals interact with other species as predators, prey, or competitors for a common resource? What specific food items or resources does an individual utilize, and how does it obtain them? In what ways do parents assist their offspring?
- *Genetics*. How much variation in morphological and physiological characteristics occurs in the population? How much of this variation is genetically controlled? What percentage of the genes are variable? How many alleles does the population have for each variable gene?

- *Distribution.* Where is the species found in the environment? Are individuals clustered together, distributed at random, or spaced out regularly? Does the species move and migrate among habitats, or to different geographical areas, over the course of a day or over a year? Does the species have difficulties moving among habitats? How efficient is the species at colonizing new habitats?
- *Environment.* What are the habitat types where the species is found, and how much area is there of each? What is the condition of each habitat? Does the environment provide enough of the resources that the species needs to survive? How abundant are the competitors, predators, and pests of the species? How variable is the environment in time and space? How frequently is the environment affected by catastrophic disturbance?
- *Demography.* How long do established adults live? What are the mortality rates of adults under favorable and unfavorable conditions? Do individuals grow quickly or slowly? What is the age and size at first reproduction? How many offspring are produced per individual? Does the population have a mixture of juveniles and adults, indicating that recruitment of new individuals is occurring?

Gathering Natural History Information

The basic information needed for an effort to conserve a species or determine its status can be obtained from three major sources.

- *Published literature.* Other people have often studied the same rare species or related species, or have investigated the species' habitat type. Library indices such as *Biological Abstracts* or *Zoological Record*, often accessible by computer, provide easy access to a variety of books, articles, and reports. Sometimes sections of the library will have related material shelved together, so finding one book leads to other books. Asking biologists and naturalists for ideas on references is another way to locate published materials. Also, once one key reference is obtained, its bibliography can often be used to discover earlier useful references. The *Science Citation Index*, available in many libraries, is another valuable tool for tracing the literature forward in time; for example, by looking in the current *Science Citation Index* for the name of W. K. Kenyon, who wrote several important papers about the Hawaiian monk seal between 1959 and 1981, recent scientific papers on the Hawaiian monk seal that cited his works can be located.
- *Unpublished literature.* A considerable amount of information

in conservation biology is contained in unpublished reports by individuals, government agencies, and conservation organizations, such as national and regional forest and park departments, government fisheries and wildlife agencies, The Nature Conservancy, the International Union for the Conservation of Nature, and the World Wide Fund for Nature. This so-called "gray literature" is sometimes cited in published literature or mentioned by leading authorities in conversations and lectures. For example, the unpublished Tropical Forest Action Plans produced by individual countries are some of the most comprehensive sources of information on conservation in the tropics. Often, a report known through word of mouth can be obtained through direct contact with the author or conservation organizations; conservation organizations are also sometimes able to supply additional reports not found in the published literature. (A list of such information sources is found in the Appendix.)

- *Fieldwork.* The natural history of a species usually must be learned through careful observations in the field. Fieldwork is necessary because only a tiny percentage of the world's species have been studied, and because the ecology of many species may change from one place to another. Only in the field can the conservation status of a species, and its relationships to the biological and physical environment, be determined. Fieldwork can be time-consuming, expensive, and physically arduous for some species, such as the polar bear, the humpback whale, or tropical canopy species, but it can also be exhilarating and deeply satisfying. There is a long tradition, particularly in Britain, of dedicated amateurs conducting excellent studies of species in their immediate surroundings with minimal equipment and financial support. While much natural history information can be obtained through careful observation, many of the techniques are technical and are best learned by studying under the supervision of an expert or by reading manuals.

The need for natural history information is highlighted by a recent paper on the conservation of the red panda (*Ailurus fulgens*). Despite the attractive appearance of this species, its unique taxonomic status, and the threat of extinction over much of its Himalayan range, there was virtually no information about the autecology of the red panda until a recent study from Lantang National Park in Nepal (Yonzon and Hunter 1991). This study showed that the red panda is a specialist on fir-jhapra bamboo forests between 2800 and 3900 meters in altitude, a habitat rare within the park. The population of red pandas is probably less than 40 in the park and is divided into at least four subpopulations. Red pandas produce only one cub per year, and these cubs suffer from a high mortality rate, mostly caused when cattle frighten or accidentally trample them. The red panda has a low-quality diet consisting mainly of bamboo leaves, seasonally supplemented with fruits and mushrooms. This combination of natural history and population information demonstrates that the precarious ex-

BOX 20 CHEESE, TOURISTS, AND RED PANDAS IN THE HIMALAYAS

Habitat degradation regularly occurs in many protected areas, as illustrated by Langtang National Park in the Nepal Himalayas (Yonzon and Hunter 1991). This park includes 1710 km² of a fragile mountain ecosystem that includes habitat for the endangered red panda and snow leopard. The human population of 30,000 consists mainly of farmers. The farmers raise livestock, principally the chauri, a hybrid between yak and hill cow, to produce milk to supplement their diet and for sale. In order to expand the cheese industry, the government has sponsored a program to encourage the farmers to increase their cattle herds to produce more milk. This program has resulted in local people being drawn into activities that are not sustainable uses of the ecosystem. First of all, the density of cattle is too high for the park, resulting in overgrazing of tree seedlings, trampled vegetation, soil compaction, and poor health for the cattle. Second, the presence of cattle, herders, and dogs has a variety of detrimental effects on red pandas and other wildlife. And

third, the fuelwood needed to process the milk into cheese exceeds the sustainable wood supply of the region, resulting in progressive deforestation. The park is capable of producing 213,000 kg of wood per year, but the annual demand by villagers for their own use is 169,000 kg, while tourists

The red panda, *Ailurus fulgens*, also known as the lesser panda. Although not as widely recognized among the general public as the giant panda (to which it is distantly related), the shy and attractive red panda is equally in danger of extinction. (Photograph by Jessie Cohen, National Zoological Park, Smithsonian Institution.)

and their porters use another 44,000 to provide cooked food, hot water, and warm rooms, and the cheese processing industry uses an additional 100,000 kg, resulting in a substantial overharvesting of the fuelwood resource.

The truly strange part of the story is that the cheese is almost all eaten by foreign tourists, since at $4.20 per kg, it is too expensive for most Nepalis. The irony is that tourists coming to see the scenic beauty of the Himalayas are contributing to the region's destruction by consuming an unsustainable food product and by insisting on hot water and warm rooms while on trekking expeditions. One possible solution is for the chauri herds to be reduced in size but the price of the cheese to be increased substantially, so that the integrity of the ecosystem can be maintained and the farmers can still earn a cash income (Yonzon and Hunter 1991). Is it such a radical idea for tourists to live as the local people do, eating less cheese and getting used to the cold, for the brief duration of their visit to this beautiful country?

istence of the red panda in the Himalayas is due to its specialized habitat requirements, low density, population fragmentation, low-quality diet, low birth rate, and vulnerability to human disturbance. The report suggests that effective conservation measures might include reducing the amount of cattle grazing in red panda areas (see Box 20), protecting the red panda itself, and protecting its specialized habitat.

Monitoring Populations

The key to learning the status of a rare species of special concern is to census the species in the field and monitor its populations over time. By repeatedly censusing a population on a regular basis, changes in the population over time can be determined. Long-term census records can help to distinguish long-term population trends of increase or decrease, possibly caused by human disturbance, from short-term fluctuations caused by variations in weather or unpredictable natural events (Pechmann et al. 1991; Hellawell 1991). Monitoring is effective at showing the response of a population to a change in its environment; for example, a decline in an orchid species was shown to be connected with heavy cattle grazing in its habitat (see below).

Monitoring studies are increasing dramatically as government agencies and conservation agencies have become more concerned with protecting rare and endangered species (Davy and Jeffries 1981; Palmer 1986, 1987; Goldsmith 1991). Some of these monitoring studies may be mandated by law as part of management efforts (Hellawell 1991). A review of projects monitoring rare and endangered plants in the United States showed a phenomenal increase in the number of research projects initiated from 1974 to 1984; only one project was initiated in the three years from 1974 to 1976, while over 120 projects were initiated from 1982 to 1984 (Palmer 1987). The most common types of monitoring projects were inventories (40%) and population demographic studies (40%), with survey studies somewhat less frequently used (20%). Statistical analysis of monitoring studies often requires special care, since many of the studies were not designed with the conservation of species as their original goal (Eberhardt and Thomas 1991; Usher 1991).

An **inventory** is a count of the number of individuals present in a population. By repeating an inventory over successive time intervals, it can be determined whether the population is stable, increasing, or decreasing in numbers (Figure 12.2). An inventory is an inexpensive and straightforward method for conservation purposes. An inventory of the only sweet bay magnolia (*Magnolia virginiana*) population in Massachusetts, at Ravenswood Park (Primack, Del Tredici, and Hen-

(A)   (B)

12.2 Monitoring populations requires specialized techniques suited to each species. (A) An ornithologist checks the health and weight of a piping plover on Cape Cod. Note the identification band on the bird's leg. (Photograph by Laurie McIvor.) (B) At Bako National Park, Sarawak, on the island of Borneo, marked trees in permanent research plots are measured and monitored for growth and survival. Here Forest Department staff measure a tree for its girth at breast height. (Photograph by R. Primack.)

dry 1986), might answer such questions as: How many plants exist at present? Has the population been stable in numbers during the period for which inventory records exist? Has the population increased during the last 15 years when the overstory pine trees were cut to increase light levels? Are the plants flowering and fruiting, and are any seedlings present?

Inventories conducted over a wider area can help to determine the range of a species and its areas of local abundance. By repeating such inventories over time, changes in the range of species can be highlighted. The most extensive inventories have been carried out in the British Isles by a large number of local amateur naturalists supervised by professional societies (Harding 1991). The most detailed mapping efforts have involved plants, lichens, and birds, with presence or absence recorded in 10-km squares covering the British Isles. The Biological Records Centre (BRC) at Monks Wood Experimental Station maintains and analyzes the 4.5 million distribution records relating to 16,000 species.

A population **survey** involves using a repeatable sampling method to estimate the density of a species in a community. An area can be divided into sampling segments and the number of individuals in each segment counted. These counts can then be used to estimate the actual population size. For example, the number of Venus's flytrap plants can be counted in a series of 100 m × 2 m transects through a North Carolina savannah to establish the overall density of plants in the savannah and then estimate the total population size. If there are an average of 30 plants in each of four such transects, the density would be 30 plants per 200 m^2. In a total area of 10,000 m^2, the population estimate would be 1500 plants. Similar methods can be used for different species in a variety of ecosystems; for instance, the number of crown-of-thorns starfish can be counted in a series of 10 m $\times$ 10 m quadrats to estimate the total starfish population on a coral reef. The number of bats caught in mist nets per hour, or the density of a particular crustacean species per liter of seawater, can also be counted. Survey methods are used when a population is very large or its range is extensive. Although these methods are relatively time-consuming, they are a methodical and repeatable way to examine a population and determine whether it is changing in size.

Survey methods are particularly valuable when there are stages in a species' life cycle that are inconspicuous, tiny, or hidden, such as the seed and seedling stages of many plants or the larval stages of aquatic invertebrates (Hutchings 1991). In the case of the magnolia population, a series of 100 m $\times$ 2 m transects through the population could be used to determine the density of seedlings on the ground. Soil samples could be taken at fixed survey points and examined in the laboratory to determine the density of magnolia seeds in the soil. Disadvantages of survey methods are their expense (chartering a vessel to sample deep-sea species), technical difficulty (identifying poorly known immature stages in the life cycle), and possible inaccuracy (sampling may miss or include infrequent aggregations of species). This is particularly true in the marine environment (Grassle 1991).

Demographic studies follow known individuals in a population to determine their rates of growth, reproduction, and survival. Individuals of all ages and sizes must be included in such a study. Either the whole population or a subsample can be followed. In a complete population study, all individuals are counted, aged if possible, measured for size, sexed, and tagged or marked for future identification; their position on the site is mapped, and tissue samples are collected for genetic analysis. The techniques used to conduct a population study vary depending on the characteristics of the species. Each discipline has its own technique for following individuals over time; ornithologists band birds' legs, mammalogists often attach tags to an animal's ear, and botanists attach aluminum tags to trees (see Goldsmith 1991). Information from demographic studies can be used in life history formulae to calculate the rate of population change and to identify critical stages in the life cycle (Caswell 1989; Menges 1986). Demographic studies provide the most information of any monitoring method, and when analyzed thoroughly can suggest ways in which a site can be managed to ensure population persistence. The disadvantages of demographic studies are that they are often time-consuming, expensive, and require repeated visits to a site and a knowledge of the species' life history. In the case of the magnolia population, demographic data can be used to predict whether the population will be present at different dates in the future and what the population size will be. If the population is predicted to go extinct, estimates can be made of the extent to which the survival and reproductive rates need to be increased through site management to maintain or enlarge the current population.

Demographic studies can provide information on the age structure of a population. A stable population typically has an age distribution with a characteristic ratio of juveniles, young adults, and older adults. An absence or low number of any age class, particularly of juveniles, may indicate that the population is in danger of declining. Similarly, a large number of juveniles and young adults may indicate that the population is stable or even expanding. However, it is difficult to determine the age of individuals of many species, such as plants, fungi, and colonial invertebrates. A small individual may be either young or just slow-growing and old; a large individual may be either old or unusually fast-growing but young. For these species, the distribution of size classes is often taken as an approximate indicator of population stability, but this needs to be confirmed by following individuals over time to determine rates of growth and mortality. For many long-lived species, such as trees, the establishment of new individuals in the population is an episodic event; an occasional year with abundant reproduction may be followed by many years of low reproduction. Careful analysis of long-term data, or of changes in the population over time, is often needed to distinguish short-term fluctuations from long-term trends.

In general, a population is stable when the growth rate is zero. This occurs when the average birth rate equals the average death rate. While a population with an average growth rate of zero is expected to be stable over time and a growth rate above zero should lead to an expanding population, random variation in population growth rates among years can lead to population decline and extinction even with a positive average growth rate, as explained in Chapter 11.

Demographic studies can also indicate the spatial characteristics of a species, which might be very important to maintaining the vitality of separate populations. The number of populations of the species, movement among populations, and the stability of these populations in space and time are all important considerations, particularly for species that occur in metapopulations. Demographic studies can identify the core sites that support large, fairly permanent populations and supply colonists to satellite areas.

Reproductive characteristics of populations such as sex ratio, percentage of breeding adults, and monogamous or polygamous mating systems will also affect the success of conservation strategies and should be thoroughly analyzed. For example, a strategy to increase genetic diversity in a highly inbred population, such as the lions inhabiting the Ngorongoro Crater of Tanzania (Packer 1992), might include introducing individuals from outside this population in the hope that they will mate with the inbred animals. If the new individuals are not selected to fit into the social dynamics of the current population, however, they may not breed and may even be driven out or killed by the native population.

Finally, demographic studies can supply clues to the carrying capacity of the environment. These studies are important in determining how large a population the environment can support before the environment deteriorates and the population declines. Nature reserves may have abnormally large populations of certain species due to the recent loss of adjoining habitat or the inability of individuals to disperse away from the reserve. With the limited space available in nature reserves, many of which are expected to support large populations over long periods of time, such data are crucial to preventing population and environmental stress, particularly in circumstances where natural population control mechanisms, such as predators or diseases, have been eliminated by humans.

Some examples of monitoring studies follow.

- *Butterflies.* Long-term trends in the abundance of butterfly species in the Netherlands were analyzed using an inventory of 230,000 butterfly records from 1901 to 1986 (van Swaay 1990). Of 63 species analyzed, 29 species (46%) have declined in abundance or gone extinct. Most of the species that have declined in this century occupied nutrient-poor grasslands that have now largely been converted to agriculture. Also, most of these declining species reproduce only once a year and overwinter as larvae, so they are vulnerable to land-use changes.
- *Hawaiian monk seals.* Population inventories of the Hawaiian monk seal, *Monachus schauinslandi*, have documented a decline from almost 100 adults in the 1950s to less than 14 in the late 1960s (Figure 12.3; Gerrodette and Gilmartin 1990). The number of pups similarly declined during this period. On the basis of these trends, the Hawaiian monk seal was declared an endangered species under the U.S. Endangered Species Act in 1976, and conservation efforts were implemented that reversed the trend (Ackerman 1992).
- *Griffon vulture.* A severe decline in the griffon vulture (*Gyps fulvus*) in northern Spain during the 1950s and 1960s led to strict protection of the species and its nesting sites, as well as

12.3 Inventories of Hawaiian monk seal populations on Green Island, Kure Atoll (black trace), and on Tern Island, French Frigate Shoals (gray trace). Population counts were plotted from either a single count, the mean of several counts, or the maximum of several counts. Note the effect on seal populations of the Coast Guard stations on the islands. (From Gerrodette and Gilmartin 1990.)

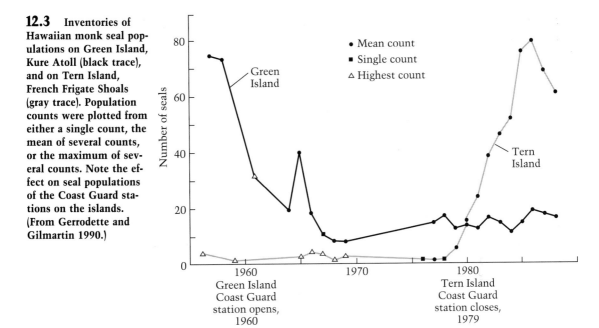

banning the use of strychnine to kill pest animals (which were then eaten by griffon vultures). Repeated inventories have shown a steady recovery of the species in northern Spain from 282 breeding pairs in 23 colonies during 1969–1975, to 1097 breeding pairs in 46 colonies by 1989 (Donázar and Fernandez 1990).

• *Marine mollusks.* In Transkeii, South Africa, coastal people collect and eat marine mollusks, such as the brown mussel, the abalone, and the turban shell, to supplement their diet (Lasiak and Dye 1989; Lasiak 1991). A survey method was used to determine whether traditional collecting methods are likely to deplete shellfish populations; the frequency and size distribution of mollusks was compared in protected and exploited rocky areas. Even though collection depleted the adult populations in exploited areas, they were quickly replaced by larvae, probably due to immigration from nearby protected areas and adjacent, inaccessible subtidal areas.

• *Early spider orchid.* The early spider orchid (*Ophrys sphegodes*) has shown a substantial decline in range during the last half century in Britain. A nine-year demographic study of this species showed that the plants were unusually short-lived for perennial orchids, with only half of the individuals surviving beyond two years (Hutchings 1987a,b). This short half-life makes the species unusually vulnerable to unfavorable conditions. In one population in which the species was declining in numbers, demographic analysis highlighted soil damage by cattle grazing as the key element in the population decline. A change in land management to sheep grazing only during times when plants were not flowering and fruiting has allowed the population to make a substantial recovery.

Monitoring studies are playing an increasingly important role in conservation biology. Monitoring, which has a long history in temperate countries, particularly Britain (Goldsmith 1991), is now being done on a more widespread basis in tropical countries. Some of the most elaborate projects involve establishing permanent research plots in tropical forests, such as the 50-ha site at Barro Colorado Island in Panama, to monitor changes in species and communities over time (Primack 1992; Condit et al. 1992; Dallmeier 1992). These studies have shown that many tropical tree and bird species are more dynamic in numbers than had previously been suspected (Primack and Hall 1992; Loiselle and Black 1992; Condit et al. 1992; Bierregaard et al. 1992).

Population Viability Analysis

Population viability analysis (PVA) is an extension of demographic analysis that focuses on determining whether a species has the ability to persist in an environment (Menges 1991; Shaffer 1991; Boyce 1992). By looking at the range of requirements that the species has and the resources available in its environment, vulnerable stages in the natural history of the species can be identified (Gilpin and Soulé 1986). PVA can be useful in considering the effects of habitat loss, habitat fragmentation, and habitat deterioration of a rare species. Though PVA is still developing as an approach for examining species persistence, and does not yet have a specific methodology or statistical framework (Thomas 1990; Shaffer 1990), its goals of systematically and comprehensively examining species data are natural extensions of autecology, natural history research, and demographic studies. In this sense, the comprehensive research efforts made already to manage populations of rare species, such as the black-footed ferret, Furbish's lousewort, and the Florida panther, could be called forerunners of PVA. Population viability analysis may be a particularly useful approach for investigations of species with fluctuating populations best studied at the metapopulation level (Murphy et al. 1990, Menges 1990). This approach is being developed to the point at which accurate predictions can be made concerning the ability of a rare species to survive in its altered environment, and population viability analysis is becoming a central part of conservation strategies.

Minimum Viable Population Size

Many species are threatened with extinction as humans increasingly dominate the landscape. The question for land managers and conservation biologists is how to prevent species from going extinct. The usual answer is that as many individuals of the endangered species should be preserved as possible within as large a reserve as possible. However, this general statement does not provide specific guidelines to assist planners, land managers, politicians, and wildlife biologists who are trying to protect species and who have to reconcile conflicting demands on finite resources ("owls versus jobs"). For example, to preserve the red-cockaded woodpecker, does long-leaf pine habitat need to be preserved for 50, 500, 5000, 50,000, or even more individuals?

It is well known that small populations are particularly vulnerable to extinction. Extending this idea leads to the concept of a minimum number of individuals necessary to maintain a population of a species. In a ground-breaking paper, Shaffer (1981) tried to provide a specific technique for determining a species' **minimum viable population (MVP):** "A minimum viable population for any given species in any given habitat is the smallest isolated population having a 99% chance of remaining extant for 1000 years despite the foreseeable effects of demographic, environmental, and genetic stochasticity, and natural catastrophes." In other words, an MVP is the smallest population size that can be predicted to have a very high chance of persisting for the next 1000 years. Shaffer emphasized the tentative nature of this definition, saying that the survival probabilities could be set at 95% or 99% and that the time frame might similarly be adjusted, for example to 100 years or 500 years. The key point is that the minimum viable population size allows a quantitative estimate to be made of how to preserve a species. Combining known distribution patterns, population turnover rates, and population data with computer simulations of extinction possibilities, biogeographical models, and models of genetic drift and inbreeding (see Chapter 11) provide valuable approaches to determining the MVP (Lande 1988a; Dobson and Lyles 1989; Harris and Allendorf 1989; Boyce 1992; Menges 1992).

Shaffer (1981) compares MVP protection efforts to flood control efforts. In planning flood control systems and in regulating building on wetlands, it is not sufficient to use average annual rainfall as a guideline. We recognize the need to plan for severe flooding, which may occur only once every 50 years. Likewise, in protecting natural systems, we understand that certain catastrophic events, such as massive hurricanes, earthquakes, forest fires, and eruptions of volcanos, may occur at even longer intervals. To plan for the long-term protection of endangered species, we not only have to provide for the requirements of the species in average years, but also for the needs of the species in exceptional years. In drought years, for instance, animals may migrate well beyond their normal ranges to obtain the water needed to survive.

Obtaining an accurate estimate of the minimum viable population size for a particular species may require a detailed demographic study of the population (Thomas 1990). Estimating the MVP also requires some educated guesswork about the probability of natural catastrophes. Collecting this data may be overly time-consuming when conservation decisions have to be made quickly. An alternative rule of thumb is to try to protect about 1000 individuals for vertebrates, as this number seems adequate to preserve genetic variability (Lande 1988a); protecting 500 individuals of a vertebrate species may be adequate to prevent the population from dropping below 100 individuals in catastrophic years and entering into an extinction vortex. For species with extremely variable population sizes, such as certain invertebrates and annual plants, Lande suggests that protecting a population of 10,000 individuals is probably an effective strategy.

More field evidence supporting the need for large populations to ensure population persistence comes from long-term studies of birds on the Channel Islands; only birds having populations of greater than 100 pairs had a greater than 90% chance of surviving for 80 years. On the other hand, there is no need to give up entirely on small populations: many populations of birds have apparently survived for 80 years with 10 or fewer breeding pairs (Jones and Diamond 1976). Northern elephant seals have recovered to a population of 30,000 individuals after being reduced by hunting to only about 20 individuals (Bonnell and Selander 1974). Exceptions notwithstanding, large populations are needed to protect most species, and species with small populations are in real danger of going extinct.

Once a minimum viable population size has been established for a species, the **minimum dynamic area (MDA),** the area of suitable habitat necessary for maintaining the minimum viable population, can be estimated. The MDA can be estimated by studying the home range size of individuals and bands of endangered species (Thiollay 1989).

Two Case Studies

Attempts to utilize population viability analysis have already begun. One of the most thorough examples of PVA, combining both ge-

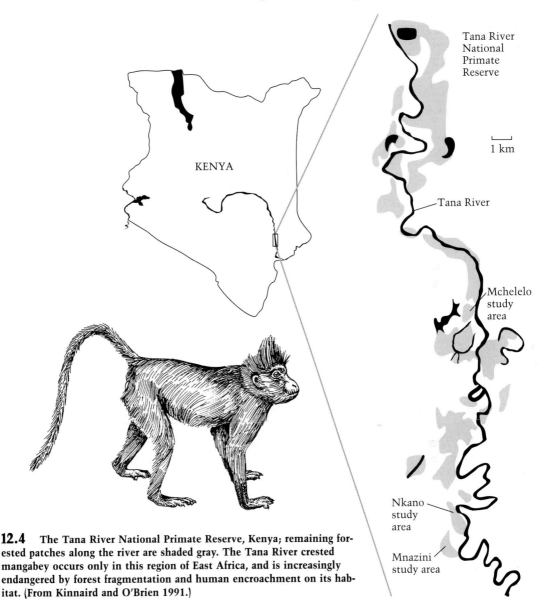

Tana River
National
Primate
Reserve

1 km

Tana River

Mchelelo
study
area

Nkano
study
area

Mnazini
study
area

KENYA

12.4   The Tana River National Primate Reserve, Kenya; remaining forested patches along the river are shaded gray. The Tana River crested mangabey occurs only in this region of East Africa, and is increasingly endangered by forest fragmentation and human encroachment on its habitat. (From Kinnaird and O'Brien 1991.)

netic and demographic analyses, is a study of the Tana River crested mangabey (*Cercocebus galeritus galeritus*), an endangered primate confined to the floodplain forests in a nature reserve along the Tana River in eastern Kenya (Figure 12.4; Kinnaird and O'Brien 1991). The species is naturally rare and restricted in range. As its habitat has become reduced in area and fragmented by agricultural activities in the last 15–20 years, the species has experienced a decline in overall population size of about 50%, as well as a decline in the number of groups. While the number of mangabeys in 1989 was about 700 individuals, the effective population size (see Chapter 11) is only about 100, due to the large proportion of nonreproductive individuals and the variation in number of offspring produced by different individuals. With such a low effective population size, the mangabey is in danger of losing significant amounts of its genetic variation. To maintain an effective population size of 500 individuals, a number considered sufficient to maintain genetic variability, a population of about 5000 mangabeys would have to be maintained. In addition, a demographic analysis of the population suggests that in the current situation, the probability of the population going extinct over the next 100 years is 40%. To assure that the population has a 95% probability of persisting for 100 years, based just on demographic factors, the population size would have to be almost 8000 individuals.

Both the genetic and the demographic analyses suggest that the long-term future of the present mangabey population is bleak. Given the restricted range and habitat of the species, increasing the population size to 5000 to 8000 individuals is probably unrealistic. A management plan that combines increasing the area of protected forests, enrichment of existing forests, and establishment of corridors to facilitate movement between forest fragments might increase the survival probability of the Tana River crested mangabey.

The grizzly bear, *Ursus arctos*, has important symbolic value in capturing the essence of wilderness, and efforts are being made to maintain populations in at least four areas of the western United States. The species presents special management problems in national parks, since it is large (up to 1800 kg) and extremely powerful. Grizzly bears will attack and even kill humans that accidently or deliberately come too close, violating the animal's "security zone"; this zone may extend up to 100 m in radius for a mother with a cub. Population viability analyses of grizzly bears have been made using estimates of environmental and demographic stochasticity (Shaffer and Samson 1985). These models predict that grizzly populations need at least 50 to 90 animals to be viable, defined in this case as a 95% chance of persisting for 100 years (Knight and Eberhardt 1985; Shaffer and Samson 1985). Efforts being made by park managers to increase the park populations will reduce the probability of extinction. Because of the wide-ranging nature of grizzly bears, however, even the largest national parks in the United States may be insufficient in area to maintain long-term populations of the species (Newmark 1985). Across its range, many small, isolated populations of grizzly bears will go extinct, particularly due to inbreeding depression and genetic drift, unless gene flow is encouraged by facilitating movement among populations (Allendorf and Servheen 1986; Maguire and Servheen 1992).

Long-term Monitoring of Species and Ecosystems

Long-term monitoring of ecosystem processes (temperature, rainfall, humidity, soil acidity, water quality, discharge rates of streams, soil erosion, etc.), communities (species present, amount of vegetative cover, amount of biomass present at each trophic level, etc.), and population numbers (number of individuals present of particular species) is necessary to protect biological diversity, since it is otherwise difficult to distinguish normal year-to-year fluctuations from long-term trends (Magnuson 1990; Primack 1992). For example, many amphibian, insect, and annual plant populations are highly variable from year to year, so many years of data are required to determine whether a particular species is actually declining in abundance over time or merely experiencing a number of low-population years that are in accord with its regular pattern of variation. In one instance, a salamander species that initially appeared to be very rare on the basis of several years of low breeding numbers turned out to be quite common in a favorable year for breeding (Pechmann et al. 1991).

Another challenge in understanding change is that effects may lag for many years behind their initial causes. For example, acid rain may weaken and kill trees over a period of decades, increasing the amount of soil erosion into adjacent streams and ultimately making the aquatic environment unsuitable for certain larvae of rare insect species. In such a case, the cause (acid rain) would have occurred decades before the effect (insect decline) actually was detectable.

Acid rain, global climate change, vegetation succession, nitrogen deposition, and invasion of exotic species are all examples of processes that cause long-term changes in biological communities but are often hidden from our short-term perspective (Figure 12.5). Some long-term records are available from weather stations, annual census counts of birds, forestry plots, water authorities, and old photographs of vegetation, but the number of long-term monitoring efforts for biological communities is inadequate for most conservation purposes. To remedy this situation many scientific research stations have begun to implement programs for monitoring ecological change over the course of decades and centuries. One such program is the system of 172 Long-Term Ecological Research (LTER) sites established by the U.S. National Science Foundation (Swanson and Sparks 1990) (Figure 12.6). Other programs include the United Nations Man and the Biosphere system of Biosphere Reserves and the increasing number of community-level permanent research plots being established in tropi-

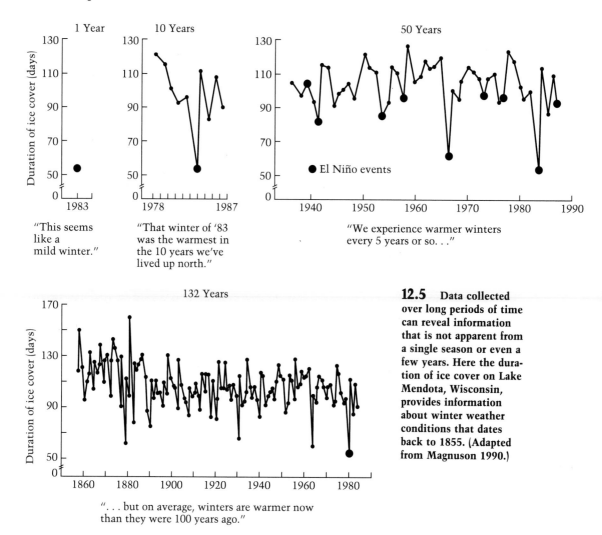

12.5 **Data collected over long periods of time can reveal information that is not apparent from a single season or even a few years. Here the duration of ice cover on Lake Mendota, Wisconsin, provides information about winter weather conditions that dates back to 1855. (Adapted from Magnuson 1990.)**

cal forests (see Chapter 20; Primack 1992; Condit et al. 1992). A major purpose of these sites is to gather essential data on ecosystem functions and biological communities that can be used to monitor changes in the natural communities. The sites also provide an early warning system for disruption or decline of ecosystem functions. Magnuson (1990) has expressed the need as follows:

> In the absence of the temporal context provided by long-term research, serious misjudgments can occur not only in our attempts to understand and predict change in the world around us, but also in our attempts to manage our environment. Although serious accidents in

12.6 The Long-Term Ecological Research (LTER) program focuses on time scales ranging from years to centuries in order to understand changes in the structure, function, and process of biological communities that are not apparent from short-term observations. (From Magnuson 1990.)

	Years	Research Scales	Physical Events	Biological Phenomena
10^5	100 Millennia			Evolution of species
10^4	10 Millennia	Paleoecology and limnology	Continental glaciation	Bog succession Forest community migration
10^3	Millennium		Climate change	Species invasion Forest succession
10^2	Century		Forest fires CO_2-induced climate warming	Cultural eutrophication Population cycles
10^1	Decade			Prairie succession
10^0	Year		Sun spot cycle El Niño events Prairie fires Lake turnover	Annual plants Seasonal migration Plankton
10^{-1}	Month		Ocean upwelling	succession
10^{-2}	Day	Most ecology	Storms Daily light cycle Tides	Algal blooms Daily movements
10^{-3}	Hour			

LTER brackets the rows from 10^2 Century through 10^{-1} Month.

an instant of mismanagement can be envisioned that might cause the end of Spaceship Earth (sensu Fuller 1970), destruction is even more likely to occur at a ponderous pace in the secrecy of the invisible present.

Summary

1. Conservation biologists often need to determine whether a population of an endangered species is stable, increasing, fluctuating, or declining. A declining population may require human intervention to save it from extinction. In some cases, a species is characterized by a shifting mosaic of temporary populations, known as a metapopulation.

2. The key to protecting and managing a rare or endangered species is understanding its natural history. The essential knowledge covers the species' morphology, physiology, behavior, genetics, distribution, environment, and demography. This information can come from published and unpublished literature and from fieldwork. Long-term monitoring of a species in the field can reveal temporal changes in population size, and can help to distinguish short-term fluctuations from long-term decline. Monitoring involves counting the population, and in more complete studies, following individuals over time. Demographic studies are particularly valuable in calculating the long-term stability of the population.

3. Population viability analysis uses demographic, genetic, environmental, and natural catastrophe data to estimate the probability of a population being able to persist in an environment. The goal of this analysis may be to determine the minimum viable population size, sometimes defined as the population size necessary to give the population a 99% chance of persisting for 1000 years. Population viability analysis is still developing needed methodologies and statistical techniques, but it shows considerable promise.

4. Long-term monitoring efforts are now being developed throughout the world to follow changes in populations and communities over the course of decades and centuries. These programs will provide an early-warning system for damage to species, communities, and ecosystem functions.

Suggested Reading

Boyce, M. S. 1992. Population viability analysis. *Annual Review of Ecology and Systematics* 23: 481–506. Review of current developments in this growing field.

Caswell, H. 1989. *Matrix Population Models: Construction, Analysis, and Interpretation.* Sinauer Associates, Sunderland, MA. Key reference for demographic analysis.

Goldsmith, B. (ed.). 1991. *Monitoring for Conservation and Ecology.* Chapman and Hall, New York. Purpose and methods of conservation monitoring, clearly explained.

Kinnaird, M. E., and T. G. O'Brien. 1991. Viable populations for an endangered forest primate, the Tana River Crested Mangabey (*Cercocebus galeritus galeritus*). *Conservation Biology* 5: 203–213. Excellent case study demonstrating application of population viability analysis.

Magnuson, J. J. 1990. Long-term ecological research and the invisible present. *BioScience* 40: 495–501. Long-term research provides important insights not apparent in short-term studies.

Menges, E. S. 1991. The application of minimum viable population theory to plants. *In* D. A. Falk and K. E. Holsinger (eds.), *Genetics and Conservation of Rare Plants*, pp. 45–61. Oxford University Press, New York. MVP theory can be used to estimate how many individuals need to be protected.

Murphy, D., K. E. Freas and S. B. Weiss. 1990. An environment-metapopulation approach to population viability analysis for a threatened invertebrate. *Conservation Biology* 4: 41–51. Excellent introduction to the metapopulation concept.

Packer, C. 1992. Captives in the wild. *National Geographic* 181 (April): 122–136. The story of the Ngorongoro Crater lions.

Primack, R. B. and P. Hall. 1992. Biodiversity and forest change in Malaysian Borneo. *BioScience* 42: 829–837. The conservation, policy, and management implications of tropical community studies. See other monitoring studies in the same issue.

Shaffer, M. L. 1981. Minimum population sizes for species conservation. *BioScience* 31: 131–134. A key paper.

PART

Practical Applications

CHAPTER 13

Establishing Protected Areas

The great majority of the world's species exist only in the wild, so preserving habitats that contain biological communities is the most effective way to preserve biological diversity. One could even argue that it is the only way to preserve species, because we do not have the resources or knowledge to maintain the majority of the world's species in captivity. Case studies in conservation research typically mention multiple threats to biological diversity; damage to endangered species and unique habitats often comes from a variety of causes (Ehrlich and Ehrlich 1981; Myers 1987). A wetland that provides habitat for unique aquatic plants and animals may be affected by encroaching development, water pollution, hunting by local people, and alteration of drainage patterns. Determining the best way to preserve the species in the wetland not only requires a knowledge of the natural history of the species and the habitat, but a clear grasp of the needs of the local human population and the conservation and development policies of the government. Protecting the habitat where threatened species occur is one of the most effective mechanisms for preserving biological diversity; however, considerable political will and financial resources are required not only to establish conservation areas but to insure that they achieve their purposes once established.

Habitat Conservation

One of the most critical steps in protecting biological communities is establishing legally designated protected areas. While legislation and purchase of land will not by themselves insure habitat preservation, they represent an important starting point. Protected areas can be established in a variety of ways, but the two most common mechanisms are government action (often at a national level, but

301

also regionally or locally) and purchases of land carried out by private conservation organizations. Protected areas can be governed by laws that allow varying degrees of commercial resource use, traditional use by local people, and recreational use. Many protected areas have been established by private conservation organizations, such as the Nature Conservancy (Box 21) and the Audubon Society. An increasingly common pattern is that of a partnership between the government of a developing country in the tropics and international conservation organizations, multinational banks, and governments of developed countries; in such partnerships the conservation organizations provide funding, training, and scientific and management expertise to assist the tropical country in establishing new national parks (see Chapter 21). The pace of this collaboration is accelerating with the recent creation of a $1.5 billion Global Environment Facility, implemented by the World Bank and agencies of the United Nations.

Protected areas have also been established by traditional societies who wish to maintain their way of life. National governments have recognized the rights of traditional societies in many countries, including the United States, Canada, Brazil, and Malaysia—though often not before conflict in the courts, in the press, and on the land. In many cases, assertions of traditional rights have involved violent confrontations with the existing authorities, sometimes with loss of life (Poffenberger 1990; Gadgil and Guha 1992).

BOX 21 ECOLOGISTS AND REAL ESTATE EXPERTS MINGLE AT THE NATURE CONSERVANCY

For nearly a century, nonprofit organizations such as the Sierra Club and the Audubon Society have sought to encourage the conservation of wild species and habitats. The number of these organizations has increased in recent decades, at the local, national, and international levels. Of the many organizations that now fight to protect biological diversity, The Nature Conservancy (TNC) is set apart by a unique approach that applies the methods of private business to accomplish the conservation of rare species and habitats. Simply put, The Nature Conservancy either buys threatened habitat outright or shows landowners how managing their land for conservation can be as profitable as developing it.

Founded in 1951, The Nature Conservancy reached a membership of over 700,000 by 1993. TNC is not as widely known as some conservation organizations, nor is it as large or as vocal. TNC advocates a nonconfrontational, business-like approach that contrasts with the methods of some high-profile, activist conservation groups such as Greenpeace and EarthFirst! (Grove 1988). Still, its methods have been quietly successful: in the United States alone, over 2.8 million hectares (6.9 million acres) have been set aside through the work of TNC, much of which has become state or national parks or wildlife refuges. Outside the U.S., 8 million hectares have been protected with TNC help.

Furthermore, TNC maintains a fund of over $150 million created from private and corporate donations with which they can make direct land purchases when necessary. Through these methods, the organization has created the largest system of private natural areas and wildlife sanctuaries in the world. One such outstanding property is the 14 hectares of land surrounding the shoals at Pendleton Island on Virginia's Clinch River (Stolzenburg 1992). These 300 meters of shoals have 45 species of freshwater mussels, and may be among the richest localities for mussels in the world. The decision of TNC in 1984 to buy this land and get involved in cleaning up the Clinch River represents an important step in recognizing the importance of aquatic invertebrates to overall biological diversity.

The Nature Conservancy's approach is creative; if TNC cannot purchase land outright for the protection of habitat, it offers alternatives to landowners that make conservation a financially attractive option. Such alternatives include enabling the landowner to obtain tax benefits in exchange for accepting legal restrictions or conservation easements pre-

venting development. In other cases, the landowner donates the land to TNC in exchange for lifetime occupancy, which in essence permits the owner to have a rent- and mortgage-free home and to receive a sizable tax deduction as well (Hoose 1981; Jenkins 1988). TNC also seeks methods that allow the preserves it owns to support themselves financially. The cost of maintaining one South Dakota prairie preserve is partially defrayed by maintaining a resident bison herd. Grazing by bison enhances biological diversity in these grasslands, provided the herd does not become too large; when the herd grows to a size at which overgrazing becomes a possibility, the excess animals are sold. The sale of bison brings roughly $25,000 annually to the preserve (Grove 1988).

In addition, The Nature Conservancy supports and encourages efforts to identify populations of rare and endangered species and to manage biological diversity in the United States (Noss 1987). Often TNC sells or donates land it has purchased to state governments or federal agencies that will protect the land. Some parcels of land are so valuable they have been designated as national wildlife refuges by the government. TNC also promotes state-level conservation through Natural Heritage Programs in all 50 states, which generally are joint ven-

tures between TNC and state governments. The Nature Conservancy provides training and staff to inventory plant and animal populations in each state. The data collected are recorded in a computerized data base located in each state and at TNC headquarters in Arlington, Virginia; with this data base, TNC biologists can keep track of the status of species and populations throughout the nation (Grove 1988). When Natural Heritage Program biologists identify populations of species that are rare, unique, declining, or threatened, the state is able to take appropriate conservation action with assistance from TNC.

The Nature Conservancy is distinguished from other conservation organizations by its determinedly apolitical stance (Grove 1988). TNC is not part of the environmental lobby at either the state or the federal level, and generally refuses to bring lawsuits against projects encroaching on the habitat of endangered species. But the Conservancy's businesslike approach is successful largely because developers, who too often have no great love for environmentalists, respect and understand the Conservancy's methods. The fundamental principle of TNC is, in essence, "Land conservation through private action." So far, the idea has proven to be a sound one.

Levels of Disturbance and Protected Areas

Biological communities vary from a few that are virtually unaffected by human influence, such as communities found on the ocean floor or in the remotest parts of the Amazon rain forest, to those that are heavily modified by human activity, such as agricultural land, cities, and artificial ponds. But even in the most remote areas of the world, human influence is apparent in the form of air pollution and rising carbon dioxide levels; and even in the most modified of human environments there are often remnants of the original biota. Habitats with intermediate levels of disturbance present some of the most interesting challenges and opportunities for conservation biology, since they often cover large areas. Considerable biological diversity may remain in selectively logged tropical forests, heavily fished oceans and seas, and grasslands grazed by domestic livestock (Webster 1989; Redford 1992). When a conservation area is established, the right compromise must be found between protecting biological diversity and ecosystem function and satisfying the immediate and long-term needs of the local human community and the national government for resources.

Once land comes under government control, decisions must be made regarding how much human disturbance will be allowed. Human population density, income levels, and past history of land use will all affect the degree to which biological diversity can be protected. The International Union for the Conservation of Nature has developed a system of classification for protected areas that ranges from minimal to intensive allowed use of the habitat by humans (IUCN 1985):

1. *Scientific reserves* and *strict nature reserves* are strictly protected areas maintained for scientific study, education and environmental monitoring. These reserves allow populations of species to be maintained and ecosystem processes to continue in as undisturbed a state as possible.
2. *National parks* are large areas of scenic and natural beauty maintained for scientific, educational, and recreational use; they are not usually used for commercial extraction of resources.
3. *National monuments and landmarks* are often smaller areas designed to preserve unique areas of special national interest.
4. *Managed wildlife sanctuaries and nature reserves* are similar to strict nature reserves, but some human manipulation may be necessary to maintain the characteristics of the community and some controlled harvesting may be permitted.

5. *Protected landscapes* allow nondestructive uses of the environment by resident people and provide opportunities for tourism and recreation.
6. *Resource reserves* are areas in which resource use is controlled in ways compatible with national policies.
7. *Natural biotic areas* and *anthropological reserves* allow traditional societies to continue to maintain their way of life without outside interference. Often these people will hunt and extract resources for their own use and practice traditional agriculture.
8. *Multiple-use management areas* allow for the sustained production of natural resources, including water, wildlife, grazing for livestock, timber, tourism, and fishing. Often the preservation of biological communities is compatible with these activities.

Of these categories, the first five can be considered truly **protected areas**, with the habitat managed for biological diversity. Areas in the last three categories are not managed primarily for biological diversity but they still may contain many or even most of their original species. These **managed areas** may be particularly significant since they are often much larger in area than protected areas, and since protected areas are often embedded in a matrix of managed areas.

Protected Areas

As of 1989 a total of 4545 protected areas had been designated worldwide, covering a total of 4,846,300 km^2 (Table 13.1; World Conservation Monitoring Centre 1989). Although this amount may seem impressive, it represents only 3.2% of the Earth's land surface. Only 2% of the Earth's surface is in the strictly protected categories of scientific reserves and national parks. The world's largest park is in Greenland and covers 700,000 km^2. Excluding this one mammoth park, only 1.6% of the Earth's surface is strictly protected. The figures for individual countries and continents are only approximate because sometimes the laws protecting national parks and wildlife sanctuaries are not actually enforced, and sometimes sections of resource reserves and multiple-use management areas are carefully protected in practice. Examples of the latter are the sections designated wilderness areas within U.S. national forests. Recent surveys produced by the World Resources Institute (1992) give slightly more encouraging figures of 6931 protected areas covering 6,512,900 km^2 as of 1990.

The momentum to establish protected areas has been increasing

TABLE 13.1

Number of protected areas and the total area of various biogeographical regions they cover

Region[b]	Scientific reserve		National park		National monument		Wildlife reserve		Protected landscape		Total	
	No.	km²	No.	km²	No.	km²	No.	km²	No.	km²	No.	km²
Nearctic	5	11,600	142	1,155,500[c]	32	64,200	259	380,700	40	113,600	478	1,725,600
Palearctic	313	273,100	204	112,300	24	2,000	649	172,800	494	171,700	1684	731,900
Afrotropical	23	17,600	152	574,300	1	<100	260	268,700	8	300	444	860,900
Indomalayan	63	27,900	158	111,300	5	300	411	180,400	39	2,900	676	322,800
Oceanian	17	25,900	10	3,300	0	0	24	19,600	1	<100	52	48,900
Australian	58	23,100	248	192,600	0	0	277	137,500	40	3,700	623	356,900
Antarctic	29	6,500	11	21,000	5	200	85	3,500	0	0	130	31,200
Neotropical	55	63,900	224	423,900	22	2,900	126	244,900	31	32,500	458	768,100
Total	563	449,600	1,149	2,549,200	89	69,600	2,091	1,408,100	653	324,800	4,545	4,846,300

Protected area designation (IUCN designations)[a]

Source: Protected Areas Data Unit, World Conservation Monitoring Centre, May 1989. Table from Reid and Miller 1989.

[a] The table includes all protected areas larger than 1000 hectares, classified in IUCN Management Categories 1 through 5, and managed by the highest accountable authority in the country.

[b] The eight biogeographic regions (Udvardy 1975) are described on page 315.

[c] The Greenland National Park—700,000 km² in size—has a significant effect on comparative statistics, since it is an order of magnitude larger than any other single site in the world.

TABLE 13.2

The rate at which new protected areas have been established throughout the world

Date established	Number of areas[a]	Total area protected (km^2)
Unknown	711	194,395
Pre-1900	37	51,455
1900–1909	52	131,385
1910–1919	68	76,983
1920–1929	92	172,474
1930–1939	251	275,381
1940–1949	119	97,107
1950–1959	319	229,025
1960–1969	573	537,924
1970–1979	1,317	2,029,302
1980–1989	781	1,068,572

Source: From Reid and Miller 1989.

[a] Includes nationally recognized protected areas over 1000 ha in size in the first five IUCN categories.

throughout the twentieth century and reached its peak in the 1970s (Table 13.2). The drop-off during the 1980s is a reflection of the fact that most of the undeveloped land that was not designated for development had already come under protection. Protected areas will never cover a large percentage of the Earth's surface—perhaps 6% or slightly more—due to the perceived needs of human society for natural resources. This limited area of protected habitat emphasizes the biological significance of the 10–20% of the land that is managed for resource production. In the United States, the Forest Service and the Bureau of Land Management together manage 23.5% of the land, while in Costa Rica about 17% of the land is managed as forests and Indian reserves.

The Effectiveness of Protected Areas

If protected areas cover only a small percentage of the total area of the world, how effective will they be at preserving the world's species? Concentrations of species occur at particular places in the landscape: along elevational gradients, at places where different geological formations are juxtaposed, in areas that are geologically old, and at places that have an abundance of critical natural resources (e.g., tropical rain forests; the edges of lakes; salt licks) (Terborgh 1986; Carroll 1992). Often a landscape contains large expanses of a fairly uniform habitat type and only a few small areas of rare habitat types. Protecting biological diversity in this case will probably not depend so much on preserving large areas of the common habitat type as on including representatives of all the habitats in a system of protected areas. Recent conservation management plans for Sarawak on the northwestern coast of Borneo have emphasized the need to distribute new national parks to cover all major vegetation types and biological communities (Kavanagh et al. 1991). The following examples illustrate the potential effectiveness of protected areas of limited extent:

- Parks and wildlife sanctuaries cover only about 8% of Thailand but include 88% of its resident forest bird species (Rand 1985, reported in Reid and Miller 1989).
- The Indonesian government plans to protect populations of all native bird and primate species within its system of national parks and reserves (IUCN/UNEP 1986a). This goal will be accomplished by increasing the coverage of protected areas from 3.5% to about 10% of Indonesia's land area.

TABLE 13.3
Percent of a country's bird species found within protected areas for selected African nations

Country	Percent of national land area protected	Number of bird species	Percent of bird species found in protected areas
Cameroon	3.6	848	76.5
Côte d'Ivoire	6.2	683	83.2
Ghana	5.1	721	77.4
Kenya	5.4	1064	85.3
Malawi	11.3	624	77.7
Nigeria	1.1	831	86.5
Somalia	0.5	639	47.3
Tanzania	12.0	1016	82.0
Uganda	6.7	989	89.0
Zaire	3.9	1086	89.0
Zambia	8.6	728	87.5
Zimbabwe	7.1	635	91.5

Source: From Sayer and Stuart 1988.

13.1 Even though the Ethiopian ground hornbill and many other bird species occur in Africa's national parks and protected areas, the long-term future of these species remains in doubt due to continuing habitat destruction, fragmentation, hunting, and other human activities. (Photograph © San Diego Zoo.)

- In most of the large tropical African countries, the majority of the native bird species have populations inside protected areas (Table 13.3). For example, Zaire has over 1000 bird species, and 89% of them occur in the 3.9% of the land area under protection. Similarly, 85% of Kenya's birds are protected in only 5.4% of the land area included in parks (Sayer and Stuart 1988).
- Thirty large protected areas distributed across tropical Africa collectively include 67 of the 70 species of kingfishers, bee-eaters, rollers, hoopoes and hornbills (Figure 13.1), with 90% of the species found in more than one park and 55% of the species found in more than five parks (IUCN/UNEP 1986b). A similarly high percentage of other vertebrate groups are protected, and over 90% of all vertebrates would be included if only a few more critical protected areas could be created (Sayer and Stuart 1988).
- A dramatic illustration of the importance of small protected areas is given by Santa Rosa Park in northwestern Costa Rica. This park covers only 0.2% of the area of Costa Rica, yet it contains breeding populations of 55% of the country's 135 species of sphingid moth. Santa Rosa Park will be included within the new 82,500-ha Guanacaste National Park, which is expected to have populations of almost every sphingid moth (Janzen 1988).

These examples clearly show that well-selected protected areas can include many if not most of the species in a country. However, the long-term future of many species in these reserves remains in doubt. Populations of many species may be so reduced in size that their eventual fate is extinction (Janzen 1986). Similarly, in isolated reserves, catastrophic events like fires, outbreaks of disease, and episodes of poaching can rapidly eliminate particular species from reserves. Consequently, while the number of species existing in a relatively new park is important as an indicator of the park's potential, the real value of the park is its ability to support viable long-term populations of species. In this regard, the size of the park and the way it is managed are critical.

Establishing Priorities for Protected Areas

In a crowded world with limited funding, priorities must be established for conserving biological diversity. While some conservationists would argue that no species should ever be lost, the reality is that species are being lost every day. The real question is how this loss of species can be minimized given the financial and human resources available. The interrelated questions that must be addressed by conservation planners are: *What* needs to be protected, *where* should it be protected, and *how* should it be protected (Johnson, in press). Three criteria can be used in setting conservation priorities:

1. *Distinctiveness.* A biological community is given higher priority for conservation if it is composed primarily of rare endemic species than if it is composed primarily of common, widespread species. A species has more conservation value if it is taxonomically unique than if it has many close relatives.
2. *Endangerment.* Species in danger of extinction are of greater concern than species that are not threatened with extinction; thus the whooping crane requires more concern than the sandhill crane (Box 22). Biological communities threatened with imminent destruction are also given priority for protection.
3. *Utility.* Species that have present or potential value to people are given more conservation value than species of no obvious use to people. For example, wild relatives of wheat, which are potentially useful in developing new improved cultivated varieties, are given greater priority than species of grass that are not known to be related to any economically important plant.

The Komodo dragon of Indonesia (Figure 13.2) is an example of a species that would be a conservation priority using all three criteria: it is the world's largest lizard (distinctive); it occurs on only a few small

13.2 The carnivorous Komodo dragon of Indonesia is the largest living monitor lizard; many people feel it has unique status, and that protecting this endangered species is a conservation priority. (Photograph by Jessie Cohen, National Zoological Park, Smithsonian Institution.)

BOX 22 CONSERVATION OF AMERICAN CRANES: DIFFERENCES BETWEEN SIMILAR SPECIES

It is often the case in conservation biology that very similar species have completely different conservation needs. A classic illustration of this principle is the difference between two American crane species, the whooping crane, *Grus americana*, and the sandhill crane, *Grus canadiensis*. Though these species are closely related, relatively small differences in their biology and behavioral patterns have resulted in disparate success for each population: while most populations of sandhill cranes are well established and thriving, whooping cranes have only a small, unstable remnant population.

The overall population of sandhill cranes is estimated at well over 500,000 birds. This estimate is somewhat deceiving, since it hides the differential success of various subspecies of sandhill crane; for example, the lesser sandhill crane numbers in the hundreds of thousands, while the Mississippi sandhill crane has only 54 individuals (Faanes 1990). However, the majority of sandhill crane subspecies, whatever their population size, have responded well to conservation efforts, particularly where habitat preservation guidelines and hunting restrictions have been effectively enforced.

In contrast, the total whooping crane population is estimated at only 155 individuals. While this total represents a substantial increase from the population low point of 15 birds in 1942 (Allen 1952; Johnsgard 1991), whooping crane conservation efforts over the past four decades have not been entirely successful despite their intensity and creativity (Doughty 1990). The rate of population increase among whooping cranes has been very slow, and the survival of young from year to year has been highly variable (Johnsgard 1983; Nedelman 1987; Dennis et al. 1991). Whooping cranes also appear to have a ten-year cycle of population fluctuation superimposed on their long-term trend of population increase (Boyce and Miller 1985).

The contrast between the sandhill cranes' positive response to comparatively limited conservation efforts and the failure of whooping cranes to rebound despite intensive efforts on their behalf raises an intriguing problem: Why should one species be successful while such a similar species is not? In general, the answer lies in the response of each species to the presence of human disturbance in its favored habitat. Whooping cranes have a number of biological and behavioral attributes that make them much more sensitive to human disturbance than sandhill cranes. First, the fact that agricultural development in the midwestern plains and prairies has greatly reduced wetlands there has affected whooping cranes more than sandhill cranes. Whooping cranes are strictly aquatic feeders and rely primarily on animal rather than plant foods, whereas sandhill cranes prefer vegetable foods and are opportunistic feeders who forage in the agricultural fields as well as wetlands (Iverson et al. 1987; Hunt and Slack 1989). Second, most of the historical nesting sites for

One conservation strategy for whooping cranes involved an experiment in which wild sandhill cranes in the Bosque del Apache Refuge of New Mexico hatched "extra" whooping crane eggs and raised the chicks. The whooping cranes grew to healthy adulthood (the white bird), but continued to associate with the gray sandhill cranes. Because of their unusual upbringing, these "fostered" birds did not learn the species-specific social behaviors needed to form mating pairs with other whooping cranes. (Photograph © Art Wolfe.)

whooping cranes are located in the midst of what is now prime agricultural and range land in the United States. Whooping cranes prefer large open areas with deep water for nesting (Folk and Tacha 1990). Most of these habitats are disappearing as wetlands and large Midwestern waterways, such as the Platte River and tributaries of the Missouri River, are drained or diverted for agriculture. Sandhill cranes, on the other hand, generally nest in remote areas of Canada, Alaska, and eastern Siberia, far from most human settlements. Third, whooping cranes apparently have a much lower rate of success in raising their young than sandhills; though both species typically

lay two eggs per nest, sandhills are usually able to raise both chicks successfully, whereas whooping crane pairs do so in only 15% of observed cases (Johnsgard 1983). Different parental responses to aggression between hatchlings might be one factor influencing this difference: sandhill cranes have been observed to intervene when their chicks fight, but whooping cranes do not, and whooping crane chicks often kill their siblings within days of hatching. Finally, the location of whooping crane nesting grounds near human population centers invites disturbance by birdwatchers and tourists eager to catch a glimpse of these spectacular birds.

This combination of factors inhibiting whooping crane reproductive success and survival requires conservation methods tailored to the specific needs of the species. Conservation methods used so far have been problematic: attempts to create a captive flock by raising chicks hatched from "second" eggs taken from whooping crane nests have not been successful. One strategy involved placing whooping crane eggs in sandhill crane nests, to be raised by sandhill crane "foster parents." Although a number of whooping crane chicks were raised successfully by these substitute parents, none of those that reached maturity took mates or bred, possibly because the different social dynamics among sandhill cranes interfered with the whooping cranes' ability to interact with one another. The project was subsequently abandoned (Johnsgard 1991).

The most important contribution to conservation of whooping cranes, however, comes in the form of wetlands preservation. The isolated and undamaged wetland habitat that whooping cranes require continues to decrease despite attempts by conservationists to prevent development of these areas. For both whooping cranes and those subspecies of sandhill crane that are in jeopardy, habitat protection is fundamental to survival.

islands of a rapidly developing nation (endangered); and it has major potential for the ecotourism industry, as well as being of great scientific interest (utility). Using these criteria, several priority systems have been developed at both national and international scales to target either species or communities (Johnson, in press). These approaches are generally complementary, with each one giving a different perspective.

Species Approaches

Protected areas can be established to conserve unique species. Many national parks have been created to protect the "charismatic megavertebrates" that capture public attention, have symbolic value, and are crucial to ecotourism. In the process of protecting these species, whole communities that may consist of thousands of species are also protected. For example, Project Tiger in India was started in 1973 after a census revealed that the Indian tiger was in imminent danger of extinction. Project Tiger has helped to provide attention, funding, and a management philosophy for national parks in India. The establishment of 18 Project Tiger reserves, combined with strict protection measures, has halted the decline in the number of tigers (Panwar 1987; Ward 1992). However, with only 3% of India's land area protected and its vast human population crowding the park edges, the long-term future of India's wildlife remains in doubt.

In the Americas, the Natural Heritage Programs (see Box 21) and Conservation Data Centers are acquiring data on rare and endangered species from all 50 U.S. states, 3 Canadian provinces, and 13 Latin

American countries (Master 1991). This information is being used to target new localities for conservation.

Another important program is the IUCN Species Survival Commission Action Plans. About 2000 scientists are organized into 80 specialist groups to provide evaluations and recommendations for mammals, birds, invertebrates, reptiles, fishes, and plants (Stuart 1987; Species Survival Commission 1990). One group produced the Action Plan for Asian primates, in which priority rankings were developed for 64 Asian primates based on degree of threat, taxonomic uniqueness, and association with other threatened primates (Eudey 1987). The plan identified 37 species that were in need of urgent conservation action, including field surveys to determine status, establishment of protected areas, and captive breeding programs (Figure 13.3).

Community and Ecosystem Approaches

A number of conservationists have argued that communities and ecosystems rather than species should be the target of conservation efforts (Scott et al. 1991; McNaughton 1989; Reid 1992). Conservation of communities can preserve large numbers of species in a self-maintaining unit, whereas species rescues are often difficult, expensive, and ineffective. Spending $1 million on habitat protection and management might preserve more species in the long run than spending the same amount of money on just one conspicuous species.

The placement of new protected areas should try to insure that representatives of as many types of biological communities as possible are protected. Determining which areas of the world have adequate conservation protection and which urgently need additional protection is critical to the world conservation movement. Resources, research and publicity must be directed to areas of the world that require additional protection (Reid and Miller 1989; Ayres et al. 1991). At this point a total of 124 countries have protected areas. While it could be argued that all countries should have at least one national park, large countries that have a rich biota and a variety of ecosystem types obviously would benefit from having many protected areas.

The need to conserve ecosystems was recognized in the 1982 IUCN Bali Action Plan. Its objective is to establish a worldwide network of national parks and protected areas covering all terrestrial ecological regions. As part of this effort, regions throughout the world are being evaluated for the current percentage of area under protection, threats, need for action, and conservation importance. Re-

13.3 The Asian proboscis monkey has been targeted for special conservation efforts due to its restricted geographical distribution and its need for mangrove vegetation, which is rapidly being destroyed. (Photograph © NYZS/The Wildlife Conservation Society.)

views have been completed for the Indomalayan Realm (MacKinnon and MacKinnon 1986a), the Afrotropical Realm (MacKinnon and MacKinnon 1986b), and Oceania (Dahl 1986).

Gap Analysis

One way to determine the effectiveness of ecosystem and community conservation programs is to compare biodiversity priorities with existing and proposed protected areas (Burley 1988); this comparison can identify "gaps" in biodiversity preservation that need to be filled with new protected areas. On an international scale, this means protecting representative examples of all of the world's biological communities. Biogeographers have divided the world into eight terrestrial biogeographic regions. Each of these regions has numerous endemic species and genera, and each has been isolated from the others for a considerable time by geographical barriers such as oceans, mountains, or deserts (Dasman 1973; Udvardy 1975; Pielou 1979). These regions include the Nearctic (North America), Neotropical (Central and South America), Palearctic (Europe, most of Asia, and North Africa), Indomalayan (India and Southeast Asia), Afrotropical (Africa south of the Sahara; Madagascar), Australian, Oceanian (New Guinea and islands to the east) and the Antarctic. These eight regions are further subdivided into 227 provinces. Although all the biogeographical regions have some protected areas (see Table 13.1), 15 of the provinces have no protected areas and 30 of the provinces have less than 1000 km^2 under protection (MacKinnon and MacKinnon 1986a,b; Dahl 1986; Reid and Miller 1989). These provinces are targets for new parks.

Another approach to establishing conservation priorities is to use systems of classification based on vegetation and climate to identify biological communities that need to be protected. The most easily recognizable component of ecosystems is the dominant plant forms that give structure to the community. One such approach used in tropical America is the Holdridge Life Zone system, while fine-grained vegetation maps are used in places such as the United States (Bailey and Hogg 1986) and Africa (White 1983).

Establishing priorities for marine conservation has lagged behind terrestrial conservation efforts. Determining biogeographic provinces for the marine environment is much more difficult than for the terrestrial environment because boundaries between realms are less sharp, dispersal of marine organisms is more widespread, and the marine environment is less well known (Grassle 1991). Attempts have been made to determine biogeographic regions using a combination of the distributions of related marine animals (coastal, shelf, ocean) and physical properties that affect ecology and distribution (currents, temperature). Using such an approach, 40 marine biogeographic provinces have been described (Hayden et al. 1984).

Urgent efforts are being made throughout the world to protect marine biological diversity in a way comparable to the terrestrial parks by establishing marine parks (Ray 1991a; Kenchington and Agardy 1990), such as the Hol Chan Marine Reserve in Belize (see Box 17 in Chapter 9). The El Nido Marine Reserve along the coast of Palawan Island in the Philippines provides protection for the sea cow (also called the dugong), the hawksbill sea turtle, and the Ridley sea turtle. About one-fourth of the 300 biosphere reserves (see Chapter 20) worldwide include coastal or estuarine habitats (Ray and Gregg 1991). Protection of the nursery grounds of commercial species and recreational diving are among the main reasons for establishing these reserves (Moyle and Leidy 1992). Unfortunately many of these reserves exist only on the map and receive little protection from overharvesting and pollution (Salm and Clark 1984).

At the national level, biological diversity is protected most efficiently by ensuring that all major ecosystem types are included in a system of protected areas (U.S. Congress, Office of Technology Assessment 1987). Ecosystem types include those that are unaffected by human activity as well as ecosystems disturbed by human activity but similar in structure and function to the undisturbed condition, such as managed rangeland and managed forest. In the United States, various federal and state agencies, often led by personnel from the Natural Heritage Programs, are involved in an intensive "bottom–up" effort to survey and classify ecosystems on a local level as part of a program to protect biological diversity.

An alternative "top–down" approach is to compare a detailed vegetation map with a map of lands under governmental protection (Crumpacker et al. 1988). In the United States, the most comprehensive system of ecosystem mapping has been the potential natural vegetation (PNV) system of Kuchler (1964), defined as "the vegetation that would exist today if man were removed from the scene and if plant succession after his removal were telescoped into a single moment." PNV can be thought of as climax vegetation types, or the types of vegetation that existed prior to European settlement. The United States has 135 PNV types, such as Spruce-Cedar-Hemlock forest and Bluestem Prairie. The distribution of these PNV types was compared to the 348 million hectares of land owned and managed by U.S. government agencies, based on a map produced by the National Geographic Society (Crumpacker et al. 1988). Nine of the 135 PNV types are not represented on federal lands and eleven others are represented only by small areas and are either naturally rare or have been largely destroyed. Most of the unrepresented ecosystems are in central or southern Texas (e.g., Mesquite savannah) and in Hawaii (e.g., Mixed Guava forest). These community types should be highlighted in conservation efforts and included if possible in new protected areas. Certain PNV types are represented by enormous areas and are not high priorities for new protected areas: Juniper-Piñon woodland (21 million ha), Cottonsedge tundra (23 million ha), Dryas meadows and Barren (25 million ha), Sagebrush steppe (27 million ha) and Spruce-Birch forest (28 million ha). No single federal agency controls a complete representation of U.S. ecosystem types, and certain government agencies that have other priorities, such as the Department of Defense and the Bureau of Land Management, may be important in efforts to preserve biological diversity (Figure 13.4). Cooperation among the diversity of federal agencies as well as among state and local governments and private land owners may represent the key to protecting biological communities (Salwasser et al. 1987).

Another example of gap analysis at a more local level (Klubnikin, cited in Scott et al. 1987) compared California vegetation maps with protected areas. The surprising result was that 95% of the alpine habitats were in reserves, but less than 1% of biologically rich riverbank

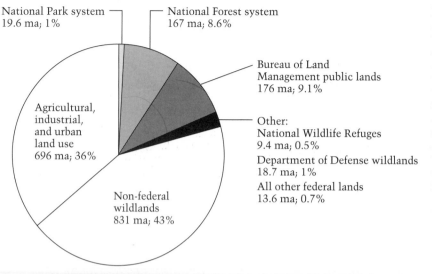

National Park system
19.6 ma; 1%

National Forest system
167 ma; 8.6%

Bureau of Land
Management public lands
176 ma; 9.1%

Agricultural,
industrial,
and urban
land use
696 ma; 36%

Other:
National Wildlife Refuges
9.4 ma; 0.5%
Department of Defense wildlands
18.7 ma; 1%
All other federal lands
13.6 ma; 0.7%

Non-federal
wildlands
831 ma; 43%

13.4   **Land ownership and use in the 48 contiguous United States. The federal government owns about 20% of the land in the U.S. (shaded), but this is divided among many agencies. Land areas are expressed in millions of acres (ma) and as a percentage of the total. (Based on 1986 data from the U.S. Department of the Interior, Council on Environmental Quality.)**

habitats were protected. In this instance, while preservation efforts had been successful at maintaining one type of habitat, another had been virtually ignored.

Geographic Information Systems (GIS) represent the latest development in gap analysis technology, using computers to integrate the wealth of data on the natural environment with information on species distributions (Scott et al. 1987, 1991). GIS analyses make it possible to highlight critical areas that need to be included within national parks and areas that should be avoided by development projects. The basic GIS approach involves storing, displaying, and manipulating many types of mapped data, such as vegetation types, climate, soils, topography, geology, hydrology, and species distributions (Figure 13.5). This approach can highlight correlations among the abiotic and biotic elements of the landscape, help plan parks that include ecosystem diversity, and even suggest potential sites to search for rare species. Aerial photographs and satellite imagery are additional sources of data for GIS analysis (Budd 1991). In particular, a series of images taken over time can reveal patterns of habitat destruction that need prompt attention.

International Approaches

In order to help establish priorities in conservation efforts, attempts have been made to identify tropical areas of the world that have great biological diversity and high levels of endemism, and are

13.5 Geographic Information Systems (GIS) provides a method for integrating a wide variety of data for analysis and display on maps. In this example, vegetation types, animal distributions, and preserved areas are overlapped to highlight areas that need additional protection. (After Scott et al., 1991.)

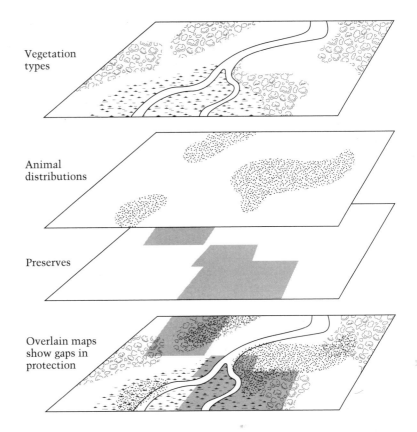

Vegetation types

Animal distributions

Preserves

Overlain maps show gaps in protection

under immediate threat of species extinctions and habitat destruction—so-called "hotspots" for preservation (Figure 13.6A; Table 13.4; NAS 1980; IUCN/UNEP 1986a,b,c). Using these criteria for rain forest plants, Myers (1988a) identified 12 hotspots that together include 14% of the world's plant species in only 0.2% of its total land surface. This analysis was later expanded (Myers 1991) to include 8 nonforest habitats—four in the tropics, and four outside of the tropics in Mediterranean-type climates (Figure 13.6B). One such notable area is southern Africa, including South Africa, Lesotho, Swaziland, Namibia, and Botswana, which has 23,200 plant species, 80% of which are endemic to the region. Another valuable approach has been to identify 12 "megadiversity" countries that together contain 60–70% of the world's biological diversity: Mexico, Columbia, Brazil, Peru, Ecuador, Zaire, Madagascar, Indonesia, Malaysia, India, China, and Australia. These countries are possible targets for increased funding and conservation attention (Table 13.5; Mittermeier 1988; Mittermeier and Werner 1990).

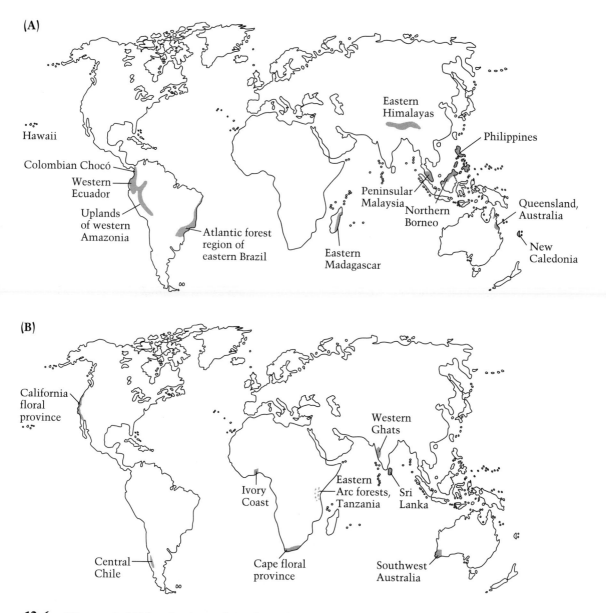

(A)

Eastern Himalayas

Hawaii

Philippines

Colombian Chocó

Western Ecuador

Peninsular Malaysia

Northern Borneo

Queensland, Australia

Uplands of western Amazonia

Atlantic forest region of eastern Brazil

Eastern Madagascar

New Caledonia

(B)

California floral province

Western Ghats

Ivory Coast

Eastern Arc forests, Tanzania

Sri Lanka

Central Chile

Cape floral province

Southwest Australia

13.6 **"Hotspots" of high endemism and significant threat of imminent extinctions.** **(A) Twelve tropical rain forest hotspots. (B) Eight hotspots in other climatic ecosystems. (Data from Myers 1988, 1991.)**

TABLE 13.4
Ten tropical rain forest "hotspots"

Hotspot	Original extent of forest (× 1000 ha)	Present extent of forest (× 1000 ha)	No. of plant species in original forest	Endemic species in original forest (percent)
New Caledonia	1,500	150	1,580	89
Madagascar	6,200	1,000	6,000	82
Atlantic forest of Brazil	100,000	2,000	10,000	50
Philippines	25,000	800	8,500	44
Eastern Himalayas	34,000	5,300	9,000	39
Northern Borneo	19,000	6,400	9,000	39
Peninsular Malaysia	12,000	2,600	8,500	28
Western Ecuador	2,700	250	10,000	25
Colombian Choco	10,000	7,200	10,000	25
Western Amazonia	10,000	3,500	20,000	25

Source: From Myers 1988.

Attempts to determine international priorities and global hotspots have had considerable overlap. There is general agreement on the need for increased conservation efforts in the following areas:

- *Latin America*: The coastal forests of Ecuador; the Atlantic coast forest of Brazil.
- *Africa*: The mountain forests of Tanzania and Kenya; the island of Madagascar.
- *Asia*: Southwestern Sri Lanka; the eastern Himalayas; Indochina (Burma, Thailand, Kampuchea, Laos, Viet Nam, and southeastern China); the Philippines.
- *Oceania*: New Caledonia.

Additional priorities include eastern and southern Brazilian Amazon, the uplands of the western Amazon, Colombia, Chocó, Cameroon, equatorial West Africa, the Sudanian zone, Borneo, Sulawesi, peninsular Malaysia, Bangladesh/Bhutan, eastern Nepal and Hawaii.

The "hotspot" targeting approach is being further refined to take into consideration the percentage of each individual country that is now protected and the degree to which habitats within each country are predicted to remain intact in the near future (Dinerstein and Wik-

TABLE 13.5

"Top ten" countries with the largest number of species of selected well-known groups of organisms

Rank	Mammals	Birds	Amphibians	Reptiles	Swallowtail butterflies	Flowering plants[a]
1	Indonesia 515	Colombia 1721	Brazil 516	Mexico 717	Indonesia 121	Brazil 55,000
2	Mexico 449	Peru 1701	Colombia 407	Australia 686	China 99–104	Colombia 45,000
3	Brazil 428	Brazil 1622	Ecuador 358	Indonesia ca. 600	India 77	China 27,000
4	Zaire 409	Indonesia 1519	Mexico 282	Brazil 467	Brazil 74	Mexico 25,000
5	China 394	Ecuador 1447	Indonesia 270	India 453	Burma 68	Australia 23,000
6	Peru 361	Venezuela 1275	China 265	Colombia 383	Ecuador 64	So. Africa 21,000
7	Colombia 359	Bolivia ca. 1250	Peru 251	Ecuador 345	Colombia 59	Indonesia 20,000
8	India 350	India 1200	Zaire 216	Peru 297	Peru 58–59	Venezuela 20,000
9	Uganda 311	Malaysia ca. 1200	U.S.A. 205	Malayasia 294	Malaysia 54–56	Peru 20,000
10	Tanzania 310	China 1195	Venezuela Australia 197	Thailand Papua New Guinea 282	Mexico 52	U.S.S.R. (former) 20,000

Source: After Conservation International; data from numerous sources. Swallowtail data from Collins and Morris 1985. Data on flowering plants from Davis et al. 1986.

[a] Numbers of species given for flowering plants are estimates.

ramanayake 1993). Further work needs to be done to assess the ability of governments to actually establish and maintain protected areas once funding becomes available.

Wilderness Areas

Large areas of wilderness are important priorities for conservation efforts. Large blocks of land that have been minimally affected by human activity, that have a low human population density, and are not likely to be developed in the near future are perhaps the only places

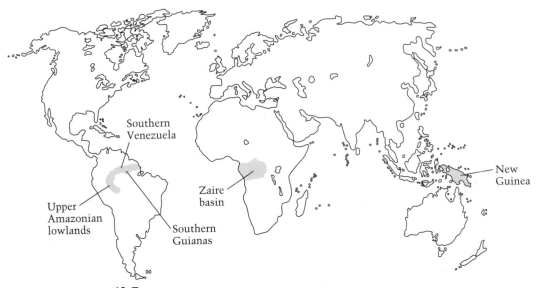

13.7 Only three major areas of tropical forest wilderness in the entire world remain essentially intact (Conservation International 1990).

on Earth where the natural processes of evolution may continue. These wilderness areas can potentially remain as controls showing what natural communities are like with minimal human influence. Three such wilderness areas have been identified and established as conservation priorities (Figure 13.7; McCloskey and Spaulding 1987; Conservation International 1990):

- *South America*. One arc of wilderness, containing rain forest, savannah, and mountains but few people, runs through the southern Guianas, southern Venezuela, northern Brazil, Colombia, Ecuador, Peru and Bolivia.
- *Africa*. A large area of equatorial Africa, centered on the Zaire basin, has a low population density and undisturbed habitat, including large portions of Gabon, the Republic of the Congo, and Zaire.
- *New Guinea*. The island of New Guinea has the largest tracts of undisturbed forest in the Asian Pacific region, despite the impacts of logging, mining, and transmigration programs. The eastern half of the island is the independent nation of Papua New Guinea, with 3.9 million people in 462,840 km^2. The western half of the island, Irian Jaya, is a state of Indonesia and has a population of only 1.4 million people in 345,670 km^2.

Centers of Diversity

Certain organisms can be used as biological diversity indicators when specific data about whole communities are absent. Diversity in birds, for example, is considered a good indicator of the diversity of a community (Bibby et al. 1992). Several analyses have put this principle into practice. Terborgh and Winter (1983) identified the areas of Colombia and Ecuador that had the greatest concentrations of bird species and proposed that these areas should be protected. The distribution and status of forest birds can demonstrate threats to a community that may not be readily apparent. Kepler and Scott (1985) plotted the distribution of endangered forest birds in Hawaii in relation to nature reserves to highlight the need for additional ecosystem protection. Collar and Stuart (1985), taking a slightly different approach, identified 75 African forests clustered in 5 distinct biogeographic areas that were important to the conservation of endangered bird species, and that should be considered in conservation efforts.

This "center of diversity" approach is being expanded in a systematic way. The IUCN Plant Conservation Office in England is identifying and documenting about 250 global centers of plant diversity with large concentrations of species (WCMC 1992). The International Council for Bird Protection (ICBP) is identifying localities with large concentrations of birds that have restricted ranges (Bibby et al. 1992). To date 221 such localities, containing 2484 bird species, have been identified. Many of these localities are islands and isolated mountain ranges that also have many endemic species of lizards, butterflies, and trees.

National Approaches

The international conservation community can help to establish guidelines and find opportunities to protect biological diversity, but in the end it is up to national and local governments to determine their own priorities. Many countries are in the process of determining their conservation priorities, or have done so recently, through the preparation of National Environmental Action Plans, National Biodiversity Action Plans, and Tropical Forest Action Plans.

The Indonesian government has recently considered the need to balance the protection of its treasurehouse of 15–25% of the world's species against the requirements of its growing population of 185 million people. The Indonesian Biodiversity Action Plan (Ministry for Population and Environment 1992) calls for the expansion of Indonesia's park system to include about 10% of the area of the country to protect all species in the wild. Additional priorities include strengthening of park managment, developing the support of local people for parks, securing stable funding for the parks, and developing new zoo and botanical garden facilities. Appropriately for a nation made up of 17,000 islands, plans are included to greatly increase the protection of marine and coastal waters, particularly mangroves.

Establishing Protected Areas with Limited Data

In general, new protected areas should encompass biological communities that are rich in endemic species, that contain community types underrepresented in other protected areas, that support threatened species, and that contain resources of potential use to people, such as species of potential agricultural or medicinal use. However, such data typically do not exist. One way of circumventing the lack of data is to base decisions on general principles of ecology and conservation biology. For example, Jared Diamond (1986) was asked by the Indonesian government to help design a national park system covering Irian Jaya, the western half of New Guinea. On the surface this seemed like an impossible task, since much of the region had never been surveyed biologically. However, Diamond was able to propose a series of reserves based on sound conservation principles (Figure 13.8) The principles include protecting elevational gradients that

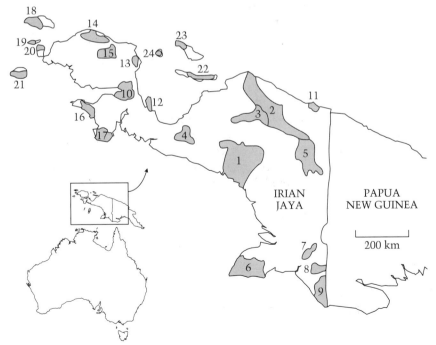

13.8   Reserves in Irian Jaya approved by the Indonesian government as of 1983. In addition to the 24 large reserves indicated, an additional 11 smaller reserves were established. Reserves were positioned to include major habitat types, centers of endemism, and islands. (From Diamond 1986.)

encompass diverse habitats, the need to have large parks to protect low-density species, the need to protect habitats in different climate zones, and the need to protect biogeographical areas that have many endemic species.

Some Conclusions

Priorities need to be established so that resources and personnel are directed to the most critical problems. Such an allocation process might reduce the tendency of funding agencies, tropical scientists, and development officers to cluster together in a few politically stable, accessible countries with high-profile projects. The decision of the MacArthur Foundation, one of the largest private sources of funds for conservation activities, to concentrate on different areas of the world for several years at a time—a "moving spotlight" approach—is a valuable counter to the tendency to concentrate all of the resources on a few well known places, such as Costa Rica, Kenya, and Brazil.

Establishing global conservation priorities is also important because the amount of money available to acquire and manage new national parks will be increasing substantially as a result of the creation of the Global Environment Facility (GEF) at the 1992 Earth Summit. The GEF will provide $1.5 billion over three years for environmental projects, with about a third of these funds allocated for biodiversity projects. Conservation biologists can play a valuable role by using their field experience to identify and recommend new areas suitable for preservation.

Establishing protected areas is only one step in the preservation of biological diversity. Biological communities inside a protected area are not immune to outside forces that may threaten their existence, such as hunting, pollution, and war. As Dasmann (1987) stated, "The Park Service is responsible for all resources within the park, but its authority ends at the park boundary. Air, water, wildlife, seeds, nutrients, energy move in and out of the park, but once across the boundary some other agency assumes responsibility." In many cases, this means that if issues outside the park boundaries are not addressed, biological diversity will continue to decline within established protected areas.

Summary

1. Protecting habitat is one of the most effective methods of preserving biological diversity. Land may be purchased and set aside by governments or by private organizations such as the Nature Conservancy. Protected habitats vary greatly in the extent of human disturbance they have experienced, and compromises have to be made between protecting biological diversity and satisfying human needs. Protected areas include nature reserves, national parks, national monuments, wildlife sanctuaries, and protected landscapes. Considerable biological diversity may still be present in unprotected multiple-use management areas.

2. About 3.2% of the Earth's land surface is covered by about 4500 protected areas, but this figure drops to a less impressive 1.6% if the world's largest park (in Greenland) is excluded. In contrast, in many countries 10–20% of the land is managed for multiple-use resource production. The area of protected habitats will probably never exceed 6% worldwide due to the perceived needs of human societies for natural resources.

3. Well-selected protected areas can initially protect large numbers of species, but the long-term future of many of these species remains in doubt. To be effective at preserving biological diversity, protected areas need to include examples of all vegetation types and biological communities.

4. Government agencies and conservation organizations are now setting priorities for establishing new protected areas based on the relative distinctiveness, endangerment, and utility of the species occurring in a place. Natural Heritage Programs and IUCN specialist groups focus on establishing parks to protect endangered species and biological communities. International approaches have involved identifying "hotspots" where there are large concentrations of species such as birds and plants. If these areas can be protected, then most of the world's biological diversity will be protected. Identifying priorities for new protected areas is important in order to insure that the funds available to protect biological diversity are used effectively.

Suggested Readings

Crumpacker, D. W., S. W. Hodge, D. Friedley and W. P. Gregg, Jr. 1988. A preliminary assessment of the status of major terrestrial and wetland ecosystems on federal and Indian lands in the United States. *Conservation Biology* 2: 103–115. Comprehensive analysis of the gaps in the protection of U.S. biological communities.

Diamond, J. 1986. The design of a nature reserve system for Indonesian New Guinea. *In* M. E. Soulé (ed.), *Conservation Biology: The Science of Scarcity and Diversity*, pp. 485–503. Sinauer Associates, Sunderland, MA. Priorities for establishing a protected area are outlined.

Grove, N. 1988. Quietly conserving nature. *National Geographic* 174(6): 818–844. The activities and philosophy of the Nature Conservancy.

Johnson, N. In press. *What to Save First? Setting Biodiversity Conservation Priorities in a Crowded World.* Biodiversity Support Program, Washington, D.C. Excellent overview of the approaches used to establish conservation priorities.

McNeely, J. A., K. R. Miller, W. V. Reid and others. 1990. *Conserving the World's Biological Diversity.* IUCN/WRI/CI/WWF-US/World Bank,

Gland, Switzerland and Washington DC. An excellent summary of the ways to preserve biological diversity.

Mittermeier, R. A. and T. B. Werner. 1990. Wealth of plants and animals unites "megadiversity" countries. *Tropicus* 4(1): 1, 4–5. Twelve countries have the majority of the world's plant and animals species.

Myers, N. 1988. Threatened biotas: "Hotspots" in tropical forests. *Environmentalist* 8: 1–20. Highlights centers of biological diversity.

Myers, N. 1991. The biodiversity challenge: Expanded "hotspots" analysis. *Environmentalist* 10: 243–256. More centers of biodiversity.

Scott, J. M., B. Csuti and F. Davis. 1991. Gap analysis: An application of Geographic Information Systems for wildife species. *In* D. J. Decker et al. (eds.), *Challenges in the Conservation of Biological Resources: A Practitioner's Guide.* Westview Press, Boulder, CO. Geographic Information Systems (GIS) represents an innovative approach to identifying gaps in conservation protection.

World Resources Institute. 1992. *World Resources 1992–1993.* Oxford University Press, New York. Assessment of the world's natural resources and protected areas, with numerous tables and summaries.

CHAPTER 14

Designing Protected Areas

The size and placement of protected areas throughout the world are often determined by the distribution of people, potential land values, and the political efforts of conservation-minded citizens. The largest parks are usually found in areas where there are few people and the land is considered unsuitable or too remote for agriculture, logging, urban development, or other human activities. The extreme example of this is Greenland National Park, a frozen land mass of 700,000 km^2. The low heath forests on nutrient-poor soils at Bako National Park in Malaysia are another example. Small conservation areas are common in large metropolitan areas and in densely settled and industrialized countries like England. Many of the conservation areas and parks in metropolitan areas of Europe and North America were formerly estates of wealthy citizens and royalty. In the midwestern United States a number of intact fragments of the prairie ecosystem are preserved on old cemeteries, railroad rights-of-way, and other odd pieces of land that were not developed into farmland.

Although most parks and conservation areas have been acquired and created in a haphazard fashion, depending on the availability of money and land, a considerable body of ecological literature has been developed addressing the most efficient way to design conservation areas to protect biological diversity (Shafer 1990). Many of the conclusions of this research have been derived from equilibrium theory, often using the island biogeography model of MacArthur and Wilson (1967) described in Chapter 4. Such models make the critical assumption that parks are "habitat islands" isolated by expanses of inhospitable terrain—which often is not the case. The guidelines developed from this research have proved to be of great interest to governments, corporations, and private landowners who are being urged and mandated to manage their properties for both the commercial production of natural resources and biological diversity. Key questions conservation biologists have attempted to address include:

1. How large should nature reserves be to protect species?
2. Is it better to have a single large reserve or many smaller reserves?
3. How many individuals of an endangered species must be protected in a reserve to prevent extinction?
4. What is the best shape for a nature reserve?
5. When several reserves are created, should they be close together or far apart, and should they be isolated from one another or connected by corridors?

Researchers working with island biogeography models have proposed some answers to these questions (Figure 14.1), but they are still being debated. The most thorough, critical look at this subject is *Nature Reserves: Island Theory and Conservation Practice* (Shafer 1990). In

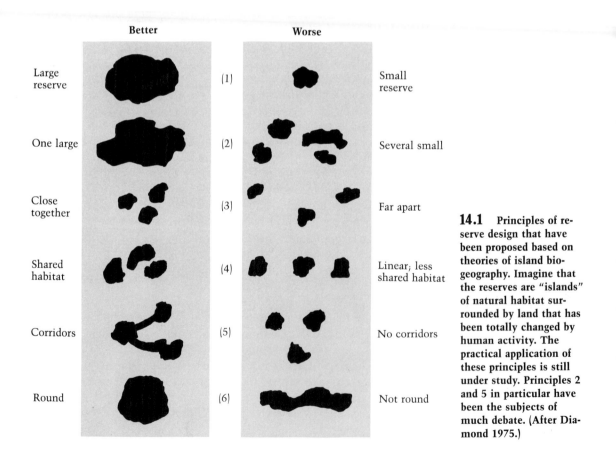

Better   **Worse**

Large reserve	(1)	Small reserve
One large	(2)	Several small
Close together	(3)	Far apart
Shared habitat	(4)	Linear; less shared habitat
Corridors	(5)	No corridors
Round	(6)	Not round

14.1 **Principles of reserve design that have been proposed based on theories of island biogeography. Imagine that the reserves are "islands" of natural habitat surrounded by land that has been totally changed by human activity. The practical application of these principles is still under study. Principles 2 and 5 in particular have been the subjects of much debate. (After Diamond 1975.)**

this book, models are examined in relation to empirical evidence and practical experience.

In describing the desire of academic conservation biologists to provide land managers ~~to~~ with simple, general guidelines for designing nature reserves, Ehrenfeld (1989) states,

> I feel obliged to point out that there is a widespread obsession with a search for general rules of scientific conservation, the genetic code of conservation so to speak, and this finds expression in very general statements about extinction rates, viable population sizes, ideal reserve designs, and so forth. . . . Yet this kind of generality is easily abused, especially when would-be conservationists become bewitched by models of their own making. When this happens, the sight of otherwise intelligent people trying to extract non-obvious general rules about extinction from their own polished and highly simplified versions of reality becomes a spectacle that would have interested Lewis Carroll. . . . We should not be surprised when different conservation problems call for qualitatively different solutions.

Commenting on this, Shafer (1990) says, "I do not believe that the above remarks were intended so much to belittle efforts at finding rules as to warn against becoming blinded by them. Modeling is badly needed and should continue." And finally, on a more positive note, Franklin (1985) states, "Any guidelines on size, shape, and other criteria must of necessity be general as each design problem is unique."

Reserve Size

An early debate within conservation biology occurred over whether species richness is maximized in one large nature reserve or several smaller ones of an equal total area (Diamond 1975; Simberloff and Abele 1976, 1982; Terborgh 1976; Terborgh and Winter 1980); this is known in the literature as the "SLOSS debate" (single *l*arge *o*r *s*everal *s*mall). For example, is it better to set aside one reserve of 10,000 ha, or four reserves of 2500 ha each? The proponents of large reserves argue that only large reserves have sufficient populations of big, wide-ranging species (such as panthers) and low-density species (such as hawks) to maintain long-term populations (Figure 14.2; Table 14.1). Also, a large reserve minimizes edge effects, encompasses more species, and has greater habitat diversity than a small reserve. These advantages of large parks follow from island biogeography theory and are demonstrated in numerous surveys of animals and plants in parks. There are three practical implications to this viewpoint. First, when a new park is being established it should be made as large as possible in order to preserve as many species as possible. Second,

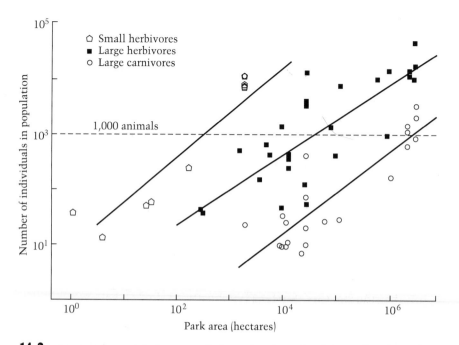

14.2　Large parks contain larger populations of each species than small parks; only the largest parks may contain long-term viable populations of many vertebrate species. Each symbol represents an animal population. If the viable population size of a species is 1000 (10^3; dashed line) individuals, parks of at least 100 (10^2) ha will be needed to protect small herbivores (e.g., rabbits, squirrels); parks of more than 10,000 ha will be needed to protect large herbivores (e.g., deer, zebra, giraffes); and parks of at least a million ha will be needed to protect large carnivores (e.g., lions, wolves). (From Schonewald-Cox 1983.)

when possible, additional land adjacent to nature reserves should be acquired in order to increase the area of existing parks. And finally, if there is a choice of creating a new small park or a new large park in similar habitat types, the large park should be created. On the other hand, once a park is larger than a certain size, the number of new species added with each increase in area starts to decline. In such a situation, creating a second large park some distance away may be an effective strategy for preserving additional species.

The extreme proponents of large reserves argue that small reserves need not be maintained, because their inability to support long-term populations gives them little value for conservation purposes. Opposing this viewpoint, other conservation biologists argue that well-placed small reserves are able to include a greater variety of habitat types and more populations of rare species than one large block of the same area (Järvinen 1979; Simberloff and Gotelli 1984).

TABLE 14.1
Estimated population size for three large carnivore species in five African protected areas

Protected area	Country	Area (ha)	Estimated population size		
			Leopard	Spotted hyena	African lion
Manyara N.P.	Tanzania	9,100	10	10	35
Nairobi N.P.	Kenya	11,500	10	12	25
Ngorongoro Crater	Tanzania	26,000	20	479	70
Kruger N.P.	So. Africa	1,908,400	650	1500	1120
Serengeti N.P.	Tanzania	2,550,000	900	3500	2250

Source: After Schonewald-Cox 1983, from various sources.

The value of several well-placed reserves in different habitats is demonstrated by a comparison of four national parks in the United States. The total number of large mammalian species in three national parks located in contrasting habitats (Big Bend in Texas; North Cascades in Washington; and Redwoods in California) is greater than the number of species in the largest U.S. national park, Yellowstone, even though Yellowstone's area is larger than the combined area of the other three parks (Quinn and Harrison 1988). Also, creating more reserves, even if they are small ones, prevents the possibility of a single catastrophic force, such as an exotic animal or a disease, destroying an entire population located in a single large reserve.

The consensus now seems to be that strategies on reserve size depend on the group of species under consideration as well as the scientific circumstances (Game and Peterkin 1984; Soulé and Simberloff 1986). It is accepted that large areas are better able than small reserves to maintain many species because of their larger population sizes and greater variety of habitats (Abele and Connor 1979). However, well-managed small nature reserves also have value, particularly for the protection of many species of plants, invertebrates, and small vertebrates (Lesica and Allendorf 1992). Often there is no choice other than to accept the challenge of managing species in small reserves when land around the small reserves is unavailable for conservation purposes. This is particularly true in places that have been intensively cultivated and settled for centuries, such as Britain, China, and Java (Figure 14.3).

Reserve Design in South Africa

The issue of nature reserve size is highlighted by conservation efforts on behalf of threatened plant species in the Cape floral province

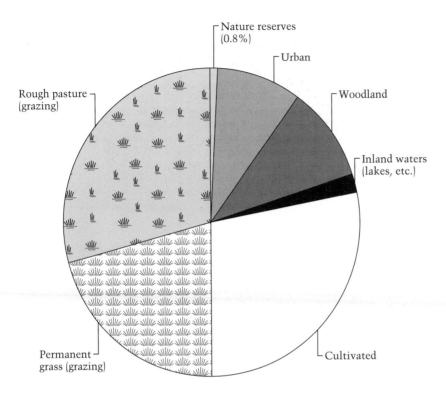

14.3   In a country like Great Britain, which has been settled and cultivated for centuries, very little land is left for nature reserves. Almost 80% of the land in Great Britain is used for agricultural purposes (farming or grazing). (From Green 1989.)

of South Africa. The Cape flora is one of the most remarkable in the world; it occupies only 90,000 km^2—about the same area as the state of Maine—but contains 8590 vascular plant species, the highest ratio of plants to area of any temperate or subtropical region in the world (Figure 14.4; Goldblatt 1978; Bond 1983). Furthermore, 68% of the species and five plant families are endemic to this small region. Part of the reason for this high species diversity is the occurrence of five distinct vegetation types in the area, a variety of soil types, and the great species richness of the fynbos vegetation type in particular (Rebelo and Siegfried 1990).

Unfortunately, this unique ecosystem is being reduced in area and damaged by agriculture, urban development, fire, and the invasion of exotic species. About 19% of the Cape flora, 1621 species, are naturally rare or threatened by human activity (Hall et al. 1984). One of the most characteristic families of the Cape flora is the Proteaceae. Plants of this family are horticulturally important for their attractive flowers arranged in condensed, colorful heads. In the Cape region, 53 species of Proteaceae are restricted to only one or two populations; each population occupies an area of less than 5 km^2 and usually con-

14.4   Species richness for flowering plants increases to an incredible level in the fynbos vegetation of the southwest corner of the Cape Floral Province in South Africa. The map shows the number of species in characteristic and endemic plant families, such as the Proteaceae, Ericaceae, and Restionaceae. (From Rebelo and Siegfried 1990, based on Oliver et al. 1983.)

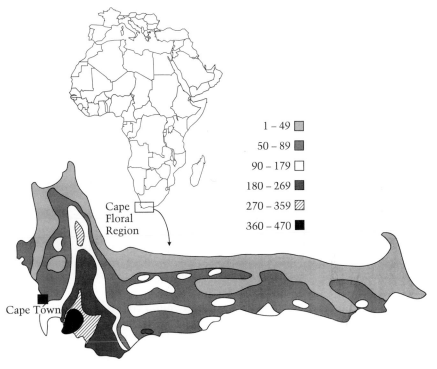

tains less than 1000 individuals. A proposed system of a few large parks of at least 100 km² in area would not protect most of these rare species. A better alternative would be a system of many smaller reserves that would protect populations of individual rare species and unique habitat types (Tansley 1988). Such a system of smaller reserves might be difficult, but not impossible, to manage.

Effective Preservation of Species

Because population size is the best predictor of extinction probability, reserves should be sufficient in area to preserve large populations of important species (rare and endangered species, keystone species, economically important species, etc.). Small populations may be vulnerable to extinction due to environmental fluctuations, demographic variation, inbreeding, and the loss of alleles, as described in Chapter 11. An isolated nature reserve that contains only four elephants will not have a self-perpetuating elephant population; an elephant population of 500 is far more likely to be self-perpetuating. The best evidence to date suggests that populations of at least several hundred reproductive individuals are needed to ensure the long-term viability of vertebrates; several thousand individuals is a desirable goal (Chapters 11 and 12). Having more than one population of a rare species within the nature reserve will also increase the probability of survival for the species.

While small populations of rare species in isolated nature reserves may not be viable, several strategies exist to facilitate their survival. Small populations in scattered reserves can be managed as one meta-population, with efforts made to encourage natural migration between the reserves (Ehrlich and Murphy 1987). Occasional individuals can be collected from one reserve and added to the breeding population of another. Such efforts may serve to bolster the numbers at a site with a declining population. Also, if even a small number of individuals move between populations, the harmful genetic effects of small population size are dramatically reduced. Transporting individuals long distances and transporting individuals between populations that appear to be genetically distinct should be done cautiously to avoid potential problems caused by outbreeding depression (the lowered survival and reproduction of offspring that follows from breeding individuals that are genetically far apart, such as sibling species or separate varieties).

A more difficult aspect of ensuring viable populations in reserves is dealing with the needs of wide-ranging species. Ideally, a nature reserve should be large enough to include a viable population of the most wide-ranging species. Protecting the habitat of these wide-ranging species, which are often large or conspicuous "flagship" species, will often provide adequate protection for the other species in the community. The large parks in eastern Africa managed for elephants, lions, and other large mammal species also protect numerous bird, insect, and plant species. Extensive areas of long-leaf pine habitat surrounding the Savannah River nuclear processing plant in South Carolina are being managed to protect the red-cockaded woodpecker (Figure 14.5) and in the process many endangered plant species are being protected.

The effective design of nature reserves requires a thorough knowledge of the ecology of the important species as well as information on the distribution of biological communities. Information on species' feeding requirements, reproductive behavior, daily and seasonal movement patterns, potential predators and competitors, and susceptibility to disease and pests is needed to plan an effective conservation strategy. This information allows the design of nature reserves that include as many as possible of the habitat requirements of each species. A balance has to be developed between focusing on the needs of the "flagship" species to the exclusion of all other species, and managing only for maximum species diversity, which may result in the loss of the "flagship" species that interest the general public.

14.5   Long-leaf pine habitat in the southeastern U.S., including areas of South Carolina, North Carolina, and Georgia, are being managed to protect the endangered red-cockaded woodpecker. Heavily logged areas lack older trees with the nesting holes that the woodpecker requires; in managed forests, artificial nesting holes are drilled in the trees. Here a young woodpecker leaves the nest for its first flight. (Photograph © Derrick Hamrick.)

Minimizing Edge and Fragmentation Effects

It is generally agreed that parks should be designed to minimize the harmful edge effects discussed in Chapter 6. Conservation areas that are rounded in shape will have the minimum edge-to-area ratio, and the center of such a park will be farther from the edge than other park shapes. Long, linear parks will have the most edge, and all points in the park will be close to the edge. Using these same arguments for parks with four straight sides, a square park is a better design than an elongated rectangle of the same area. These ideas have rarely, if ever, been implemented. Most parks have irregular shapes because land acquisition is typically a matter of opportunity rather than a matter of completing a geometric pattern.

As was discussed in Chapter 6, internal fragmentation of reserves by roads, fences, farming, logging, and other human activities should be avoided as much as possible, because fragmentation often divides a large population into two or more smaller populations, each of which is more vulnerable to extinction than the large population (Schonewald-Cox and Buechner 1992). Fragmentation also provides an entry point for exotic species that may harm native species, creates further undesirable edge effects, and creates barriers to dispersal, thus reducing the probability of colonization of new sites and decreasing the gene flow necessary to maintain genetic variability.

The forces promoting fragmentation are powerful, since protected areas are often the only undeveloped land available for new roads, power lines, agriculture, dams, and residential areas. In the eastern United States many parks near cities are crisscrossed by roads, railroad tracks, and power lines that divide large areas of habitat like pieces of a roughly cut pie.

Coordinating New and Existing Reserves

Strategies exist for aggregating small nature reserves into larger conservation blocks. Nature reserves are often embedded in a larger matrix of habitat managed for resource extraction such as timber forest, grazing land, and farmland. If the protection of biological diversity can be included as a secondary priority in the management of the production areas, then larger areas can be included in conservation management plans. Nature reserves should be managed as a regional matrix to facilitate gene flow and migration among populations and to ensure adequate representation of species and habitats (Figure 14.6). Cooperation among public and private landowners is particularly important in developed metropolitan areas, where there are often numerous small, isolated parks under the control of a variety of different government agencies and private organizations (Salwasser et al. 1987).

Linking Nature Reserves with Habitat Corridors

One intriguing suggestion for managing a system of nature reserves has been to link isolated protected areas into one large system through the use of **habitat corridors**: strips of land running between the reserves (Simberloff and Cox 1987; Simberloff et al. 1992). Such

14.6   Often national parks (shaded areas) are adjacent to other areas of public lands (unshaded) that are managed by different agencies. The United States government is considering managing large blocks of land that include national parks, national forests, and other federal lands as networks of natural areas in order to maintain populations of large and scarce wildlife. This illustration shows four of ten such networks that have been proposed. Privately owned land is shown in black. (Modified from Salwasser 1987.)

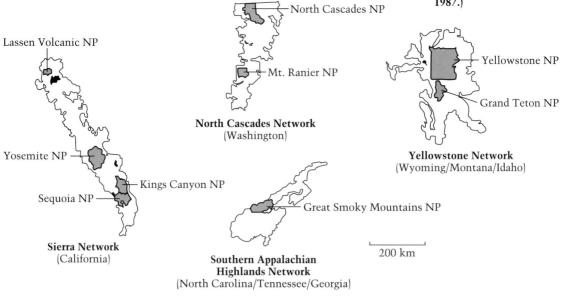

North Cascades NP

Lassen Volcanic NP

Mt. Ranier NP

North Cascades Network
(Washington)

Yellowstone NP

Grand Teton NP

Yellowstone Network
(Wyoming/Montana/Idaho)

Yosemite NP

Kings Canyon NP

Sequoia NP

Great Smoky Mountains NP

200 km

Sierra Network
(California)

**Southern Appalachian
Highlands Network**
(North Carolina/Tennessee/Georgia)

habitat corridors, also known as conservation corridors or movement corridors, could allow plants and animals to disperse from one reserve to another, facilitating gene flow and colonization of suitable sites. Corridors might also help to preserve animals that must migrate seasonally among a series of different habitats to obtain food; if these animals are confined to a single reserve, they could starve. The idea of corridors has been embraced with enthusiasm by some park managers as a strategy for managing wide-ranging species. For example, the Florida Natural Areas Inventory has proposed a 2400-ha corridor to link three large publicly owned areas, at a cost of $5 million (Simberloff et al. 1992).

Corridors that facilitate natural patterns of migration will probably be the most successful at protecting species. For example, large grazing animals often migrate in regular patterns across a rangeland in search of water and the best vegetation. In seasonally dry savannah habitats, animals often migrate along the forests that grow along streams and rivers. In the temperate zone, many bird and mammal species regularly migrate to higher elevations during the warmer months of the year. This principle was put into practice in Costa Rica to link two wildlife reserves, the Braulio Carillo National Park and La Selva Biological Station. A 7700-ha corridor of Costa Rican forest several kilometers wide, known as La Zona Protectora, was set aside to provide an elevational link that allows at least 35 species of birds to migrate between the two large conservation areas (Wilcove and May 1986).

Although the idea of corridors is intuitively appealing, there are some potential drawbacks (Simberloff et al. 1992; Simberloff and Cox 1987). Corridors may facilitate the movement of pest species and disease, so that a single infestation can quickly spread to all of the connected nature reserves and cause the extinction of all populations of a rare species. Also, animals dispersing along corridors may be exposed to greater risks of predation, since human hunters as well as animal predators tend to concentrate on routes used by wildlife. At the present time, the empirical evidence to support the value of corridors is very limited, despite a lot of enthusiasm for the idea. The value of habitat corridors probably needs to be evaluated on a park-by-park basis. In general, maintaining existing corridors is probably worthwhile. For one thing, many of the corridors that currently exist are along water courses, and these may themselves be biologically important habitats. Also, when new parks are being carved out of large blocks of undeveloped land, incorporating corridors among the parks is also probably worthwhile. In some cases, leaving small clumps of original habitat between large conservation areas may also serve to facilitate movement by a "stepping-stone" process.

A Proposed Wildlife Corridor in Tanzania

In Tanzania, large herds of wildebeest, zebras, and other grazing animals migrate seasonally between Tarangire National Park and Lake Manyata National Park in search of food and water (Mwalyosi 1991). Lake Manyata National Park is notable for its scenic lake and rift escarpment and the high density and diversity of animal life attracted by its lush vegetation and reliable source of water. Because of these attractions, and despite its small size, Lake Manyata National Park has the third largest number of visitors per year and the second highest amount of tourist revenue of any park in Tanzania. Managing wildlife in the area is a problem because the park is only 110 km^2 and is increasingly becoming surrounded by agricultural settlements. Originally the park was surrounded by the rough rangeland of the Masai Ecosystem, across which wildlife migrated freely (Prins 1987). However, in the last few decades, humans and their cattle have increased dramatically in numbers as irrigation has allowed farming to expand in the area. Agricultural acreage in the vicinity of the park increased from 218 ha in 1957 to 1093 ha in 1980. While the total cultivated area is not great, some of the cultivated fields are within the traditional migratory routes of animals, resulting in damage to crops and killing of animals by farmers, and repeated conflicts between farmers and conservation authorities. If farmland continues to expand this conflict will only get worse.

To deal with this situation, a proposal has been made to establish a defined migratory route between Tarangire and Lake Manyata National Parks and to buy the farmland along this route and return it to rangeland (Figure 14.7; Mwalyosi 1991). If animals do wander off their migratory routes and damage crops, farmers will be compensated for their losses from revenue paid by tourists. In this way the value of preserving animals will be demonstrated to the people. In addition, barriers are being constructed to keep animals within the confines of the parks, and arrangements are being made with local people to allow them to continue to obtain the firewood, sand, and fish that they have traditionally collected in the park. The implementation of this proposal will be closely followed by the conservation community.

Corridors in Louisiana Wetlands

The potential of corridors for protecting biological diversity is also highlighted in the case of the Tensas River basin in northeastern Louisiana. The Tensas River basin originally contained one million hectares of flat, poorly drained forested land characteristic of the Missis-

14.7 A game corridor has been proposed to allow game herds to migrate between Lake Manyara and Tarangire National Parks in northeastern Tanzania. Current cropland is indicated by diagonal hatching. The horizontally hatched area between the lake and the proposed corridor is an area used by tribespeople for dry-season grazing of their herds. A proposed "buffer zone" (dark gray) would allow additional areas for grazing. (From Mwalyosi 1991.)

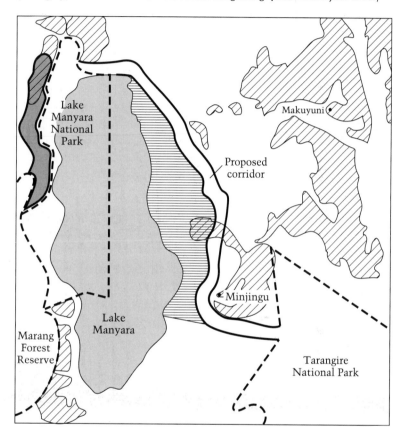

sippi River floodplain (Gosselink 1990; Burdick et al. 1989). These floodplain forests are among the most productive in the United States for fish and wildlife, and support large populations of migratory and resident birds. The Tensas River basin is now being fragmented by logging and conversion to agricultural fields for soybean, cotton, rice, and corn production. In 1957, 560,000 ha of forest remained, with the two largest blocks containing a total of 326,500 ha. By 1990, only 157,000 ha remained, with four large patches (ranging in area from 10,000 to 30,000 ha) in protected areas. The remaining forest is in smaller patches scattered throughout the basin. The Tensas River basin originally supported populations of the red wolf (*Canis lyceon*), the Florida panther (*Felis concolor coryi*), and the ivory-billed woodpecker (*Campephilus principalis*), none of which are now found in the area; significant declines have also been observed in about one-third of the forest-dependent bird species. Along with habitat destruction has come a decline in water quality, an increase in soil erosion, and changes in the cycling and flow of the water in the region.

One strategy that has been proposed for protecting the biological diversity of the Tensas basin is a system of corridors that will link together the separate forest blocks into larger units (Gosselink 1990). Adding 400 hectares of forest corridor would increase the size of the largest complex of forest from 50,000 ha to 100,000 ha. Adding 600 hectares of corridors in the western edge of the basin would link together several large fragments in the 3,000- to 10,000-ha range, forming a 63,000-ha forest complex. These large forest blocks might be of sufficient size to protect some of the large wildlife species in the region, such as the black bear.

Present and Future Human Use of Reserves

Human use of the landscape is a reality that must be dealt with in park design. People have been a part of virtually all the world's ecosystems for thousands of years, and excluding humans from nature reserves could have unforeseen consequences; for example, a savannah protected from fires set by people may change to forest, with a subsequent loss of the savannah species. People who have traditionally used products from inside a nature reserve and are suddenly not allowed to enter the area will suffer from their loss of access to basic resources that they need to stay alive. They will be understandably angry and frustrated, and people in such a position are unlikely to be strong supporters of conservation; this issue is discussed further in Chapter 15.

The United Nations Educational, Scientific, and Cultural Organization (UNESCO) has pioneered one approch to this situation with its Man and the Biosphere (MAB) Program. This program has designated a number of Biosphere Reserves worldwide in an attempt to integrate human activities, research, and protection of the natural environment at a single location (Dyer and Holland 1991). The concept involves a core area in which biological communities and ecosystems are strictly protected, with a surrounding buffer zone in which human activities are monitored and nondestructive research is carried out. Then there is a transitional zone in which research and sustainable development are allowed (Figure 14.8).

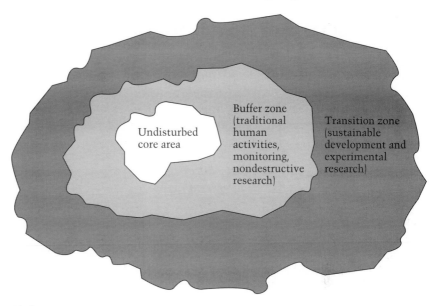

14.8   The general pattern of an MAB reserve includes a core protected area surrounded by a buffer zone in which human activities are monitored and managed and research is carried out; and a transition zone of sustainable development and experimental research.

The general principle of surrounding core conservation areas with buffer and transition zones in which the traditional extraction of products occurs can have several desirable effects. First, the good will of local people may be maintained. Second, certain desirable features of the landscape created by human use may be maintained. And third, buffer zones create a transition between highly protected core conservation areas and human-dominated areas, which may facilitate animal movement and dispersal (Figure 14.9). Without a buffer zone, the effective area of a park may be greatly reduced by human activity, since fewer species may be able to survive in parks that are immediately surrounded by human-dominated areas. In the words of Lamprey (1974), "In the experience of many African national park authorities, the presence of intensive settlement on the boundaries produces a *de facto* zone of 'limited conservation' inside the park, within which some poaching, tree-felling, grazing, grass burning, and other illicit activities may occur, necessitating constant policing." The human population is expected to continue to increase dramatically in coming decades, and resources such as firewood and wild meat will become harder to find. Consequently, managers of nature reserves need to anticipate ever-greater demand for use of the remaining patches of natural habitat.

Opening conservation to man

Is the best way to protect a natural area to seal it off in a "closed jar" from the outside human world? Sooner or later such a policy can destroy the area it was intended to protect. Ecological and sociological pressures - both inside and outside - eventually may shatter the reserve.

Almost all natural areas have been modified by man: creating a reserve by excluding man can upset the ecological balance. Boundaries may not coincide with territorial areas and feeding grounds. Pressure builds up within the reserve. Jammed inside, some animals overbreed, others "eat themselves to starvation".

In some cases, nature reserves are created by excluding the local inhabitants from their traditional grazing and hunting areas. They have difficulty in accepting that these areas are only accessible to tourists. Gradually illicit hunting, grazing and cropping may encroach upon and eradicate the reserve.

MAB emphasizes man's partnership with nature. A reserve is open and interacts with its region. The local people can be its guardians.

It is not suggested that the traditional policy of conservation should be changed everywhere. Certainly some areas must remain untouched. But there are fewer and fewer natural areas left to conserve and certain reserves are being destroyed by these internal and external pressures. Opening conservation to man does not only apply to the Kenyan situation here but to many other countries. It may be a longer term solution.

MAN AND THE BIOSPHERE (MAB) PROGRAMME, UNESCO

The diagram (right) illustrates how a Kenyan specialist envisages integrating wildlife conservation, tourism and traditional land use through zonation into different use areas, research for rational management and participation of the local population. The term "biosphere reserve" was coined to identify reserves putting the "open" concept into practice.

Summary

1. Guidelines for protecting biological diversity have been developed using island biogeography models. These guidelines have been used to analyze and design isolated protected areas on the assumption that these areas have island-like characteristics.

2. In general, a large park will have more species than a small park in an equivalent habitat. Conservation biologists have debated whether it is better to create a single large park or several small parks comprising equivalent area; convincing arguments have been presented on both sides. Parks must be sufficient in area to include viable populations of important species. The numbers needed appear to be at least several hundred individuals, and preferably several thousand, for many vertebrate species. However, small populations can also be managed to increase their probability of survival. Some of the most difficult species to protect are wide-ranging species that require a large area.

3. Parks should be designed to minimize harmful edge effects. To achieve this, the ideal shape for a park of a given area would be rounded. The tendency to fragment parks with roads, fences, and other human developments should be avoided insofar as is possible. When possible, government authori-

14.9 Past policies have often attempted to protect natural areas by sealing them off from outside influences. Such policies may fail to recognize ecological and social forces that threaten the ecosystem. The Man and the Biosphere Program is an attempt to integrate the needs and cultures of local people in park planning and protection. (Poster from "Ecology in Action: An Exhibit," UNESCO, Paris, 1981.)

ties and private owners should try to coordinate their activities and manage adjoining small parcels of land as one unit.

4. Habitat corridors have been proposed for linking together isolated conservation areas. These corridors would allow the movement of animals between protected areas, potentially facilitating gene flow as well as dispersal and the colonization of new sites. The concept of habitat corridors has been embraced by many land managers, but the empirical evidence supporting their value is still limited. Habitat corridors will probably be most effective when they protect existing routes of migratory animals.

5. The effects of humans on protected areas are often important and need to be considered in park design and management. Restricting human activities to buffer zones that surround protected core areas represents one possible strategy.

6. Guidelines developed on the basis of the island biogeography model have had only limited predictive power in dealing with actual problems in park design. Combining these guidelines with common sense and experience may provide the best strategies for conservation.

Suggested Readings

Diamond, J. M. 1975. The island dilemma: Lessons of modern biogeographic studies for the design of natural reserves. *Biological Conservation* 7: 129–146. An early attempt to apply the island biogeography model to nature reserve design.

Game, M. and G. F. Peterkin. 1984. Nature reserve selection strategies in the woodlands of central Lincolnshire, England. *Biological Conservation* 29: 157–181. Practical guide to land acquisition decisions.

Salwasser, H., C. M. Schonewald-Cox and R. Baker. 1987. The role of interagency cooperation in managing viable populations. *In* M. E. Soulé (ed.), *Viable Populations for Conservation*, pp. 159–173. Cambridge University Press, Cambridge. Land managers need to coordinate their activities on a regional scale to protect biodiversity.

Shafer, C. L. 1990. *Nature Reserves: Island Theory and Conservation Practice.* Smithsonian Institution Press, Washington D.C. A comprehensive, well-illustrated review of the theories of reserve design that also presents evidence and counter-arguments.

Simberloff, D. and J. Cox. 1987. Consequences and costs of conservation corridors. *Conservation Biology* 1: 63–71. Arguments for and against habitat corridors.

Simberloff, D., J. A. Farr, J. Cox and D. W. Mehlman. 1992. Movement corridors: Conservation bargains or poor investments? *Conservation Biology* 6: 493–505. A critical examination of the conservation value of habitat corridors.

Tansley, S. A. 1988. The status of threatened Proteaceae in the Cape flora, South Africa, and the implications for their conservation. *Biological Conservation* 43: 227–239. Numerous rare plant species occupy limited areas that could be protected by small nature reserves.

CHAPTER **15**

Managing Protected Areas

Once a protected area has been legally established, it must be effectively managed if biological diversity is to be maintained. The conventional wisdom that "nature knows best" and that there is a "balance of nature" leads some people to the conclusion that biodiversity is best served when there is no human intervention. The reality is often very different: in many cases humans have already modified the environment so much that the remaining species and communities need human intervention in order to survive (Pimm 1991; Buttrick 1992; Blockhus et al. 1993). The world is littered with parks that have been created on paper but not effectively managed. These parks have gradually—or sometimes rapidly—lost species and their habitat quality has been degraded. The crucial point is that parks must sometimes be actively managed to prevent deterioration. However, decisions on park management can usually be made most effectively when information is provided by a research program.

It is also true that sometimes the best management involves doing nothing; management activities are sometimes ineffective or even detrimental (Chase 1986). For example, the U.S. government spent $180 million unsuccessfully fighting the 1988 fires in Yellowstone National Park, even after it became apparent that the forests were going to burn regardless of human intervention. Long-term fire suppression policies had allowed a buildup of dead wood over a large area, creating ideal conditions for a massive burn. As another example, active management to promote the abundance of game species, such as deer, has frequently involved eliminating top predators, such as wolves and cougars. Game populations (and, incidentally, rodents) once freed from their predators, may increase explosively in numbers, resulting in overgrazing, habitat degradation, and a collapse of the animal and plant communities (Redford 1992). Overenthusiastic park managers who remove dead trees and underbrush to "improve" a park's appearance may remove a critical resource needed by certain animal species for nesting and overwintering.

Many of the best examples of park management come from Britain, where there is a history of scientists and volunteers successfully monitoring and managing small reserves, such as the Monks Wood and Castle Hill Nature Reserves (Usher 1975, 1991; Steele and Welch 1973; Duffey et al. 1974; Peterken 1982; Pollard 1991). At these sites, the effects of different grazing methods (sheep vs. cattle, light vs. heavy grazing) on populations of wildflowers, butterflies, and birds are closely followed. In a symposium volume entitled *The Scientific Management of Animal and Plant Communities for Conservation* (Duffey and Watt 1971), Morris concluded, "There is no inherently right or wrong way to manage a nature reserve . . . the aptness of any method of management must be related to the objects of management for any particular site. . . . Only when objects of management have been formulated can results of scientific management be applied."

Government agencies and conservation organizations have clearly articulated the protection of rare and endangered species as their top priority. These priorities are often outlined in mission statements, which allow managers to defend their actions. For example, at the Cape Cod National Seashore in Massachusetts, protecting tern nesting habitat has been given priority over the use of the area by off-road vehicles and the "right" of sport fishermen to fish (Figure 15.1).

15.1 Tern nesting habitat in the Cape Cod National Seashore and at nearby beaches is extremely vulnerable to the "wear and tear" that is inevitable in a heavily visited recreation area. (Photograph by David C. Twichell.)

Dealing with the Threats to Parks

Effective management of parks must take into account factors that threaten the biological diversity and ecological health of the park. These include many of the threats detailed in Chapters 6 and 7, including exotic species, low population size among rare species, habitat destruction and degradation, and human use. In 1990, the World Conservation Monitoring Centre and UNESCO conducted a survey of 89 World Heritage sites to gain an idea of the management problems there (World Resources Institute 1992). The results showed a wide range of management problems as well as significant differences among continents (Table 15.1). Threats to protected areas were generally greatest in South America and least in Europe. The most serious management problems in Oceania were introduced plant species, while illegal wildlife harvesting, fire, grazing, and cultivation were the major threats in both South America and Africa. Inadequate park management was a particular problem in the developing countries of Africa, Asia, and South America. The greatest threats faced by parks in industrialized countries were internal and external threats associated with economic developments such as mining, logging, agriculture, and water projects. Although these general patterns give an overview, any single park has its own unique problems—such as illegal logging and hunting in many Central American parks, or the vast numbers of tourists who crowd into Yellowstone National Park in July and August.

Assessing the threats to parks does not necessarily mean attempting to eliminate their presence; in many cases it is nearly impossible to do so. For example, exotic species may already be present inside the park, and new exotic species may be invading along the park boundaries. However, if these species are allowed to increase unchecked, native species and even entire communities may be eliminated from the park. Where an exotic species threatens native species it should be removed if possible, or at least reduced in frequency (Temple 1990). An exotic species that has just arrived and has known noxious tendencies should be aggressively removed while it is still at low densities. European purple loosestrife, which invades North American wetlands, is one such species. Purple loosestrife can outcompete many native plants, often forming pure stands along river and pond edges and in marshes. Purple loosestrife has a detrimental effect on wildlife since it is not eaten by most waterfowl, yet it crowds out beneficial species. Once such an exotic species becomes established in an area, it may be difficult (if not impossible) to eliminate it.

TABLE 15.1
Problems faced by World Heritage sites

Continent	Number of sites	Development[a]	Tourism	External threats[b]	Grazing and cultivation	Illegal wildlife harvesting	Fire, natural threats	Introduced species	Insufficient management[c]
Africa	25	48	16	36	56	68	52	8	52
Asia	10	40	50	50	40	40	40	10	70
Europe	11	45	18	18	27	9	18	27	0
Oceania	10	70	30	10	40	10	40	60	10
South America	8	38	63	63	75	63	88	25	63
North and Central America	21	57	33	43	29	33	24	43	10

Source: World Resources Institute 1992.

[a] Development includes logging, mining, and other human activities originating within the site.
[b] External threats originate outside of site boundaries.
[c] Includes lack of staff, funding, equipment, and training.

The role of exotic species may occasionally be beneficial, as the example of the monarch butterfly illustrates. The monarch (*Danaus plexippus*) is one of the most beautiful and well-known butterfly species of North America. Monarchs that breed west of the Rocky Mountains migrate to overwintering sites spread 500 miles along the Pacific coast, from Marin County to Baja California (Weiss et al. 1991). These sites are forest groves located within 1 kilometer of the ocean, protected from the wind and characterized by moderate temperatures. Most of the aggregations of monarchs occur in forest groves dominated by blue gum trees (*Eucalyptus globulus*), a species introduced into California from Australia about a century ago (Figure 15.2). The flowers of this tree are an important winter nectar source for the monarchs. Protecting these butterfly wintering sites from development has proved to be a controversial issue, because the blue gum itself is an exotic species. Even though blue gum forests are im-

15.2 In California, monarch butterflies aggregate in groves of blue gum eucalyptus trees. Should these trees be protected, despite the fact that they are themselves an exotic species from Australia? (Photograph by Ken Lucas/Biological Photo Service.)

portant for monarchs, they harbor few other species of insects and small mammals. Blue gum forests also pose fire hazards because of the buildup of dead brush. The complexity of these issues creates a difficult problem for those responsible for deciding whether or not to protect blue gum forests.

In the case of rare species whose populations inside a park are too small to be viable (see Chapters 11 and 12), thorough studies and observations of their natural history need to be carried out (Hutchings 1991; Pollard 1991; Baillee 1991). Such information can be used to determine whether the habitat can be managed to increase the species' population size and whether additional populations of the species can be established elsewhere in the park. It might also be possible to periodically augment the population by the release of individuals raised in captivity or by other techniques (see Chapter 18).

Habitat Management

The habitats of a park may have to be aggressively managed to insure that the original habitat types are maintained. Many species occupy only a specific habitat or a specific successional stage of a habitat. When land is set aside as a protected area, the pattern of disturbance and human use may change so markedly that many species previously found on the site fail to persist (Gomez-Pompa and Kaus 1992). Natural disturbances, including fires, grazing, and tree-falls, are key elements in the presence of certain rare species (Box 23). In small parks, the full range of successional stages may not be present and many species may be missing for this reason; for example, in an isolated park dominated by old-growth trees, species characteristic of the early successional herb and shrub stage may be missing (Figure 15.3). If this small park is swept by a fire or a windstorm, the species characteristic of old-growth forest may be eliminated. In many isolated protected areas in metropolitan areas, frequent fires started by people and other human disturbance may eliminate many of the plant and animal species that require the environmental condition of older vegetation. However, the early successional species may also be missing if they are not present in adjacent sites that can serve as colonization sources.

Park managers sometimes must actively manage sites to ensure that all successional stages are present. One common way to do this is to periodically set localized, controlled fires in grassland, shrublands, and forests to re-initiate the successional process. In some wildlife sanctuaries, open fields are maintained by mowing, disking, or grazing livestock in order to maintain a variety of habitat types.

BOX 23 HABITAT MANAGEMENT: THE KEY TO SUCCESS IN THE CONSERVATION OF ENDANGERED BUTTERFLIES

In 1980 the heath fritillary butterfly, *Mellicta athalia*, had the dubious honor of being closer to extinction than any other butterfly species in England. The distribution of the species had declined steadily for 70 years. However, since the butterfly became protected by law in 1981, the number and size of fritillary colonies have stabilized or increased in southern England (Warren 1990).

The short-term success of these conservation efforts is the product of a plan based on detailed ecological studies of the species' habitat requirements. Areas with heath fritillary populations were managed to encourage the habitat types, such as newly felled woodlands and unimproved grasslands, that the species prefers. This task was made particularly difficult because these habitats are ephemeral and patchy by nature, requiring limited disturbance or grazing to maintain populations of the butteflies' food plants. Assessment of the fritillary's progress after nearly a decade of active habitat management maintaining the early-successional food plants demonstrates that human intervention was a significant factor in the success of the colonies. Where habitat management did not occur, the majority of colonies became extinct (Warren 1990).

Although habitat management has been generally successful in preventing the fritillary's extinction, the practice raises the disturbing issue of the extent to which endangered species depend upon human action. The heath fritillary now appears to be utterly dependent upon human intervention for survival; the fate of many other species probably rests entirely in our hands.

The problem faced by the heath fritillary is similar to that of a number of butterfly species that prefer specialized, ephemeral habitats. The silver-studded blue butterfly, *Plebejus argus*, found in the heathlands of East Anglia; the Karner blue, *Lycaeides melissa samuelis*, found in lupine patches of New England; and the bay checkerspot, *Euphydryas editha bayensis*, found in serpentine grasslands in the San Francisco Bay area, are three species that share the fritillary's difficulties. Long-term studies of the checkerspot have been particularly valuable in establishing key factors in the conservation and management of butterfly populations in habitat patches (Ehrlich and Murphy 1987; Murphy and Weiss 1988; Dirig 1988). Many of these species have subtle, specialized habitat require-

The bay checkerspot butterfly, *Euphydryas editha bayensis*. Studies done on this species have been extremely useful in providing principles for managing other endangered butterflies. (Photograph courtesy of Dennis Murphy, Stanford University.)

ments that are best satisfied by extensive and diverse habitat, in which natural processes can create favorable conditions. In the case of the silver-studded blue butterfly, the species is only found in young stands of bell heather (*Erica cinera*), with adults feeding on nectar and larvae feeding on leaves (Ravenscroft 1990). The specialization goes even further because the larvae must be tended by a certain type of black ant (*Lasius* sp.) in order to survive, and the distribution of these ants is variable. When the butterflies' habitats are fragmented, these natural patterns are often interrupted. Species may be unable to locate new suitable habitat due to limited dispersal abilities; the silver-studded blue butterfly in particular seems to be unable to disperse more than 1 km from existing populations. Experimental attempts to establish new populations by carrying adults to unoccupied sites have had some degree of success. Humans can simulate natural processes, as the case of the heath fritillary demonstrates, but in the absence of human intervention these species will most likely become extinct.

But the survival or loss of a single species is not the sole issue. These butterflies are often among the most sensitive indicators of ecological health in remnant habitats (Murphy et al. 1988; Hafernik 1992); if they require management, it is a fairly safe assumption that their habitat has been heavily damaged, to the extent that it is no longer viable without significant restoration. The decline of these species is a warning flag: the habitat is at a point of crisis (Lawton 1991). Habitat management is a solution for short-term population recovery, but it cannot take the place of habitat preservation in the long term. Ultimately, the real key to preventing extinctions among rare butterflies is the prevention of habitat fragmentation and, where possible, the restoration of damaged areas.

For example, many of the unique wildflowers of Nantucket Island off the coast of Massachusetts are found in the scenic heathland areas. These heathlands were previously maintained by sheep grazing; they must now be burned every few years to prevent scrub oak forest from taking over the area and shading out the wildflowers (Figure 15.4A). In other situations, parts of protected areas must be carefully managed to minimize human disturbance and provide the conditions required by old-growth species (Figure 15.4B).

Fire appears to be a major factor maintaining species diversity in Mediterranean-climate shrubland ecosystems, such as the chaparral of California and the the fynbos of South Africa (Kruger et al. 1977, 1984; Christiansen 1985). Speciation in response to the effects of fire appears to be a major factor in the development of the rich flora found in many shrubland communities. As humans have settled in such areas, they have altered the natural fire regime. Frequent intentional fires to encourage the growth of young vegetation, combined with overgrazing, led to severe soil erosion and desertification in many Mediterranean countries. However, the complete suppression of fires also has negative effects, since a buildup of dead wood can result in infrequent accidental fires that are widespread, very hot, and

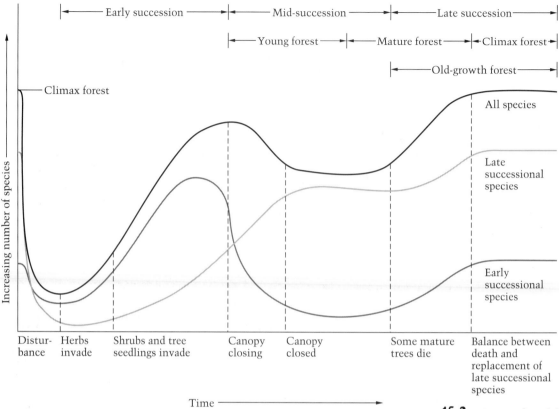

15.3 A general model of the change in species diversity during succession following a major disturbance such as a fire, hurricane, or clear-cut logging. Early successional species are generally fast-growing and intolerant of shade; late successional species grow more slowly and are shade-tolerant. The full successional time span covers many years. (After Norse et al. 1986.)

very destructive. A regular schedule of controlled burns appears to be the best strategy for maintaining species diversity and vegetation structure, and minimizing fire damage in shrubland vegetation (Figure 15.5; Brown et al. 1991).

Wetlands management is a particularly crucial issue. The maintenance of wetlands is necessary for populations of waterbirds, fish, amphibians, aquatic plants, and a host of other species (Moyle and Leidy 1992). Yet parks may become direct competitors for water resources with irrigation projects, flood control schemes, and hydroelectric dams in places such as the floodplains of India (Pandit 1991) and the Florida Everglades. Only a finite amount of water is available and wetlands are often interconnected, so a decision affecting water in one place has repercussions on other areas. Park managers may have to become politically sophisticated and effective at public relations to ensure that the wetlands under their supervision continue to receive the water they need to survive.

(A)

15.4 Conservation management: intervention versus leave-it-alone. (A) Heathland in protected areas of Cape Cod is burned on a regular basis in order to maintain the open vegetation habitat and protect wildflowers and other rare species. (Photograph by P. Dunwiddie.) (B) Sometimes management involves keeping human disturbance to an absolute minimum. This old-growth stand in the Olympic National Forest is the result of many years of solitude. (Photograph by Thomas Kitchin/Tom Stack & Associates.)

(B)

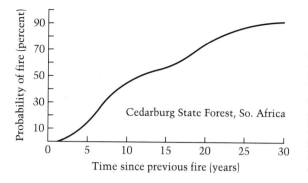

15.5 In many areas with a Mediterranean climate, such as South Africa, the probability of a fire having burned the shrubland vegetation increases with each year since the previous fire. In any 15-year period, a patch of shrubland has about a 60% chance of being burned. Preventing fires in such areas is unnatural and, indeed, almost impossible. In managed shrubland, controlled burning is often used to maintain the ecosystem. (From Brown et al. 1991.)

Case Studies: Two Fens

The need for managing habitat to maintain populations of rare species is illustrated by two examples from fens (a type of wet meadow), one in the United States and one in England. Crystal Fen in northern Maine is recognized for its numerous rare plant species (Jacobson et al. 1991). An apparent drying of the fen and an increase in woody vegetation was attributed by some biologists to the construction of a railroad in 1893 and a drainage ditch in 1937; there was concern that the fen community might be lost. Subsequent studies using aerial photography, vegetation history, and dated fossil remains from peat layers collectively showed that the construction of the railroad bed had allowed the wetland to *expand* in area by impeding drainage. The fen also increased in area following fires started by cinder-producing locomotives. Today, the large area of fen in which rare plants occur is primarily a product of human activity. The construction of the drainage ditch and the decrease in fires following the change to diesel-powered engines are allowing the vegetation to return to its original state. If the goal is to maintain the current extent of the fen and the populations of rare species, management practices involving periodic burning, removing woody plants, and manipulating drainage patterns are necessary.

Controlled disturbance is also necessary to maintain some species in English fens. The fen violet, *Viola persicifolia*, has declined rapidly in Europe as its habitat of open, alkaline, peaty areas such as fens has been drained and altered. In England the species is known from only two sites in Cambridgeshire (Pullin and Woodell 1987). At Woodwalton Fen National Nature Reserve, the species had apparently disappeared, but it was found again after considerable soil disturbance caused by commercial digging of peat and scrub clearance. The ability of populations to reappear after several decades suggests that the species has long-lived seeds that germinate opportunistically when soil is brought to the surface after trees are blown over. A management policy involving destruction of vegetation and soil disturbance appears to be necessary for the continued existence of the fen violet in England.

The type of controlled disturbance that provides optimal results can be determined through field experiments. For example, chalk grasslands in Britain require specific management measures to maintain a biologically rich community. Experiments have shown that the number of species, the particular species present, and the species' relative abundance are determined by the management regime: whether the grassland is grazed, open, or burned; the time of year of the management; and whether the management is carried out continuously, annually, or rotationally (Morris 1971). Certain management regimes favor certain groups of species over others.

Landscape Ecology and Park Design

The interaction of actual land use patterns and conservation theory is evident in the discipline of **landscape ecology**. Landscape ecology investigates patterns of habitat types on a regional scale and their influence on species distribution and ecosystem processes (Naveh and Lieberman 1984; Forman and Godron 1986; Urban et al. 1987). A landscape is defined by Forman and Godron (1981) as an "area where a cluster of interacting stands or ecosystems is repeated in similar form" (Figure 15.6). Landscape ecology has been more intensively studied in the human-dominated environments of Europe than in North America, where research in the past has emphasized single habitat types.

Landscape ecology is important to the protection of biological diversity because many species are not confined to a single habitat, but move between habitats or live on borders where two habitats meet. For these species, the patterns of habitat types on a regional scale are of critical importance. The presence and density of many species may be affected by the size of habitat patches and their degree of linkage. For example, the population size of a rare animal species will be different in two 100-ha parks, one with an alternating checkerboard of 100 patches of field and forest, each 1 ha in area, the other with a checkerboard of 4 patches, each 25 ha in area (Figure 15.7). These alternative landscape patterns may have very different effects on the microclimate (wind, temperature, humidity, and light), pest outbreaks, and animal movement patterns, as described in Chapter 6. Different land uses often result in dramatically contrasting landscape patterns. Forest areas cleared for shifting agriculture, suburban development, intensive subsistence agriculture, and plantation agriculture

(A) Scattered patch landscapes

Open clearings
in a forest

Groves of trees
in a field

(B) Network landscapes

Network of roads
in a large plantation

Riparian network of
rivers and tributaries
in a forest

(C) Interdigitated landscapes

Tributary streams
running into a lake

Shifting
forest-grassland
borders

(D) Checkerboard landscapes

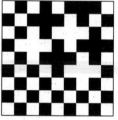

Farmland under
cultivation for
different crops

Lots in a residential
development

15.6   **Four renditions of landscape types where interacting ecosystems or other land uses form repetitive patterns. The discipline of landscape ecology focuses on such interactions rather than on a single habitat type. The first pattern (forest clearings) is illustrated in Figure 16.2. (From Zonneveld and Forman 1990.)**

have totally different appearances from the air, and differing distributions and sizes of remnant forest patches.

A landscape of large patch sizes and minimal edge effects is generally favored by conservation biologists; this pattern minimizes habitat disturbance and protects species that depend on old-growth vegetation. In contrast, wildlife managers interested in increasing species such as deer have observed that the overall abundance of birds and mammals is *greatest* at the edges of habitats (Noss 1983). At edges, foods from two habitat types are present, primary productivity (the amount of plant growth per year) is often high, and certain species from both communities live close together. In addition, many animal species appear to be specialists on conditions at the edges of habitats.

To increase the number and diversity of animals, wildlife managers often attempt to create the greatest amount of landscape variation possible within the confines of their management unit (Yahner 1988).

15.7 Two square nature reserves, each 100 ha in area (1 km on a side). They have equal areas of forest (shaded) and pasture (unshaded), but in very different size patches. Which landscape pattern benefits which species? This is a question managers will encounter.

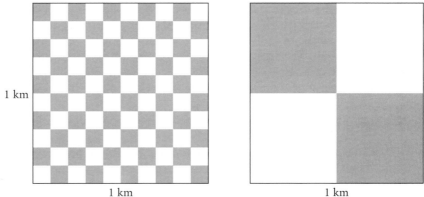

1 km

1 km   1 km

Fields and meadows are created and maintained, small thickets are encouraged, fruit trees and crops are planted on a small scale, small patches of forests are periodically cut, small ponds and dams are developed, and numerous trails and dirt roads meander across and along all of the patches (Noss 1983). The result is a park transformed into a mass of edges, where transition zones abound. In one textbook of wildlife management, managers are advised to "develop as much edge as possible" because "wildlife is a product of the places where two habitats meet" (Yoakum and Dasmann 1971). Other wildlife biologists have suggested that "a diverse wildlife population will require a planning approach that ensures a diverse environment" (Sideritis and Radtke 1977).

However, the conservation biologist's goal of maintaining biological diversity is not just to include as many species as possible within small nature reserves, but also to protect those species most in danger of extinction as a result of human activity. Small reserves broken up into many small habitat units within a compressed landscape may have a large number of species, but these are likely to be principally "weedy" species—species that depend on human disturbance—and non-native species (Yahner 1988; Noss 1983). A park that contains the maximum amount of edge may lack the truly rare species that inhabit large blocks of undisturbed habitat; these interior species may be unable to survive due to the harmful edge effects discussed in Chapter 6. The net result is that many parks intensively managed for wildlife and habitat diversity may lack any species of true conservation significance.

To remedy this localized approach, biological diversity needs to be managed on a regional landscape level in which the size of the landscape units more closely approximates the natural units prior to human disturbance (Noss 1983; Salwasser et al. 1987). An alternative to creating a miniature landscape of contrasting habitats on a small scale is to link all parks in an area in a regional plan in which larger habitat units could be created—an approach that was discussed in Chapter 14. Some of these larger habitat units would then be large enough to protect rare species that are not able to tolerate human disturbance.

Keystone Resources

In managing parks, attempts should be made to preserve and maintain **keystone resources**—sources of food, water, minerals, protection, etc. on which many species depend. If it is not possible to keep these keystone resources intact, attempts might be made to reconstruct them. For example, an artificial salt lick could be built in place of one that was destroyed, or artificial pools could be built in stream beds to provide replacement water supplies. Keystone resources and keystone species (see Chapter 2) could conceivably be enhanced in managed conservation areas to increase the populations of species whose members have declined. For example, by planting fig and fruit trees, building an artificial pond, and providing salt licks, it might be possible to maintain vertebrate species in a smaller conservation area at higher densities than would be predicted based on studies of species distribution in undisturbed habitat. Another example is the provision of nesting boxes for birds as a substitute resource when there are few dead trees with nesting cavities. In this way a viable population of a rare species could be established, whereas without such interventions the population size of the rare species might be too small to persist. In each case it has to be determined where the balance lies between establishing nature reserves free from human influence and creating seminatural gardens in which the plants and animals are dependent on people.

Park Management and People

The use of parks by local people and outside visitors must be a central part of any management plan, both in developed and developing countries (Mackinnon et al. 1992). Many parks flourish or are destroyed depending on the degree of support, neglect, hostility, or exploitation they receive from the humans who use them. If the purpose of a protected area is explained to local residents and most residents agree with the objectives and respect the rules of the park, then the area may maintain its natural communities. In the most positive scenario, local people become involved in park management and planning, are trained and employed by the park authority, and benefit from the protection and regulation of activity within the park. At the other extreme, if there is a history of bad relations and mistrust between local people and the government, or the purpose of the park is not explained adequately, the local people may reject the park concept and ignore park regulations. In this eventuality, the local people will come into conflict with park personnel to the detriment of the park.

Regulating Activities inside Parks

Certain human activities are incompatible with maintaining a protected conservation area. If these activities continue, eventually the biological communities will be destroyed. Some of these include:

- *Commercial harvesting of animals*, including hunting using high-powered guns, fishing using destructive methods such as poisons and dynamite, and commercial trapping. Some regulated hunting and fishing for personal consumption and for sport may be acceptable, but harvesting for commercial sale almost invariably leads to the elimination of species. Regulated hunting may be necessary to control exotic animals and herbivore populations in areas with reduced populations of carnivore species.
- *Intensive harvesting of natural plant products*, such as fruits, fibers, and resins. Again, collection for personal use may be acceptable, whereas commercial harvesting is detrimental. Even personal collecting may be unacceptable in national parks with tens of thousands of visitors per year, or where the local human population is large in relation to the area of the park.
- *Illegal logging and farming*. These activities degrade the habitat and eliminate species. Where these activities are large-scale, commercial in nature, and controlled by outside interests, they must be eliminated if possible. However, when logging and farming are done by local people in order to supply their basic needs, stopping them becomes very difficult and may be counterproductive.
- *Fires* set accidentally or deliberately by local people to open up farmland, to provide forage for livestock and wildlife, and to reduce undesirable species. While occasional fires of limited area may help to create a variety of successional stages, fires that are more frequent than would occur naturally can dry out a habitat, cause soil erosion, and eliminate many native species.
- *Recreational activities* popular in some developed countries can eliminate sensitive plants and animals from protected lands. Activities such as hiking off trails, camping outside designated

areas, and riding motorcycles, off-road vehicles, and mountain bikes must be controlled and restricted to specified areas by park managers. In open, sandy terrain such as dunes and deserts, off-road vehicles have damaged fragile vegetation and shorebird nesting sites. In heavily used parks, frequent traffic by hikers wearing heavy boots has degraded vegetation along trails (Bratton 1985). Redwood trees in California parks may be harmed by too many park visitors walking around the trunks, compressing the soil too tightly.

Even in a well-regulated park, outside influences from air pollution, acid rain, water pollution, global climate change, and changing composition of atmospheric gases will influence natural communities, causing some species to increase and others to decrease and be eliminated. Unfortunately, natural history studies show that invasive exotic species are likely to be the main beneficiary of an altered environment, since they tend to be adaptable, efficient dispersers able to tolerate disturbance. The ability of park managers to deal with the major externally driven alterations in ecosystem processes is rather limited. Experiments are being attempted in which basic compounds (such as lime) are added to water bodies to counter the effects of acid rain (Moyle and Leidy 1992). However, such measures will never take the place of needed environmental reforms to human production and consumption patterns.

National Parks and Local People

Park managers throughout the world frequently cite conflicts with local people as their most serious problem. In a survey of 98 national parks throughout the world, about half of the parks reported problems associated with illegal collection of wildlife, removal of plant materials, poor relations with local people, and conflicting demands on park resources (Machlis and Trahnell 1985). In the developing world, local people typically obtain the products they need—including food, fuelwood, and building materials—from their environment (MacKinnon et al. 1992). Without these products, some local people may not be able to survive. If the park provides overall benefits to local communities in terms of employment, revenue sharing, and regulated access to natural products, then the community may accept and support the park. However, when a new park is created, or if the boundaries of an existing park are rigidly enforced, people may be denied access to a resource that they have traditionally used. Most local people will react cautiously and even antagonistically when their traditional rights are curtailed. It is to be expected that when the existence of a park prevents local people from obtaining resources they need to survive, they will resist park regulations (Clay 1991; Dasmann 1991). In order to survive, local people will violate the park boundaries, sometimes resulting in confrontation with park officers. In effect, the creation of a national park often makes local people into poachers, even though they have not changed their behavior. Even worse, if local people feel that the park and its resources no longer belong to them, but rather to an outside government, they may begin to exploit the resources of the park in a destructive manner.

An extreme example of such a conflict occurred in 1989, when angry members of the Bodo tribe in Assam, India, killed 12 employees of the Manas National Park and opened the area for farming and hunting (McNeely et al. 1990). The Bodos justified their action on the basis that they were reclaiming their traditional lands that had been stolen from them by the British and not returned to them by the modern Indian government. The fact that Manas has been designated a World Heritage Site and contains such endangered species as the Indian rhinoceros and the pygmy hog was not relevant to the Bodos; the advantages of the national park were not apparent to them.

In the developing world, a rigid separation between lands used by local people to obtain natural resources and strictly protected national parks is often not possible (McNeely and Miller 1984; McNeely 1993a,b). Many examples exist in which people are allowed to enter protected areas periodically to obtain natural products. In Biosphere reserves, local people are allowed to use resources from designated buffer zones. In another example, local people are allowed to collect canes and thatch from Chitwan National Park in Nepal (Figure 15.8; Lehmkuhl et al. 1988). Large game animals are harvested for meat in many African game reserves (Lewis et al. 1990). Through such compromises, local people are considered in local management plans, to the benefit of both the people and the reserve.

Local people sometimes even take the lead in protecting biological diversity from destruction by outside influences such as mining and dam construction. The destruction of communal forests by government-sanctioned logging operations has been a frequent target of protests by traditional people throughout the world (Poffenberger 1990). In India, followers of the Chipko movement hug trees to prevent logging (Guha 1989a; Gadgil and Guha 1992). In Borneo, the Penans, a small tribe of hunter-gatherers, have attracted worldwide attention by their blockades of logging roads entering into their traditional forests. In Thailand, Buddhist priests are working with vil-

15.8 Local residents collect cane grass and thatching materials from Chitwan National Park in Nepal. Park officials weigh the bundles in order to keep the harvest at a sustainable level. (Photograph © John F. Lehmkuhl.)

lagers to protect communal forests and sacred groves from commercial logging operations (Figure 15.9). As stated by a Tampon leader in Thailand, "This is our community forest that was just put inside the new national park. No one consulted us. We protected this forest before the roads were put in. We set up a roadblock on the new road to stop the illegal logging. We caught the district police chief and arrested him for logging. We warned him not to come again" (Alcorn 1991).

An Experiment in Zambia

Africa is known for its abundant and spectacular wildlife. African wildlife has declined dramatically during the last several decades, despite attempts by national governments to impose a conservation policy and to establish effective national parks. There is now increas-

15.9 Buddhist priests in Thailand offer prayers and blessings to protect communal forests and sacred groves from commercial logging operations. (Photograph by Project for Ecological Recovery, Bangkok.)

ing recognition that involvement of local people is the crucial, missing element in conservation management strategies. "Top-down" strategies, in which governments try to impose conservation plans, need to be integrated with "bottom-up" programs, in which villages and other local groups are able to formulate and reach their own development goals (Clay 1991). As explained by Lewis et al. (1990):

> If any lesson can be learned from past failures of conservation in Africa, it is that conservation implemented solely by government for the presumed benefit of its people will probably have limited success, especially in countries with weakened economies. Instead, conservation for the people and by the people with a largely service and supervisory role delegated to the government could foster a more cooperative relationship between government and the residents living with the resource. This might reduce the costs of law enforcement and increase revenues available to other aspects of wildlife management, which

could help support the needs of conservation as well as those of the immediate community. Such an approach would have the added advantage of restoring to local residents a greater sense of traditional ownership and responsibility for this resource. Convincing proof that such a partnership is possible has yet to be demonstrated and has therefore been more theoretical than practical.

In Africa, some local residents and conservation officers are working together to increase the level of community involvement in national park wildlife management. An example of the possible effectiveness of this approach is a new program initiated in the Luangwa River valley in eastern Zambia.

The Luangwa River valley, which supports a world-famous concentration of wildlife, contains four national parks in which people are not allowed to live or hunt (Lewis et al. 1990). Surrounding the parks, and occupying a much greater total area, are game management areas, in which tribal people live and in which hunting is allowed only with permits. However, the hunting licenses are so expensive that local residents cannot afford them, so licenses are only taken out by international safari companies. Local residents resent the licensing system that prevents them from obtaining the meat that they need to feed their families but permits sport hunters to take trophies. From a national economic perspective, the single greatest revenue earner in the game management areas is the license fees paid by international safari companies—but less than 1% of this revenue is returned to the economy of the local village where the hunting takes place. Not surprisingly, local residents have come to resent the park, the tourists and the safaris. Chief Matama summarized the feelings of his people in a speech: "Tourists come here to enjoy the lodges and to view the wildlife. Safari companies come here to kill animals and make money. We are forgotten.... Employment here is too low. Luangwa Lodge employs only about four people, and safari hunting employs no one. How can you ask us to cooperate with conservation when this is so?" (quoted in Lewis et al. 1990).

As a consequence of these attitudes, illegal poaching of elephants, rhinos, and other animals increased dramatically during the 1970s and 1980s and wildlife populations began to plummet. Personnel from the National Parks and Wildlife Service (NPWS) were ineffective at stopping the poaching; since they were not from the local area, they were unwelcome in local villages and could not patrol the game management areas actively. Local people perceived the NPWS personnel as acting only in the interests of the safari companies, with little regard for local concerns; poaching at least resulted in having meat to eat and the chance to make some money.

To remedy this deteriorating situation, an experimental program was initiated in the Lupande Game Management Area, which occupies 4849 km^2 and has 20,000 residents. The key elements of the program were:

1. Local residents were hired and trained as scouts. These scouts remained in their own village areas to act as wildlife custodians.
2. All workers hired for building construction, wildlife maintenance, and seasonal work came from the local villages.
3. Revenues were provided to the program by allowing safari companies to bid for hunting rights, with 40% of this revenue returned to local villages for community development projects and 60% used for carrying out wildlife management. Additional revenue would be provided by sustained-yield harvesting of hippos and commercial sale of meat, hides, and teeth.
4. Decisions on wildlife management and employment practices were made following discussions with tribal chiefs and village leaders. Village leaders decided for themselves how best to spend the community development funds.

In the three years since its initiation, this program appears to have been successful at addressing many of the earlier problems of wildlife management (Lewis et al. 1990). Village scouts patrol actively, catching poachers and seizing firearms; as a result, illegal killing of elephants and black rhinos has declined by 90%. Local people have begun to appreciate the employment opportunities and revenue provided by the program. Poachers from outside the area are now being reported to the village scouts, since the poachers are now seen as a threat to a community resource. The income generated through regulated, sustainable safari hunting is apparently sufficient for the expenses of the program.

The Luangwa Valley example demonstrates that involving local people in wildlife management and revenue sharing offers a potentially effective way of reconciling conflicts between local people and conservation policies. However, it remains to be seen whether the village scouts can continue to enforce conservation laws if their own relatives and fellow villagers take up poaching during times of unemployment and hunger. The conflict between park employment and kin responsibilities may become impossible to reconcile.

Summary

1. Protected areas often must be managed in order to maintain their biological diversity because the original conditions of the area have been altered by human activities. In some cases the best management involves doing noth-

ing. Effective management begins with a clearly articulated statement of priorities.

2. Parts of protected areas may have to be periodically burned, dug up, or otherwise disturbed by people to maintain the habitat types and successional stages that certain species need. Such management is crucial, for example, to some endangered butterfly species that need early successional food plants to complete their life cycle.

3. In the past, wildlife biologists advocated creating a mosaic of habitats with abundant edges. While this landscape design often increases the number of species and the overall abundance of animals, it does not generally favor the species of greatest conservation value, which typically occupy large blocks of undisturbed habitat.

4. Keystone resources may need to be preserved, restored, or even added to protected areas in order to maintain populations of some species.

5. Local residents and outside visitors are key elements in park management. Protected areas may flourish or be destroyed depending on how people view them. A crucial element is finding the compromise between the extremes of banning human use of park resources and allowing unlimited use, a compromise that allows people to use park resources in a sustainable manner without harming biological diversity. Allowing limited harvesting of resources, hiring local residents as park employees, and returning revenues from park visitors to the local community are some ways to enlist cooperation.

Suggested Readings

Chase, A. 1986. *Playing God in Yellowstone: The Destruction of America's First National Park*. Atlantic Monthly Press, Boston. A case study of overmanagement of a protected area.

Duffy, E. and A. S. Watt (eds.). 1971. *The Scientific Management of Animal and Plant Communities for Conservation*. Blackwell Scientific Publications, Oxford. Examples of intensive management of small conservation areas.

MacKinnon, J., G. Child and J. Thorskell (eds.). 1992. *Managing Protected Areas in the Tropics*. IUCN, Gland, Switzerland. An excellent introduction to management issues.

Naveh, Z. and A. S. Lieberman. 1984. *Landscape Ecology: Theory and Applications*. Springer-Verlag, New York. A good introduction to the subject.

Peterken, G. F. 1982. *Woodland Conservation and Management*. Chapman and Hall, New York. Some carefully studied examples from Britain.

Poffenberger, M. (ed.). 1990. *Keepers of the Forest*. Kumarian, West Hartford, CT. Local people often protect biological diversity; governments often favor overexploitation of resources.

Poore, D. and J. Sayer. 1991. *The Management of Tropical Moist Forest Lands: Ecological Guidelines*. IUCN, Gland, Switzerland. A comprehensive guide to management issues.

Salm, R. and J. Clark. 1984. *Marine and Coastal Protected Areas: A Guide for Planners and Managers*. IUCN, Gland, Switzerland. Special methods are needed in these environments.

Warren, M. S. 1990. The successful conservation of an endangered species, the Heath Fritillary Butterfly *Mellicta athalia*, in Britain. *Biological Conservation* 55: 37–56. Case study of management for species preservation.

Yahner, R. H. 1988. Changes in wildlife communities near edges. *Conservation Biology* 2: 333–339. Managing for the most species in contrast to managing for rare species.

Zonneveld, I. S. and R. T. Forman (eds.). 1990. *Changing Landscapes: An Ecological Perspective*. Springer-Verlag, New York. Articles by leading authorities.

CHAPTER 16

Outside Protected Areas: Working with People and Restoring the Environment

A danger of relying on parks and reserves to protect biological diversity is that they can create a "siege mentality," in which species inside the parks are to be rigorously protected while species outside can be rapidly exploited. As McNeely (1989) remarks:

> By their very nature as legally established units of land management, national parks have boundaries. Yet nature knows no boundaries, and recent advances in conservation biology are showing that national parks are usually too small to effectively conserve the large mammals or trees that they are designed to preserve. The boundary post is too often also a psychological boundary, suggesting that since nature is taken care of by the national park, we can abuse the surrounding lands, isolating the national park as an 'island' of habitat which is subject to the usual increased threats that go with insularity.

This sentiment is amplified by Van Tighem (1986), who feels that national parks

> have not drawn us into a more thoughtful relationship with our habitat. They have not taught us that land is to be used frugally, and with good sense. They have encouraged us to believe that conservation is

merely a system of trading environmental write-offs against large protected areas. They have more than failed, in fact; they have become a symptom of the problem.

The strategy of establishing large national parks to protect biological diversity and establishing biosphere reserves to protect traditional hunter–gatherer societies has readily been adopted by countries such as Brazil and Malaysia to deflect international criticisms of their intensive land development policies. However, if these parks are not managed effectively and the areas outside the parks are degraded, then the biological diversity within the protected areas will decline (Table 16.1). In such cases, new national parks are just bones thrown to the international conservation community; they are answers to an immediate criticism, not solutions to the long-term problem.

A crucial element in conservation strategies must be the protection of biological diversity outside as well as inside protected areas. As stated by Western (1989), "if we can't save nature outside protected areas, not much will survive inside." More than 90% of the world's land will remain outside of protected areas, according to even the most optimistic predictions. The majority of these unprotected lands are not used intensively by humans, and still harbor some of their original biota. Strategies for reconciling human needs and conservation interests in these unprotected areas are critical to the success of conservation plans. Since the majority of the land area in

TABLE 16.1

The number of large herbivore species currently in some East African national parks, and the number expected to remain if areas outside the parks become unavailable to wildlife

National park	Area (km^2)	Number of species in park	
		Now	If areas outside parks exclude wildlife[a]
Serengeti, Tanzania	14,504	31	30
Mara, Kenya	1,813	29	22
Meru, Kenya	1,021	26	20
Amboseli, Kenya	388	24	18
Samburu, Kenya	298	25	17
Nairobi, Kenya	114	21	11

Source: Data from Western and Ssemakula 1981.

[a] Estimated number of species that will remain if areas outside the protected parks exclude wildlife due to agriculture, hunting, herding, or other human activities.

most countries will never be in protected areas, numerous rare species will inevitably occur outside protected areas. In Australia, for example, 79% of the endangered and vulnerable plant species occur outside protected areas (Leigh et al. 1982).

Conservation strategies in which private landowners are educated and encouraged to protect rare species are obviously the key to the long-term survival of many species. Government endangered species programs in many countries inform road builders and developers of the locations of rare species, and assist them in modifying their plans to avoid damage to the sites. Forests that are selectively logged on a long cutting cycle or are used for traditional shifting cultivation by a low density of farmers may still contain a considerable percentage of the original biota (Johns 1985, 1987; Thiollay 1992). In Malaysia, the majority of bird species are still found in rain forests 25 years after selective logging when undisturbed forest is available nearby to act as a source of colonists (Wong 1985).

Native species can often continue to live in unprotected areas, especially when those areas are set aside or managed for some other purpose that is not harmful to the ecosystem. Security zones surrounding government installations are some of the most outstanding natural areas in the world. In the United States, excellent examples of natural habitat occur on military reservations such as Fort Bragg in North Carolina; nuclear processing facilities such as the Savannah River site in South Carolina; and watersheds adjacent to metropolitan water supplies, such as the Quabbin Reservoir in Massachusetts. Although dams, reservoirs, canals, dredging operations, port facilities, and coastal development destroy and damage aquatic communities, some species are capable of adapting to the altered conditions, particularly when the water itself is not polluted. In estuaries and seas managed for commercial fisheries, many of the native species remain, since commercial and non-commercial species alike require that the chemical and physical environment not be damaged.

Other areas that are not protected by law may retain species because the human population density and degree of utilization is typically very low. Border areas such as the demilitarized zone between North and South Korea often have an abundance of wildlife because they remain undeveloped and depopulated. Mountain areas are often too steep and inaccessible for development. These areas are often managed as watersheds by governments for their value in producing a steady supply of water and preventing flooding, yet they also harbor natural communities. Likewise, desert species may be at less risk than other unprotected communities because desert regions are considered marginal for human habitation and use. In many parts of the world, wealthy individuals have acquired large tracts of land for their personal estates and for private hunting. These private estates frequently are used at very low intensity, often in a deliberate attempt by the landowner to maintain large wildlife populations. Some estates in Europe have been owned and protected for hundreds of years by royal families, protecting unique old-growth forests.

Large parcels of government-owned land in many countries are designated for multiple use. In the past, these uses have included logging, mining, grazing, wildlife management, and recreation. Increasingly, multiple-use lands are also being valued and managed for their ability to protect species and biological communities (Norse et al. 1986; Thomas and Salwasser 1989; Salwasser 1991a; Poore and Sayer 1991). Laws and court systems are being used to halt activities that are harmful to the survival of endangered species.

Questions have been raised about the way in which the U.S. Forest Service has been applying the multiple-use concept to the Nicolet and Chequamegon National Forests in Wisconsin. These forests have been managed by the Forest Service for timber production and deer hunting. Deer often increase in numbers following the fragmentation of forest blocks by logging because of vigorous plant growth along roads and in cut-over areas (Alverson et al. 1988). Deer populations are further encouraged when the Forest Service makes "wildlife openings" in the forest—which in effect are deliberate fragmentations of the habitat. As a result of such management, deer populations in these national forests have increased from an original density of 2–4 animals per km^2 to 5–12 animals per km^2. Since deer can forage several kilometers into the forest from an opening, the entire forest has been turned into edge habitat. This excessive deer population is now overgrazing the understory throughout the entire forest, preventing the regeneration of many woody plant species. In addition, this overgrazing threatens the existence of some 20 rare plant species that are protected under the 1976 National Forest Management Act, which established protection of biodiversity as one of the goals of national forest management policy (Mlot 1992).

About ten years ago, a group of Wisconsin botanists advised the Forest Service that the best way to maintain biological diversity in the Nicolet and Chequamegon National Forests would be to set aside 200–400 km^2 blocks of land surrounding the habitats of the rare plant species, excluding logging, road construction, and wildlife openings in these areas. When the Forest Service did not accept this suggestion and continued with business as usual, the botanists felt they had no choice but to bring a lawsuit against the Forest Service (Mlot 1992). Once the suit was filed, private conservation groups such as the Sierra Club joined the case on the side of the botanists, while a coalition of interests representing logging, hunting, and snowmobiling groups organized to oppose the botanists. The case is now being decided in a U.S. District Court.

Some North American Case Studies

The snail kite. The snail kite (or Everglades kite), *Rostrhamus sociabilis*, is a rare bird species found in southern Florida; it is protected under the U.S. Endangered Species Act (Takegawa and Beissinger 1989). In recent years the number of snail kites has fluctuated between 250 and 670 individuals, after reaching a low point of 25–60 birds during the 1960s. In most years, the kites feed on snails in large wetlands in Loxahatchee National Wildlife Refuge. However, in exceptionally dry years, the kites are forced to leave the refuge for canals, flooded fields, and small permanent marshes where they are vulnerable to shooting and accidents. These wetland habitats outside of the system of protected areas are being rapidly developed, posing a serious threat to the long-term survival of the snail kite (Graham 1990b).

Mountain sheep. These animals often occur in isolated populations in steep, open terrain surrounded by large areas of unsuitable habitat (Bleich et al. 1990). Since mountain sheep were considered to be slow colonizers of new habitat, conservation efforts have focused on the protection of known mountain sheep habitat and the release of sheep into areas that were occupied in the past. However, recent studies using radio telemetry have revealed that mountain sheep often move well outside their normal territories and even show considerable ability to move across inhospitable terrain between mountain ranges. The isolated mountain sheep populations are really part of a large metapopulation that occupies a much greater area (see Figure 12.1). These observations emphasize the need to protect not only the land occupied by mountain sheep but the habitat between populations that can act as "stepping stones" for dispersal, colonization and gene flow.

The Florida panther. The Florida panther, *Felis concolor coryi*, is an endangered subspecies of puma in South Florida with probably no more than 50 individuals (Maehr 1990). The panther was designated the Florida state animal in 1982 and has since received a tremendous amount of government and research attention. It is an example of a large, wide-ranging species that is difficult to preserve within a protected conservation area. Half of the present range of the panther is in private hands, and animals tracked with radio collars have all spent at least some of their time on private lands (Figure 16.1). In ad-

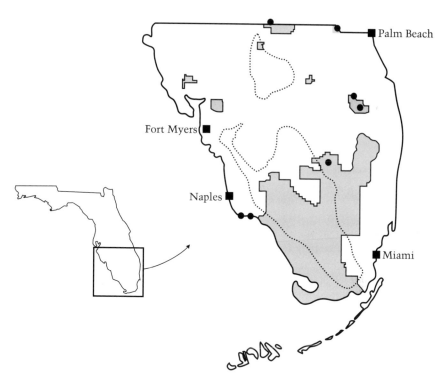

16.1 The Florida panther is found on both public and private lands in South Florida. The dotted lines enclose areas known to be used by radio-collared panthers; black dots represent sightings or other signs of uncollared panthers. Public lands are shaded gray. (From Maehr 1990.)

dition, the private lands are typically on better soils than the state's public lands and support more prey species, so panthers spending most of their time on private lands have a better diet and are in better condition than panthers on public land. Acquiring the 400,000 hectares of private land occupied by the panther would be financially and politically difficult, and even slowing down the pace of land development may not be possible, as South Florida is one of the most rapidly growing areas of the United States. Two other possibilities are educating private owners on the value of conservation, and payments to landowners willing to practice management options that allow the continued existence of panthers.

Managed Coniferous Forests

The coniferous forests of the Pacific Northwest of the United States are managed for timber production, but also contain numerous unique species (Hansen et al. 1991; Franklin and Forman 1987). In this ecosystem, the issue of timber production versus the conservation of biodiversity has become a highly emotional and political issue, cast in terms of "owls versus jobs" (see Box 33 in Chapter 19).

However, research on the ecology of these forests has indicated possible compromise solutions to this apparently insurmountable dilemma.

Following fires, windstorms, or other disturbances, forests in the Pacific Northwest pass through four distinct successional stages: an early successional stage (0–20 years), dominated by herbs and shrubs; young forest (20–80 years), characterized by vigorously growing trees; mature forest (80–200 years) with a declining growth rate and closed canopy; and old-growth forest (more than 200 years) in which shade-tolerant conifers replace the dead and dying pioneer trees (see Figure 15.4B). Old-growth forests are periodically damaged by fire and windstorms, and the damaged patches return to the early successional stage. Despite the great difference in vegetation structure among these four stages, taxonomic studies have shown that there are relatively few differences in animal species composition and numbers of species among them. The main difference is that several species of bats and cavity-nesting birds, such as the northern spotted owl, are confined to old-growth forests. The reason for the similarity is that all four of these forest types typically have at least a few old, large trees surviving, some dead standing trees, and some fallen trees that remain even after fires and storms. These resources are sufficient to support a complex community of plants and animals. However, current logging practices remove the living and dead trees of all ages in order to maximize wood production, which reduces structural complexity in the next forest cycle. In managed forests in the Pacific Northwest, the current practice of clear-cutting staggered patches of

16.2 Staggered harvesting of trees in managed forests of the Pacific Northwest produces a striking mosaic landscape of forest fragments. Within each patch, all vegetation is at the same successional stage. (Photograph by Al Levno.)

16.3 (A) The conventional clear-cutting illustrated in Figure 16.2 involves removing all trees from an area on a 70-year cycle, thus reducing the structural diversity of the forest. (B) Proposed new practices would better maintain structural diversity by leaving behind some old trees, standing dead trees, and fallen trees. (From Hansen 1991.)

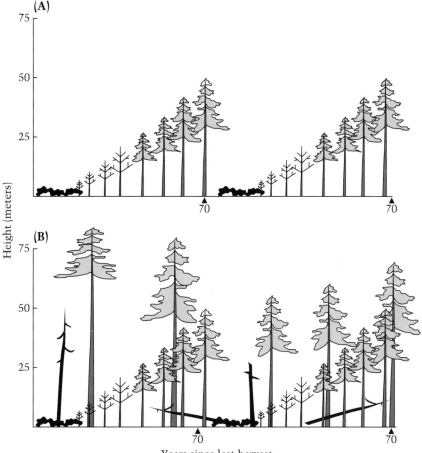

timber produces a landscape pattern that is a mosaic of forest fragments of different but uniform ages (Figure 16.2).

Current research suggests that managed forests can contain considerable biological diversity and suggests specific ways in which conifer forests can be managed to both produce timber and maintain a greater number of species. These lessons are being incorporated into the concept of the "new forestry" being advocated for the Pacific Northwest (Franklin 1989; Gillis 1990). This interesting but still untested method involves clear-cutting but leaving a low density of large live trees, in addition to standing dead trees and some fallen trees, to serve as habitat for animal species in the next forest cycle (Figure 16.3). This practice would particularly benefit cavity-nesting birds such as the spotted owl. Forest fragmentation might also be reduced by harvesting in patterns that maintain forest in larger blocks than previously. These alternative harvesting methods have policy and economic implications. If forests can be harvested in a way that allows all of the original species to be retained, then this would result in an improvement in current logging practices. However, the "new forestry" requires a reduced harvest of timber at the time of cutting and a somewhat longer cutting cycle, resulting in a lower profitability for the timber industry. The "new forestry" also fails to satisfy strict environmentalists who want to set aside large landscapes of old-growth forest as wilderness areas. United States citizens and their government will have to decide on the compromises to be made between preserving these forests and human use of natural resources (Salwasser 1991a,b, 1992).

African Wildlife Outside Parks

East African countries such as Kenya are famous for the large wildlife species found in their national parks, which are the basis of a valuable ecotourism industry. Despite the fame of these parks, about three-fourths of Kenya's 2 million large animals live in rangelands outside of the national parks (Western 1985, 1989). The rangelands of Kenya occupy 700,000 km², or about 40% of the country. Among the well-known species found predominantly outside the parks are the giraffe (89%), the impala (72%), Grevy's zebra (99%), the oryx (73%), and the ostrich (92%). Only the rhinoceros, the elephant, and the wildebeest are found predominantly inside the parks; rhinos and elephants are concentrated in parks because poachers seeking ivory, horns, and hides have virtually eliminated external populations of these animals. The large herbivores found in the parks often graze seasonally outside of the parks; many of these species would be unable to persist if they were restricted to the limits of the parks by fencing, poaching, and agricultural development.

Often the areas surrounding national parks are used as rangeland for domestic cattle. It may seem intuitively obvious that the cattle compete with wild animals for range, water, and vegetation. However, studies have shown that the main factor determining the productivity and number of Kenyan wildlife species is not competition from livestock, but the amount of rainfall (Figure 16.4). The productivity of the rangelands, as measured by the weight of animals produced per km² per year, increases in a linear fashion with rainfall, while the number of species on the land is highest with intermediate amounts of rainfall. The presence of livestock outside of parks does not affect the number of wildlife species present and has only a slight

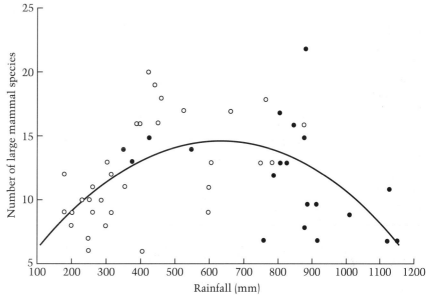

16.4 The number of large mammal species in East African ecosystems is apparently determined by annual rainfall, with the greatest number of species occurring in regions of intermediate rainfall. Ecosystems dominated by livestock (open circles) are no poorer in species than national parks, where livestock is banned (black circles). (From Western 1989.)

effect on productivity. It would appear that limited grazing of livestock may be compatible with wildlife conservation in some circumstances, and that commercial range may extend the effective area of a wildlife preserve. Even in areas in which there is commercial ranching there appears to be little change in the types of wildlife species present. To support this point, Western (1989) has pointed out that human pastoralists, such as the Masai in Tanzania, have lived in East Africa for over 3000 years without even one large herbivore going extinct.

The main factors affecting the continued existence of wild species in unprotected African lands appear to be a stable social structure and secure land tenure in the local human population. These factors tend to be characteristic of both traditional societies and highly developed societies. In these situations, use of resources is highly regulated by a recognized authority and current needs can be deferred to enhance future production of resources. Countries in which there is migration, poverty, unclear ownership of resources, and a breakdown of authority are likely to experience the greatest environmental deterioration and destruction of biological communities, since in these circumstances people must focus on their immediate needs, with the future value of the resources of little concern. In some unstable countries, there is an unregulated proliferation of guns in rural areas and often uncontrollable poaching. In a study of factors affecting the current status of the African elephant in various countries, the single most important factor was the extent of civil disruptions and, to a lesser extent, the socioeconomic status of the people (Burrill et al. 1986). At the present time, elephant populations in politically and socially stable countries are increasing at 2.5% per year, while in unstable countries they are declining by 16% annually. Literacy, annual income, and conservation measures also correlate positively with elephant population increases.

In East Africa, a number of additional factors contribute to the persistence of wildlife in unprotected areas with rural human populations (Western 1989). Many wildlife species are valued for their meat, so that their presence on rangeland is encouraged. Private ranching in which wildlife and livestock are managed together is more profitable than when livestock is maintained alone. Wildlife species that do not compete with livestock, are at very low numbers, or are elusive are ignored by ranchers. Some species, such as elephants, are tolerated since they open up woody vegetation for grassland and enhance the habitat for livestock. Some areas containing wildlife are not used by people at all because of lack of water, warfare, disease, and inaccessibility. In such areas wildlife can exist without interference. Finally, some species are protected by laws against hunting and trading them, which are enforced by government agents; others persist simply because people enjoy them, finding them beautiful or amusing, and so tolerate their presence.

Traditional Societies and Biological Diversity

A great deal of biodiversity exists in places where people have lived for many generations, using the resources of their environment in a sustainable manner (Bedigian 1991; Gomez-Pompa and Kaus 1992). Local people practicing a traditional way of life in rural areas, with relatively little outside influence in terms of modern technology, are variously referred to as tribal people, indigenous people, native people, or traditional people (Dasmann 1991). These people often have established local systems of rights to natural resources, which are sometimes recognized by their governments. These established indigenous people need to be distinguished from settlers who have arrived more recently and may not be as closely linked to the land.

People have lived in every terrestrial ecosystem of the world for thousands of years as hunters, fishermen, farmers, and gatherers. Even remote tropical rain forests that are designated as "wilderness" by governments and conservation groups often have a small, sparce human population. In fact tropical areas of the world have had a particularly long association with human societies, since the tropics have been free of glaciation and are particularly amenable to human settlement. The great biological diversity of the tropics has coexisted with human societies for thousands of years, and in most places, humans did not substantially damage the biological diversity of their surroundings (Gomez-Pompa and Kaus 1992). The present mixture and relative densities of plants and animals in many biological communities may reflect the activities of people in the area, such as selective hunting of certain game animals, fishing, and planting of useful plant species (Posey 1983; Dufor 1990). The commonly practiced agricultural system, known variously as swidden agriculture, shifting cultivation, and slash-and-burn agriculture, also affects forest structure and species composition by creating a mosaic of forest patches of different ages (see Figure 6.7A). In this system, the trees in an area are cut down, the fallen plant material is burned, and crops are planted in the nutrient-rich ash. After one or several harvests, the nutrients are washed out of the soil by the rain; the farmer then abandons the field and cuts down a new patch of forest for planting. This system works well and does not degrade the environment as long as human population density is low and there is abundant forest land available.

There is a popular misconception that complex societies did not develop in tropical rain forest areas. This is not borne out by the available information. Traditional societies utilizing innovative irrigation methods and a mixture of crops were able to support relatively high human population densities without destroying the environment or the surrounding biological communities. For example, the present-day Maya lowlands in Mexico, Belize, Guatemala, and Honduras are lightly settled, with only about 5 people per square kilometer; the area is covered by forests and has many unique species and biological communities. However, during the height of Maya civilization 1000 years ago (Figure 16.5), the region had densities of up to 500 people per square kilometer and the range of species was apparently still maintained (Turner 1976; Gomez-Pompa and Kaus 1988, 1992). The low population densities among traditional societies of many Neotropical rain forest areas today are an artifact of the repeated episodes of disease, exploitation, and fighting during the 500 years since the arrival of Europeans. Indigenous populations have been reduced to less than 10% of their original density (Dufor 1990). Traditional peoples have been an integral part of these forests for thousands of years. The greatest threat to the rain forests today, as described in Chapter 6, is from large numbers of outsiders, particularly landless farmers and ranchers coming into a region along new roads and practicing destructive and inappropriate farming methods.

Traditional peoples have been viewed in a variety of perspectives

16.5 A thousand years ago, Mayan farms and cities occupied a wide area of the Central American lowlands, with no apparent loss of species. Today the ruined cities are overgrown by tropical forests. (Photograph by R. Primack.)

by Western civilization. At one extreme, local people are viewed as destroyers of biological diversity who cut down forests and overharvest game. This destruction is accelerated when they acquire guns, chainsaws, and outboard motors. At the other extreme, traditional peoples are viewed as "noble savages" living in harmony with nature and minimally disturbing the natural environment. An emerging middle view is that traditional societies are highly varied, and there is no one simple description of their relationship to their environment that fits all groups (Alcorn 1993).

Many traditional societies do have strong conservation ethics that are more subtle and less clearly stated than Western conservation beliefs, but which affect the actions of people's day-to-day lives (Gomez-Pompa and Kaus 1992; Posey 1992). One well-documented example of such a conservation perspective is that of the Tukano Indians of northwest Brazil (Chernela 1987). The Tukano live on a diet of root crops and river fish (see Figure 8.1); they have strong religious and cultural prohibitions against cutting the forest along the Upper Río Negro, which they recognize as important to the maintenance of fish populations. The Tukano believe that these forests belong to the fish and cannot be cut by people. They have also designated extensive refuges for fish, and permit fishing along less than 40% of the river margin. Chernela observes, "As fishermen dependent upon river systems, the Tukano are aware of the relationship between their environment and the life cycles of the fish, particularly the role played by the adjacent forest in providing nutrient sources that maintain vital fisheries."

Another example of a "traditional" conservation ethic is that of the Patzcuaro Indian communities of central Mexico (Toledo 1991). The resources of Lake Patzcuaro are regulated at the community level and shared equally. Since the community as a whole benefits from these restrictions, it has been able to resist outside influences that have damaged many other nearby lakes.

Local people can also manage the environment to maintain biological diversity, as shown by the traditional agroecosystems and forests of the Huastec Indians of northeastern Mexico (Alcorn 1984). In addition to their permanent agricultural fields and swidden agriculture, the Huastec maintain managed forests—known as *te'lom*—on slopes, along watercourses, and in other areas that are either fragile or unsuitable for intensive agriculture (Figure 16.6). These forests

16.6 A Huastec Indian woman at a *te'lom*, an indigenous managed forest in northeastern Mexico. Here she collects sapote fruit (*Manilkara achras*) and cuttings of a frangipani tree (*Plumeria rubra*) for planting. (From Alcorn 1984; photograph by Janis Alcorn.)

contain over 300 species of plants, from which the people obtain food, wood, and other needed products. Species composition in the forest is altered in favor of useful species by planting and periodic selective weeding. These forest resources provide Huastec families with the means to survive the failure of their cultivated crops. Comparable examples of intensively managed village forests exist in traditional societies throughout the world (Oldfield and Alcorn 1991; Redford and Padoch 1992; Nepstad and Schwartzman 1992).

Biological Diversity and Cultural Diversity

Biological diversity and cultural diversity are often linked (Dasmann 1991; Bedigian 1991). Rugged tropical areas of the world where the greatest concentrations of species are found are frequently the areas where people have the greatest cultural and linguistic diversity. The geographical isolation by mountain ranges and complex river systems that favors biological speciation also favors the differentiation of human cultures. The cultural diversity found in places such as Central Africa, Amazonia, New Guinea, and Southeast Asia represents one of the most valuable resources of human civilization, providing unique insights into philosophy, religion, music, art, resource management, and psychology (Denslow and Padoch 1988). The protection of these traditional cultures within their natural environment provides the opportunity to achieve the dual objectives of protecting biological diversity and preserving cultural diversity (Oldfield and Alcorn 1991). In the words of Toledo (1988):

> In a country that is characterized by the cultural diversity of its rural inhabitants, it is difficult to design a conservation policy without taking into account the cultural dimension; the profound relationship that has existed since time immemorial between *nature* and *culture*. . . . Each species of plant, group of animals, type of soil and landscape nearly always has a corresponding linguistic expression, a category of knowledge, a practical use, a religious meaning, a role in ritual, an individual or collective vitality. To safeguard the natural heritage of the country without safeguarding the cultures which have given it feeling is to reduce nature to something beyond recognition, static, distant, nearly dead.

Cultural diversity is strongly linked to the genetic diversity of crop plants. In mountainous areas in particular, the inaccessible terrain often leads to the development of diverse local languages and tribes. These people in turn develop local plant varieties known as **landraces**; these cultivars are adapted to the local climate, soils, and pests, and satisfy the tastes of the local people (Figure 16.7). The genetic variation in these landraces has global significance to modern agriculture for the improvement of crop species (see Chapter 17).

16.7   A typical Huastec farm. Such polyculture fields, where numerous species of wild and cultivated plants grow together, are reserves of tremendous genetic diversity that are potentially of crucial importance to world agriculture. (From Oldfield and Alcorn 1987; photograph by Janis Alcorn.)

Individual landraces are typically named and have many identifying characteristics. For example, in the rugged Nuba mountains of Sudan, the Nuba people are divided in 62 distinct language groups and grow dozens of landraces of sesame that are identified with particular tribes and places (Bedigian 1991). These diverse landraces of sesame are starting to decline in importance and their continued existence is in doubt as a cash economy displaces the traditional way of life.

Involving Traditional Societies in Conservation Efforts

Several strategies exist for integrating the protection of biological diversity, the customs of traditional societies, and the genetic variation of traditional crops.

Man and the Biosphere reserves. UNESCO's Man and the Biosphere Program (MAB) includes among its goals the maintenance of "samples of varied and harmonious landscapes resulting from long-established land use patterns." (UNESCO 1974, 1984, 1985; Gregg 1991). The MAB Program recognizes the role of people in shaping the natural landscape, as well as the need to find ways in which people can sustainably use natural resources without degrading the environment

(Chapter 14). The MAB research framework, used at its worldwide network of designated Biosphere reserves (see Chapter 20), integrates natural science and social science research. It includes investigations of how biological communities respond to different human activities, how humans respond to changes in their natural environment, and how degraded ecosystems can be restored to their former condition.

One valuable example of a Biosphere reserve is the Kuna Yala Indigenous Reserve on the northeast coast of Panama (Houseal et al. 1985; Clay 1991; Gregg 1991). In this protected area of 60,000 hectares of tropical forest, 30,000 Kuna people in 60 villages practice traditional medicine, agriculture, and forestry, with documentation and research undertaken by scientists from outside institutions. The Kuna carefully regulate the levels of scientific research in the reserve, insisting on local training, presentation of reports before scientists leave the area, payment of research fees, and having local guides accompany the scientists. The Kuna people even control the type and rate of economic development in the reserve and have their own outside, paid advisors. The level of empowerment of the Kuna people is unusual, and it illustrates the potential for traditional people to take control of their destiny, way of life, and environment.

In situ agricultural conservation. In many areas of the world, local farmers cultivating locally adapted varieties of crop plants can preserve genetic variability in these species. For example, there are thousands of distinct varieties of potatoes grown by Andean farmers in South America. Often these farmers will grow many varieties in one field to minimize the risk of crop failure and for the different uses of each variety (Figure 16.8). Similarly, traditional farmers in the Apo Kayan of Borneo may grow more than 50 varieties of rice. These local varieties often have unique genes for dealing with disease, nutrient deficiences, pest resistance, drought tolerance, and other environmental variations (Browning 1991). Moreover, these local varieties continue to evolve new genetic combinations, some of which may be effective in dealing with looming global environmental threats. However, farmers throughout the world are abandoning their traditional forms of agriculture with local races to grow high-yielding varieties using capital-intensive methods including fertilizer and pesticides. In countries such as Indonesia, Sri Lanka and the Philippines, over 80% of the farmers have adopted modern varieties (Brush 1989).

While an increased yield may be better in the short term for the individual farmer and his society, the long-term health of modern agriculture depends on the preservation of the genetic variability represented by local varieties (see Chapter 17). One innovative suggestion has been for an international agricultural body to subsidize villages to be in situ, or in place, "landrace custodians" (Nabhan 1985, 1986;

16.8 A sampling of the many varieties of potatoes grown by farmers in the Andes of South America; growing multiple varieties in a single field minimizes the risk of losing the entire crop to a varietal-specific disease or pest. (Courtesy of International Potato Center, Lima.)

Altieri and Merrick 1988; Wilkes 1991; Altieri and Anderson 1992). One hundred landrace zones, each 5 km × 20 km, would be established in areas of high genetic variation throughout the world. Villagers in these zones would be paid to grow their traditional crops in a traditional manner, providing a crucial source of genes for modern crop improvement programs. The benefits would be increased even further because many of these agricultural areas contain wild varieties of crop plants and species related to crop plants that may have significant breeding value. The cost of subsidizing several hundred villages to maintain the genetic variability of major crops would be a relatively modest investment in the long-term health of world agriculture.

Programs based on the in situ idea have already been initiated in some places (Figure 16.9). In Mexico in particular, a number of development programs are attempting to integrate traditional agriculture, conservation and research (Gliessman 1991; Toledo 1991). A slightly different approach that has been tried in arid regions of the American Southwest involves linking traditional agriculture and genetic conservation (Nabhan 1985). A private organization, Native Seeds/SEARCH, collects the seeds of traditional crop cultivars for long-term preservation. The organization also encourages farmers to grow traditional crops, provides them with the seeds of traditional cultivars, and buys the farmers' unsold production.

Countries have also established special reserves to conserve areas containing wild relatives of crops. In the former Soviet Union, 127

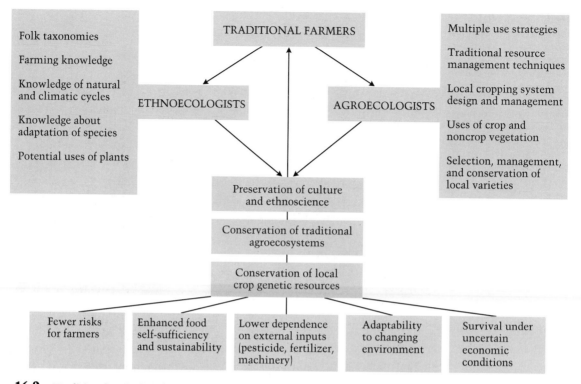

16.9   Traditional agricultural practices can be viewed both from a human cultural and an agricultural perspective. A synthesis of these viewpoints can lead to theoretical and methodological approaches toward the conservation of the environment, the culture, and the genetic variation found in these traditional agroecosystems. (Modified from Altieri and Anderson 1992.)

such reserves were created. Species reserves protect the wild relatives of wheat, oats, and barley in Israel and of corn in Mexico (Browning 1991).

Extractive reserves.   In many areas of the world, people have extracted products from natural communities for decades and even centuries. The sale and barter of these natural products serve as a major part of the livelihood of the people. The right to continue collecting natural products from the surrounding countryside is a major concern of local people. The establishment of national parks that exclude the traditional collection of products will meet with as much resistance from the local community as will a land-grab that involves exploitation of the natural resources and conversion to other uses.

The Brazilian government is trying to address the legitimate demands of local citizens through a new type of protected area known as an **extractive reserve**, in which settled people collect natural products such as rubber, resins, and nuts in a way that minimizes damage to the forest ecosystem (Fearnside 1989). In such areas, the ability of the people to continue their way of life is guaranted against the threat of conversion to cattle ranching and farming. At the same time, the government protection afforded to the local population also serves to protect the biological diversity of the area, since the ecosystem remains basically intact (Nepstad et al. 1992).

Extractive reserves appear to be appropriate in the Amazon rain forests, where about 68,000 rubber-tapper families live. The rubber-tappers live at a density of only about one family per 300–500 ha, of which they clear a few hectares for growing food. Commercial rubber-tapping has been going on in the Amazon for over 100 years, and rubber-tapping areas presently occupy 4–7% of the Amazon area. The efforts of Chico Mendes and his subsequent assassination in 1988 drew worldwide attention to the plight of the rubber tappers (see Box 31 in Chapter 19; Schwartzman 1992). In response to both local and international concern, the Brazilian government established the extractive reserves in rubber-tapping areas. The reserves make sense because the rubber collection system appears to be economically viable. The rubber-tappers themselves have a strong vested interest in preventing habitat destruction, because it would destroy their livelihood. The real challenge for the rubber-tappers and their Brazilian and international allies is to develop other natural products that can be collected and sold at a good market price (Clay 1992).

The Brazilian experiment has indicated that extractive reserves are a possible mechanism to preserve biological diversity, but the rubber-tapping example also displays a number of limitations (Browder 1992). First, the extractive reserves occupy only a small percentage of the Amazon; conservation efforts aimed at protecting the rain forest need to concentrate on reducing the rates of deforestation caused by ranching and farming activities, which already occupy 24% of the northern Amazon region. Second, these reserves provide occupations for only a tiny percentage of the millions of Brazilians who need a livelihood. And, finally, the economics are chancy: transportation to market is difficult, and any increase in the amount of product coming to market could drive the price down (Tremaine 1993; Salafsky et al. 1993). If such events cause the market to fail, the rubber-tappers could be forced to cut down their forests for timber and agriculture out of economic desperation.

Restoration Ecology

An important opportunity for conservation biologists is the chance to participate in the restoration of damaged or degraded ecosystems (Jordan et al. 1990). **Ecological restoration** is defined as "the process of intentionally altering a site to establish a defined, indigenous, historic ecosystem. The goal of this process is to emulate the structure, function, diversity and dynamics of the specified ecosystem" (Society of Ecological Restoration 1991). Restoration ecology has its origins in older applied technologies that restore ecosystem functions of known economic value: wetland replication to prevent flooding, mine site reclamation to prevent soil erosion, range management to ensure the production of grasses, and forest management for timber and amenity value (Bradshaw and Chadwick 1980; Bradshaw 1983, 1990; Kusler and Kentula 1990). However, these technologies sometimes produce only simplified communities, or communities that cannot maintain themselves. With the emergence of biological diversity as an important societal concern, the protection of species and communities has been included as a goal in restoration plans, and the input of conservation biologists is needed to make these efforts successful.

Ecosystems can be damaged by natural phenomena such as lightning-caused fires, volcanos, and storms, but they typically recover to their original biomass, community structure, and even a similar species composition through the process of succession. However, some ecosystems are so degraded by human activity that their ability to recover is severely limited. Recovery is unlikely when the damaging agent is still present in the ecosystem. For example, restoration of degraded savannah woodlands in western Costa Rica and the western United States is not possible as long as the land continues to be overgrazed by introduced cattle; reduction of the grazing pressure is obviously the key starting point in restoration efforts. Recovery is also unlikely when many of the original species have been eliminated over a large area, so that there is no source of colonists. For example, prairie species were eliminated from huge areas of the midwestern United States when the land was converted to agriculture. Even when an isolated patch of land is no longer cultivated, the original community does not become reestablished, since there is no source of seeds or colonizing animals of the original species (Cottam 1990). In addition, recovery is unlikely when the physical environment has been so altered that the original species can no longer survive there; examples of this situation include mine sites, where the restoration of natural communities may be delayed by decades or even centuries due to the poor structure, heavy metal toxicity, and low nutrient status of the soil (Figure 16.10; Bradshaw 1990).

In certain cases entirely new environments are created by human activity, such as reservoirs, canals, landfills, and industrial sites. If these sites are neglected, they often become dominated by exotic and weedy species, resulting in biological communities that are unpro-

16.10   To speed the recovery of this devastated coal mine site in Wyoming, crews planted 120,000 shrubs. Mining sites often need a great deal of human help in order to recover even a semblance of biodiversity. (From Jordan et al. 1990.)

ductive, not typical of the surrounding areas, valueless from a conservation perspective, and aesthetically unappealing. If these sites are properly prepared and native species are reintroduced, native communities can potentially be successfully restored.

Restoration ecology provides theory and techniques to address these various types of degraded ecosystems. Four main approaches are available in restoring biological communities and ecosystems (Figure 16.11; Cairns 1986; Bradshaw 1990).

1. *No action* because restoration is too expensive, or because previous attempts at restoration have failed, or because experience has shown that the ecosystem will recover on its own. The last approach is typical of old fields in eastern North America, which return to forest within a few decades after being abandoned for agriculture.
2. *Restoration* of the area to its original species composition and structure by an active program of reintroduction, in particular planting and seeding of the original species.
3. *Rehabilitation* of at least some of the ecosystem functions and some of the original species, such as replacing a degraded forest with a tree plantation.

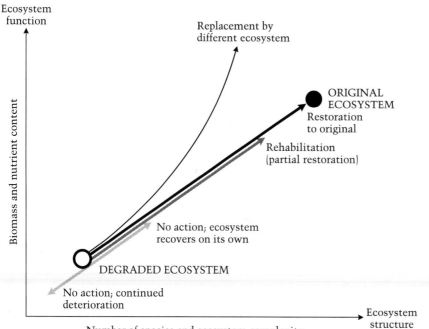

16.11 Degraded ecosystems have lost their structure (in terms of species and their interactions with the physical and biological environments) and their function (the accumulation of biomass, and soil, water, and nutrient processes). Decisions must be made as to whether the best course of action is to restore, rehabilitate, or replace the degraded site, or whether the best course is no action. (Modified from Bradshaw 1987.)

4. *Replacement* of a degraded ecosystem with another productive ecosystem type, for example replacing a degraded forest area with a productive pasture.

Civil engineers and others involved in major projects deal with the restoration of degraded habitats in a practical, technical manner. Their goals are to find economical ways to permanently stabilize land surfaces, to prevent soil erosion, to make the site look better to neighbors and the general public, and if possible to restore the productive value of the land (Bradshaw 1990). Ecologists contribute to these restoration efforts by developing ways to restore the original communities in terms of species diversity, species composition, vegetation structure, and ecosystem function. To be practical, restoration ecology must also consider the speed of restoration, the cost, the reliability of the results, and the ability of the final community to persist with little or no further maintenance. Practitioners of restoration ecology must have a clear grasp of how natural systems work and what methods of restoration are feasible. Considerations of the cost and availability of seeds, when to water plants, how much fertilizer to add, and how to prepare the surface soil may become paramount in determining the success of a project. Dealing with such practical details has not generally been attractive to academic biologists in the past, but they must be dealt with in restoration ecology.

Restoration ecology is valuable to the science of ecology because it provides a test of how well we really understand a biological community, and demonstrates how well we can reassemble it from its component parts (Diamond 1990). As Bradshaw (1990) has said, "Ecologists working in the field of ecosystem restoration are in the construction business, and like their engineering colleagues, can soon discover if their theory is correct by whether the airplane falls out of the sky, the bridge collapses, or the ecosystem fails to flourish." In this sense, restoration ecology can be viewed as an experimental methodology that complements existing programs of basic research on intact systems. Restoration ecology provides an opportunity to completely reassemble communities in different ways, to see how well they function, and to test ideas on a larger scale than would otherwise be possible (Diamond 1990; Gilpin 1990; Jordan et al. 1990).

Restoration ecology will play an increasingly valuable role in the conservation of biological communities if degraded lands can be restored to their original species composition and added to the limited existing area of protected conservation areas. Since degraded areas are unproductive and of little economic value, governments may be willing to restore them and increase their productive and conservation value. Restoration ecology is almost certain to become one of the major growth areas in conservation biology. However, conservation biologists in this field must be careful not to allow themselves to be used as a convenient public-relations cover by ecosystem-damaging industrial corporations interested only in continuing business as usual (Falk and Olwell 1992).

Efforts to restore ecological communities have focused extensively on lakes, prairies, and forests. These environments have suffered severe alteration from human activities and are good candidates for restoration work.

Lakes

Limnologists involved in multimillion-dollar efforts to restore lakes are already gaining valuable insights into community ecology and trophic structure that otherwise would not be possible (Welch and Cooke 1990). One of the most common types of damage to lakes and ponds is cultural eutrophication (discussed in Chapter 6), a change caused by excess mineral nutrients that enter the water from human activity. The signs of eutrophication include increases in the algae population (particularly surface scums of blue-green algae), lowered water clarity, lowered oxygen content of the water, fish die-offs, and an eventual increase in the growth of floating plants and other water weeds.

Attempts to restore eutrophic lakes have not only provided practical management information but have also provided insight into the basic science of limnology (the study of the chemistry, biology, and physics of fresh water). In many lakes, reducing the mineral nutrients entering the water—through better sewage treatment, or by diverting polluted water—leads to a reversal of the eutrophication process and a restoration of the original conditions; this approach is known as "bottom-up" control. In other lakes, this improvement does not occur, suggesting that there are internal mechanisms within the lake that are recycling nutrients from the sediment to the water column and keeping the nutrient levels artificially high. One possible mechanism for the return of phosphorus to the water column is the role of fish, such as the carp (*Cyprinus carpio*) and the brown bullhead (*Ictalurus nebulosus*), that eat organic matter and excrete phosphorus (Keen and Gagliardi 1981). This hypothesis is supported by declines in phosphorus concentrations after carp populations are reduced in eutrophic lakes (Shapiro et al. 1982). The composition of the fish community can also affect the eutrophication process through predation relationships. In some eutrophic lakes, planktonic invertebrates (such as the crustacean *Daphnia*) that eat algae are intensively eaten by the fish, allowing the algae to grow unchecked. If predatory fish (which feed on other fish) are added to the lake, the population of zooplankton-eating fish often drops; an increasing crustacean population then reduces the abundance of algae, and water quality improves. Such improvements in water quality achieved through manipulations of fish populations are referred to as "top-down" control.

One of the most dramatic and expensive examples of lake restoration is that of Lake Erie (Makarewicz and Bertram 1991). Lake Erie was the most polluted of the Great Lakes in the 1950s and 1960s, and was characterized by deterioring water quality, extensive algal blooms, declining indigenous fish populations, the collapse of commercial fisheries, and oxygen depletion in deeper waters. To address this problem the United States and Canadian governments have invested more than $7.5 billion since 1972 in wastewater treatment facilities, reducing the annual discharge of phosphorus into the lake from 15,260 tons in 1972 to 2449 tons in 1985. Once water quality began to improve in the mid-1970s and 1980s, stocks of the native commercial walleye (*Stizostedion vitreum vitreum*), a predatory fish, began to increase on their own, and other predatory fish species were added to the lake by state agencies. As a result, both "bottom-up" and "top-down" control agents worked to improve lake quality.

The 1980s have seen a continued improvement in Lake Erie water

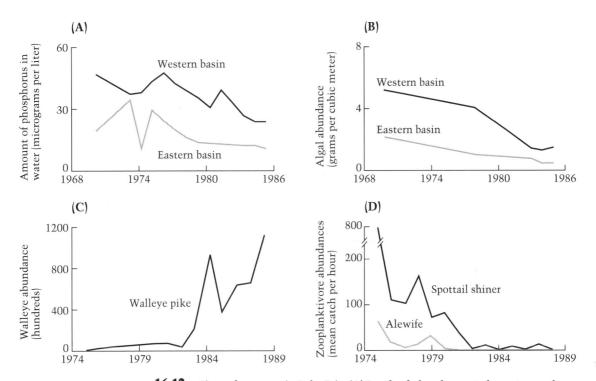

16.12 Signs of recovery in Lake Erie. (A) Levels of phosphorus at the eastern and western ends of the lake. Phosphorus levels are lowered by treating the sewage and other human effluents that enter the lake. (B) Algal abundance. Algae flourish in the presence of phosphorus; algal blooms lower the water quality and eliminate other species. (C) Walleye pike abundance, as measured by the sport fisherman catch. Walleye are predatory fish that feed on zooplanktivorous fish such alewife and spottail shiners; adding walleye to the lake is one way to increase the population of crustaceans and other zooplankton, which in turn feed on the algae. (D) The abundance of alewife and shiners, measured by the catch per hour in Lake Erie fishing trawlers. When populations of these fish reach overly high levels, they decimate the zooplankton population; with fewer zooplankton to feed on them, algal blooms flourish even more. Alewife and shiner populations declined after the introduction of pike and other predatory fish. (After Makarewicz and Bertram 1991.)

quality, as shown by lower concentrations of phosphorus, lower phytoplankton (algal) abundance, and a shift in the trophic community toward higher relative numbers of algal-feeding zooplankton and predatory fish, with lower numbers of zooplankton-feeding fish (Figure 16.12). There is even some evidence of improvement in oxygen levels at the lower depths of the lake. Even though the lake will probably never return to its original condition, the combination of bottom-up and top-down controls—and the investment of billions of dollars—has resulted in a significant degree of restoration in this large ecosystem.

Prairies

Many small parcels of former agricultural land in North America have been restored as prairies (Box 24; Kline and Howell 1990). Prairies represent ideal subjects for restoration work, since they are spe-

BOX 24　RESTORATION OF THE ILLINOIS PRAIRIE

Scientists in northern Illinois have undertaken a project designed to renew one of the most severely damaged ecosystems in the United States: the prairie grasslands that originally covered much of Illinois, Kansas, and other Midwestern states. In Illinois, the damage to this ecosystem is extensive; less than 0.01% of the original landscape remains unaltered, and many of the native species are all but extinct.

Biologists are attempting to reestablish prairie ecosystems in designated areas such as the Nachusa Grasslands, a 700-acre reserve owned by the Nature Conservancy (Schmidt 1992). Restoration of the prairie involves both conservation of the native species and reestablishment of some of the ecological relationships between species. Biologists have combed railroad abutments, cemeteries, and other areas that were never subjected to plowing in order to find remnant populations of the plants that made up the original prairie flora, and have collected numerous seeds and insects to introduce to the restoration sites. Some of the animals most critical to successful pollination and seed dispersal have been captured and released at the sites to improve the ability of the plant populations to reproduce and expand; others have begun to return of their own accord.

Restoring the animals and plants is one step in the process of recreating the prairie. Another is the reestablishment of ecological processes that encourage growth of plant and animal populations in the reserve. For example, biologists periodically set fire to the grasslands in the reserve to initiate the natural process of renewal so vital to the prairie ecosystems. Since many of the native species are fire-adapted while most introduced competitors are not, the fires have the added advantage of flushing out non-native species.

The damage to this ecosystem has been greater than apparent to the naked eye, however; the transformation of the land from prairie to agricultural fields has changed the composition of the prairie soils as well as the flora and fauna. Agricultural fields plowed for over a century leave behind a fairly simple soil ecology that bears little resemblance to the diversity of the original prairie soils. The complex soil system developed by deep-rooted prairie plants and the fungi, microbes, and other soil fauna that coexist with these plants may take decades, if not centuries, to re-create (Schmidt 1992). Naturally, biologists cannot hope to replace all of the original species and relationships that made up the original prairie; they are only just beginning to understand how all of the plants, animals, and microorganisms fit into their ecosystem. However, they can provide the grasslands with a jump-start to recovery by conserving the species and habitats that make up the prairie. The procedure will teach biologists a tremendous amount about grassland ecology for use in future restoration projects.

cies-rich, have many beautiful wildflowers, and can be restored within a few years (Jackson 1992). Also, the technology used for prairie restoration is similar to that of gardening and agriculture and is well suited to incorporating volunteer labor.

The most extensive research on the restoration of prairies has been carried out in Wisconsin, starting in the 1930s (Cottam 1990). A wide variety of techniques have been used in these prairie restoration attempts: the basic method involves a light site preparation of disking, burning, and raking if prairie species are present, or elimination of all vegetation by plowing or herbicides if only exotics are present. Native plant species are then established by transplanting in prairie sods obtained elsewhere, planting individuals grown from seed, or scattering prairie seed collected from the wild or from cultivated plants (Figure 16.13). The simplest method is gathering hay from a native prairie and sowing it on the prepared site. In summarizing the five decades of Wisconsin experiments, Cottam (1990) observes:

> All of these methods work, but the success rate is highly variable and depends on the quantity of weeds present, the amount and timing of precipitation, the way the seeds are stratified, and a number of other variables both known and unknown.... Native prairies are usually very heterogeneous, with masses of one species growing together in one place and other species growing together in another place. Why the plants distribute themselves as they do is not easily discerned, so at best a lot of guesswork and intuition goes into the actual planting of the species, and a large amount of background information increases the chance of success. There is room for error, however, because prairie plants generally have a broad range of tolerance. If they are established within their optimum habitat, the species will interact and ultimately sort themselves out into a reasonable approximation of a native prairie.

A particular value of prairie restoration projects is their usefulness for educational demonstration purposes in areas covered by agricultural landscapes and their ability to excite the imagination of urban dwellers eager to be involved in conservation efforts. In concluding his essay, Cottam says, "Prairie restoration is an exciting and rewarding enterprise. It is full of surprises, fantastic successes, and abysmal failures. You learn a lot—usually more about what not to do than what to do. Success is seldom high, but prairie plants are resilient, and even a poor beginning will in time result in a beautiful prairie."

One of the most ambitious and controversial proposed restoration schemes involves re-creating a prairie ecosystem, or "buffalo commons," on about 380,000 km^2 of the American Plains states, from the Dakotas to Texas and from Wyoming to Nebraska (Popper and Popper 1991; Mathews 1992). This land is currently used for environ-

(A)

(B)

16.13 (A) In the late 1930s, members of the Civilian Conservation Corps (one of the organizations created by President Franklin Roosevelt in order to boost employment during the Great Depression) participated in a University of Wisconsin project to restore the wild species of a Midwestern prairie. (B) The prairie as it looked 50 years later. (Photograph from the University of Wisconsin Arboretum and Archives.)

mentally damaging and often unprofitable agriculture and grazing supported by government subsidies. The human population of this region is declining as farmers and townspeople go out of business and young people move away. From an ecological, sociological, and even an economic perpective, the best long-term use of much of the region might be a restored prairie ecosystem. The human population of the region could potentially stabilize around non-damaging core industries such as tourism, wildlife management, and low-level grazing, leaving only the best lands in agriculture.

Restoration of Tropical Dry Forest in Costa Rica

Throughout the world, tropical forests are being degraded by logging, grazing, fire, shifting cultivation, and collection of fuelwood. These lands often become degraded to the point that they have few remaining trees and little value to the local human population. In order to reverse these disastrous trends, goverments, local people, and private organizations are involved in planting hundreds of millions of tree seedlings per year and in protecting remaining forests.

An exciting experiment in restoration ecology is currently taking place in northwestern Costa Rica. The tropical dry forests of Central America have suffered from large-scale conversion to cattle ranches and farms. Cattle grazing, fire, and clearing have reduced this diverse community to a few fragments. Even in these fragments, exotic grasses and hunting pressure threaten remaining native species. This destruction has gone on largely unnoticed as international scientific and public attention has focused on the more glamorous rain forests elsewhere. To reverse this destruction of tropical dry forests, the American ecologist Daniel Janzen has been working with the Costa Rican government and local people to restore 75,000 hectares of land in Guanacaste National Park (Figure 16.14; Janzen 1988b; Allen 1988). The plans for restoration include planting native trees, controlling fires and banning hunting. Livestock grazing will be reduced to levels necessary for controlling exotic grasses, which fuel fires and prevent the regeneration of native plant species. The goal is to eliminate exotic species and reestablish a forest ecosystem within the next 100 to 300 years.

One innovative aspect of this restoration effort is the incorporation of local people into many aspects of park management, and the intended role of the park in the cultural and educational life of the people. Many of the farmers and ranchers living within the park borders were given the opportunity to be trained as park employees and to use their skills and knowledge of the area to develop the park. Those individuals showing initiative and ability are being trained as park managers and biologists.

A key element in the restoration plans is what has been termed *biocultural restoration*, meaning that the park will serve as a center for teaching the 40,000 local residents about natural history and the principles of ecology and conservation. Janzen believes that in rural areas such as Guanacaste, providing an opportunity for learning about nature can be one of the most valuable functions of national parks:

> The public is starving for and responds immediately to presentations of complexity of all kinds—biology, music, literature, politics, educa-

(A)

(B)

16.14 (A) The proposed Guanacaste National Park is an experiment in restoration ecology—an attempt to restore the devastated and fragmented tropical dry forest of Costa Rica. This view of the forest shows an area of Santa Rosa National Park that serves as the nucleus for the experiment. (B) Daniel Janzen, an ecologist from the United States, is the driving force behind the restoration project in Guanacaste. Here he inspects moth specimens from the study area. (Photographs by William H. Allen.)

tion, et cetera.... The goal of biocultural restoration is to give back to people the understanding of the natural history around them that their grandparents had. These people are now just as culturally deprived as if they could no longer read, hear music, or see color. (Janzen, quoted in Allen 1988)

To achieve this goal, educational and research programs have been designed to include local students at grade schools, high schools, and universities, as well as citizen groups. By educating the local community about natural history and teaching them the value of the park, the hope is that the people will become advocates both locally and nationally for the conservation of natural resources.

> The most practical outcome is that this program will begin to generate an ongoing populace that understands biology. In 20 to 40 years, these children will be running the park, the neighboring towns, the irrigation systems, the political systems. When someone comes along with a decision to be made about conservation, resource management, or anything else, you want that person to understand the biological processes that are behind that decision because he or she knew about them since grade school. (Janzen, quoted in Allen 1988)

On a practical level, funding for land purchases and park management comes from the Costa Rican government and private international foundations. In the future, operating income will increasingly come from fees paid by foreign and Costa Rican scientists working at the biological field stations. Also, the proximity of the park to the Pan American Highway makes it an ideal location for ecotourism. Employment in these expanding research, tourist, and educational facilities will provide a significant source of employment for the local community, particularly for those who are interested in nature and education. A key element in the future success of Guanacaste National Park is that the plan for park development and management provides the proper integration of community needs and restoration needs in a way that satisfactorily fulfills both objectives. In its final analysis, this restoration effort has been so successful and has attracted so much media attention because of the efforts of a single individual—Daniel Janzen—committing all his time and resources to a cause he passionately believes in. His enthusiasm and vision have aroused many other people to join his cause. This is a classic example of how a single dedicated individual or small group of people can be a potent force for the conservation of a habitat or a species.

The Fine Points of Restoration Ecology

Efforts to restore degraded terrestrial communities have emphasized the establishment of the original plant community. This emphasis is appropriate because the plant community typically contains the majority of the biomass and provides a structure for the rest of the community. However, more attention needs to be devoted to the other major components of the community. Myccorhizal fungi (see Box 15 in Chapter 9) and bacteria play a vital role in soil decomposition and nutrient cycling (Danielson 1985; Miller 1990); soil invertebrates are important in creating soil structure; herbivorous animals are important in reducing plant competition and maintaining species diversity; and many vertebrates have vital functions as seed dispersers, insect predators, and soil diggers. Many of these non-plant species can be transferred to a restored site in sod samples, while large animals and above-ground invertebrates may have to be deliberately caught in sufficient numbers and then released onto restored sites to establish new populations. If an area is going to be destroyed and then restored later, as might occur during strip mining, the top layer of soil which contains the majority of buried seeds, soil invertebrates, and other soil organisms could be carefully removed and stored for later use in restoration efforts (Putwain and Gilham 1990; Miller 1990).

Restoration efforts could also be used to re-create a biological community that is threatened somewhere else. For example, degraded rangelands in Texas could be restored using species from a threatened savannah ecosystem in Africa. If the black rhinoceros and other wildlife are going to be hunted to extinction in Africa, then their only hope might be preserving them in restored ecosystems elsewhere. This idea is not as far-fetched as it seems; rhinos occurred in Texas until about 10,000 years ago, when they were presumably eliminated by human activities. Whole groups of plants and animals could be moved from one continent to another, although this would have to be done cautiously; introduced species are, after all, one of the problems conservation biologists must fight, and it is important to ascertain that none of the transported species has the potential to become invasive pests in their new home.

Summary

1. Considerable biological diversity exists outside of protected areas, particularly in habitat managed for multiple-use resource extraction. Such unprotected habitats are vital for conservation because protected areas account for only a small percentage of the area of any country. Also, animal species living in protected areas often forage and migrate onto unprotected land where they are vulnerable to hunting and other threats from humans. Governments are increasingly including the protection of endangered species as one of the priorities of multiple-use land.

2. In temperate forest ecosystems, biological diversity can be enhanced if logging operations harvest in larger blocks to minimize fragmentation, and if such operations leave some late-successional individuals, including living trees, standing dead trees, and fallen trees. Such trees are important resources for animal species, especially cavity-nesting birds.

3. In Africa, many of the characteristic large animals are found predominantly in rangeland outside of the parks. Limited grazing by cattle on East African rangeland appears to be compatible with maintaining the variety of wild species and their numbers. The most important factor allowing the persistence of wildlife on unprotected rangeland is a stable social structure and secure land tenure for the local human population. War and other social disruptions are major causes of decline for many African animal species.

4. Local people practicing traditional ways of life are found in every terrestrial ecosystem. Often the present mixture of plants and animals in a biological community is influenced by the activities of these residents. Some traditional societies have strong conservation ethics and management practices that are compatible with the protection of biological diversity. The protection of traditional cultures within their natural environment provides the opportunity to achieve the dual objectives of protecting biological diversity and preserving cultural diversity.

5. Restoration ecology provides methods for reestablishing species and whole communities in degraded habitat. Restoration ecology provides an opportunity to enhance biological diversity in habitats that have little other value to humans. Restoration ecology builds on applied technologies that attempt to reestablish ecosystem functions in damaged wetlands, mine sites, lakes, and degraded rangelands. Restoration projects begin by eliminating or neutralizing any factors that prevent the system from recovering. Then combinations of site preparation, habitat management, and reintroduction of original species gradually allow the community to return. Restoration ecology has practical value in restoring the ecosystem functions and natural appearance of a location, but it also provides unusual insight into community ecology as communities are reassembled from their components.

Suggested Readings

Alcorn, J. B. 1993. Indigenous peoples and conservation. *Conservation Biology* 7: 424–426. This article and others in the same issue explore different perceptions of indigenous people.

Gomez-Pompa, A. and A. Kaus, 1992. Taming the wilderness myth. *BioScience* 42: 271–279. Traditional people often have their own approaches to preserving biodiversity.

Gradwohl, J. and R. Greenberg. 1988. *Saving the Tropical Forest*. EARTH-SCAN Ltd., London. Innovative approaches to protect tropical biodiversity.

Hansen, A. J., T. A. Spies, F. J. Swanson and J. L. Ohmann. 1991. Conserving biodiversity in managed forests. *BioScience* 41: 382–392. New logging techniques can minimize the damage to biodiversity.

Janzen, D. H. 1988. Tropical ecological and biocultural restoration. *Science* 239: 243–244. Unique integration of ecology and public education.

Jordan III, W. R., M. E. Gilpin and J. D. Aber (eds.). 1990. *Restoration Ecology: A Synthetic Approach to Ecological Research*. Cambridge University Press, Cambridge. Papers outlining case studies and general approaches to restoration ecology.

Mathews, A. 1992. *Where the Buffalo Roam*. Grove Weidenfeld, New York. Superb popular account of the "buffalo commons" proposal and controversy.

Mlot, C. 1992. Botanists sue Forest Service to preserve biodiversity. *Science* 257: 1618–1619. Scientists insist that the protection of biodiversity be part of forest management.

Oldfield, M. L. and J. B. Alcorn (eds.). 1991. *Biodiversity: Culture, Conservation, and Ecodevelopment*. Westview Press, Boulder, CO. The preservation of biodiversity and culture are sometimes linked.

Redford, K. and C. Padoch (eds.). 1992. *Conservation of Neotropical Rainforests: Working from Traditional Resource Use*. Columbia University Press, New York. Case studies in which traditional resource use patterns also protect biodiversity.

Western, D. and M. Pearl (eds.). 1989. *Conservation for the Twenty-First Century*. Oxford University Press, New York. Essays by leading authorities, many of which are related to conservation outside of protected areas.

CHAPTER 17

Ex Situ Conservation Strategies

As the last few chapters have shown, the best strategy for the long-term protection of biological diversity is the preservation of natural communities and populations in the wild, known as **in situ** or **on-site preservation**. Only in natural communities are populations truly large enough to prevent genetic drift. Species are able to continue the process of evolutionary adaptation to a changing environment within their natural communities. However, for many rare species, in situ preservation is not a viable option in the face of increasing human disturbance. Species may decline and go extinct in the wild for any one of the reasons discussed previously: genetic drift and inbreeding, demographic and environmental variation, habitat loss, deteriorating habitat quality, competition from exotic species, disease, or overexploitation. If a remnant population is too small to persist, or if all the remaining individuals are found outside of protected areas, then in situ preservation may not be effective. In such circumstances it is likely that the only way a species can be prevented from going extinct is to maintain individuals in artificial conditions under human supervision (Conway 1980; Dresser 1988; Seal 1988; Cohn 1991a). This strategy is known as **ex situ** or **off-site preservation**. Already a number of species are extinct in the wild but survive in captive colonies, such as the Père David's deer (*Elaphurus davidianus*) and Przewalski's horse (*Equus caballus przewalski*) (Figure 17.1). The beautiful Franklin tree (see Figure 4.1 and Box 6) grows only in cultivation and is no longer found in the wild.

(A)

17.1   (A) Père David's deer (*Elaphurus davidianus*) has been extinct in the wild since about 1200 B.C. The species remained only in managed hunting reserves kept by Chinese royalty. (B) Przewalski's horse (*Equus caballus przewalski*) does well in captivity but is now probably extinct in the wild. This species was once abundant in Central Asia and is the last living species of wild horse. (Photographs by Jessie Cohen, National Zoological Park, Smithsonian Institution.)

(B)

Examples of ex situ facilities for animal preservation include zoos, game farms, aquaria, and captive breeding programs, while plants are maintained in botanical gardens, arboreta, and seed banks. An intermediate strategy that combines elements of both ex situ and in situ preservation is the monitoring and management of populations of rare and endangered species in small, protected areas; such populations are still somewhat wild, but human intervention may be necessary on occasion to prevent population decline.

Ex situ and in situ conservation strategies are complementary approaches (Kennedy 1987; Robinson 1992). Ex situ methods can help in several ways to preserve a species in the wild. First, individuals from ex situ populations can be periodically released into the wild to maintain numbers and genetic variability in natural populations (Powell and Cuthbert 1993; Cade and Jones 1993). Second, research on captive populations can provide insight into the basic biology of the species and suggest new conservation strategies. Third, ex situ populations that are self-maintaining can reduce the need to collect individuals from the wild for display and research purposes. And last, individuals on display can help to educate the public about the need to preserve the species, and so protect other members of the species in the wild. In situ preservation of species, in turn, is vital to the survival of species that are difficult to maintain in captivity, such as the rhinoceros (Figure 17.2), as well as to the continued ability of zoos, aquaria, and botanical gardens to display new species. Ex situ conservation should be seen as an important part of an integrated conservation strategy to protect endangered species.

Ex situ conservation efforts have certain basic limitations in comparison with in situ preservation (Conway 1988, Ledig 1988):

- *Population size.* To prevent genetic drift, ex situ populations of at least several hundred individuals need to be maintained. In any one zoo, only a few vertebrate species can be maintained at such numbers. In botanical gardens, only one or a few individuals are typically maintained for most species, particularly for trees.

17.2 The rhinoceros is an example of an animal that is not amenable at present to ex situ conservation strategies; rhinos tend not to reproduce once in captivity. Virtually all the animals seen in zoos, such as this pair of black rhinoceros, were captured in the wild. (Photograph by Jessie Cohen, National Zoological Park, Smithsonian Institution.)

- *Adaptation.* Ex situ populations may undergo genetic adaptation to their artificial conditions. For example, animal species kept in captivity for many generations may be selected for changes in mouthparts and digestive enzymes due to the diet of zoo food; when the animals from this altered population are returned to the wild, they may have difficulty eating their natural diet.
- *Learning skills.* Individuals of ex situ populations may lose their knowledge of their natural environment and may no longer be able to survive in the wild. For example, captive-bred animals may no longer recognize wild foods as edible or be able to locate water sources if they are released back into the wild. This problem is most likely to occur among social mammals and birds in which juveniles learn skills from adult members of the population.
- *Genetic variability.* Ex situ populations may represent only a limited portion of the gene pool of the species. For example, a captive population started with individuals collected from a warm lowland site may be unable to adapt physiologically to colder highland sites formerly occupied by the species.
- *Continuity.* Ex situ conservation efforts require a continuous supply of funds and a steady institutional policy. While this is true to some extent for in situ conservation efforts, interruption of care in a zoo, aquarium, or greenhouse lasting only days or weeks can result in considerable losses, both of individuals and species. The breakup of the former Soviet Union, the deterioration of the Russian economy, and civil wars in the outlying states provide abundant examples of how rapidly conditions can shift in a country and what such events can do to captive animal and plant populations.
- *Concentration.* Since ex situ conservation efforts are sometimes concentrated in one relatively small place, there is a danger of an entire population being destroyed by a catastrophe, such as a fire, hurricane, or epidemic.

In spite of these limitations, ex situ conservation strategies may prove to be the best, or the only, alternative when in situ preservation of a species is difficult or impossible. As Michael Soulé says, "There are no hopeless cases, only people without hope and expensive cases" (Soulé 1987).

Zoos

Zoos have traditionally focused on large vertebrates, particularly mammals, since these species have the greatest interest for the general public. This emphasis on "charismatic megavertebrates" tends to ignore the enormous threats to the huge numbers of insects and other invertebrates that form the majority of the world's animal species. However, it does energize public opinion for conservation purposes (Cohn 1988c). Zoos are increasingly emphasizing ecological themes and the threats to endangered species in their public displays and research programs (Fackelmann .1984; Robinson 1988, 1992). Educational programs at zoos, articles written about zoo programs, and zoo field projects all direct public attention toward animals and habitats of conservation significance. If the general public becomes interested in protecting gorillas and pandas after seeing them in zoos and reading about them, then money will be donated, pressure will be exerted on governments, and eventually mountains in Africa and China will be set aside as protected areas for these species. In the process thousands of other plant and animal species occupying these environments will be protected.

Zoos presently maintain over 500,000 individuals of terrestrial vertebrates, representing 3000 species of mammals, birds, reptiles, and amphibians (Conway 1988). While this number of captive animals may seem impressive, it is trivial in comparison to the numbers of domestic cats, dogs, and fish kept by people as pets. In the United States alone, about 50 million cats are kept as pets, 100 times more than the world's total of zoo animals.

A current goal of most major zoos is to establish captive breeding populations of rare and endangered animals. Zoos, along with affiliated universities, government wildlife departments and conservation organizations, are the logical organizations to develop viable captive populations of these species because they have the needed knowledge and experience in animal care, veterinary medicine, animal behavior, reproductive biology, and genetics. Only about 10% of the 274 species of rare mammals kept by zoos worldwide currently have self-sustaining captive populations (Ralls and Ballou 1983). In the United States, zoos have self-maintaining populations of about 100 species, only a small percentage of the number of species on display. Zoos still collect much of their stock from wild populations. To remedy this situation, zoos and affiliated conservation organizations have embarked on a major effort to build the facilities and develop the technology necessary to establish breeding colonies of rare and endangered animals, and to develop new methods and programs for reestablishing species in the wild (Dresser 1988; Foose 1983; Benirschke 1983).

For common animals, such as the raccoon and the white-tailed deer, there is no need to establish breeding colonies in zoos, since individuals of these common species can be readily obtained from the

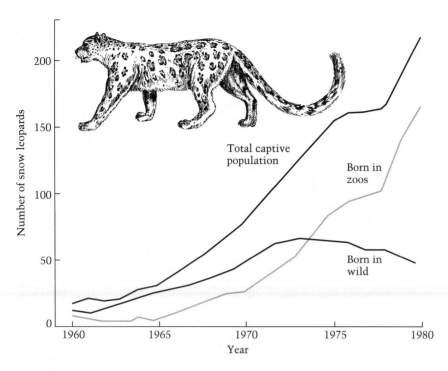

17.3   Snow leopards (*Panthera uncia*) reproduce well in captivity. Maintaining breeding colonies of these animals can reduce the zoos' need to capture individuals from the declining wild population. Since 1974, the majority of snow leopards in zoos have been born in captivity. (Data from Foose 1983.)

wild. The real need is for zoos to establish viable populations of rare species that can no longer be readily captured in the wild, such as the orangutan, Chinese alligator, and snow leopard (Figure 17.3). Colonies in zoos may represent many species' only chance of survival if their natural environments are severely damaged by human activity (Box 25). Also, if a species goes extinct in the wild, captive breeding colonies could be a source of individuals to reestablish natural populations in the wild. If a concerted effort were made to establish breeding colonies of 100 to 150 individuals per mammal species, then about 900 species could be maintained in captivity with current zoo facilities, with that number declining to 200 to 300 species if colony size were increased to 500 individuals (Conway 1988).

Most vertebrate species can be bred in captivity on a long-term basis using modern methods described by the IUCN/SSC Captive Breeding Specialist Group (Seal 1988). These methods include a diet carefully tailored to the nutritional requirements of the species, with the addition of vitamin and mineral supplements. Anesthetic techniques can be used to immobilize animals to reduce stress during transport and medical procedures. Knowledge is accumulating about vaccinations and antibiotics that can prevent the spread of disease

BOX 25   THE CALIFORNIA CONDOR RETURNS

The California condor, *Gymnogyps californianus*, is a large, vulture-like bird with an average wingspan of nearly 3 meters (10 feet). Its plight illustrates both the problems and the potential of conservation efforts in the United States during the past decade. The condor has been experiencing a gradual decline in numbers for the last 200 years (Kiff 1990; Snyder and Snyder 1990; Girdner 1992). Even in the late nineteenth century, the California naturalist James Cooper had observed that "there can be little doubt that unless protected our great vulture is doomed to rapid extinction." By 1985 condor numbers had dropped to the point of virtual extinction; the entire breeding population of the species consisted of only 6 birds in the wild and an additional 21 birds in captivity (Crawford 1986). In December of 1985, the U.S. Department of the Interior announced that, as a last-ditch measure to save the condor, the remaining birds would be captured and placed in a captive breeding program run by the San Diego and Los Angeles zoos. This decision sparked a heated debate between the Department of the Interior and the National Audubon Society, which sought a court order to block the capture on the grounds that the Department of the Interior was not developing a plan for the

eventual return of the species to the wild.

Availability of habitat free from human influence appears to be the major factor affecting the survival of the condor. The decline of the condor in the past three decades apparently is the result of a combination of habitat loss, shooting, and pollution. The species' original modern range encompasses parts of California that are now heavily urbanized, including the greater Los Angeles metropolitan area. Hunters in the area often shot at the large, soaring birds. Also, as human pressure reduced their natural prey, the birds frequently fed on the carcasses of animals killed with lead shot, which resulted in fatal lead poisoning. In addition to lead, condors have been exposed to pesticides, such as the insecticide product DDE, which caused the shells of their eggs to become fragile and break (Di Silvestro 1987; Kiff 1990; Girdner 1992) (see Box 7). These factors represented a dual crisis: high mortality among breeding adults and reproductive failure.

The National Audubon Society's suit against the Department of the Interior charged that the latter was violating environmental law by failing to acquire habitat critical to the maintenance of the species in the wild. The federal government initially refused to purchase a block of land called

the Hudson Ranch, 5300 hectares of open land that was considered by some conservationists to be vital condor habitat. The Hudson Ranch property contains one of the largest foraging areas in the condor's range that could potentially be free of hunting and pesticide use. Though the Audubon Society lost its 1986 lawsuit, the Fish and Wildlife Service reversed its original policy and purchased the Hudson Ranch—now renamed the Bitter Creek National Wildlife Refuge—with the hope of providing foraging habitat for an expanding condor population in the future (Di Silvestro 1987). The limitations of habitat preservation as a sole strategy for preserving the California condor are illustrated by the fact that the last free-living condor foraged over an area of 2.4 million hectares. In such a case, species management becomes critical, since not enough land can ever be acquired to totally protect a wild population of this species from human influence.

The captive breeding experiment has provoked controversy for several reasons. First, when the last condors were captured in 1986 and 1987, biologists were not certain that condors would breed in captivity. A related concern was that captivity might alter the birds' behavior, rendering them incapable of surviving in the wild. The zoos have avoided this problem by isolating the birds from human contact as much as possible; condors are not displayed to the public, and hand-feeding of chicks is done through blinds, with condor-shaped puppets covering the feeders' hands (Crawford 1986; Girdner 1992). A second problem was the small size of the population. Though biologists could and did prevent closely related individuals from breeding, such a small population provides a limited supply of potential mates. At least half of the current condor population stems from only three clans with 14 founder lines (Kiff, personal communication); a significant degree of inbreeding is therefore unavoidable. Furthermore, questions have been raised concerning the whole rationale of spending over $20 million on a single rare species unlikely ever to be reestablished in the wild (Willwerth 1992). However, the 100,000 hectares of California lands now protected for future condor use will certainly be vital to the preservation of other, less conspicuous species and communities. Also, the new techniques of captive rearing developed for the condor are now being applied to other species (Kiff 1990).

As part of the project, Andean condors are being experimentally released in California to develop conservation methods that can be applied to the California condor (Kiff 1990). Only females are used to pre-

California condor chicks raised in captivity are fed by researchers using puppets that look like adult birds. Conservation biologists hope that minimizing human contact with the birds will improve their chances of survival when they are returned to the wild. (Photograph by Mike Wallace, The Los Angeles Zoo.)

vent a wild population from becoming established. Andean condors breed readily in zoos, and are common in captivity. It has been found that Andean condors can adapt to artificial cave structures in the wild and learn to take contaminant-free food from feeding stations. They learn to soar without human assistance, traveling great distances before returning to their nests. While such a life might seem overly artificial, it may be the only possible way in which the California condor can return to its natural habitat and live in association with humans.

Captive breeding has thus far proven successful. By early 1992, the total population of California condors in captivity had doubled; the Fish and Wildlife Service has released one pair into the wild and plans to monitor their progress (Girdner 1992). Despite this preliminary success, the species' future is not yet certain; the ability of the pair to survive and, more importantly, for released birds to eventually reproduce in the wild will be the critical test of the captive breeding program. The success or failure of the initial pair may determine whether or not the California condor's time has truly passed.

within colonies of captive species. Central data bases of breeding records are being developed to prevent genetic drift and inbreeding.

Some rare species do not reproduce well in captivity. New techniques are being developed to enhance the low reproductive rates of these species (Dresser 1988; Seal 1988). **Cross-fostering**, for example, can increase the reproductive success of certain species. If animal mothers of a rare species are not able to raise their own young, their offspring can sometimes be raised by parents of another species (Powell and Cuthbert 1993). Many bird species, such as the bald eagle, normally lay only one clutch of eggs per year, but if the clutch is removed by biologists, the mother bird will lay and raise a second clutch of eggs. If the first clutch is given to another bird of a related species, two clutches of eggs will be produced per year for each rare female. This technique, known informally as "double-clutching," potentially doubles the number of offspring one female can produce.

Another aid to reproduction, similar to cross-fostering, is **artificial incubation**. If the mother does not adequately care for her offspring, or the offspring are readily attacked by predators, parasites, or disease, the offspring may be cared for by humans during their vulnerable early stages. This approach has been tried extensively with sea turtles, birds, fishes, and amphibians. Eggs are collected and placed in ideal hatching conditions, the hatchlings are protected and fed during their vulnerable early stages, and the young are then released into the wild or raised in captivity. This approach is sometimes called a "Head Start Program."

Artificial insemination may be used when a zoo has only one or few individuals of a rare species, such as the giant panda. When an isolated female animal comes into breeding condition, either on her own or by chemical induction, no male of her species may be present at the zoo to mate with her; also, some animal species lose interest in mating while in captivity. In such cases, sperm can be collected from suitable males, stored until needed at low temperatures, and then used for artificial insemination of a receptive female. While artificial insemination is done routinely with many animal species, exact techniques of sperm collection, sperm storage, recognition of female receptivity, and sperm delivery have to be worked out for each species.

Embryo transfer can be used to increase the reproductive output of some rare animals such as the bongo, gaur, and Przewalski's horse. Superovulation, or production of multiple eggs, is induced using fertility drugs, and the extra eggs are collected, fertilized with sperm, and surgically implanted into surrogate mothers of a related common species. The surrogate mother will then give birth to the offspring of the rare species some time later (Figure 17.4).

17.4   This gaur calf was produced via an embryo transfer using a domestic Holstein cow as a surrogate mother at the Kings Island Wild Animal Habitat. The gaur is an endangered species. (Photograph by Betsy L. Dresser.)

A wide range of additional innovative techniques are being developed to maintain and increase populations of captive species (Dresser 1988; Woolf 1986). Some of these come directly from human and veterinary medicine, while others are novel methods developed for particular species. Some promising new techniques include rearing species from a single sex (when only one individual remains); cross-species hybridization (when the remaining members of a species cannot breed among themselves); induced hibernation and diapause as a way of maintaining dormant populations; and biochemical and surgical sexing of animals that have no external sex differences. One of the most unusual techniques involves freezing eggs, sperm, embryos, and other tissue of species on the verge of extinction, with the hope that these tissues can be used to reestablish the species at some time in the future (Woolf 1986; Cade 1988; Wildt 1992).

Managers of captive breeding programs are also now much more careful than in the past about assigning mates so as to avoid genetic problems. A careful analysis of zoo breeding records for 44 mammalian species, including 16 ungulates, 16 primates, and 12 other small mammals, revealed that juvenile mortality was higher among the offspring of closely related animals (such as fathers mated with their daughters) than among the offspring of unrelated animals (Ralls and Ballou 1987; Ralls, Harvey, and Lyles 1986). "It was like re-inventing the wheel," Ballou stated (in Tangley 1988c). "Anyone who has taken an introductory genetics course knows about the potential problems of inbreeding." As a result of such studies, zoos now track the genetic lineages of captive animals carefully, using global computerized data bases to help prevent pairing of related animals and avoid inbreeding depression.

BOX 26 IS DOMESTICATION THE WAY TO SAVE THE GREEN IGUANA?

Species sometimes continue to be overharvested to the point where they become extinct because of the simple principle of supply and demand: as a species becomes scarce, its price keeps rising. One solution is to cultivate or domesticate a species, cause a price reduction, and reduce the demand for wild products. One of the more novel conservation efforts is recommending just this strategy: herpetologists in Panama and Costa Rica are promoting the creation of "iguana farms" in an attempt to increase the population of the green or common iguana, which is rapidly becoming endangered throughout most of Central America (Miller 1984; Cohn 1989; Werner 1991).

The iguana, intensively hunted for its delicate, tasty meat, is prized above other reptile species for its supposed aphrodisiac and curative properties (Vietmeyer 1989). Overhunting and rapid destruction of the iguana's rain forest habitat have combined to reduce

its numbers in recent decades. Yet the iguana reproduces rapidly and easily; researchers have discovered that adults breed yearly after reaching maturity at age 2–3 years, and female iguanas are not overly selective about where they build their nests (Rand and Greene 1982; Werner and Miller 1984). One significant reason that the iguana's high reproductive rate has not counteracted the species' decline is that over 95% of newly hatched wild iguanas are lost to predators. In an effort to improve the survivorship of iguana hatchlings, scientists initiated a program in which they collected both hatchlings and unhatched eggs from iguana nests and brought them into a protected enclosure (Cohn 1989).

The initial effort to protect hatchlings in Panama led to the discovery in 1984 that iguanas could be easily bred in captivity. Though wild-collected adult iguanas refused to mate in captivity, iguanas that had been captured as hatch-

lings, as well as iguanas hatched from eggs collected from wild nests, bred enthusiastically during the first three years of the project. The population of captive iguanas increased rapidly: the initial 700 iguanas hatched in 1984 ballooned to over 8000 iguanas by 1988 (Cohn 1989). Project scientists concluded that if local farmers could be persuaded to raise iguanas instead of cattle, the species, the farmers, and the rain forest would all benefit. Pound for pound, iguanas utilize the land far more efficiently than cattle (Vietmeyer 1989): unlike cattle, iguanas do not require large, cleared spaces, so a "herd" of 60–100 iguanas can thrive on a relatively small tract of enclosed, forested land (Miller 1984). The work required to clear pasture is eliminated, as is the damage done to the soil through erosion following clearing. To promote this idea, project scientists encouraged local people to visit the captive breeding site and spon-

Cages and enclosures housing a captive iguana breeding colony at Soberiana National Park, Panama. (Photographs by Karen M. Allen, Conservation International.)

sored lectures, films, and other educational activities. Farmers willing to attempt iguana breeding were given materials for the construction of a simple enclosure and instructed in the care of the animals (Cohn 1989).

The project has been extremely successful thus far. By encouraging involvement of local people, project scientists have effectively demonstrated the simplicity and cost-effectiveness of iguana farming while simultaneously increasing the local population of iguanas. Over 70,000 iguanas had been released as of 1993 on farms in Panama and at an additional site in Costa Rica, where an iguana farming project was begun in 1988 (Werner, personal communication). The 80% survival rate for the iguanas is very high, suggesting that the behavioral and genetic effects of captive breeding have been minimal. In addition, the educational program in Panama has reduced hunting; many former poachers who learned about the program began to bring iguanas to the project's directors rather than killing them for sale in the market (Cohn 1989). Other Central American countries may soon follow the lead of Panama and Costa Rica in developing iguana farming programs. Still, one troubling question remains: though the program is undeniably successful in creating a population explosion of green iguanas, it amounts to a virtual domestication of the species. Do these methods really help reduce hunting pressure on wild iguana populations, or do they just add another species to the list of animals and plants that human beings have taken under our control? It may be that domestication of the species is the trade-off for saving it, but the consequences, both to the species and to the ecology of its rain forest habitat, are yet to be seen.

These ex situ techniques represent technological solutions to problems caused by human activities. The most natural solution is to protect the species and its habitat in the wild so that it can recover naturally. When this solution is not possible, artificial methods are available to support those species that will become extinct without human intervention (Box 26). When scientists decide to use these methods, they face a series of ethical questions that need to be addressed. First, how necessary and how effective are these methods for a particular species? Is it better to let the last few individuals of a species live out their days in the wild, or to start a captive population that may be unable to readapt to wild conditions? Second, does a population of a rare species that has been raised in captivity and does not know how to survive in its own natural environment really represent survival for the species? Third, are species held in captivity for their own benefit or for the benefit of zoos?

Even when the answers to these questions indicate a need for ex situ management, it is not always feasible to create ex situ populations of rare animal species (Box 27). A species may have been so severely reduced in numbers that there is a low breeding success and a high infant mortality due to inbreeding depression. Certain animals, particularly marine mammals, are so large or require such specialized environments that the facilities for maintaining and handling them

BOX 27 CONSERVATION OF LAND TORTOISES IN THE INDIAN OCEAN

Conservation of endangered land tortoises on islands off the coast of Africa is complicated by two factors: first, these animals are often so rare that they have not been well studied, and second, they are subject to frequent harassment by collectors, thieves, and poachers. Two of these tortoises, the angonoka (*Geochelone yniphora*) and the Aldabra giant tortoise (*Geochelone gigantea*), are different in terms of population size, natural history, and conservation needs,

but the survival of both in the wild is in doubt.

The severely threatened angonoka, also known as the plowshare tortoise, is found only in a small patch of secondary forest and savannah habitat in northwestern Madagascar. Only 400 individuals are thought to exist in the wild, and approximately 50 are in captivity, giving this species the dubious distinction of being the world's rarest tortoise. Studies indicate that this species apparently has always

had a small range, and has never been very numerous (Burke 1990); however, its numbers may have been reduced in recent years by a combination of introduced pigs, which prey on eggs and young tortoises, and disturbance by human beings. Though the angonoka is not considered good to eat, these tortoises are commonly kept as pets by local people because of their attractive shell patterns, a practice which seemingly does no harm to the individual animal but which does reduce its chances to breed (Burke 1990).

The Aldabra giant tortoise is the only remaining representative of the 15 species of giant tortoises that formerly occurred on the islands of the western Indian Ocean. In contrast to the angonoka, the Aldabra giant tortoise, which can reach up to 120 kg and live to be 65 to 90 years old, has reasonably healthy numbers of over 100,000 animals. However, the only self-maintaining population of this rare species is found on the Aldabra atoll; the species is thus susceptible to disease or other natural disasters (Samour et al. 1987). Large colonies of endemic birds and the presence of coral reefs make the atoll a tourist attraction, a fact that could affect the breeding success of the tortoises. Like the angonoka, the Aldabra tortoise is attractive to collectors; it is also hunted illegally for food, its eggs are eaten, and introduced animals such as pigs, goats, and rats threaten its young.

Several different methods have been attempted in order to conserve these two tortoise species. For the angonoka, a clear priority was to increase the overall numbers of the animal. Captive breeding was initially unsuccessful; only five of these tortoises survived away from Madagascar, and only one hatchling was produced in all of the breeding attempts (Burke 1990). Studies showed that the single male

that provided sperm for artificial insemination was producing low-quality sperm; further attempts to collect sperm from this male apparently led to his death. Subsequent attempts at captive breeding have incorporated information from the study of the angonoka and other tortoise species to improve their success: for example, a successful breeding program is located at a station built on Madagascar, near the species' normal range, to eliminate the possibility of environmental variables preventing successful reproduction. This station also incorporates multiple individuals into the breeding group, based upon research on other tortoise species that indicates a need for multiple males, and possibly male combat, to produce offspring.

The Aldabra tortoise required a different approach. The genetic diversity and population size of the species are not yet a problem; the priority in this case was to extend the range of the species to guard against disease or disaster. To this end, a number of giant tortoises was released on Curieuse Island in the Seychelles (Samour et al. 1987). The colony was established between 1978 and 1982, and censused in 1986. The results of the survey were somewhat disheartening: less than half of the introduced tortoises were present, despite the fact that earlier surveys had shown that the population was both

healthy and growing. Curieuse Island had been chosen because there were no natural predators, and exotic species were kept largely under control by the few human inhabitants; the tortoises had adapted well to the island and were breeding there. The disappearance of the tortoises was therefore thought to be primarily due to theft and poaching.

Collection of both species for pets is a problem of considerable concern to biologists. Collectors do not feel that what they are doing is wrong, since they are not physically harming the animals; in fact, one might argue that the individual animal's chances of survival are improved in captivity, since it is protected from predators or natural disasters and is always well fed. What is good for the individual, however, is bad for the species: the removal of each individual animal, particularly if the animal is immature, decreases the gene pool of the population and increases the chance of extinction. In the case of the angonoka, which are spread out over a fairly large area, removal of even a single individual may have dramatic repercussions for the entire population; not only will that individual have no chance to mate, but several other wild angonoka may also be deprived of the opportunity because they are too far from other individuals to attract or find a mate.

are prohibitively expensive. Many invertebrates have complex life cycles in which their diet changes as they grow and in which their environmental needs vary in subtle ways. Many of these species are not possible to raise with our present knowledge. Finally, certain species are simply difficult to breed, despite the best efforts of scientists. Two prime examples of this are the giant panda (Box 28) and the Sumatran rhino, which have low reproductive rates in the wild and do

BOX 28 LOVE ALONE CANNOT SAVE THE GIANT PANDA

The giant panda is one of the most easily recognized endangered species in the world. It is so well known, and so well beloved by millions of people, that its image is the symbol for the World Wide Fund for Nature, also known as the World Wildlife Fund, a prominent international conservation organization. Nevertheless, the panda's future is in serious jeopardy. As with many endangered species, pressure upon the panda's habitat by human populations is the most significant threat to its survival. However, human pressures appear to exacerbate some of the unusual traits of the panda's physical and behavioral makeup, making this species particularly vulnerable to extinction.

One of the more bizarre features of panda biology is the species' diet of bamboo. Pandas are related to carnivores,

Among the most beloved of all endangered species, the panda has become a symbol of conservation efforts. (Photograph by Jessie Cohen, National Zoological Park, Smithsonian Institution.)

and they lack many of the anatomical adaptations—such as elongated digestive tracts—that enable herbivores to use plant foods efficiently. Most herbivores also have symbiotic bacteria in their digestive systems that assist in breaking down cellulose, further improving digestive efficiency; pandas lack these organisms (Dolnick 1989). Consequently, pandas must eat continually in order to absorb sufficient nutrients to survive.

To further complicate matters, pandas periodically must change their feeding habits in response to cyclical bamboo die-offs. Bamboo species reproduce in long-term cycles of anywhere from 15 to over 100 years; typically, all individuals in a given species within a certain area will produce flowers and seeds in a single season, then die (Machlis and Johnson 1987; Roberts 1988; Dolnick 1989). Two to three years are generally required before new shoots appear from the seeds. Though in a certain place pandas may prefer one particular species of bamboo above all others, during die-off events they switch to other species. Frequently, this change requires them to migrate from the high-altitude regions they prefer to lowland areas, especially on those uncommon occasions when two or three bamboo species flower simultaneously. The bamboo die-offs trigger a direct conflict between the needs of the pandas and human populations (Johnson et al. 1988; Schaller et al. 1989). Pandas are solitary and shy; they will not go into the human-populated lowland areas. Cut off from the lowland regions by humans, the pandas have no recourse when die-offs occur. In the 1970s, when three species of bamboo flowered simultaneously, at least 138 pandas starved—almost 14% of the current panda population (Reid et al. 1989).

Following this catastrophe, in 1983 the Chinese government instituted a "rescue" policy of searching for starving pandas during bamboo die-offs. The policy is not without its drawbacks; rescued animals frequently end up in zoos, depleting the wild population further (Roberts 1988). Although one might argue that a live panda in captivity is better than a dead one in the wild, attempts to establish a self-sustaining captive breeding colony have not been successful. Pandas are extremely selective in choosing mates, and often pandas paired by zoos will prove incompatible. Artificial insemination can circumvent the pandas' choosiness, but even using this technique, pandas rarely give birth to live young, and those often do not survive more than a few days. Between 1963, when China first began to breed captive pandas, and 1989, only 90 cubs were born; of these, only 37 survived for

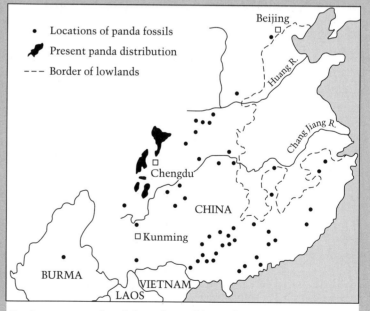

Pandas were once found throughout China and even into Burma; they are now restricted to a few localities in the vicinity of the city of Chengdu. (After Roberts 1988.)

more than 6 months (Dolnick 1989). However, the birth of 11 pandas in captivity in China in 1992 is a hopeful sign that some of these problems can be eventually overcome.

The difficulties encountered in breeding pandas are only partially understood. Some of the problems are purely a consequence of panda physiology, and are commonly found in other mammals as well: females go into heat only once every year, and are fertile for only two or three days. Cubs are born singly or in pairs, and are remarkably small and helpless, weighing just four ounces. A panda female will usually raise only one cub at a time, even if she gives birth to two live cubs. Moreover, the five-month pregnancy is followed by five months of nursing; since the female does not generally go into heat again immediately after she stops nursing, she misses a year's breeding season. In addition, young pandas are dependent on the care of their mothers for up to 22 months after birth, so that injury, sickness, or death of the mother may mean the loss of the cub as well. A panda female will produce at most one cub every other year, so the rate of population growth is very slow even under the best of conditions. Would-be panda breeders worldwide are attempting to work out why captive pandas refuse to mate—or in the case of many females, do not even come into heat (Dolnick 1989).

Getting pandas to reproduce in captivity is only one of several major obstacles facing this species. The Chinese government has put significant financial resources into setting aside habitat for the remaining wild pandas, but it will be difficult for the reserves to withstand the pressure of China's immense human population. (Machlis and Johnson 1987; DeWolf et al. 1988). Moreover, poaching of pandas for their skins has become increasingly common despite stiff penalties imposed by the Chinese government. Given these circumstances, the odds that this popular species will escape extinction in the wild seem poor.

not reproduce well in captivity despite a considerable effort by some of the best scientists to find effective methods.

Aquaria

Public aquaria have traditionally been oriented toward the display of unusual and attractive fish, supplemented with performances by seals, dolphins, and other marine mammals. However, as concern for the extinction of species has increased, conservation has developed into a major educational theme in aquaria. The need is great, since hundreds, if not thousands, of fish species are threatened with extinction. In North America alone, 24 species are known to have gone extinct since the arrival of European settlers, and 63 species are now classified as endangered (Ono et al. 1983; Courtney and Stauffer 1984; Williams and Nowak 1986). The rich fauna of the Mississippi basin and the unique desert pupfish of the southwestern United States are in particular danger. Extinctions of fishes are occurring worldwide in places such as the African Rift lakes and Andean lakes (Parenti 1984). Already 15 of the 18 endemic fishes in Lake Lanao in the Philippines are extinct (Kornfield and Carpenter 1984). Freshwater mollusks are also a priority for preservation because of their vulnerability to changes caused by water pollution and dams.

To deal with this threat to species, ichthyologists, marine mammalogists, and coral reef experts who work for public aquaria are increasingly linking up with colleagues in marine research institutes, government fisheries departments, and conservation organizations to develop programs for the conservation of rich natural communities and species of special concern.

Major efforts are presently being made to develop breeding techniques so that rare species can be maintained in aquaria, sometimes for release back into the wild, and so that aquatic species do not have to be collected in the wild (Kaufman 1988). Many of the techniques used in fish breeding were originally developed by fisheries biologists for large-scale stocking operations involving trout, bass, salmon, and other commercial species. Other techniques were discovered in the aquarium pet trade, as dealers attempted to propagate tropical fish for sale. These techniques are now being applied to such endangered freshwater fauna as the desert pupfish, stream fishes of the Tennessee River Basin, and cichlids of the African Rift lakes. Programs for breeding endangered marine fishes and coral species are still in an early stage, but this is an area of active research at the present time.

Some of the most serious problems faced by marine species are the destruction of tropical coral reefs by water pollution, overcollection of fish for the worldwide tropical fish trade and shells for the shell market, and dynamiting of reefs for collection of coral fragments for the international market. These problems are particularly severe in the Philippines, where there is an unfortunate combination of rich coral reefs, extensive rural poverty, and an ineffective government. Throughout the world, aquatic conservation biologists and aquarium managers are developing techniques for more reliable breeding success and legislation for better habitat protection.

Aquaria have a particularly important role to play in the conservation of endangered cetaceans. Aquarium personnel often respond to public requests for assistance in dealing with whales stranded on beaches or disoriented in shallow waters. The aquarium community potentially can use the lessons learned from working with common species to develop programs to aid endangered species. Extensive experience with captive populations of the bottle-nosed dolphin, the most popular aquarium species, may eventually be applied to other species (Figure 17.5; Ames 1991). Researchers are now able to maintain colonies, perform artificial insemination, raise calves, and release captive-born animals back into the natural environment. Tech-

17.5 Breeding bottle-nosed dolphins in captivity has provided aquarium personnel with valuable experience that can be applied to endangered species. Shown here are a mother and calf. (Photograph courtesy of Sea World.)

niques learned with dolphins may eventually be applied to such other endangered cetaceans as the Chinese Yangtze River baiji, the Gulf of Mexico vaquita, the Mediterranean striped dolphin, and the harbor porpoise (Ames 1991).

A practical problem with establishing populations of captive marine mammals and large fish is that they require enormous volumes of water. One possible approach is to use small, protected natural bodies of water to create habitats that are somewhat intermediate between wild and artificial conditions. Such an approach is being used to protect the baiji (*Lipotes vexillifer*) in the Yangtze River. This species has undergone a precipitous decline and now has only about 200 individuals remaining in the wild. The most obvious causes of the decline are dams and floodgates that reduce fish populations and interfere with migration patterns, and accidental injuries caused by commercial fishing activities and boat propellers. Water pollution may also disrupt the baiji's reproductive physiology, and the noise from motorboats and other industrial activity may interfere with the echolocation used by the baiji to find food and mates and avoid danger. To protect the baiji, Chinese scientists have established experimental breeding reserves in oxbow lakes. While this species is normally found in moving river water rather than lakes, at least the baiji are protected from the most damaging aspects of human activity within the lakes.

Botanical Gardens and Arboreta

Gardening is a hobby of millions of people worldwide, and has a history going back thousands of years. Household gardens have long provided a source of vegetables and herbs. In ancient times, doctors and healers kept gardens of medicinal plants to treat their patients. In more recent centuries, royal families established large private gardens for their personal enjoyment, and nations established gardens for the public. While the display of beautiful plants was the major purpose of many of these large gardens, an additional purpose was to illustrate the diversity of the living world and to assist in the dissemination and propagation of plants that could be used in horticulture, agriculture, forestry, landscaping, and industry. In recognition of these vital roles of plants in the economic activities of society, many European countries set up botanical gardens throughout their colonial empires. The world's 1500 botanical gardens now contain major collections of living plants and represent a crucial resource in plant conservation efforts (Figure 17.6). The botanical gardens of the world are currently growing at least 35,000 species of plants, approximately 15% of the world's flora (Raven 1981; IUCN/WWF 1989), with perhaps even double that number of species being grown in greenhouses, subsistence

17.6 The New England Wild Flower Society's Garden in the Woods in Massachusetts. Such botanical gardens offer great pleasure and enjoyment to visitors as well as serving as preservers of rare plant species. (Photograph by John A. Lynch.)

17.7 Endangered species of plants can often be propagated in large numbers using modern tissue culture techniques. A new plant is growing inside each plastic container. The plants will later be transferred outside or into pots in a greenhouse. (Photograph by John A. Lynch.)

gardens, hobby gardens, and other such situations. The world's largest botanical garden, the Royal Botanical Gardens of England at Kew, has an estimated 25,000 species of plants under cultivation, about 10% of the world's total, of which 2700 are listed as threatened under the IUCN categories (Reid and Miller 1989). In addition to growing plants, botanical gardens and research institutes have developed collections of seeds, sometimes known as **seed banks**, collected from the wild and from cultivated plants. These seed banks represent a crucial backup to the cultivated plants.

Many botanical gardens specialize in particular types of plants. The Arnold Arboretum of Harvard University grows hundreds of different temperate tree species. The New England Wild Flower Society has a collection of hundreds of perennial temperate herbs at its Garden in the Woods location. In California, a specialized pine arboretum grows 72 of the world's 110 species of pines, while in South Africa, the leading botanical garden has 25% of South Africa's plant species growing in cultivation (Raven 1981).

In many ways, plants are easier to maintain in controlled conditions than animals are (Ashton 1988). Plants can be readily established from seeds, cuttings, and rhizomes, and through tissue culture techniques (Figure 17.7). Most plants have the same basic needs of light, water, and minerals, which can be readily supplied in greenhouses and gardens. Adjusting light, temperature, and humidity levels to suit individual species is the main concern, but their needs are often readily determined through knowledge of the plant's natural growing conditions. Since plants do not move, they often can be grown at high densities. Also, plants can be pruned to a small size if space is a limiting factor. As a result, adequate population samples can be established for many species. Plants can often be maintained out of doors in gardens, where they need minimal care and weeding to survive; unlike animals, plants do not have to be caged and regularly fed. Most plant species are bisexual, so that fewer individuals are needed to maintain a population, and many plant species readily produce seeds on their own. Wind, insects, and other animals cross-pollinate many plants in botanical gardens, while other species naturally self-pollinate. Hand pollination to produce seed is also a simple procedure in most plant species. Many plants, particularly those found in the temperate zone and dry climates and those found growing in disturbed conditions, have seeds that can lie dormant for years and even decades in cool, dry conditions. Some perennial plants, particularly shrubs and trees, are long-lived, so that individuals can be kept alive for decades and even centuries once they are beyond the seedling stage.

Botanical gardens are in a unique position to contribute to conservation efforts because their living collections and their associated herbaria of dried plant collections represent one of the best sources of information on plant distribution and habitat requirements. The staff of botanical gardens are often recognized authorities on plant identification, distribution, and conservation status. Expeditions sent out by botanical gardens discover new species and determine the distribution and status of known species, while over 250 botanical gardens maintain nature reserves that serve as important conservation areas in their own right. In addition, botanical gardens are in a position to educate the public about conservation issues, since an estimated 150 million people per year visit them (IUCN/WWF 1989). As stated by Ashton (1984): "Botanic gardens have an opportunity, indeed an obligation which is open to them alone, to bridge between the traditional concerns of systematic biology and the returning needs of agriculture, forestry and medicine for the exploration and conservation of biological diversity."

The conservation of endangered species is becoming one of the major goals of botanical gardens as well as zoos. In the United States, conservation efforts by a network of 25 botanical gardens are being coordinated by the Center for Plant Conservation based at the Missouri Botanical Garden (Mlot 1989b; Falk 1991). The United States has over 4000 plant species that are threatened in some way, with perhaps 700 of these species in danger of going extinct within 5 to 10 years. Through the efforts of these botanical gardens, more than 450 of the threatened species are now being grown in cultivation. A major goal of such programs is the reintroduction of these species into the wild.

At the international level, the Botanical Gardens Conservation Secretariat (BGCS) of IUCN is involved in organizing and coordinating conservation efforts by the world's botanical gardens (BGCS 1987). Priorities of this program involve a worldwide data base system for coordinating collecting activity and identifying important species that are underrepresented or absent from living collections. A problem with the distribution of botanical gardens is that most of them are located in the temperate zone, even though most of the world's plant species are found in the tropics. While a number of major gardens exist in such places as Singapore, Sri Lanka, Java, and Colombia, establishing new botanical gardens in the tropics should be a priority for the international community, along with training local plant taxonomists to fill staff positions.

Seed Banks

The seeds of most plant species can be stored in cold, dry conditions in seed banks for long periods of time and then later germinated to form new plants. At low temperatures, the metabolism of seeds is reduced and the food reserves of the embryo are depleted very slowly. This property of seeds is extremely valuable for ex situ conservation efforts, since it allows seeds of large numbers of rare species to be frozen and stored in a small space, with minimal supervision, and at a low cost (Figure 17.8). One such seed bank facility is the United States Department of Agriculture (USDA) National Seed Storage Laboratory (NSSL) at Fort Collins, Colorado, in which some seeds are stored at temperatures as low as $-196°C$. More than 50 other major seed banks exist in the world, many of them in developing countries. The focus of most of these seed banks is on preserving genetic variability in crop species. While seed banks have great potential for conserving species, they have certain problems as well. If power supplies fail or equipment breaks down, the entire frozen collection may be damaged. Even in cold storage, seeds gradually lose their ability to germinate, due to an exhaustion of energetic reserves and an accumulation of harmful mutations. Old seed supplies simply may not germinate. To overcome this gradual deterioration of seed quality, seed samples must be periodically germinated, adult plants grown to maturity, and new seed samples stored. For seed banks with large collections, this testing and rejuvenation of seed samples can be a formidable task. For large, long-lived species such as trees, renewing seed vigor may be extremely expensive and time-consuming.

Approximately 15% of the world's plant species have "recalci-

17.8 At the U.S. National Seed Storage Laboratory, seeds are stored in hermetically sealed packets at −20°C; sometimes even lower temperatures are used. (Photograph courtesy of U.S. Department of Agriculture and the Center for Plant Conservation.)

trant″ seeds that either lack dormancy or do not tolerate low-temperature storage conditions, and consequently cannot be stored in seed banks. Seeds of these species must germinate right away or die. Species with recalcitrant seeds are much more common in the tropical forest than in the temperate zone, and the seeds of many economically important tropical fruit trees, timber trees, and plantation crops, such as cocoa, rubber, and Asian dipterocarp trees, cannot be stored (BGCS 1987). Intensive investigations are under way to find ways of storing recalcitrant seeds; one possible means may be storing only the embryo after removing the surrounding seed coat, endosperm, and other tissues. Plant species can also be maintained in tissue culture in controlled conditions or propagated by cuttings from a parent plant, though these processes are currently more expensive than growing plants from seeds.

Seed Sampling Strategies for Wild Species

Strategies for collecting seeds from wild plants are influenced by the distribution of genetic variability, since species that are genetically variable may require more extensive sampling to acquire the majority of their alleles than species that are more genetically uniform. Genetic variability includes both the percentage of genes that are variable and the number of alleles per polymorphic gene (see Chapter 2). The most extensive information on genetic variability in plants comes from accumulated studies of allozyme variation in over 400 species (Hamrick et al. 1991; Hamrick and Godt 1989). These studies show that the most important factor in determining the total amount of genetic variability in a plant species is the geographical range of the species; widespread species have more than twice as much genetic variability as species of restricted range.

Genetic variability can be further partitioned into the amount of variability within individual populations and the amount of variability among populations of a species. On average, each plant species has 78% of its genetic variability within populations and 22% of its variability among populations. Outbreeding, wind-pollinated species have proportionately more genetic variability within their individual populations than do self-pollinating species and animal-pollinated species with mixed mating systems. Variation among populations shows quite a different pattern: self-pollinating species have proportionately more genetic variability among populations than wind-pollinated, outcrossing species. Annual species have proportionately more genetic variability among populations than long-lived woody perennial species (Hamrick et al. 1991; Hamrick and Godt 1989). Overall, these results show that species with the greatest potential for long-distance pollen movement (i.e., outcrossing species and wind-pollinated species) have more genetic uniformity across their populations than species with less potential for gene flow.

These results demonstrate that with information on plant distribution, breeding system, pollination ecology, seed dispersal, growth, habitat, and other botanical characteristics, it is possible to make an initial prediction of the amount of genetic variability that a species possesses and how it is distributed within and among populations. This information in turn suggests strategies for protecting the gene pool of a rare or endangered species by both in situ protection of carefully selected populations and a well-planned collection program for ex situ preservation.

Using information on patterns of genetic variability, the Center for Plant Conservation (1991) has developed a set of seed sampling guidelines for conserving the genetic variability of endangered plant species; these guidelines could be modified for other groups of species, including animals.

1. The highest priority for collecting should be species (*a*) that are in danger of extinction, that is, species showing a rapid decline in number of individuals or number of populations; (*b*) that are evolutionarily or taxonomically unique; (*c*) that can be reintroduced into the wild; (*d*) that have the potential to be preserved in ex situ situations; (*e*) that have potential economic value for agriculture, medicine, forestry, or industry.

2. Samples should be collected from up to five populations per species to ensure a sampling of the genetic variability contained among populations. Where possible, populations should be selected to cover the geographical and environmental range of the species. All populations should be sampled for the 70% of endangered species that have five or fewer populations.

3. Samples should be collected from 10 to 50 individuals per population. Sampling fewer than 10 individuals may miss alleles that are common in the population. Sampling more than 50 individuals may not result in obtaining enough new alleles to justify the effort. In general, sample sizes should be at the high end of the range when the population appears to be phenotypically variable, the site is heterogeneous, and the plants are outcrossing.

4. The number of seeds (or cuttings, bulbs, etc.) collected per plant is determined by the viability of the species' seeds. If seed viability is high, then only a few seeds need to be collected per individual; if seed viability is low, then many seeds have to be collected per individual.

5. If individual plants of a species have a low reproductive output, collecting many seeds in one year may have a negative effect on the sampled populations. This is particularly true for annuals and other short-lived plants. In such cases, a better strategy would be to spread the collecting over several years.

As the CPC (1991) concludes: "Conservation collections are only as good as the diversity that they contain. Thus the forethought and methods that go into sampling procedures play a critical role in determining the ultimate quality of the collection, as well as its usability for purposes such as reintroduction and restoration. In the long run, the real significance of collections in biological conservation is their role in reinforcing the management and maintenance of natural populations. Collectors should view themselves not as 'preserving' static entities, but as providing a stepping-stone on the pathway to survival and evolution."

Agricultural Seed Banks

Seed banks have been embraced by the international agricultural community as an effective way of preserving the genetic variability that exists in agricultural crops (Figure 17.9). Often resistance to particular diseases and pests is found in only one variety of a crop that is grown in only one small area of the world. This genetic variability is often crucial to the agricultural industry in its efforts to maintain and increase the high productivity of modern crops and to respond to

17.9 The genetic variation in corn (*Zea mays,* also known as maize) is evident in the variety of cob shapes, seed (kernel) shapes, and color patterns. (Photograph © Steven King.)

changing environmental conditions, such as acid rain, changing weather patterns, and soil erosion. Researchers are in a race against time to preserve this genetic variability because traditional farmers throughout the world are abandoning their local crop varieties in favor of standard, high-yielding varieties (Wilkes 1991; Altieri and Anderson 1992). This world-wide phenomenon is illustrated by Sri Lankan farmers, who grew 2000 varieties of rice until the late 1950s, when they switched over to five high-yielding varieties (Rhoades 1991).

The value of agricultural seed banks is illustrated by one classic example. Rice crops in Africa were being devastated by grassy stunt virus strain 1. In an attempt to find a response to this problem, agricultural researchers grew rice plants from thousands of seed collections of wild and cultivated rice from around the world (Khush and Ling 1974; Vaughan and Sitch 1991; Lin and Yuan 1990). One sample of wild rice from Gonda, Uttar Pradesh, India, was found to contain a gene for resistance to the disease. These wild plants were immediately incorporated in a major breeding program to transfer the gene for virus resistance into high-yielding varieties of rice. If the sample of wild rice had not been collected, or had died out before being discovered, the future of rice cultivation in Africa would have been uncertain.

Agricultural researchers have been combing the world for local varieties of crop plants, known as landraces (see Chapter 15), that can be stored in seed banks and later be hybridized with modern varieties in crop improvement programs (Wilkes 1977, 1987, 1989; Rhoades 1991). So far over 2 million collections of seeds have been acquired by agricultural seed banks (Peeters and Williams 1984). Many of the major food crops, such as wheat, maize (corn), oats, and potatoes, are well represented in seed banks, and other important crops such as rice, millet, and sorghum are being intensively collected as well (Plucknett et al. 1987; Williams 1984).

Despite their obvious successes in collecting and storing material, agricultural seed banks are not wholly satisfactory. Collections are often poorly documented regarding the locality of collection and original growing conditions. Many of the seeds are of unknown quality, and may not be able to germinate. Seed collections have focused on major food crops; crops of only regional significance, medicinal plants, fiber plants, and other useful plants are not as well represented in seed banks. Species with recalcitrant seeds, such as rubber, cacao, palms, and many tropical fruit trees, are not represented in seed collections, yet these have major significance to the economies of tropical countries. One of the few ways to preserve the genetic variability in these species is to establish special botanical gardens, which require considerable area and expense. Also, root crops, such as cassava (manioc), yams, and sweet potatoes, are not well represented in seed banks because they often do not form seeds. To be preserved, these species must be propagated vegetatively in special gardens (Figure 17.10), though new techniques are being developed to produce and store seeds of root crops. This effort is crucial, as these root crops are very important in the diets of people in developing tropical countries. Among the root crops, only potatoes are well represented in gene banks (Gulick et al. 1983). An alternative method of conserving this genetic variability involves in situ preservation of traditional agricultural practices (see Chapter 16).

17.10 The International Potato Centre maintains a living collection of 5000 samples of potatoes growing outdoors at its facility in Peru. (Photograph courtesy of the International Potato Centre.)

One of the most important sources of genetic variability for use in breeding programs is the wild relatives of crop species. For example, over 20 wild species of potatoes have been used in the development of modern potato varieties. However, only about 2% of the collections in agricultural seed banks come from wild relatives of crop plants (Hoyt 1988). Only the wild relatives of wheat and potatoes are well represented in seed banks. For such major crops as rice and cassava, the majority of the genetic variability found in their wild relatives still remains to be collected.

Seed banks are coordinated by the Consultative Group on International Agricultural Research (CGIAR) and the International Board for Plant Genetic Resources (IBPGR). One of the largest seed banks is maintained by the International Rice Research Institute (IRRI), which has 86,000 rice collections (Rhoades 1991). Other examples of specialized seed collections are The International Maize and Wheat Improvement Center (Centro Internacional de Mejoramiento de Maíz y Trigo) in Mexico with 12,000 samples of maize and 100,000 samples of wheat, and a center for apples in Geneva, New York.

A major controversy in the development of agricultural seed banks is who owns and controls the genetic resources of crop plants (Shulman 1986; Kloppenburg and Kleinman 1987). The genes of local landraces of crop plants and wild relatives of crop species represent the building blocks needed to develop advanced "elite" high-yielding varieties suitable for modern agriculture. An estimated 96% of the genetic variability necessary for modern agriculture comes from the developing countries of the world, such as India, Ethiopia, Peru, Mexico, Indonesia, and Egypt (Figure 17.11), yet the breeding programs for "elite" strains frequently take place in the industrialized countries of North America and Europe. In the past, the staff of international seed banks freely collected seeds and plant tissue from developing countries and gave them to research stations and seed companies. Genetic material was perceived as free for the taking. Yet, once seed companies developed new "elite" strains through sophisticated breeding programs and field trials, they would insist on being able to sell their seeds at a high price to make a profit.

Developing countries are now questioning this system, arguing that it is not equitable and possibly even a "holdover colonial mentality of maintaining ignorance" in which "dependent nations are robbed of diversity" (Goldstein in Shulman 1986). From their perspective, the developing countries question why they should share their genetic resources freely but then have to pay for advanced seeds based on those genetic resources. One solution proposed is for developed countries and seed companies to pay for genetic resources that they obtain from developing countries (Kloppenburg and Kleinman 1987). Another alternative is that all seed samples, including those

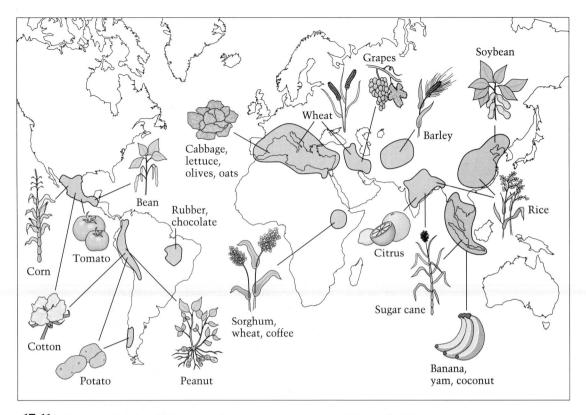

17.11 Crop species show high genetic diversity in certain areas of the world. These areas are often where the species was first domesticated, or where the species is still grown in traditional agricultural settings. (Courtesy of Garrison Wilkes.)

developed by seed companies, should be shared freely. Understandably, the industrialized countries and seed companies resist this suggestion, arguing that they are taking genetic resources that cost the developing countries nothing, investing money in their improvement, and creating a product that has value. As the late William Brown of Pioneer Hi-Bred International of Iowa has said: "To ask that an elite strain costing companies hundreds of thousands of dollars be exchanged with a primitive cultivar is simply not reasonable." A possible solution might be increased international investment in plant breeding capabilities in developing countries, so that these countries can produce and sell their own improved seeds using their own genetic variability as the starting point (Wilkes 1987).

This unresolved issue has the familiar North-South division common in many environmental debates, in which discussions of the immediate issues are tainted by present economic and technological inequalities and past injustices. Trying to strike a middle position, one conference participant pointed out (Shulman 1986):

> The great irony in the germplasm controversy is that in a world economic system where private property dominates, each side in the debate wants to define the other side's possessions as 'common heritage.' . . . In a global economy in which the Third World has precious little to offer save labor and natural resources, might it not be equitable to acknowledge this.

Conservation of Tree Genetic Resources

Forestry is a huge, global industry that depends on the genetic variability found in trees for its long-term success (Ledig 1988). As with crop plants, trees are genetically variable and often adapted to local weather conditions and pests. The success of a forestry program depends on obtaining a good sample of seeds for establishing tree plantations. The results of seed selection may not be known until years or decades later, with a poor initial seed sample resulting in slow-growing, misshapen, disease-ridden trees with poor wood quality. Relying on wild-collected seeds for establishing plantations has its dangers, since selective logging often removes the superior trees and leaves the inferior trees behind.

To conserve genetic variation, foresters have established plantations of superior genetic varieties, known as "clone banks," for long-term maintenance and research of commercially important tree species. For just loblolly pine (*Pinus taeda*), 8000 clones are being grown in clone banks in the southeastern United States (McConnell 1980). Selected trees are then used to establish seed orchards for producing commercial seed. Storage of seeds is difficult for many important genera of trees, such as oaks (*Quercus*) and poplars (*Populus*). Even pine seeds cannot be stored indefinitely and must eventually be grown out as trees. Preservation of natural areas where commercial species occur is increasingly being considered as a way of protecting the genetic variation needed for forestry (Figure 17.12). International cooperation is needed in forestry research and conservation because commercial species are often grown far from their countries of origin; for example, loblolly pine and slash pine (*Pinus elliottii*) from the United States are planted on 2.8 million ha of land in Africa, Asia, and Latin America, and Monterey pine (*Pinus radiata*) from the United States is planted on 3 million hectares of land in Chile, New Zealand, Australia, and Spain.

17.12 A sugar pine stand in Baja California with unique genetic characteristics. (From Ledig 1988.)

Summary

1. Some species that are in danger of going extinct in the wild can be maintained in artificial conditions under human supervision; this is known as ex situ or "off-site" conservation. These captive colonies can be used later to reestablish species in the wild.

2. Zoos are developing self-maintaining populations of many rare vertebrates, often using modern techniques of veterinary medicine to increase their reproductive rates. Managers of zoos are developing breeding programs so as to maintain the genetic variability of the captive colony and prevent inbreeding depression. Certain species, such as the giant panda, may initially be hard to maintain and breed in captivity, but these problems can usually be overcome with hard work and experience.

3. Aquaria are important in the protection of fishes, marine mammals, and aquatic invertebrates. The preservation of coral reef species and marine mammals are important scientific challenges that also appeal to the general public.

4. The world's 1500 botanical gardens and arboreta are now collecting rare and endangered plant species as a priority. Plants are generally easier to maintain than animals because their needs are more basic and they require less frequent care. The seeds of most species of plants can be stored for long periods of time under cold conditions in seed banks. Sampling strategies for seed collections attempt to preserve the range of genetic variability present in a species. Seed banks often specialize in the seeds of major crop species, commercial timber species, and their close relatives in order to preserve material for genetic improvement programs.

Suggested Readings

Ashton, P. S. 1988. Conservation of biological diversity in botanical gardens. *In* E. O. Wilson and F. M. Peter (eds.), *Biodiversity*, pp. 269–278. National Academy Press, Washington, D.C. The new role of botanical gardens.

Dresser, B. L. 1988. Cryobiology, embryo transfer, and artificial insemination in ex situ animal conservation programs. *In* E. O. Wilson and F. M. Peter (eds.), *Biodiversity*, pp. 296–308. National Academy Press, Washington, D. C. Modern veterinary techniques used to enhance reproduction in captive animals.

Falk, D. A. and K. E. Holsinger. 1991. *Genetics and Conservation of Rare Plants*. Oxford University Press, New York. Authoritative articles on seed banking and sampling strategies.

Plucknett, D. L., N. J. H. Smith, J. T. Williams and N. M. Anishetty. 1987. *Gene Banks and the World's Food*. Princeton University Press, Princeton, N.J. Gene banks are central to ensuring the security of the world's food supply and the continuing vitality of modern agriculture.

Ralls, K., J. D. Ballou and A. Templeton. 1988. Estimates of lethal equivalents and the cost of inbreeding in mammals. *Conservation Biology* 2: 185–193. Inbreeding in small captive populations lowers the rate of reproduction.

Rhoades, R. E. 1991. World's food supply at risk. *National Geographic* 179(April): 74–105. A beautifully illustrated popular account of the decline of traditional agricultural varieties and the need for seed banks.

Robinson, M. H. 1992. Global change, the future of biodiversity and the future of zoos. *Biotropica* 24: 345–352. In a special issue of the journal, this article focuses on the modern role of zoos in conservation and education.

Seal, U. S. 1988. Intensive technology in the care of ex situ populations of vanishing species. *In* E. O. Wilson and F. M. Peter (eds.), *Biodiversity*, pp. 289–295. National Academy Press, Washington, D.C. Describes technological advances in maintaining captive animals.

Snyder, H. A., and N. F. Snyder. 1990. The comeback of the California Condor. *Birds International* 2: 10–23. Excellent account of the California condor program, with beautiful pictures.

Tarpy, C. 1993. Zoos: Taking down the bars. *National Geographic* 184(July): 2–37. A great article on the new role of zoos in animal species conservation.

U.S. Office of Technology Assessment. 1987. *Technologies to Maintain Biological Diversity*. U.S. Government Printing Office, Washington, D.C. A summary of the methods needed to accomplish ex situ conservation.

CHAPTER **18**

Establishing New Populations

Some exciting conservation methods being developed as part of ex situ programs involve the establishment of new wild and semi-wild populations of rare and endangered species, and increasing the size of existing populations. These experiments are important in allowing species living only in captivity to regain their ecological and evolutionary roles within the biological community. Also, populations in the wild may have less chance of being destroyed by catastrophes (such as epidemic disease or war) than confined captive populations. Further, increasing the number and size of populations for a species will generally lower the probability of its extinction.

Such establishment programs are unlikely to work effectively, however, unless the factors leading to the decline of the original wild populations are clearly understood and then eliminated, or at least controlled (Campbell 1980; Kleiman 1989; Gipps 1991). For example, bird species have been eliminated from some Pacific islands because of predation by the introduced brown tree snake (see Figure 7.1). In a successful reestablishment program, the snake would have to be removed from the island, or bird nests would have to be protected in some way from the snake. Alternatively, the birds could be introduced onto another island where there are no snakes. Or if a marine turtle species has been hunted nearly to extinction in the wild by local villagers and its nesting beaches have been damaged by development, these social and economic issues would have to be addressed as an integral part of a reestablishment program. Simply releasing captive-bred turtle hatchlings into the wild without discussions with local people and a change in land use patterns would result in a recurrence of the original situation (Frazer 1992).

Three basic approaches have been used to establish new animal populations. A **reintroduction program** involves releasing captive-bred animals or wild-collected animals into an area of their historic range where the species no longer occurs. The principal objective of a reintroduction program is to create a new population in the original environment (Box 29). Frequently animals are released at the site where they or their ancestors were collected to ensure genetic adaptation to the site. Animals are also sometimes released elsewhere within the range of the species when a new protected area has been established, when an existing population is under a new threat and will no longer be able to survive in its present location, or when natural or artificial barriers to the normal dispersal tendencies of the species exist. Unfortunately there is confusion about the terms denoting the reintroduction of populations, and sometimes these programs are also called "reestablishments," "restorations," or "translocations."

BOX 29 WOLVES RETURN TO A COLD WELCOME IN YELLOWSTONE

When conservationists speak of saving endangered animal species, the creatures that come to mind are usually those whose very existence on the planet hangs in the balance. Species such as the California condor, with less than 100 remaining individuals, or the giant panda, whose numbers are estimated at less than 1100, receive the majority of the headlines because of the immediacy of the danger to the species. Frequently, other species with healthier populations get lost in the shuffle until some conflict or controversy erupts over their management.

Such is the case with gray wolf (*Canis lupus*) populations in the United States. Wolves were systematically exterminated throughout the United States as a matter of federal policy since the late 1800s (Lipske 1991; Begley et al. 1991). Yellowstone National Park in particular was pinpointed for wolf extermination, as the predators were thought to pose a threat to the herds of elk, deer, and other game animals inhabiting the park (McNamee 1986). When the U.S. Fish and Wildlife Service adopted a plan in 1987 for the recovery of the gray wolf, which included the reintroduction of the species into Yellowstone National Park and surrounding government lands, controversy and conflict were the natural results. Ranchers in Montana, Wyoming, and Idaho argued that the wolves posed a threat to livestock, and possibly to humans as well. Hunters objected that the wolves would severely reduce the supply of game animals, while logging and mining companies were concerned that the presence of wolves would limit their ability to utilize re-

Gray wolves may be reintroduced into Yellowstone National Park to restore the ecological equilibrium of predators and herbivores. (Photograph © NYZS/The Wildlife Conservation Society.)

sources on federal lands (Cohn 1990a). Underlying all of their objections is the argument that the wolf, with an estimated population of 50,000 in Canada alone, is in no immediate danger of extinction. Furthermore, wolves have been steadily recolonizing the northern states, including Wisconsin and Michigan, and individual animals have been sighted in Idaho, Washington, and Montana. Since the survival of the species is not at issue, from the ranchers' perspective there is no compelling reason for the species to be reintroduced to the area.

From the perspective of conservation biologists, however, there is a rationale for the reintroduction of wolves to Yellowstone: The wolf is the single "missing link" in the Yellowstone ecosystem, and is necessary to the restoration of ecological balance in the park. Wolves are vital to preventing overpopulation of elk, moose, and deer in the park; without these predators, herbivore herds become so dense that hundreds starve each winter (Begley et al. 1991). Wolves maintain the health of the herds by removing old or sick animals and keeping population levels below the environmental carrying capacity. Furthermore, fears that wolves will attack livestock or people appear to be exaggerated. In Minnesota, until recently the only state in which wolves remained, few domestic animals are killed by wolves annually. Ranches that surround Yellowstone have a far lower density of cattle and sheep than Minnesota; livestock losses due to wolf predation probably will not be very high. Creating a fund to pay for damages caused by wolves could help to overcome the objections of the ranchers. Wolf attacks on humans are virtually nonexistent, except when the animals are cornered or provoked (Cohn 1990a; Begley et al. 1991; Lipske 1991).

The reintroduction plan may not work in any case because the Yellowstone area is fragmented by roads, which the wolves may avoid (Schonewald-Cox and Buechner 1992). Also, the high volume of tourists visiting the park may frighten the wolves away.

The controversy surrounding the reintroduction of the wolf reflects a debate central to conservation biology: how to balance environmental needs and human needs. When survival of a species is at issue, conservationists have an urgency to their arguments that tips the balance in favor of the environment. In the case of the wolf, however, the problem is not so clear-cut. People living and working in the Yellowstone area who will be directly affected by the reintroduction plan are unwilling to accommodate the wolf without a compelling reason for the inconvenience; ecological equilibrium as yet does not rank so high.

Two other, distinct types of release programs are also being used. An **augmentation program** involves releasing animals into an existing population to increase its size and gene pool (Powell and Cuthbert 1993). These released animals may be wild animals caught elsewhere or animals raised in captivity. An **introduction program** involves moving animals to areas outside their historic range (Conant 1988). Such an approach may be appropriate when the environment within the known range of a species has deteriorated to the point where the species can no longer survive there. Within its historical range, the species may be either extinct or in severe decline; reintroduction may be impossible if the factor causing the original decline is still present. The only alternative may be to introduce wild-caught or captive-raised individuals into areas outside the historic range in the hope of establishing new populations. An introduction of a species to new sites needs to be carefully thought out to be sure that the species does not damage its new ecosystem or harm populations of any local endangered species (Conant 1988). Care must be taken that released animals have not picked up any diseases while in captivity that could spread to and decimate wild animal populations. Also, a species may adapt genetically to its new environment in ways that make it different from the original population.

Social Behavior of Released Animals

Successful reintroduction, augmentation, and introduction programs need to consider the social organization and behavior of the animals that are being released. When social animals, particularly mammals and some birds, grow up in the wild, they learn about their environment and how to interact socially from other members of their species. In the wild, these animals learn, often from their family members, how to search their environment for food and how to recognize, capture, gather, and consume the food. For carnivores such as lions and wild dogs, hunting techniques are complex, subtle, and require considerable teamwork. Herbivores like hornbills and gibbons must learn seasonal migration patterns over a wide area to obtain the variety of food items necessary to stay alive and reproduce. When mammals and birds are raised in captivity, their environment is limited to a cage or pens, so exploration is impossible. Searching for food and learning about new food sources is not necessary, since the same food items arrive day after day on schedule. And social behavior may become highly distorted if the animals are raised alone or in unnatural social groupings. In such cases, animals may lack the skills to survive in their natural environment, as well as the social skills necessary to cooperatively find food, sense danger, find mating partners, and raise young.

To overcome these socialization problems, captive-raised mammals and birds may require extensive training before as well as after release into the environment. They must learn how to find food and shelter, avoid predators, and interact in social groups. Training techniques have been developed for several mammals and a few birds. Captive chimps, for instance, have been taught how to use twigs to feed on termites and how to build nests in captivity (Carter 1988). Red wolves are taught how to kill live prey. Golden lion tamarins are given complex food boxes to gain skills that will be useful for opening wild fruit (Box 30; Kleiman 1989). Captive animals are taught to fear potential predators by being frightened in some way when a dummy predator is shown.

Social interaction is one of the most difficult behaviors to teach

BOX 30 REINTRODUCTION OF THE GOLDEN LION TAMARIN: PROBLEMS, METHODS, AND SUCCESSES

Golden lion tamarins (*Leontopithecus rosalia*) are beautiful, small primates that have declined precipitously in numbers as their Atlantic coastal rain forest habitat has become fragmented by human activities. Since tamarins breed well in captivity, they are an excellent choice for a reintroduction program. However, the reintroduction of golden lion tamarins into remaining fragments of the Brazilian rain forest is a complicated process. Captive-bred tamarins cannot simply be dumped into the forest and left to fend for themselves; unused to living in the wild, they would be unable to survive very long, and each individual animal is simply too valuable to the survival of this rare species to be wasted. To increase the likelihood of survival among released animals, biologists with the Golden Lion Tamarin Conservation Program in Brazil have developed a set of procedures to identify and minimize possible problems facing the animals in their new habitat (Kleiman 1989).

To be successful, a reintroduction program must consider a range of variables. First, biologists need to be certain that the captive population is truly stable before removing individuals for release into the wild; they must make sure that the genetic diversity of the captive population is maintained and also that the reproductive potential of the captive population is not severely disrupted by the removal of the animals to be released. This concern extends to the wild population as well: haphazard introductions could lead to social stress among wild groups. Too great a difference in genetic makeup between the wild and captive-bred animals could counteract important microenvironmental adaptations in isolated wild populations and could negatively effect reproductive potential. Second, the presence of protected, suitable habitat is of primary importance. There is little point in releasing animals into the wild if their habitat will be destroyed shortly afterward; effective legal protection of the release site is therefore a prerequisite for re-establishment. In addition, the release site must be large enough to accommodate the released animals, since conflicts caused by overcrowding or territorial disputes are potential hazards to both the introduced animals and the remaining wild population. A third variable affecting the program is the ability of biologists to teach captive-bred animals the behaviors that enable their wild counterparts to find food, shelter, and mates. Captive-reared tamarins have no experience in finding what they need to survive, and without some training will be unable to fit into either their habitat or the arena of tamarin social behavior (Kleiman 1989). Finally, the survival of the released tamarins depends upon the biologists' ability to eliminate or reduce the factors that led to the original decline of the golden lion tamarin in the wild. Since a principal cause of the decline of the tamarin was habitat destruction, releasing the animals on protected lands partially alleviates the problem; however, further fragmentation of unprotected forest lands is still a potential threat to tamarins in the wild.

When reintroduction of the golden lion tamarin was first attempted in the early 1980s, the first variable affecting the program's success was of little concern to biologists: the captive population of golden lion tamarins was clearly stable and thriving. Today over 500

The logo used for the Poço das Antas Biological Reserve in Brazil was designed to increase public awareness of the Golden Lion Tamarin Project. (From Kleiman 1989.)

golden lion tamarins live in zoos throughout the world, with their breeding and genetics managed by an international team of experts (Kleiman et al. 1990). The second variable, however, was somewhat more complicated. Brazil's Reserva Biológica Poço das Antas was the central site for the reintroduction effort, but the Conservation Program's personnel felt that additional public support was necessary to assure the stability of the protected lands. Through public service announcements on radio and television, educational programs in public schools, production of audiovisual educational materials, and other methods, the biologists increased public awareness not only of the plight of the tamarin, but also of the broader conservation efforts centered around the reserve. The tamarin, an animal with an attractive and charismatic appearance, served as a focal point for garnering public support for the reserve, and even served as the logo for the conservation program. The heightened public awareness also helps conservationists to deal with the potential threats to tamarins outside the reserve. Landlords whose past activities have damaged unprotected forest lands are encouraged by these educational programs, as well as by direct discussions with biologists, to find alternative uses for their land that will not degrade the forest.

The trickiest problem of

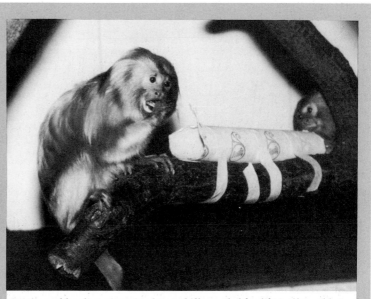

Captive golden lion tamarins learn skills needed for life in the wild; in this case, the animals must find food inside a complicated puzzle box. (Photograph by Jessie Cohen, National Zoological Park, Smithsonian Institution.)

all, however, is preparing the captive animals for life in the wild. Biologists developed a series of methods to promote behaviors similar to those of wild tamarins in several important areas: gathering of food, finding shelter, avoiding predators, navigation and locomotion, and social interaction with other tamarins. For example, increasingly complex simulated environments were used to teach the captive tamarins to be both tenacious and inventive in food gathering and locomotion (Kleiman 1989). The effectiveness of these programs has been uncertain. Captive tamarins are now being released with relatively little prerelease training, but are given food at feeding sta-

tions and other support for their first 18 months in the wild as they learn to forage and move through the forest on their own (Kleiman et al. 1990).

Pairing captive-born animals with tamarins captured from the wild is also effective, especially with regard to social interaction. The captive-bred animal adopts many of the behaviors of its wild partner, and is therefore better equipped to survive (Kleiman 1989). Between 1984 and 1991, 91 golden lion tamarins were released into the Poço das Antas Reserve, of which 33 were surviving as of June 1991. A total of 57 infants had been born to released tamarins, of which 38 survived (Beck et al. 1991).

The young tamarins born in the wild are better able to find food and orient in the forest than their captive-born parents, suggesting that the tamarin population may eventually become independent of people.

A particular behavior that is difficult for biologists to teach to captive tamarins is avoidance and fear of human beings; surrounded by humans since birth, the captive animals are much less inclined to steer clear of human activities than their wild counterparts. Unfortunately, avoidance of people may be a key to continued survival, since 20% of the losses of released tamarins are attributable to humans (Beck et al. 1991). Biologists must hope that the remaining wild tamarins can teach the captive-bred animals to avoid humans.

captive-bred mammals and birds because for most species, the subtleties of social behavior are poorly understood by humans. Nevertheless, some successful attempts have been made to socialize captive-bred mammals. In some instances, wild individuals are used as "instructors" for captive individuals of the same species (Kleiman 1989). For instance, wild-bred golden lion tamarins are caught and held with captive-bred tamarins to form social groups that are then released together, in the hope that the captive-bred tamarins will learn from the wild ones. In other cases, humans mimic the appearance and behavior of the wild species. This method is particularly important when dealing with very young animals; for example, captive-bred California condor hatchlings were originally unable to learn the behaviors of their wild relatives because they had imprinted on their human keepers. Newly hatched condors are now fed with condor puppets and kept from the sight of zoo visitors so that they learn to identify with their own species rather with than a foster species or with humans (see Box 25). When captive-bred animals are released into the wild, they sometimes join existing social groups or mate with wild animals and thereby gain some knowledge of their environment. The development of social relationships with wild animals may be crucial to the success of the captive-bred animals once they have been released. Failure to associate with wild birds during migration appears to be one of the reasons for the high mortality of captive-raised bald ibis (*Geronticus eremita*) (Akçakaya 1990). On the other hand, killdeer chicks (*Charadrius vociferus*) raised in captivity by humans and cross-fostered by a related plover species in the wild did not behave appreciably differently from chicks reared by wild killdeer parents when they were released into a natural population (Powell and Cuthbert 1993).

Considerations for Successful Programs

Programs to establish new populations are often expensive and difficult; they require a serious, long-term commitment. The programs to capture, raise, monitor, and release California condors, peregrine falcons, and black-footed ferrets, for instance, have cost millions of dollars and required years of work. When the animals are long-lived, the program itself may have to last for many years before its outcome is known. Decisions on initiating reintroduction programs can also become highly emotional public issues, as evidenced by the California condor, the black-footed ferret, the grizzly bear, and the gray wolf programs (Luoma 1992; Cohn 1990a; Lipske 1991). Programs can be attacked as a waste of money ("Millions of dollars for a few ugly birds!"), unnecessary ("Why do we need wolves here when there are so many elsewhere?"), poorly run ("Look at all of the ferrets that died of disease in captivity!"), or unethical ("Why can't the last animals just be allowed to live out their lives in peace without being captured and put into zoos?"). The answer to all of these criticisms is straightforward: well-run, well-designed captive breeding and reintroduction programs are the best hope for preserving a species that is about to go extinct in the wild or is in severe decline.

The selection of animals for reintroduction programs has an important genetic component. Captive populations may have lost much of their genetic variability. Gene frequency changes may have occurred in populations that have been raised for several generations in captive conditions, such as the Pacific salmon (Waples and Tel 1990). These gene frequency changes may have resulted in the species having a lower ability to survive in the wild following their release. To ensure that the released population does not suffer from inbreeding depression, individuals have to be carefully selected to produce the most genetically diverse release population (Haig et al. 1990).

Monitoring of released animals is necessary to determine whether the new population is self-sustaining. Released animals may require special care and assistance during and immediately after release; this approach is known as "soft release." Animals may have to be fed and sheltered at the release point until they are able to subsist on their own, or they may need to be caged temporarily at the release point and released gradually, so they can become familiar with the area. Social groups abruptly released from captivity (a "hard release") may explosively disperse in different directions and away from the protected area, resulting in a failed establishment effort (Zwank and Wilson 1987). Intervention may be necessary if animals appear to be unable to survive, particularly during episodes of drought or low food abundance. In such cases a decision has to be made whether it is better to give the species occasional temporary help to get established or to force the species to survive on its own. It is difficult to know what is the right decision beforehand; at the two extremes, the results will be a highly dependent population congregated around a feeding station with little ability to survive on its own, or a population forced to survive on its own in which the individuals starve to death and are killed by predators.

Successful reintroduction programs often have considerable educational value. One highly visible success is the reintroduction of native red squirrels into London's Regent's Park after they were apparently displaced by North American gray squirrels. In Brazil, efforts to protect golden lion tamarins through conservation and reintroduction have become a rallying point in attempts to protect the last remaining fragments of the Atlantic coastal forest. In Oman, captive-bred Arabian oryx were successfully reintroduced into desert areas, creating an important national symbol and a source of employment for local Bedouins who run the program (Figure 18.1; Stanley-Price 1989).

Establishment programs involving common game species have always been widespread and have contributed a great deal of knowledge to the new programs being developed for threatened and endangered species. A detailed study examined 198 bird and mammal establishment programs conducted between 1973 and 1986 and found a number of significant generalizations (Griffith et al. 1989). The reported success of programs in establishing new populations was:

- Greater for game species (86%) than for threatened, endangered, and sensitive species (44%)
- Greater for release in excellent quality habitat (84%) than in poor quality habitat (38%)
- Greater in the core of the historic range (78%) than at the periphery of and outside the historic range (48%)
- Greater with wild-caught (75%) than with captive-reared animals (38%)
- Greater for herbivores (77%) than for carnivores (48%)

For these bird and mammal species, the probability of establishing a new population increases with the number of animals being released up to about 100 animals. Releasing more than 100 animals does not further enhance the probability of success (Griffith et al. 1989).

18.1   The Arabian oryx (*Oryx leucoryx*), almost extinct in the wild, is being reintroduced to places in its former range, such as Oman. (Photograph by Ron Garrison, © San Diego Zoo.)

Case Studies

The following are brief descriptions of some recent attempts to establish new populations of threatened species.

- Red wolves (*Canis rufus*) have been reintroduced in the Alligator River National Wildlife Refuge in northeastern North Carolina through the release of 42 captive-born animals. Survival of adults is about 50% after three years, and 23 pups have been born (Cohn 1987; Phillips 1990; Parker and Phillips 1991). Animals in the program have established packs, and survive by hunting deer and raccoons and eating carrion (Phillips and

Henry 1992). The Red Wolf Recovery Program appears to be successful.

- Florida gopher turtles (*Gopherus polyphemus*) have greatly declined in numbers throughout their range. An attempt was made to establish a new population by transferring 85 wild-caught turtles to a new site; two years later, 35 adults were present with some hatchlings (Burke 1989).

- The Mauritius kestrel (*Falco punctatus*) had only two breeding pairs remaining in the 1970s, earning it the distinction of being the most endangered bird of prey in the world (Cade and Jones 1993). Degradation of its forest habitat and introduced mammalian predators were largely to blame for the decline. An active program has been initiated involving the augmentation of the existing small population with captive-reared young and the reintroduction of the species into other parts of the island. Young have been raised in captivity using eggs collected in the wild and laid by adults in the captive colony. A total of 235 young kestrels have been released back into the wild. As a result of natural population increase and these augmentation efforts, the wild population has now risen to 30 breeding pairs and more than 170 birds, distributed in four forest areas.

- The kakapo (*Strigops habraptilus*) is not only the largest parrot in the world, it is also flightless, nocturnal, and solitary. The New Zealand kakapo was believed to be extinct due to introduced mammalian predators, but a few individuals were discovered in the late 1970s (Triggs et al. 1989). Twenty-two kakapos were collected in the wild and released on Little Barrier Island, which lacks most predators (Moorhouse and Powlesland 1991). While the kakapos have survived for several years at this new site, they still have not formed any nests, so the success of the experiment is unknown.

- Sri Lankan fishes that inhabit streams are potentially endangered by rain forest destruction. An introduction experiment demonstrated that new populations of rare fish species could be established in other Sri Lankan streams beyond their known range (Wikramanayake 1990).

- Coral reefs on the Pacific coast of Costa Rica have been destroyed in places by human activities and natural disturbances. Experimental attempts at restoration have involved attaching live pieces of coral from nearby living reefs onto the dead reef. Survival of these transplants is high, resulting in a large increase of new colonies (Guzman 1991).

- St. Catherine's Island is a 14,000-acre conservation area off the Georgia coast that is managed for conservation, research, and education by a private foundation in cooperation with the Bronx Zoo and the American Museum of Natural History (Cohn 1990c). The native coastal vegetation provides habitat for such native wildlife as alligators, wild turkeys, white ibises, and wood storks. In addition, endangered animals that do not do well in captivity or need large areas for social interactions have been introduced to establish breeding populations. Examples of such species are the red-fronted macaw, known only from one valley in Bolivia; lion-tailed macaques from India; and Madagascar's angonoka tortoise, which is very rare in the wild. A captive group of 12 Madagascan ring-tailed lemurs was released in 1985; this group has survived and grown to number 17 individuals with the birth of 11 new lemur infants on the island.

Establishment of New Plant Populations

Efforts at establishing new populations of rare and endangered plant species are fundamentally different from attempts using terrestrial vertebrate animal species. Animals can disperse to new locations and actively seek out the microsite conditions that are most suitable for them. In the case of plants, seeds are dispersed to new sites by such agents as wind, animals, and water (Guerrant 1992; Primack and Miao 1992). Once the seeds land on the ground they are unable to move farther, even if a suitable microsite exists just a few centimeters away. The immediate microsite is crucial for plant survival: if the environmental conditions are unsuitable in any way—too sunny, too shady, too wet, or too dry—either the seed will not germinate, or if it does germinate, the resulting seedling will die (Harper 1977). Seedlings are also vulnerable to predation by insects and other animals and to attack by fungi and nematodes. As a result of these subtle ecological requirements, plant species may become established at a new site only if many seeds arrive there and land by chance in just the right conditions. Disturbance in the form of fire or blowdowns may also be necessary for seedling establishment in many species (Hobbs 1989; DeSteven 1991; Facelli and Picket 1991). As a result, a site may be suitable for seedling establishment only once every several years.

Plant populations typically fail to become established from seeds at most sites that appear to be suitable for them. In one study, large numbers of seeds of four species of annual plants were introduced at

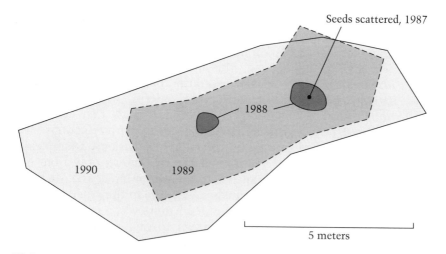

18.2   Sometimes a new plant population can be established by the introduction of seeds. In September 1987, 100 seeds of *Impatiens capensis*, an annual species of jewelweed, were introduced into an unoccupied site in the Hammond Woods, near Boston, Massachusetts. The seeds were scattered within a meter of a stake (black dot). In 1988, two groups of plants separated by several meters had established themselves (darkest gray areas). The populations continued to expand, as shown by the limits in 1989 (dashed lines) and 1990 (solid lines). By 1991, population size had reached 750 individuals that collectively produced approximately 56,000 seeds. (After Primack and Miao 1992.)

34 apparently suitable sites where the species did not grow naturally. Of these 34 introductions, only three resulted in established populations, each persisting through four generations (Primack and Miao 1992). Only at one site has the population expanded substantially in numbers and area occupied (Figure 18.2). Subsequent attempts to establish new populations of perennial herbs by sowing seeds at apparently suitable sites have had an even lower rate of success (Primack, unpublished).

To increase their chances of success, botanists often germinate seeds in controlled environments and grow the young plants in protected conditions. Only after the plants are past the fragile seedling stages will they be transplanted into the field. While this method has a better chance of ensuring that the species survives at a new location, it does not mimic a natural process, and the new population often fails to produce the seeds and seedlings needed to form the next generation (Hall 1987). However, the failure of such an attempt is generally not irreparably damaging to the species, as seeds from the original plants are still available for future attempts. Plant ecologists are currently trying to work out new techniques to overcome these difficulties in establishing new plant populations, such as fencing to exclude animals, removing some of the existing vegetation to reduce competition, and mineral nutrient additions.

Reestablishment Programs and the Law

Reintroduction, introduction, and augmentation programs will increase in the coming years as the biological diversity crisis eliminates more species from the wild. Some of the reintroduction programs for endangered species will be mandated by official recovery plans set up by national governments. However, establishment programs, as well as research in general on endangered species, are increasingly being affected by endangered species legislation that restricts the possession and use of endangered species (New England Wild Flower Society 1992). If government officials rigidly apply these laws to scientific research programs, the creative insights and new approaches coming out of these programs could be stifled (Ralls and Brownell 1989). New scientific information is central to establishment programs and other conservation efforts. Government officials that block reasonable scientific projects may be doing a disservice to the citizens of their country. The potential harm to endangered species caused by carefully planned scientific research is relatively insignificant in comparison with the actual massive loss of biological diversity being caused by habitat destruction and fragmentation, pollution, and overexploitation.

Experimental populations of rare and endangered species successfully created by introduction and reintroduction programs are sometimes given a degree of legal protection (Falk and Olwell 1992). Legislators and scientists alike must understand that the establishment of new populations through reintroduction programs in no way reduces the need to protect the original populations of the endangered species; the original populations are more likely to have the most complete gene pool of the species and the most intact interactions with other members of the biological community.

Summary

1. New populations of rare and endangered species can be established in the wild using either captive-raised or wild-caught animals. A reintroduction program involves releasing individuals within the historic range of the species; an introduction involves releasing individuals at a site outside of the historic range of the species; an augmentation involves releasing individuals into an existing population to increase population size and genetic variability.

2. Mammals and birds raised in captivity may lack the skills needed to survive in the wild. Some species require social and behavioral training before release, and often some degree of maintenance after release. Establishment of new populations of rare bird and mammal species is enhanced when the release occurs into excellent habitat within the historic range of the species, and uses large numbers (up to 100) of animals.

3. Reintroduction of plant species requires a different approach because of their specialized environmental requirements at the seed and seedling stages. Reintroduction efforts involving plants appear to have a lower chance of success than attempts with animals. Site disturbance to reduce plant competition and using adult transplants may increase the success of plant reintroduction efforts.

4. Conservation biologists involved in establishing new populations of endangered species must be careful that their efforts do not weaken the legal protection currently given to natural populations of those species.

Suggested Readings

Beck, B. B., D. G. Kleiman, J. M. Dietz, I. Castro, C. Carvalho, A. Martins and B. Rettberg-Beck. 1991. Losses and reproduction in reintroduced golden lion tamarins *Leontopithecus rosalia. Dodo: Journal of the Jersey Wildlife Preserve Trust* 27: 50–61. The latest word on this large, well-documented project.

Cade, T. J. and C. G. Jones. 1993. Progress in the restoration of the Mauritius kestrel. *Conservation Biology* 7: 169–175. Case study of a successful program.

Conant, R. 1988. Saving endangered species by translocation. *BioScience* 38: 254–257. The benefits and dangers of establishing new populations outside of a species' historic range.

Falk, D. A. and P. Olwell. 1992. Scientific and policy considerations in restoration and reintroduction of endangered species. *Rhodora* 94: 287–315. Clear cautionary statements about the implications of reintroduction efforts.

Frazer, N. B. 1992. Sea turtle conservation and halfway technology. *Conservation Biology* 6: 179–184. Attempts to help endangered species must first deal with the human activities leading to population declines.

Gipps, J. H. W. (ed.). 1991. *Beyond Captive Breeding: Reintroducing Endangered Species through Captive Breeding.* Zoological Society of London Symposium 62. Clarendon Press, Oxford. Essays on reintroduction by leading authorities.

Griffith, B., J. M. Scott, J. W. Carpenter and C. Reed. 1989. Translocation as a species conservation tool: Status and strategy. *Science* 245: 477–480. Analysis of 198 establishment programs.

Kleiman, D. G. 1989. Reintroduction of captive mammals for conservation. *BioScience* 39: 152–161. Outstanding outline of mammal reintroduction efforts, with many examples.

Primack, R. and S. L. Miao. 1992. Dispersal can limit local plant distribution. *Conservation Biology* 6: 513–519. The special requirements of plant reintroduction efforts.

Stanley-Price, M. R. 1989. *Animal Re-introductions: The Arabian Oryx in Oman*. Cambridge University Press, Cambridge. History of a successful example of reintroduction.

PART VI

Conservation and Human Societies

CHAPTER 19

How Are Species and Habitats Legally Protected?

At local, national, and international levels, there is a tension between forces acting to exploit natural resources and forces acting to preserve them. A traditional view of conservation is that resources are either to be exploited or protected; areas inside parks and nature reserves are to be protected, while areas outside parks are available for any type of use (Western 1989). The Resource Conservation Ethic (see Chapter 1) and, more recently, the policy of sustainable development, have been advocated by conservationists to encourage the utilization of resources only at a level that can be maintained indefinitely and which does no harm to either the biological community or to the ecosystem (WCED 1987; IUCN 1980; IUCN/UNEP/WWF 1991). In most efforts to preserve species and habitats, the situation comes down to concerned citizens, conservation organizations, and government officials taking the initiative to act (Caldwell 1985; Gross et al. 1991). This action may take many forms, but it begins with individual and group decisions to prevent the destruction of habitats and species in order to preserve something of perceived value (Figure 19.1). As Peter Raven, Director of the Missouri Botanical Garden, emphasized regarding the accelerated loss of biological diversity: "You can think about it on a worldwide basis, and then it becomes discouraging and insoluble, or you can think about it in terms of specific opportunities, seize those opportunities, and reduce the problem to a more manageable size" (quoted in Tangley 1986).

457

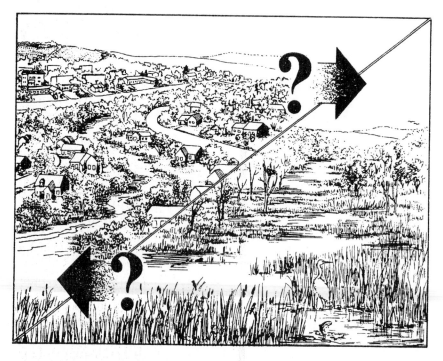

19.1   Decisions have to be made concerning compromises between development for human needs and the preservation of the natural world. (From Gersh and Pickert 1991; drawing by Tamara Sayre.)

Local Conservation Legislation

In traditional societies the use of natural resources is often regulated through communal rules enforced by common social pressures or by designated authorities. In modern societies, local (city and town) and regional (county, state, provincial) governments pass laws to provide effective protection for species and habitats. Such laws were passed in the first place because citizens and political leaders felt that they represented the will of the majority and provided long-term benefits to the society.

Conservation laws cover a number of activities that directly affect species and ecosystems. The most prominent of these laws are rules governing when and where hunting can occur, the size and number of animals that can be taken, the types of weapons and traps and other equipment that can be used, and the species of animals that can be taken. In some settled areas and protected areas, there is an absolute ban on hunting and fishing. Hunting and fishing restrictions are enforced through licensing requirements and patrols by game wardens. Related legislation includes prohibitions on trade in wild-collected animal and plant products. Certification of origin of biological products may be required to ensure that wild populations are not illegally depleted. These restrictions have long applied to animals, such as trout and deer, and plants of horticultural interest, such as rhododendrons, azaleas, and Venus's flytrap.

Laws that control the manner in which land is used are another means of protecting biological diversity. These laws include restrictions on amount of land use or access, type of land use, and pollution generated. For example, vehicles and even people on foot may be restricted from habitats and resources that are sensitive to damage, such as bird nesting areas, wildflower patches, and sources of drinking water. Uncontrolled fires may severely damage habitats, so practices contributing to accidental fires, such as campfires, are often rigidly controlled. Habitat that is less sensitive may still have use restrictions placed on it, including regulations on the harvest of natural products and grazing by domestic animals. Zoning laws sometimes prevent construction in sensitive areas such as barrier beaches and floodplains. Even where development is permitted, building permits increasingly are reviewed to ensure that damage is not done to endangered species or to wetlands. For major regional and national projects, environmental impact statements must be prepared which describe the damage that the project could cause. Governments regulate manufacturing, mining, transportation, waste disposal, and other human activities to prevent air and water pollution that can damage biological communities, natural resources, and human health.

Passage and enforcement of conservation-related laws on a local level can become an emotional issue that divides a community and can even result in life-threatening situations (Box 31). Conservationists must be able to articulate the point that using a resource in an intelligent and sustainable manner creates the greatest long-term benefit for the society. The general public must be encouraged to look beyond the immediate benefits that come with rapid and destructive exploitation. For example, towns often need to restrict development in watershed areas to protect water supplies; this may mean that houses and businesses are not built, and the owners of the land may have to be compensated. Off-road vehicles ("dune buggies") may need to be restricted to particular areas of coastal parks to prevent damage to ground-nesting birds and fragile dune vegetation; such restrictions limit the enjoyment of popular recreational activities. The ability to negotiate, compromise, and explain positions are important qualities for conservation activists—a fervent belief in one's own cause is not enough (Lowenthal 1991).

The rhetoric of conservation biology can also enter local debate in ways that may be inappropriate. Citizens seeking to resist any development near their own homes may use conservation arguments to

BOX 31 THE DANGERS OF ENVIRONMENTAL ACTIVISM

The past decade has witnessed a tremendous increase in popular awareness of environmental issues. Many conservation organizations, such as the Sierra Club, the World Wildlife Fund, the Nature Conservancy, and Greenpeace, to name only a few, have gained hundreds of thousands of new members. The surge in environmental awareness, however, has recently triggered a disturbing backlash from industries, business interests, and labor organizations, and even some governments, which resent and fear the new power of the conservation movement. Conservation of natural resources may be linked to a short-term loss in profits and job opportunities; local people may be afraid of losing their jobs and feel anger toward environmental activists, particularly in a slow economy. Incidents of intimidation, threats, and physical harassment, sometimes frighteningly violent, have been reported worldwide by environmental activists.

Perhaps the best-known violent incident occurred in 1988, when Chico Mendes, a Brazilian activist organizing rubber tappers to resist the encroachment of cattle ranching and logging in the Amazon rain forest, was assassinated by ranchers. Mendes' martyrdom created a worldwide uproar and focused tremendous attention upon the destruction of the rain forest. The conviction of the rancher who allegedly ordered Mendes' murder was an initial victory for his supporters; before this incident, large landholders, loggers, and miners could—and did—act with impunity against activists, labor leaders, native tribespeople, and anyone else who stood in the way of the development of Brazil's forests (Schwartzman 1992; Toufexis 1992). The victory, however, was short-lived; the conviction was overturned on appeal and the rancher will be retried. Moreover, many other activists in Brazil have been shot at, beaten, or killed, both before and since Mendes' death. Activists claim that hit lists circulate openly among developers, loggers, and ranchers, none of whom government officials wish to offend.

Activists in many countries often face arrest and abuse at the hands of their own governments. They have been branded "subversives" or "traitors" for fighting against government policies that promote development at the expense of the environment. Based upon these and other charges, some of these activists have been interrogated, jailed, beaten, and tortured (Toufexis 1992). Such incidents occur even in countries considered progressive with regard to conservation.

The persecution and injustices faced by environmentalists are most commonly found in developing nations, but such problems are by no means limited to the Third World. Individual activists fighting against industrial pollution in the United States have also been victims of persecution, ranging from arson to assault to attempted murder (Toufexis 1992). On occasion, authorities responsible for investigating the crimes have responded with either indifference or overt antagonism toward the victims. In one particular case, an EarthFirst! activist, severely injured when a bomb in her car exploded, was herself arrested in connection with the case. Police claimed that the woman had been deliberately transporting the bomb when it went off, but charges were never filed against her; she has since filed civil suits against both the police department and the FBI (Toufexis 1992). In another famous incident, French government agents were convicted in the 1985 bombing in New Zealand of the Greenpeace flagship *Rainbow Warrior*, in which a crew member died. The ship was being readied to head into the South Pacific to protest the French nuclear weapons testing program.

The increased opposition to environmentalism represents an unwillingness to accept the limitations that must be imposed upon human consumption if the biosphere is to survive. Though sometimes opposition is subtle, all too frequently the backlash against conservation has become overt and violent. Business organizations and conservative elements in the labor movement are increasingly lobbying and forming action groups to counter conservation groups. In an interesting twist, these pro-development groups often use environmental rhetoric to argue for the "wise use" of natural resources.

justify their objections. For example, a citizens' group in Boston attempted to support their objections to a new school on the basis that the cars dropping off students would create excess carbon dioxide and contribute to global warming. Though the argument had little basis in fact, the use of this rationale by the citizens' group was detrimental to conservation efforts in one respect: when conservation issues are misused in this manner, valid conservation initiatives may suffer by association. Frivolous use of conservation arguments may cause community leaders to take genuine conservation issues less seriously, or to suspect the intentions of conservationists. Real concern for endangered species and habitats may be ridiculed as "hysteria" by policymakers, or treated as a manifestation of a hidden political or business agenda.

One of the most powerful strategies to protect biological diversity at the local level is the acquisition of intact biological communities as nature reserves (Bickford 1991; Land Trust Exchange 1988; Mackintosh 1990). Government bodies buy land as local parks for recreation, as conservation areas and wildlife refuges for biological diversity, as forests for timber production and other uses, and as watersheds to protect water supplies. Private conservation organizations often acquire land solely for preservation purposes. In some cases, land is purchased outright, but land is often donated to conservation organizations by public-spirited citizens. Many of these citizens receive significant tax benefits from the government to encourage these land donations.

In many countries, private conservation organizations are among the leaders in acquiring land for conservation efforts (Land Trust Exchange 1988; Elfring 1989). In the United States alone, over 800,000 hectares of land have been protected at a local level by land trusts, which are private, nonprofit corporations established to protect land and natural resources. At a national level, major land trusts such as the Nature Conservancy and the Audubon Society have protected an additional 3 million hectares. Jean Hocker, executive director of the

Land Trust Exchange, an association of land trust organizations, explains that:

> Different land trusts may save different types of land for different reasons. Some preserve farmland to maintain economic opportunities for local farmers. Some preserve wildlife habitat to ensure the existence of an endangered species. Some protect land in watersheds to improve or maintain water quality. Whether biologic, economic, productive, aesthetic, spiritual, educational, or ethical, the reasons for protecting land are as diverse as the landscape itself. (Elfring 1989)

In addition to outright purchase of land, both governments and conservation organizations protect land through **conservation easements**. Landowners will often be willing to give up the right to develop, build on, or subdivide their property in exchange for a sum of money, a lower real estate tax, or a tax benefit. For many landowners accepting a conservation easement is an attractive option, because they receive a financial advantage while still owning their land, and they feel that they are assisting conservation objectives. Because of these considerations, many landowners will voluntarily accept conservation restrictions without compensation. Another option that land trusts use is **limited development**, in which a landowner, a property developer, and a conservation organization reach a compromise involving part of the land being commercially developed and the remainder being protected by a conservation easement. Limited development projects are often successful since the developed lands typically have their value enhanced by being adjacent to conservation land. Limited development also allows the construction of necessary buildings for an expanding human society.

Local efforts by land trusts to protect land are sometimes criticized as being elitist because they remove land from productive use and often lower the revenue collected from land taxes. However, the loss of tax revenue from land being acquired by a land trust is often offset by the increased value of property adjacent to the conservation area. In addition, nature reserves, national parks, wildlife refuges, and other protected areas generate revenue throughout the local economy, which benefits the community (Power 1991). Finally, by preserving important features of the landscape and the natural communities, local nature reserves enhance the significant and important cultural features of the local community.

National Legislation

Throughout much of the modern world, national governments and national conservation organizations play a leading role in conservation activities (Bean 1983). The establishment of national parks is a common conservation strategy. National parks are the single largest source of protected lands in many countries. For example, Costa Rica's national parks protect over half a million hectares, or about 8% of the nation's land area (Tangley 1986; World Resources Institute 1992). Outside of the parks deforestation is proceeding rapidly, and soon the parks may represent the only undisturbed habitat and source of natural products, such as timber, in the whole country. The U.S. National Park system, with 357 sites, covers 32 million hectares (Pritchard 1991).

Government agencies are the principal instrument for developing national standards on environmental pollution. Laws regulating aerial emissions, sewage treatment, waste dumping, and development of wetlands are often enacted to protect human health as well as resources such as drinking water, forests, and commercial and sport fisheries (Untermaier 1991). The effectiveness with which these laws are enforced determines a nation's abilities to protect its citizens and natural resources (Tobin 1990). At the same time, these laws protect biological communities that would otherwise be destroyed by pollution. The air pollution that exacerbates human respiratory disease, kills commercial forests, and ruins drinking water also kills terrestrial and aquatic species.

National governments, through their control of their borders, ports, and commerce, can have a substantial effect on the protection of biological diversity. To protect forests and regulate their use, governments can ban logging, as was recently done in Thailand; restrict the export of logs, as was done in Indonesia; and even ban the import of forestry equipment. To prevent the exploitation of rare species, governments can restrict the possession of certain species and control all imports and exports of the species. For example, the export of ivory from the rare rhinoceros hornbill bird, a valuable international commodity used for carving, is strictly controlled by the Malaysian government.

Finally, national governments can identify endangered species within their borders and take steps to conserve them, such as acquiring habitat for the species, controlling use of the species, developing a research program on the species, and implementing in situ and ex situ recovery plans (de Klemm 1990).

The Endangered Species Act of the United States

In the United States, the principal conservation law protecting species is the Endangered Species Act of 1973 (U.S. Congress 1973; Endangered Species Coalition 1992). This legislation represents one

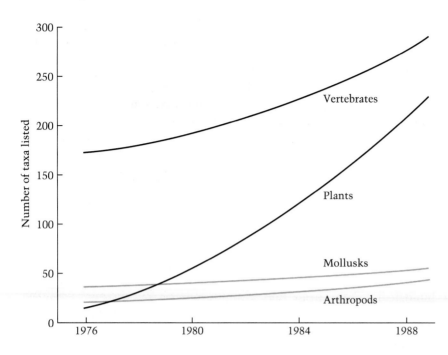

19.2 Plants and vertebrates have received the most attention under the U.S. Endangered Species Act since it went into effect in 1976. Comparatively few mollusks and arthropods have been listed, despite the large numbers of species in these groups that are in danger of going extinct. (From Hafernik 1992.)

of the most effective means of protecting species in the United States, and it has been a model for other countries (Rohlf 1989, 1991; Webster 1987; Simon 1988).

The Endangered Species Act was created by the U.S. Congress to "provide a means whereby the ecosystems upon which endangered species and threatened species depend may be conserved (and) to provide a program for the conservation of such species." Species are protected under the Act if they are on the official list of endangered and threatened species (see Chapter 5). As defined by law, endangered species are those likely to go extinct, as a result of human activities and natural causes, in all or a major portion of their range, while threatened species are likely to become endangered in the near future. The Secretary of the Interior, acting through the U.S. Fish and Wildlife Service, and the Secretary of Commerce, acting through the National Marine Fisheries Service (NMFS), can add and remove species from the list based on information available to them. In addition, a recovery plan is required for each listed species. More than 800 U.S. species have been added to the list, and around 500 species from elsewhere in the world (Figure 19.2). The Act requires all U.S. government agencies to consult with the Fish and Wildlife Service and the NMFS to determine whether their activities will affect listed species, and prohibits activities that will harm these species and their habitat. The law also prevents private individuals, businesses, and local governments from harming or "taking" listed species, and prohibits all trade in listed species.

In the two decades since its enactment, the Endangered Species Act has become increasingly important as a conservation tool. The Act has provided the legal basis for protecting some of the most significant animal species in the United States, such as the grizzly bear, the bald eagle, the whooping crane, and the gray wolf. It has also become a source of contention between conservation and business interests in the United States. The conservation opinion is expressed by Bill Reffalt of the Wilderness Society: "The Endangered Species Act is a safety net for species we've put in jeopardy while we get our act together to take care of the planet. Ultimately we've got to convince people that human progress running counter to the existence of species simply is not sustainable" (Horton 1992). The best hope of long-term success for the protection of species is to prevent the habitat loss that caused species to be endangered in the first place. This point is emphasized by Randall Snodgrass of the National Audubon Society: "If the Act is truly to prevent further species loss it's got to evolve into a next generation wildlife law instead of [an emergency] law that just recovers species from the brink."

The protection afforded to listed species is so strong that business interests often lobby vigorously against the listing of species in their area. At the present time 3600 species are candidates for listing; while awaiting official decisions, some of these species have probably gone extinct (Horton 1992). Business leaders are reluctant to allow new species to be added to the list in part because of the difficulty of rehabilitating species to the point where they can be removed from the list, known as "de-listing." These people worry that once a species is protected under the Act, it will be protected forever. So far only 5 of the 749 listed species have been de-listed, with the most notable successes being the brown pelican and the American alligator (Figure 19.3). The difficulty of implementing recovery plans is to a large extent financial; the U.S. Fish and Wildlife Service annually spends less than $50 million on activities related to the Act, but a recent estimate suggests that over $4 billion is needed to remove the threat of extinction from all listed species.

The economic implications of protecting listed species can be staggering. Pro-business groups have been formed that use environmental-sounding rhetoric to argue against the Endangered Species Act and for the "wise use" of natural resources. In an attempt to find compromises between the economic interests of the country and conservation priorities, the Endangered Species Act was amended in 1978 to allow a Cabinet-level committee, the so-called "God Squad,"

19.3 The American alligator is one of only five species in the United States that has been "delisted" under the Endangered Species Act—in other words, it is judged no longer to be threatened or in need of strict protection, a conservation "success story." (Photograph by Brian Parker/Tom Stack & Associates.)

to exclude areas from protection (Box 32). This issue is illustrated dramatically by the 2.8 million hectares of old-growth forest in the Pacific Northwest that have been designated as critical protected habitat for the northern spotted owl. Limitations on logging in this region have been fiercely resisted by business and citizen groups in the region, as well as by many politicians (Box 33).

BOX 32 THE SNAIL DARTER, THE TELLICO DAM, AND THE ENDANGERED SPECIES ACT

The Endangered Species Act is a keystone of United States environmental law, but like the species it is supposed to protect, the law faces a variety of threats to its very existence. A case in point is the controversy surrounding the construction of the Tellico Dam in Tennessee. In 1973, when the $116 million dam was about half complete, construction of the dam was halted when zoologist David Etnier discovered an unknown species of perch in the Little Tennessee River near the dam site (UNESCO 1980; Norman 1981). The fish, known scientif-ically as *Percina tanasi*, and more familiarly as the snail darter, was not known to inhabit any other locations; under the provisions of the Endangered Species Act, the construction of the dam could not continue because it jeopardized a critical population of an endangered species (UNESCO 1980).

What followed was an uproar: a multimillion dollar public works project that was supposed to benefit one of the poorest regions of the United States might be scrapped because of a three-inch, light brown fish of no known value (UNESCO 1980). Litigation over the dam's completion went all the way to the Supreme Court, which ruled that the court order halting construction of the dam was both valid and correct. Because of the Endangered Species Act, the Tennessee Valley Authority—the agency that had planned and supervised the construction of the Tellico Dam—could not complete the dam. However, the fight did not end there: local politicians and business interests appealed to Congress, and in 1978 legislation was passed to exempt the project from the Endangered Species Act (Cahn 1979a; Norman 1981). Foreseeing similar conflicts between public projects and endangered species in the future, Congress added an amendment to the Endangered Species Act, permitting a government committee, popularly dubbed the "God Committee" or the "God Squad," to decide the fate of an endangered species when overriding public interest dictates that habitat vital to the species be developed (Cahn 1979a). The amendment has caused considerable consternation for environmentalists, who feel that in essence, it undermines the intended purpose of the Endangered Species Act by allowing development interests the opportunity, literally and figuratively, to bulldoze endangered species. Though the Act still prevents developers from destroying species at whim, if political pressure on the side of development is strong enough, then the endangered species must give way (Cahn 1979b).

For the snail darter, the ending of this story was not as disastrous as feared. Three additional populations of the species were discovered in creeks and tributaries of the Tennessee River in 1980 and 1981, after the Tellico Dam was completed. The story of the tiny brown fish that stopped the dam (temporarily) does not, however, have a happy ending, in spite of the species' apparent survival. The ultimate effect of the conflict has been to weaken the legislation that was intended to provide the strongest possible protections for endangered species in the United States. While the "God Squad" has seldom been activated, the recent furor over the northern spotted owl in Oregon serves as a reminder of how fragile the protections of the Endangered Species Act have become. The snail darter did not fall victim to the amendment, but other species certainly will.

The Endangered Species Act has often forced business and conservation groups to develop compromise habitat conservation plans that reconcile both conservation and business interests. In one such case, an innovative program in Riverside County, California, allows developers to build within the historic range of the endangered Stephen's kangaroo rat if they contribute to a fund that will be used to buy wildlife sanctuaries. Already, more than $25 million has been raised for the program, which has an eventual goal of $100 million. In this particular case, the result is a compromise in which development proceeds but pays a higher cost to support conservation activities.

An analysis of the U.S. Endangered Species Act shows a number of revealing trends. The great majority of species listed under the Act are plants and vertebrates, despite the fact that most species are insects and other invertebrates. Over 40% of the 300 mussel species found in the U.S. are extinct or in danger of extinction, yet only a handful are listed under the Act (Stolzenburg 1992). Clearly, greater efforts must be made to study the various invertebrate groups and extend listing to endangered species whenever necessary. Another study of species covered by the Act has shown that animals only have about 1000 remaining individuals at the time of listing and plants

BOX 33 OWLS VERSUS JOBS

The recent furor over the northern spotted owl (*Strix occidentalis caurina*) demonstrates just how volatile conservation issues can be when the species in question resides in habitat with high economic potential. The spotted owl is able to survive only in substantial tracts of old-growth conifer forest (Blondin 1989; Lamberson et al. 1992; Thomas et al. 1990) (see Chapter 16). Conservative estimates place the owls' needs at between 320 and 800 hectares per pair of owls, at bare minimum (Doak 1989). Fragmentation of the forest by logging limits the ability of the owls to disperse to new habitats and to find mates, potentially leading to a population collapse (Lande 1988b). Most of the remaining old-growth forests in the species' range are federal government properties in Oregon, Washington, and northern California. Since it has been determined that the spotted owl is in imminent danger of extinction should its remaining habitat be destroyed (USDA 1988; Doak 1989), its presence on federally owned property would seem to dictate that these lands be set aside for conservation in accordance with the Endangered Species Act. These old-growth forests, however, are of considerable economic value: a single acre contains roughly $4000 in timber (Simberloff 1987), a com-

The northern spotted owl of the Pacific Northwest old-growth forests sits at the center of a controversy over the survival of species and ecosystems. (Photograph by Greg Vaughn/Tom Stack & Associates.)

modity that is a mainstay in the economies of the states of Oregon and Washington. Conservation of the roughly 3000 pairs of owls in these states (Simberloff 1987; Dawson et al. 1987) thus would require the loss of billions of dollars of private timber sales, federal and state tax revenues from those sales, and employment that the timber industry would provide: truly a staggering sacrifice on behalf of a single species of bird. Some residents and politicians of these states are understandably reluctant to make that sacrifice.

The bird's importance, however, is greater than might at first be apparent. The spot-

ted owl is a Pacific Northwest Region "management indicator species"; that is, the health of the owl population can be an indication of the health of the forest ecosystem as a whole (Simberloff 1987; USDA 1988; Doak 1989). Though the arguments between conservationists and timber interests tend to focus upon this species, many other species in the old-growth forests of the Pacific Northwest will also be affected by logging activities, and some of these may pass into extinction without anyone raising a fuss—or, indeed, without anyone being aware that they're gone at all. Because the spotted owl requires the preservation of such large areas of old-growth forest, conservation of this species would in essence protect many other species as well. For example, the Western or Pacific yew, a species commonly found in these forests, would also be protected if adequate land were set aside for the spotted owl. Since this species was only recently discovered to have medicinal value in the treatment of cancer, protection of the owl could have significant secondary benefits for human beings by preventing the destruction of this tree and others like it which may have undetermined medicinal value (see Box 18).

Despite the importance of the species, conservation of the spotted owl is a monumental challenge. In the face of the jobs and income that conservation could cost the local economies, arguing the importance of nebulous "possible" benefits is something of a losing proposition. In mid-1992, a high-level Federal government committee, the so-

called "God Squad," voted to override the owl's protected status and permit logging on 680 hectares of land set aside as owl habitat (Gup 1992). This marks only the second time that the Endangered Species Act has been overridden, but it is nevertheless an unwelcome second precedent: the difference between this case and the earlier one, in which construction of the Tellico Dam was permitted to continue despite the presence of the endangered snail darter (see Box 32), is that scientists are reasonably certain that continued logging in the forests would lead to the extinction not only of the owl, but of many other species dependent upon old-growth forests (Lamberson et al. 1992). In 1979, so little was known about the snail darter and its habitat in the Tennessee River that no one could say for sure what would happen to the species if the dam were finished; as it turned out, other populations of the fish were located and it

is no longer endangered. Such is not the case with the spotted owl, which has been thoroughly studied by biologists. Without the old-growth forests, the owl will become extinct.

The battle over the spotted owl and its old-growth forest habitat has recently taken a turn in favor of the owls. Environmental legislation designed to protect the old-growth forests was recently passed by Congress and implemented by the U.S. Forest Service. This legislation includes provisions for retraining timber cutters—who, environmental activists point out, would be unemployed anyway once the trees are gone—making it more acceptable to local residents who rely upon timber companies for jobs (Gup 1992). Conservationists hope that their efforts on behalf of the spotted owl have succeeded in the long-term preservation of the old-growth forests and the many species that depend upon these ecosystems.

have fewer than 120 individuals remaining (Wilcove et al. 1993). Populations of such species may encounter the genetic and demographic problems associated with small population size that can prevent recovery. At the extreme were 39 species listed when they had 10 or fewer individuals remaining, and a freshwater mussel that was listed when it had only a single remaining population that was not reproducing. Endangered species probably should be given protection under the Act before they decline to the point where recovery becomes difficult. An earlier listing of a declining species might often allow it to recover and become a candidate for de-listing more quickly. Clearly, there is potential for the Endangered Species Act to be used more widely and more effectively to protect biological diversity.

Summary

1. Biological diversity can be protected by laws and conservation activities at local, national, and international levels. Local conservation laws regulate such activities as hunting, fishing, and gathering of plants; they also restrict pollution. Land can be acquired for conservation by purchasing it directly as well as by buying conservation easements. Local conservation efforts sometimes encounter considerable opposition from the local community, business interests, and the government.

2. National governments can protect biodiversity by establishing national parks, controlling imports and exports at their borders, and creating regulations for air and water pollution. One of the most effective laws in the United States is the Endangered Species Act of 1973. The protection afforded under the law is so strong that pro-business and development groups often lobby against the listing of new species. An amendment to the Act allows economic development to proceed at the expense of an endangered species if there is an overwhelming public interest involved. The controversy over the northern spotted owl is a recent and highly publicized example of the difficulties of finding acceptable compromises between conservation and development.

Suggested Readings

Bean, M. J. 1983. *The Evolution of National Wildlife Law.* 2nd ed. Praeger Publishers, New York. Conservation law at the national level.

Gross, D. W., B. T. Wilkins, R. R. Quinn and A. E. Zepp. 1991. Local land protection and planning efforts. *In* D. J. Decker, M. E. Krasny, G. R. Goff, C. R. Smith and D. W. Gross (eds.), *Challenges in the Conservation of Biological Resources: A Practitioner's Guide*, pp. 355–366. Westview Press, Boulder, CO. Local efforts are critical to the success of conservation.

Hafernik, J. E., Jr. 1992. Threats to invertebrate biodiversity: Implications for conservation strategies. *In* P. L. Fiedler and S. K. Jain (eds.), *Conservation Biology: The Theory and Practice of Nature Conservation, Preservation and Management*, pp. 171–195. Chapman and Hall, New York. Invertebrates have not received enough attention, and many species are in danger of extinction.

Hoose, P. M. 1991. *Building an Ark: Tools for the Preservation of Natural Diversity through Land Protection.* Island Press, Covelo, CA. A practical guide to legal procedures.

Horton, T. 1992. The Endangered Species Act: Too tough, too weak, or too late? *Audubon* (March/April): 68–74. A critical look at the Endangered Species Act.

Lamberson, R. H., R. McElvey, B. R. Noon and C. Voss. 1992. A dynamic analysis of Northern Spotted Owl viability in a fragmented forest landscape. *Conservation Biology* 6: 505–512. The latest in a series of articles on this controversial species, explaining why large stands of old-growth forest are needed. Earlier articles by Doak (1989) and Simberloff (1987) are also good accounts.

Land Trust Exchange. 1988. *Land Trust Standard Practices*. Land Trust Exchange, Alexandria, VA. How to establish a land trust.

Rohlf, D. L. 1989. *The Endangered Species Act: A Guide to its Protections and Implementation*. Stanford Environmental Law Society, Stanford, CA. How the Endangered Species Act works.

Rohlf, D. L. 1991. Six biological reasons why the Endangered Species Act doesn't work—and what to do about it. *Conservation Biology* 5: 273–282. Some weaknesses of the Endangered Species Act.

Tobin, R. 1990. *The Expendable Future: U.S. Politics and the Protection of Biological Diversity*. Duke University Press, Durham, NC. Political and economic factors strongly influence the enactment and enforcement of laws relating to biological diversity.

Toufexis, A. 1992. A new endangered species: Human protectors of the planet put their lives on the line. *Time* 139(17): 48–50. The personal dangers faced by environmental activists around the world.

CHAPTER **20**

International Agreements

The protection of biological diversity needs to be addressed at multiple levels of government. While the major control mechanisms that presently exist in the world are based within individual countries, agreements at international levels are increasingly being used to protect species and habitats (de Klemm 1990, 1993). International cooperation is an absolute requirement for several crucial reasons. First, species have no regard for international borders. Conservation efforts must protect species at all points in their ranges; such efforts in one country will be ineffective if critical habitats are destroyed in a second country to which an animal migrates. For example, efforts to protect migratory bird species in northern Europe will not work if the birds' overwintering habitat in Africa is destroyed. Species are particularly vulnerable when they are migrating, as they may be more conspicuous, more tired, or in more desperate need of food and water.

Second, there is international trade in biological products. A strong demand for a product in a wealthy country can result in the overexploitation of the species by people in a poor country to supply this demand. To prevent overexploitation, control and management of the trade are required at both the points of export and import.

Third, the benefits of biological diversity are of international importance. The community of nations benefits from the species and varieties that can be used in agriculture and medicine, the ecosystems that help regulate climate and control flooding, and the national parks and biosphere reserves of international scientific and tourist value. Wealthy countries of the temperate zone that benefit from tropical biological diversity need to be willing to help the less wealthy countries of the world that preserve it.

Finally, many of the factors that threaten ecosystems are international in scope and require international cooperation. Such threats include overfishing and hunting, atmospheric pollution and acid rain, pollution of lakes, rivers, and oceans, global climate change, and ozone depletion.

Agreements for the Protection of Species

The single most important treaty protecting species at an international level is the Convention on International Trade in Endangered Species (CITES), established in 1973 in association with the United Nations Environmental Program (UNEP) (Faure 1989; de Klemm 1990, 1993; Wijnstekers 1992). The administration of CITES is based in Geneva, Switzerland, with eight full-time staff. The treaty is currently endorsed by 118 countries. CITES establishes lists of species whose international trade is to be controlled; the member countries agree to restrict trade in and destructive exploitation of these species (Fitzgerald 1989). Appendix I includes 406 animals and 146 plants whose commercial trade is prohibited, and Appendix II includes about 2500 animals and 25,000 plants whose international trade is regulated and monitored. For plants, Appendixes I and II cover such important horticultural species as orchids, cycads, cacti, carnivorous plants, and tree ferns; increasingly they cover timber species as well. For animals, closely regulated groups include parrots, large cat species, whales, birds of prey, rhinos, bears, primates, species collected for the pet, zoo, and aquarium trades, and species harvested for their fur, skin, or other commercial products.

International treaties such as CITES are implemented when a country signing the treaties passes laws making it a criminal act to violate them (de Klemm 1993). Once CITES laws are passed within a country, police, customs inspectors, wildlife officers, and other government agents can arrest and prosecute individuals possessing or trading in CITES listed species and seize the products or organisms involved. In one recent case in Florida, an individual was sentenced to 13 months in jail for attempting to smuggle an orangutan into the United States. The CITES Secretariat periodically sends out bulletins aimed at halting specific illegal activities. In recent years these bulletins have even specified that *all* wildlife trade with countries such as Italy and Thailand be temporarily halted because of their unwillingness to restrict the illegal export of wildlife from their countries. Member countries establish their own management and scientific authorities to implement their CITES obligations, with technical advice coming from the International Union for the Conservation of Nature (IUCN) Wildlife Trade Specialist Group, from the World Wildlife Fund (WWF) TRAFFIC Network, and from the World Conservation Monitoring Centre (WCMC) Wildlife Trade Monitoring Unit. CITES is particularly active in encouraging cooperation among countries as well as conservation efforts by development agencies. The CITES treaty has been instrumental in restricting the trade in certain endangered wildlife species, with its most notable success being a ban on the ivory trade, which was causing severe declines in African elephant populations (Box 34).

The Convention on Conservation of Migratory Species of Wild Animals was signed in 1979 with a primary focus on bird species. This convention serves as an important complement to CITES by encouraging international efforts to conserve bird species that migrate across international borders, and by emphasizing regional approaches to research, management, and hunting regulations. The problem with this convention is that only 36 countries have signed it and its budget is very limited. It also does not cover migratory marine mammals and fish.

Other international agreements that protect species include:

- The Convention on Conservation of Antarctic Marine Living Resources
- The International Convention for the Regulation of Whaling, which established the International Whaling Commission.
- The International Convention for the Protection of Birds and the Benelux Convention on the Hunting and Protection of Birds
- The Convention on Fishing and Conservation of Living Resources in the Baltic Sea and the Belts
- Miscellaneous agreements protecting specific groups of animals, such as prawns, lobsters and crabs, fur seals, Antarctic seals, salmon, and vicuña

BOX 34 THE WAR TO SAVE THE ELEPHANT

Conservationists must occasionally take radical steps to save an overexploited species from extinction. For those concerned with the fate of the African elephant, the measures employed to preserve the species have sometimes amounted to actual warfare. Park rangers who wanted to prevent the elephant's extinction literally protected the animals with drawn weapons. The ivory ban imposed by CITES in 1989 stopped the world trade in ivory in a last-ditch effort to end poaching of elephants (Cohn 1990b; Jones 1990; Milner-Gulland and Mace, 1991). The CITES ban may have been a key to saving rapidly dwindling elephant populations in Kenya, Tanzania, Uganda, and Zaire; however, the drastic measures employed by wildlife officers in those countries, most notably Richard Leakey of Kenya, were the first steps in preventing the animal's extinction. Moreover, the battle is by no means ended; several southern African nations that have healthy elephant populations want to end the ivory ban, as do some of the countries that are the main importers of ivory, such as Japan and Hong Kong (Dobson and Poole 1992). The CITES ban could be reversed at any time, in which case poaching would again become a major problem.

At the center of the conflict is the market for elephant ivory, which grew rapidly in volume during the 1970s and early 1980s. Over 800 tons of ivory was required annually to meet market demands (Jones 1990). In 1989, before the worldwide trade was halted, raw ivory sold for as much as $120 per pound. A villager might realize only $6 of that price per pound, but elephant tusks weigh an average of 10 pounds apiece (Cohn 1990b; Milner-Gulland and Mace 1991). In impoverished East Africa, the ivory of just three elephants could provide a family in Kenya with a year's income (Jones 1990). Given the uncertainty of agriculture in drought-plagued East Africa, hunting elephants for ivory was a more lucrative way, or perhaps the only way, for individual families to obtain income, as well as meat. However, most elephant hunting was not done by impoverished small-time hunters, but by organized bands of poachers. The poachers carried automatic weapons, including AK-47 assault rifles—more than sufficient firepower to take down an entire family group of elephants, and certainly heavy enough armament to withstand any group of wildlife officials trying to prevent them from poaching. Frequently the bands had ties to military or government forces, and they did not hesitate to cross national borders in pursuit of elephant herds; for instance, the United States' 1989 ban on ivory imports was precipitated by the capture of Somalian sol-

Wholesale slaughter of elephants for their tusks has led to an international ban on trade in ivory. (Photograph by Karen Allen, Conservation International.)

diers who were poaching elephants and rhinos in Kenya. In some cases, the poachers were the same people whose job it was to protect the animals: the game wardens themselves.

Under these circumstances it is hardly surprising that the total elephant population dropped from 1.3 million in the late 1970s to under 600,000 by the late 1980s (Cohn 1990b; Caughly et al. 1990). Poaching accounted for 60% of elephant mortality in Kenya's Tsavo ecosystem during the late 1970s (Ottichilo 1987), and probably increased in the 1980s as the price of ivory rose in world markets. Even that number is not entirely indicative of the extent of the slaughter, since a significant portion of the remaining elephants are part of the large,

well-managed herds of Zimbabwe, Botswana, South Africa, Malawi, and Namibia (Jones 1990). East Africa's population was decimated: Kenya lost an estimated 85% of its elephant herd, Uganda nearly 90%—some 150,000 animals in less than a decade (Cohn 1990b). Kenya's president, Daniel arap Moi, decided that drastic action was required to stop the illegal killing of elephants. In 1989, he appointed world-famous paleoanthropologist Richard Leakey to head the new, semi-autonomous Kenya Wildlife Service. Moi and Leakey instituted a harsh policy toward poachers: they would be shot on sight, no questions asked. Game wardens and park personnel were supplied with vehicles and arms, provided with training, and given incentives, including higher pay, to increase their commitment to the job. At the same time, the East African countries joined together to ask the member nations of CITES to halt ivory imports. Even though the ivory trade was officially regulated by the CITES treaty, ivory from countries with an export ban freely passed to neighboring countries, where it was re-exported with official permits. It has been estimated that more than 80% of the ivory being exported from Africa came from elephants killed by poachers (Dobson and Poole 1992). When the ban was finally instituted in 1989, the price of ivory dropped dramatically, and the pace of the poaching declined with it.

The damage done to the East African elephant herds by three decades of unrestricted hunting is more than a matter of mere numbers. First, elephants are social animals, with complex behaviors that are taught to younger elephants by their elders. Because the poachers selectively killed the elephants with the largest tusks—in other words, the older elephants, generally between 30 and 60 years of age—the transmission of knowledge from the mature animals to the next generation has been disrupted (Cohn 1990b; Jones 1990). The remaining population of elephants is essentially composed of young animals, in their teens or twenties, which lack the experience to know where food and water may be available in times of drought. In addition, male elephants under normal circumstances rarely breed before age 30; adult male elephants, prized by poachers for their large tusks, are now virtually nonexistent in East Africa.

Second, elephants have a profound effect upon the development of microhabitats on which many other animals depend (Cohn 1990b). Elephants strip leaves, upset trees, and trample brush as they feed, opening up habitat for other kinds of vegetation. The elephants' foraging patterns can initiate successional phases in the West and Central African rain forest and the East African bush, opening up areas for grazing animals such as gazelle, zebra, and wildebeest, and, in West and Central Africa, encouraging the growth of vegetation favored by gorillas. With fewer elephants available to perform this service, less open habitat is created, and the other species suffer as a consequence, including those which, like the gorilla, are already hard-pressed by habitat destruction. Finally, elephants learn to avoid areas where the herd has suffered heavy poaching losses. This restriction on the elephants' range came on top of the loss of habitat due to human activities (Chadwick 1991). Game wardens in several countries have observed that the elephants tend to cluster near park headquarters, where they are safest from poaching gangs. The concentration of elephants in these small areas strains the capacity of the region to support the population (Jones 1990).

The recent efforts to save the elephant have had a significant, positive effect. The ivory ban appears to be working— the price of ivory dropped precipitously, making poaching much less attractive (Dobson and Poole 1992; Graham 1990). Furthermore, efforts to hire, train, and supply a military-style cadre of dedicated park rangers have created a deterrent to further poaching. Yet the elephant is not entirely safe. Countries in southern Africa with stable elephant herds have been pressing for at least a partial lifting of the ivory trade ban, claiming that the sale of elephant products provides financial support for their successful elephant management programs (Cohn 1990b). Herds in Zimbabwe, Botswana, and South Africa are so healthy that they require annual culling to prevent habitat damage; they are in no danger of becoming extinct on a local scale. For the greater good of the species throughout its range, however, the ban needs to remain in place (Dobson and Poole 1992). There is too much danger that reestablishing a legal trade in ivory will result in a renewal of massive uncontrolled poaching of elephants to supply a new black market.

Agreements for the Protection of Habitats

Habitat conventions complement species conventions by emphasizing unique ecosystem features that need to be protected. Within these habitats, multitudes of individual species can be protected. Three of the most important conventions are the Ramsar Convention on Wetlands of International Importance Especially as Waterfowl Habitat, the Convention Concerning the Protection of the World Cultural and Natural Heritage, and the UNESCO Biosphere Reserves Programme (McNeely et al. 1990).

The Ramsar Convention on Wetlands was established in 1971 to halt the continued destruction of wetlands, and to recognize the ecological, scientific, economic, cultural, and recreational values of wetlands (Koester 1987; Kusler and Kentula 1990; Untermaier 1991). The value of wetlands for migratory waterfowl is given particular emphasis. This convention is associated with the IUCN and is overseen by a staff of five in the United Kingdom. The Ramsar Convention covers freshwater, estuarine, and coastal marine habitats, and includes more than 400 sites with a total area of over 30 million ha. The Ramsar Convention includes sites that are already designated by national legislation, and assists in their protection by giving them international status. The 61 signing countries agree to conserve and protect their wetland resources and designate for conservation purposes at least one wetland site of international significance.

The Convention Concerning the Protection of the World Cultural and Natural Heritage is associated with UNESCO, IUCN, and the International Council on Monuments and Sites (Hales 1984; Slayter 1983; Thorsell and Sawyer 1992). This convention has received unusually wide support, with 109 countries participating, among the most of any conservation convention. The goal of the convention is to protect natural areas of international significance through its World Heritage Site Program. The convention is unusual because it emphasizes the cultural as well as biological significance of natural areas and recognizes that the world community has an obligation to support the sites financially. Limited funding for World Heritage Sites comes from a special World Heritage Fund, which also supplies technical assistance. As with the Ramsar Convention, this convention seeks to give international support to parks that are created initially by national legislation. Included in the list of World Heritage Sites are some of the world's premier conservation areas: Serengeti National Park (Tanzania), Sinharaja Forest Reserve (Sri Lanka), Iguaçu (Brazil), Manu National Park (Peru), Queensland Rain Forest (Australia), and Great Smokies National Park (U.S.).

An international network of Biosphere Reserves was established by UNESCO's Man and the Biosphere Program (MAB) in 1971. Biosphere Reserves are designed to be models demonstrating the compatibility of conservation efforts and sustainable development (see Chapters 14 and 16). The concept of the Biosphere Reserve places an emphasis on scientific research and international cooperation, monitoring the environment, educating and training local people in ecological concepts, and developing sustainable human activities (UNESCO 1974, 1985, 1987; Batisse 1986). Countries can apply to have national parks and other natural areas of international significance designated as Biosphere Reserves. As of 1990, a total of 283 reserves had been created in 72 countries, covering about 1.5 million km^2, and including 43 reserves in the United States (Figure 20.1). The success of the Biosphere Reserve concept will depend on whether the sites can be successfully organized into a network that can address larger ecosystem and biodiversity questions at a regional and landscape level (Dyer and Holland 1991).

Regional agreements to protect unique ecosystems and habitats cover the Western Hemisphere, the Antarctic flora and fauna, the South Pacific, Africa, and European wildlife and natural habitat (World Resources Institute 1992).

International Agreements to Control Pollution

International agreements have been signed to prevent or limit pollution that poses regional and international threats to the environment (McNeely et al. 1990; World Resources Institute 1992). The Convention on Long-Range Trans-Boundary Air Pollution in the European Region recognizes the role that long-range transport of air pollution plays in acid rain, lake acidification, and forest dieback. More recently, the Convention on the Protection of the Ozone Layer was signed in 1985 to regulate and discourage the use of chlorofluorocarbons, which have been linked to the destruction of the ozone layer and a resulting increase in the levels of harmful ultraviolet light.

Marine pollution is another key area of concern because of the extensive areas of international waters not under national control and the ease with which the pollutants released in one area can spread to another area. Agreements covering marine pollution include the Convention on the Prevention of Marine Pollution by Dumping of Wastes and Other Matters, and the Regional Seas Conventions of the United Nations Environmental Program (UNEP) (World Resources Institute 1992). Regional agreements cover the northeastern Atlantic, the Baltic, and other specific locations, particularly in the North Atlantic region.

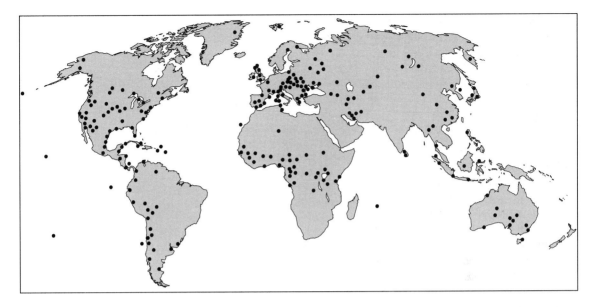

20.1   The locations of recognized Biosphere Reserves as of 1988. A lack of reserves is apparent in such biologically important regions as New Guinea, the Indian subcontinent, South Africa, Amazonia, and Southeast Asia. (Data from Gregg et al. 1988.)

The Earth Summit

For 12 days in June 1992, world attention was focused on a major conference in Rio de Janeiro, Brazil, known officially as the United Nations Conference on Environment and Development (UNCED) and also as the Earth Summit, Eco '92, and the Rio Summit. Representatives from 178 countries including over 100 heads of state, plus leaders of the United Nations and major nongovernment and conservation organizations, met to discuss ways of combining increased protection of the environment with more effective economic development in less wealthy countries (United Nations 1993a,b). The conference was successful in heightening awareness of the seriousness of the environmental crisis and placing the issue at the center of world attention (Haas et al. 1992; Alyanak 1992). A noteworthy feature of the conference was the clear linkage established between the protection of the environment and Third World poverty. Whereas the wealthy countries of the world have the resources to provide for their citizens and protect the environment, most poor countries see the immediate use of their natural resources as the key to raising the standard of living for their impoverished populations. At the Earth Summit, the wealthy countries were collectively made to understand that they would have to assist the poor countries in order to protect the global environment and tropical biodiversity.

The conference participants discussed and eventually signed five major documents described below, and initiated many new projects (Parson et al. 1992). Aside from these specific achievements, the central achievement of the Earth Summit was the willingness of the participants to keep working together on long-term goals. As stated in the opening address by UNCED Secretary-General Maurice F. Strong,

> The Earth Summit is not an end in itself, but a new beginning. The measures you agree on here will be but the first steps on a new pathway to our common future. Thus, the results of this conference will ultimately depend on the creditability and effectiveness of its follow-up.

The Rio Declaration. The Declaration provides general principles to guide the actions of both wealthy and poor nations on issues of the environment and development. The right of nations to exploit their own resources is recognized, as long as the environment elsewhere is not harmed. At the same time, the need to stop unsustainable activities and protect the environment is also recognized. The Declaration affirms the "polluter pays" principle, in which companies and governments take financial responsibility for the environmental damage that they cause.

Convention on Climate Change. An agreement was reached that requires industrialized countries to reduce their emissions of carbon dioxide and other greenhouse gases and to make regular reports on their progress. While specific emission limits were not decided upon, the convention states that greenhouse gases should be stabilized at levels that will not interfere with the Earth's climate. Funding for implementing this convention will come from the Global Environment Facility (see Chapter 21).

Convention on Biodiversity. The Convention on Biodiversity has three objectives: protecting biological diversity, using it sustainably, and sharing the benefits of new products made with wild and domestic species. While the first two objectives are straightforward, the last point recognizes that developing countries should receive fair compensation for the use made of species collected within their borders. In the past, industrialized countries have developed new crops, medicines, and other biotechnology products based on tropical species without returning any of the resulting technology, new products, or profits to the countries in which the wild species were originally found. The treaty, signed by 153 nations, affirms that countries have certain rights over species occurring within their borders. The United States at first refused to sign because of what were perceived to be potential restrictions on its enormous biotechnology industry. The United States finally signed the convention in early 1993, in the aftermath of the Clinton/Gore victory in the 1992 election. Funding for this convention has been set initially at $200 million, administered by the Global Environment Facility.

Statement on Forest Principles. An agreement on the management of forests proved to be difficult to negotiate, with strong differences of opinion between tropical and temperate countries. This final, nonbinding treaty calls for the sustainable management of forests without making any specific recommendations.

Agenda 21. This 800-page document is an innovative attempt to describe in a comprehensive manner the policies needed for environmentally sound development (United Nations 1993a). Agenda 21 shows the linkages between the environment and other issues that are often considered separately, such as child welfare, poverty, women's issues, technology transfer, and unequal divisions of wealth. Plans of action are described to address problems of the atmosphere, land degradation and desertification, mountain development, agriculture and rural development, deforestation, aquatic environments, and pollution. Financial, institutional, technological, and legal mechanisms for implementing these action plans are also described.

The most contentious issue was deciding how to fund the programs of Agenda 21. The cost of these programs was estimated to be about $600 billion per year, of which $125 billion would come from the developed countries as overseas development assistance (ODA). Since existing levels of ODA amount to $60 billion per year for all activities, this means that implementing Agenda 21 would require a tripling of the present foreign aid commitment. This increase in funding was not agreed to by the major developed countries, known as the Group of Seven, whose economies were suffering from the global recession. As an alternative proposal, the Group of 77, a group of 128 developing countries, suggested that industrialized countries increase their level of foreign assistance to 0.7% of their Gross National Product by the year 2000. While the richer countries agreed in principle to this figure, no schedule was set for implementing it. The frustration of the developing countries was eloquently summarized by Dr. Mahathir bin Mohamed, Prime Minister of Malaysia:

> The poor countries of the world have been told to preserve their forests and other genetic resources on the off chance that at some future

date something is discovered which might prove useful to humanity. But now they are told that the rich will not agree to compensate the poor for their sacrifices, arguing that the diversity of genes stored and safeguarded by the poor are of no value until the rich, through their superior intelligence, release the potential within.

In the end, industrialized countries did announce $6 billion of new contributions. Considering that this was the moment when the world's leaders and the public's attention were focused on environmental issues and there was maximum pressure to get something accomplished, the inability of the major industrial countries to allocate enough funds for Agenda 21 was disappointing to many conference participants. Raising funds for the implementation of Agenda 21 will be part of the ongoing process that the Earth Summit initiated.

The Value of International Agreements

International agreements are signed voluntarily by the countries involved, and implemented by the passage of supporting laws at the national level (de Klemm 1990, 1993). Member countries that are in violation of these conventions are often persuaded or embarrassed into compliance by the other countries, as well as by international nongovernmental organizations and the media. A weakness of these international treaties is that countries can withdraw from the convention to pursue their own interests when they find the conditions of compliance too difficult. This occurred recently when several countries walked out on the International Whaling Commission because of its ban on hunting (Ellis 1992). The crucial test of these international agreements is their ability to allocate funds to support positive steps and to stop destructive activities. The wealthy nations of the world are indicating that they are willing to increase substantially the amount of funds available to international conservation activities, but it remains to be seen whether they will invest enough to meet the challenge. It also remains to be seen whether the countries who receive this funding can channel the funds into preservation and conservation, and not find it disappearing into other political and economic uses. Finding the political will to enforce laws and prosecute violators represents an additional challenge.

Summary

1. International agreements and conventions on the protection of biological diversity are needed because species migrate across borders, because there is an international trade in biological products, because the benefits of biological diversity are of international importance, and because the threats to diversity are often international in scope and require international cooperation.

2. The Convention on International Trade in Endangered Species (CITES) was enacted to prevent destructive trade in endangered species. CITES prohibits trade in some species and regulates and monitors trade in others. Enforcement of such treaties involves individual countries implementing their own legislation. The most notable success of CITES was stopping elephant poaching by banning all trade in ivory. Other important conventions protect habitats, species, and cultural sites of international importance.

3. Five major environmental documents were signed at the 1992 Earth Summit, attended by over 100 heads of state. Implementing and funding these new treaties could prove vital to international conservation efforts.

Suggested Readings

Batisse, M. 1986. Developing and focusing the biosphere reserve concept. *Nature and Resources* 22: 1–10. The philosophy behind the UNESCO Man and the Biosphere Program's system of reserves.

de Klemm, C. 1993. *Guidelines for CITES Implementation Legislation.* IUCN, Gland, Switzerland. Establishing and enforcing CITES laws at the national level.

Dobson, A., and J. H. Poole. 1992. Ivory: Why the ban must stay! *Conservation Biology* 6: 149–151. Arguments against lifting the ban on the ivory trade.

Fitzgerald, S. 1989. *International Wildlife Trade: Whose Business Is It?* World Wildlife Fund, Washington, D.C. An overview of international wildlife trade and the effectiveness of CITES.

Haas, P. M., M. A. Levy and E. A. Parsons. 1992. Appraising the Earth Summit: How should we judge UNCED's success? *Environment* 34(8): 7–35. A summary of the accomplishments of the Earth Summit.

Jones, R. F. 1990. Farewell to Africa. *Audubon* 92: 50–104. An excellent extended essay on the problems faced by wildlife in Africa.

Thorsell, J. and J. Sawyer. 1991. *World Heritage: The First Twenty Years.* IUCN, Gland, Switzerland. An overview of the program with 16 case study reports on World Heritage sites.

TRAFFIC USA. World Wildlife Fund, Washington, D.C. *TRAFFIC USA* is an informative newsletter covering the international trade in wildlife and wildlife products, with an emphasis on CITES activities.

United Nations. 1993. *The Global Partnership for Environment and Development: A Guide to Agenda 21.* United Nations Publications, New York. Explains Agenda 21, and how countries can protect the environment and still allow economic development.

Untermaier, J. 1991. *Legal Aspects of the Conservation of Wetlands.* IUCN, Gland, Switzerland. The effectiveness of the Ramsar convention is discussed along with case studies.

CHAPTER 21

International Funding

Biological diversity is concentrated in the tropical countries of the developing world, most of which are relatively poor and are experiencing rapid rates of development, population growth, and habitat destruction. The developed countries of the world depend on the biological diversity of the tropics to supply genetic material and natural products for agriculture and industry. The general public in the developed world is also anxious to preserve the unusual species, such as elephants, rhinos, and lemurs, that are so interesting to see in zoos, read about, and watch on television. Developing countries are often willing to preserve biological diversity, but they may be unable to pay for the needed research, habitat preservation, management, and infrastructure required for the task. Increasingly, groups in the developed countries are realizing that if they want to preserve biological diversity in species-rich but cash-poor countries, they must pay for it themselves.

Institutions within the United States represent some of the largest sources of assistance to developing countries for the conservation of biological diversity. A comprehensive examination of U.S. conservation activities and research in the developing world was recently undertaken by the World Resources Institute to determine patterns in funding activity (Abramovitz 1991). In 1989, a total of 1093 projects were identified in 127 developing countries, for a total investment of $62.9 million. This compares with 873 projects costing $37.5 million in 1987, a whopping 68% increase in just two years (Figure 21.1). These projects were funded and carried out by government agencies (e.g., U.S. Fish and Wildlife Service, National Science Foundation, Agency for International Development, the Peace Corps), charitable foundations (e.g., Pew Charitable Trusts, MacArthur Foundation), nongovernment conservation organizations (e.g., Wildlife Conservation International, World Wildlife Fund), museums (e.g.,

21.1 Funding for United States biodiversity research and conservation efforts in developing countries, by type of funding organization. The first three categories include charitable foundations and trusts, organizations and divisions of the U.S. Government, and nongovernment organizations such as the World Wildlife Fund. The final bars show funding that came from multiple sources or from miscellaneous institutions (universities, zoos, museums, etc.) (From Abramovitz 1991.)

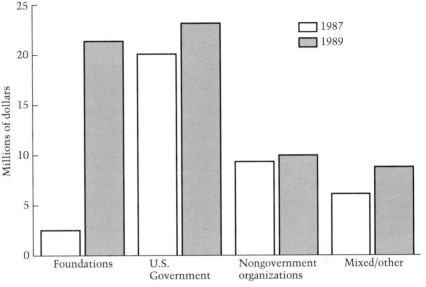

Field Museum of Natural History, Smithsonian Institution), botanical gardens (e.g., Missouri Botanical Garden, New York Botanical Garden), zoos (e.g., The New York Zoological Society) and universities. Of the total funding, the U.S. government and the foundations each gave about one-third. The most dramatic change was a sevenfold increase in funding between 1987 and 1989 by large foundations; tropical conservation has clearly been targeted as a funding priority by the major foundations, such as the Mellon Foundation, W. Alton Jones Foundation, and Pew Charitable Trusts. Over half of the funding from foundations came from the MacArthur Foundation alone.

In allocation of funds, the major activities receiving U.S. funds were research (38%), site and species management (25%), institutional strengthening (15%), policy planning and analysis (11%), and education (8%) (Abramovitz 1991). The projects were overwhelmingly concentrated in Latin America, which received 68% of the funds. Costa Rica, Mexico, and Brazil each received over $5 million (Table 21.1). Funding was much lower in other regions of the world, with only four countries in Africa and Asia—Indonesia, Madagascar, Uganda, and Kenya—receiving over $1 million per year. While funding levels in developing countries are increasing substantially, the amount of money being spent is still inadequate to protect the great storehouse of biological riches needed for the long-term prosperity of human society. In comparison with the vast sums being considered and allocated for other large U.S. science projects, such as the human

TABLE 21.1
Estimated U.S. funding for tropical biodiversity in 1989[a]

Country	1989 Funding ($U.S.)	Land area (× 1000 ha)	Dollars per 1000 ha
Costa Rica	6,214,897	5,106	1217
Mexico	5,528,809	190,869	29
Brazil	5,483,535	845,651	6
Ecuador	3,257,943	27,684	118
Madagasgar	2,835,649	58,154	49
Kenya	2,101,170	56,697	37
Peru	1,923,445	128,000	15
Colombia	1,450,650	103,870	14
Indonesia	1,394,244	181,157	8
Guatemala	1,240,995	10,843	114
Belize	1,199,342	2,280	526
Jamaica	1,141,076	1,083	1054
Uganda	1,020,701	19,955	51

Source: Funding data from World Resources Institute Biodiversity Projects Database, in Abramovitz 1991.

[a]Table includes only those countries that received more than $1,000,000.

genome project ($3 billion), the space station ($13 billion), and the superconducting supercollider ($8 billion), the amount being spent on biological diversity is tiny.

The $23 million in funds annually spent on biodiversity by the United States government is less than 1% of the $9 billion spent each year in U.S. foreign assistance. Within the U.S. government, the leading agency in economic and humanitarian aid is the Agency for International Development (AID), with a budget of over $5 billion per year. With programs and missions in about 60 developing countries, AID is a dominant presence in foreign assistance activities. To reflect growing environmental concerns, the U.S. Congress amended the Foreign Assistance Act in 1983 to make the conservation of biological diversity and the protection of endangered species an integral part of AID foreign assistance activities (Shaffer and Saterson 1987; Brady 1988). The amendment requires AID to spend money each year on biodiversity projects, in particular supporting local conservation efforts, education, and ecosystem protection. To provide guidance in this effort, a panel of government and nongovernment representatives was convened to formulate an official U.S. policy on biodiversity, released as *U.S. Strategy on the Conservation of Biological Diversity:*

An Interagency Task Force Report to Congress (AID 1985). This report is being used to develop and guide programs in the AID and other federal agencies.

A major new source of funds for conservation and environmental activities will be the Global Environment Facility (GEF), created in 1990 by the World Bank along with the United Nations Development Program (UNDP) and the U.N. Environmental Program (UNEP). The GEF was established as a three-year pilot program with a budget of $1.3 billion to be used for funding projects relating to global warming, biodiversity, international waters, and ozone depletion (Haas et al. 1992). At the June 1992 Earth Summit, it was suggested that the GEF be expanded to fund the Agenda 21 programs, but no additional funds were allocated for this purpose. The Group of 77 developing countries have expressed distrust of the GEF and the World Bank, which they regard as dominated by the major industrialized countries and too assertive in dictating the domestic policies of recipient countries. As a result, the GEF has been restructured to give the recipient countries a greater voice in the decision-making and voting processes.

Linkages among Conservation Organizations

When a conservation need is identified, such as protecting a species or establishing a nature reserve, this often begins a complex process of project design, proposal writing, fundraising, and implementation that involves different types of conservation organizations. Conservation foundations (e.g., the MacArthur Foundation) and government agencies (e.g., AID) often provide money for conservation programs through direct grants to the institutions (e.g., Colorado State University, Missouri Botanical Garden) that implement the projects. In some cases, the major foundations and government agencies give money to major conservation nongovernment organizations (NGOs) (e.g., World Wildlife Fund, Wildlife Conservation International), which in turn provide grants to local conservation organizations. The major international conservation organizations are often active in establishing, strengthening, and funding local organizations in the developing world that can run conservation programs. From the perspective of an international conservation organization like the World Wildlife Fund, working with local organizations in developing countries is an effective strategy because it trains and supports groups of citizens within the country, who can then be advocates for conservation for many years.

A common pattern is that an active local conservation program in a tropical country will have funding from one or more conservation foundations and foreign governments, scientific links to international conservation NGOs, and affiliations with local and overseas research institutions. In such a manner, the world conservation community is knit together through networks of money, expertise, and mutual interests. The Program for Belize (PFB) is a good example of this situation. At first glance the PFB is a Belizean organization, staffed by Belizean personnel, with the main purpose of managing a Belizean conservation facility, the Río Bravo Conservation and Management Area. However, there are additional facets to this organization that extend beyond its basic purpose: the PFB has an extensive network of research, institutional, and financial connections to government agencies and private organizations in other countries.

International Development Banks and Tropical Deforestation

The rate of tropical deforestation and habitat destruction has sometimes been accelerated by poorly conceived projects financed by the four major **multilateral development banks (MDBs)**: the World Bank, which lends to all regions of the globe, and the regional MDBs, which include the Inter-American Development Bank (IDB), the Asian Development Bank (ADB), and the African Development Bank (AFDB). The MDBs annually loan more than $25 billion per year to 151 countries to finance economic development projects (Rich 1990). The impact of the MDBs is actually even greater, since their funding is often linked to financing from donor countries, private banks, and other government agencies; the $25 billion funding of the MDBs attracts about another $50 billion in loans, which makes the MDBs major players in the developing world.

While the goal of the MDBs is ostensibly economic development, the effect of many of the projects they support is to exploit natural resources to create exports for international markets. In many cases, these MDB-funded projects have resulted in the destruction of ecosystems over a wide area (Norse 1987; Rich 1990). Since the MDBs are controlled by the governments of the major developed countries, such as the United States, Japan, Germany, the United Kingdom, and France, the policies of the MDBs can be scrutinized by the elected representatives of the MDB member countries, the national media, and conservation organizations. In particular, as some of the ill-conceived projects of the World Bank have been publicly criticized, the World Bank has reacted by making the conservation of biological diversity part of its assistance policy and requiring new projects to be more environmentally responsible (Goodland 1987, 1989, 1992; Pearl

1989). However, it remains to be seen whether the MDBs will actually change their practices. As Rich (1990) points out:

> Real reform in the MDBs will not occur without steady and increased political pressure. What has been won so far is an unprecedented and undeniable place for citizen activism, the only force that can bring accountability to the agencies controlling the international development agenda. But the fact that the World Bank and the IDB have undertaken some bureaucratic reforms does not mean that environmentalists can assume that their case is won, or even that their ideas will get a sympathetic hearing. New posts have been created in the past without disrupting 'business as usual.' Environmentalists should remember that for any bank or bureaucracy, let alone the MDBs, nothing is cheaper than words.

The most highly publicized examples of environmental destruction resulting from World Bank lending are the transmigration program in Indonesia, road construction, agricultural development, and industrialization projects in Brazilian Amazonia, and large dams.

Indonesia

From the 1970s to the late 1980s, the World Bank loaned $560 million to the Indonesian government to resettle millions of people from the densely populated inner islands of Java, Bali, and Lombok on the sparsely inhabited, heavily forested outer islands of Borneo (Kalimantan), New Guinea (Irian Jaya) and Sulawesi (Rich 1990). These farmers were supposed to raise crops to feed themselves as well as cash crops, such as rubber, oil palm, and cacao, that could be exported to pay off the World Bank loan. This transmigration program has been an environmental and economic failure because the poor tropical forest soils on the outer islands were not suited to the intensive agriculture practiced by the farmers (Whitten 1987). As a result, many of the farmers have become impoverished and are forced to practice shifting agriculture. The production of export crops to pay off the World Bank loans has not materialized. In addition, at least 2 million and possibly up to 6 million hectares of tropical rain forest has been destroyed by the transmigration settlers. While this amount of land is enormous, it still represents less than 1% of the forested area of the outer islands.

Brazil

Many of the large Brazilian projects financed by the World Bank and other MDBs have failed to take into account the loss of biologi-

cal diversity that results from their activities (Fearnside 1987, 1990;
Cavalcanti 1991). Agricultural, industrial, and transportation projects
have consistently been launched without environmental impact
studies or land use studies to determine their feasibility. Decisions
about these projects are typically made at a high level in the government without adequate knowledge of realities on the ground, and
generally benefit urban elites rather than the rural poor. Promises to
protect biological and Amerindian reserves have not been kept.
Often, the banks express a willingness to acknowledge past mistakes,
but do not institute changes in the way projects are implemented. As
a result the mistakes are repeated in other places.

Several examples illustrate the effects of these projects upon Brazil's natural resources. For example, the World Bank and the Inter-
American Development Bank have made hundreds of millions of dollars in loans since 1981 to support the Northwest Development Program (Polonoroeste) in Brazil (Figure 21.2). The eventual cost of the
program will be over $1.6 billion (Fearnside 1987, 1990; Anderson

21.2   The locations of
the Polonoroeste, Acre,
and Grande Carajás proj-
ects in Brazil. Highways
are indicated by dashed
lines; the effect of such
highways was illustrated
in Figure 6.8. (From
Fearnside 1987.)

1990). Most of these funds were spent on the construction of a 1500-km section of road connecting Cuiaba, capital of Mato Grosso, to Porto Velho, capital of Rondonia. The remaining funds were mainly used for building secondary and feeder roads and settlement areas, with only 3% of the budget allocated for biological and Amerindian reserves and 0.5% for research. Once the highway was opened, farmers from southern and northeastern Brazil, who had been displaced from their land by increasing mechanization and land ownership laws that favored the wealthy, flocked to Rondonia seeking free land. As a result, during the 1980s, Rondonia had one of the most rapid rates of deforestation in the world. At the peak of deforestation in 1987, 20 million hectares—2.5% of Brazil's total land area—was burned in one of the world's most massive episodes of environmental devastation (Pinto 1990). Much of the land was unsuitable for agriculture, but was cleared to establish land claims, leading to fragmentation of the forest (Hecht and Cockburn 1989; Anderson 1990). In its haste to develop the region, the Brazilian government also built roads across Amerindian reserves and biological reserves that were supposed to be completely protected, effectively opening up even these areas to deforestation. As one example, the Ianomãmi Indians were given legal rights to only 30% of the land that they occupied, and this holding was eventually fragmented into 19 separate pieces by roads and other developments (Costa 1989). In general, the cattle ranches and tree plantations that were supposed to pay for the loan have failed after colonists abandoned their plots, resulting in increased indebtedness for the Brazilian government (Hecht et al. 1988).

The IDB is also providing $58 million to assist in constructing a road from Porto Velho in Rondonia to Rio Branco in Acre. Work on this road started before adequate studies were undertaken on the feasibility of agriculture in the region and the effects of the road on Amerindians and biological communities (Fearnside 1987; Anderson 1990). Rapid colonization by farmers, deforestation, and violent confrontations between settlers and local people were the logical consequences of the construction of this road (Schwartzman 1992). In both the Acre and the Polonoroeste projects, road construction was the crucial point of control; the Brazilian government and the MDBs decided whether and when to build the road, and how development should proceed. However, once the road was constructed, migration, settlement, and deforestation occurred as a result of thousands of uncontrollable acts by land speculators and desperately poor farmers. Also, because the pace of road construction was so rapid, research by biologists, anthropologists, and other scientists could not be an effective part of the planning process, but only played a role in dealing with problems created after the road was completed (Fearnside 1987).

A third major project in Brazil is the Grande Carajás Program in eastern Amazonia. This project principally involves mining, processing, and exporting the vast mineral deposits in the region (Oren 1987). The project covers an area of 900,000 km^2, larger than Texas and Oklahoma together, and is financed in part by a $300 million loan from the World Bank. Associated with this project are several massive pig-iron factories, which were not financed by the World Bank, but are clearly part of the overall development plan for the area. These factories require at least one million and possibly up to two million tons of charcoal each year. At the time the pig-iron projects were approved, it had still not been decided where these enormous quantities of charcoal were going to come from. Since tree plantations on the scale needed to supply this much charcoal to the factories—an estimated 700,000 ha—have not been planned or planted, the only source of charcoal will be rain forests; in effect, the factories will gradually consume all of the rain forests in their vicinity and then create an ever-widening demand for wood that will continue to devastate Brazil's rain forests (Mahar 1988). It has been estimated that once fully operational, these factories will require the destruction of 74,000 ha (183,000 acres) of forest per year (Fearnside 1988; Cavalcanti 1991). Even though the World Bank did not finance the pig-iron factories, and may have even tried to discourage their construction, the World Bank was certainly involved in the many phases of the project leading up to this decision.

Dam Projects

Ironically, recent research appears to indicate that protection of biological diversity may be a key to the success of some of the large international projects that have devastated tropical rain forests. A major class of projects financed by MDBs are dam and irrigation systems that provide water for agricultural activities and generate hydroelectric power (Figure 21.3; McNeely 1987; Goodland 1990b). Protecting the forests and other natural vegetation in the watersheds is now widely recognized as an important and relatively inexpensive way to ensure the efficiency and longevity of these water projects, while at the same time preserving large areas of natural habitat. The loss of plant cover on the slopes above water projects often results in soil erosion and siltation, with resulting loss of efficiency, higher maintenance costs, and damage for irrigation systems and dams (Vaux and Goldman 1990). In one study of irrigation projects in Indonesia, it was found that the cost of protecting watersheds ranged from only 1% to 10% of the total cost of the project, in contrast to an estimated 30% to 40% drop in efficiency due to siltation if the for-

21.3   A hydroelectric dam on the Volta River in Ghana. The watersheds around such dams must be protected if the dams are to operate efficiently. (Photograph courtesy of FAO.)

ests were not protected (MacKinnon 1983). One of the most successful examples of an effective environmental investment was the $1.2 million irrigation sector loan made by the World Bank to assist in the development and protection of Dumoga-Bone National Park in northern Sulawesi, Indonesia (McNeely 1987). A 278,700-ha primary rain forest, which included the catchment area on the slopes above a $60 million irrigation project financed by the World Bank, was converted into a national park (Figure 21.4). In this particular case, the World Bank was able to protect its original investment through environmental funding representing less than 2% of the project's cost, and create a significant new national park in the process.

Changing the Funding Process

If many large international development projects are so economically and environmentally unsuccessful, why do host countries want them, and why do the MDBs agree to finance them? Projects are often funded because economists make overly optimistic predictions on production schedules and prices of commodities, and minimize potential problems. Surveys and pilot studies are not undertaken, or their results are minimized; comparable projects elsewhere are not

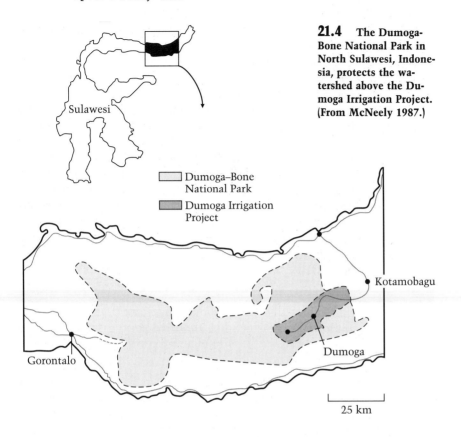

21.4 **The Dumoga-Bone National Park in North Sulawesi, Indonesia, protects the watershed above the Dumoga Irrigation Project. (From McNeely 1987.)**

Sulawesi

▫ Dumoga–Bone
National Park

▪ Dumoga Irrigation
Project

Kotamobagu

Dumoga

Gorontalo

25 km

evaluated. The environmental costs of projects are often ignored or minimized because these variables are considered external to the economic analysis (see Chapters 8 and 9). However, an accurate analysis of a project would include all of its costs and benefits, including the effects of soil erosion, the loss of biological diversity, the impact of water pollution on the health and diets of local people, and the loss of income associated with the destruction of renewable resources (Repetto 1990a,b; Repetto et al. 1989; El Serafy 1989; Daly and Cobb 1989).

Host governments often want large projects to proceed despite unfavorable reviews because the projects provide temporary jobs and temporary economic prosperity, and as a result some release from social tensions during the duration of the project. Local business leaders, especially those with close links to the government, may endorse the projects because they can make large profits on project contracts (Hecht and Cockburn 1989). Industrialized countries that support the banks may encourage these loans to stabilize governments in the host countries that are friendly to their interests but which may lack popular support. The banks themselves make loans because that is their defined reason for existing, and they continue to make loans for projects that experience should tell them are not economically profitable and not environmentally sound.

How could the multilateral development banks operate more responsibly? First, they could stop making loans for environmentally destructive projects (Ledec 1989; WRI/IUCN/UNEP 1992). This step would require the banks to use economic cost–benefit models for the analysis of development projects that would include the environmental and ecological effects of projects. Also, banks need to encourage open public discussion among all groups in a country before projects are implemented (Goodland 1992). In particular, the banks should allow examination, independent evaluations, and discussions of environmental impact reports before a project is approved for funding.

Second, the MDBs could offer incentives to countries to encourage land reform, so that landless people have the opportunity to own good farmland rather than being displaced into shifting agriculture. At the present time, 4.5% of Brazil's landowners own 81% of the country's farmlands, and 70% of rural families own no land at all (Caulfield 1984). In addition, the banks could support research identifying local methods of sustainable agriculture that small farmers can be encouraged to use. MDBs could also assist in the establishment of national parks to protect biological diversity and extractive reserves that can be used sustainably, a goal that could potentially be accomplished by the purchase of international conservation easements. The banks could also support the development of markets for natural products that come from these reserves.

Finally and most importantly, the MDBs could play a role in reducing or forgiving international debts owed by the governments of the poor countries of the world to the governments and banks of wealthy countries. This action would reduce the desperate need of many countries to exploit their natural resources just to pay off foreign debt.

Debt-For-Nature Swaps

Many of the countries in the developing world have accumulated huge international debts. Collectively, these countries owe about $1.3 trillion to international financial institutions, which represents 44% of their collective Gross National Products (Dogsé and von Droste 1990; Hansen 1989). This money was borrowed in the hopes of stimulating economic development, which would generate the income necessary to pay off the loans. However, combinations of low prices for the exports of developing countries, high fuel costs, mismanagement of projects, and overpopulation have made repayment of these loans difficult or impossible for many countries. Some developing countries have rescheduled their loan payments, or unilaterally reduced and even stopped making their payments. As a result, the commercial banks that hold these debts are selling the debts at a steep discount on the international secondary debt market. For example, Brazilian debt has traded for about 22% of its face value, Costa Rican for 14–18%, and Peruvian for as little as 5%. In Peru's case this would mean that a $1 million loan owed to a bank could be purchased for $50,000 on the international secondary debt market.

The inability to repay loans on schedule has serious negative consequences for all parties involved, as well as for the environment. The developing countries are hurt because they are unable to borrow new money to finance government activities, import foreign goods, and develop their economies. The profits of the banks are hurt because their debts are not being repaid as contracted. The people of the developing countries are hurt because their standard of living is being reduced by the austerity measures their governments must impose to be able to pay cash for needed imports and to pay off at least some of the debts. These economic problems and austerity measures can lead to an uncertain financial and political climate in the country, which discourages outside investment in the country's industries. The developing countries of the world are now paying $50 billion per year more in interest payments back to the international banks than they are receiving in new investment loans (Figure 21.5). The result is that capital is actually being extracted from poorer countries to create income in the wealthier nations of the world.

One consequence of Third World debt is that many governments are attempting to extract and sell their natural resources as rapidly as possible to pay off these debts. Environmentally destructive practices associated with logging activities, mining, and ranching are all justi-

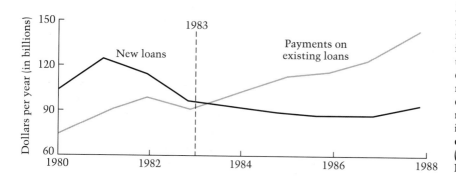

21.5 Prior to 1983, more money was being invested in the developing world than was being used to pay international debts. After 1983, payments on existing loans exceeded the amount of new investments, resulting in a flow of money out of these countries. (After Dogsé and van Droste 1990.)

fied on the basis of the need to earn export exchange (Gullison and Losos 1993). In addition, because of the inability of the weak economies of these countries to absorb a growing labor force, large numbers of people are forced into marginal economic activities that are destructive to the environment, such as shifting agriculture. To deal with this combination of problems, an innovative new idea has been proposed to use Third World debt as a vehicle for financing projects to protect biological diversity, so-called **debt-for-nature swaps** (Lovejoy 1984; Hansen 1988). These debt swaps appear to offer great opportunities to help all participants, and they have great public relations appeal (Debt-for-Development Coalition 1990).

A debt-for-nature swap is relatively simple in theory. A nongovernment conservation organization, such as Conservation International in Washington, D.C., cooperates with the government of a debtor country, such as Bolivia, in developing a proposal involving an environmental activity. This activity could involve land acquisition for conservation purposes, park management, development of park facilities, conservation education, or a sustainable development project. The international NGO then negotiates with a local NGO or local government agency that is willing to actually implement the environmental program. The international NGO then finds a bank that has a loan from the debtor country that it is willing to donate or sell to the NGO at a large discount. This debt must be for an amount and of a form that is acceptable to the debtor country. After the international NGO has purchased the loan, the loan is returned to the debtor country so that no more payments have to be made on the debt. In exchange, the debtor country agrees to supply local currency for the agreed-upon conservation activities, often by issuing bonds that pay a fixed annual amount for the project. In summary, debt-for-nature swaps involve an international NGO relieving a developing country of its obligation to repay an international debt in exchange for the country agreeing to increase its funding efforts in conservation.

The first debt-for-nature swap was initiated in 1987 in Bolivia, with other swaps arranged later with Costa Rica, the Philippines, Madagascar, Zambia, Ecuador, and other countries (Dogsé and von Droste 1990; WRI 1992). The major NGOs involved have been the World Wildlife Fund, Conservation International, and The Nature Conservancy. The total amount of debt involved in nature swaps is approximately $100 million, or about 0.01% of Third World debt. The majority of these funds have been used in Costa Rica.

The first debt swap provided many lessons on the potential problems of this method (Dogsé and von Droste 1990). This debt swap involved Conservation International buying $650,000 of Bolivian debt for $100,000. The debt was given to the government of Bolivia in exchange for the government establishing a local currency fund worth $250,000. $150,000 of this fund was actually donated by the U.S. Agency for International Development using excess pesos accumulated by the U.S. government. The fund was to be used by the Bolivian government for managing and protecting the Beni Biosphere Reserve. The government agreed to give maximum protection to the Reserve and to establish three large buffer zones around the park, totaling over 1,000,000 ha. This first debt swap proved to be extremely complex and difficult to negotiate, due to the novelty of the negotiations, the weak financial condition of the Bolivian government, and a difficult political climate. In particular, there was a perception within Bolivia that the country would lose sovereignty over the lands governed under the agreement.

The Bolivian agreement was supposed to begin in 1987 but was not actually implemented until 1989, and has proved problematic from the start for reasons unrelated to the debt swap itself. First, legal and illegal harvesting of mahogany trees in the buffer zone has increased dramatically, with little concern for sustainable harvesting or reforestation (Collett 1989). Second, Indian groups numbering 25,000 people that live in and around the reserve have claimed 200,000 ha of the forest for themselves, some of which is in areas where logging is occurring. The Indians were not party to the debt swap agreement and have opposed it by staging a protest march to La Paz (Pearce 1990; Dudley 1992). Third, illegal coca plantations for cocaine production and drug trafficking have increased in the area. Not only does the growing and processing of coca cause environmental damage to forests and watersheds, but it creates instability in the area, which makes local management difficult. A more positive development is the fact that considerable resources from both the Bolivian government and Conservation International have been focused on developing management, training, infrastructure, and protection for the reserve.

Costa Rica has taken the lead in debt swaps (Dogsé and von Droste 1990; WRI 1992). International conservation organizations have spent $12 million to purchase more than $79 million of foreign debt, which has then been exchanged for $42 million in Costa Rican bonds for use in conservation activities at La Amistad Biosphere Reserve, Braulio Carillo National Park, Corcovado National Park, Guanacaste National Park, Tortuguero National Park, and Monteverde Cloud Forest, a private reserve. The interest on the bonds has been used to establish a fund administered by the Costa Rican government and several local NGOs, including the Costa Rican Parks Foundation. This money has been used in acquiring land for reserves, developing and instituting a plan for managing these reserves, and establishing sustainable development projects. Even in Costa Rica, which has signed the majority of the existing debt swap agreements, only 5% of its international debt has been released by these agreements.

Debt-for-nature swaps have been undertaken primarily with Latin American countries that have large debts. However, innovative conservation efforts are also needed in African and South Asian countries, which are even poorer and have less international debt. Also, with the breakup of the Soviet Union, it has become apparent that its constituent countries have massive environmental problems that will require international expertise and financial support to solve (Box 35). Debt-for-nature swap agreements might be one mechanism for accomplishing this.

BOX 35 THE DEATH OF THE ARAL SEA: UNFORESEEN DISASTER IN A PLANNED ECONOMY

The case of the Aral Sea is a prime example of the massive scale of the damage that was done to the environment by poorly planned exploitation of natural resources in the former Soviet Union (Feschbach and Friendly 1992). The Aral Sea is a large, river-fed lake bordering the newly independent republics of Kazakhstan and Uzbekistan, in what was formerly Soviet Central Asia. Once one of the largest lakes in the world, the Aral has been drained to the extent that it has decreased in area by 29,000 square kilometers in the past three decades (Ellis 1990). The waters of the Aral's feeder rivers have been tapped for irrigation since the late 1930s with the goal of increasing agricultural productivity. As the volume of water in the lake decreased, many negative

The water is gone from more than 40% of the original area of the Aral Sea, leaving fishing boats stranded on the dry lakebed. (Photograph © David C. Turnley/Black Star.)

environmental, economic, and social consequences began to appear in the 1960s. First, the salinity of the lake's water rose as the inflow of fresh water from the two rivers feeding the lake, the Amu Darya and the Syr Darya, decreased. The 24 species of fish native to the lake died off, with some species going extinct while others survived in rivers and lakes elsewhere (Birstein; personal communication). The local fishing industry, dependent upon pike, perch, and other species, died with them (Ellis 1990). The Aral Sea and its associated rivers previously had four species of sturgeon, considered to be the most primitive living bony fish; biologists believe that studies of these species may contribute unique insights into evolution, physiology, and biogeography. Three of the four sturgeon species are now extinct in the wild, and the fourth is extremely endangered. The Aral ship sturgeon (*Acipinser nudiventris*), extinct in the wild, now survives only as six captive individuals in the Moscow Aquarium.

The receding waters of the Aral Sea left behind them an expanse of bare, dry lakebed too saline to support much vegetation. Dust and crystalline salt from the lakebed—an estimated 43 million tons each year—are blown into the surrounding countryside, causing a sharp increase in the occurrence of respiratory ailments, eye diseases, and throat cancer among the people of northern Uzbekistan (Ellis 1990). The salt carried by these dust storms settles in the soils of regions far from the Aral, altering their ecology and decreasing their agricultural productivity. Furthermore, regional climatic conditions have become more extreme since the 1960s: without the ameliorating influence of the lake, the winters are colder, the summers hotter. The quality of water in the area south of the lake has worsened as the Aral's water level has dropped, leaving local inhabitants susceptible to diseases transmitted by impure water.

The problem facing the new republics following the breakup of the Soviet Union is extremely difficult to resolve: how to undo the environmental damage inflicted by Soviet agricultural policies that mandated the diversion of the Amu Darya and the Syr Darya in order to increase the production of cash crops (Ellis 1990). Official recognition of the problem occurred only in the later years of Gorbachev's governance, over two decades after it first became apparent; methods of stabilizing or partially restoring the Aral were under debate prior to the Soviet Union's dissolution. Options for even partial restoration of the lake are now further complicated by international politics, since the republics that benefit from the irrigation system that drains the Aral may not be willing to lower their agricultural productivity on behalf of their neighbors. Yet if the Amu Darya and the Syr Darya continue to be exploited at the current rate, the Aral Sea will shrink even further, resulting in a body of water roughly one-tenth its present size with a salinity four times that of ocean water. Such a scenario would be disastrous, since the environmental problems caused by dust, salt, and water loss would be substantially increased. The overuse of the Aral has already devastated the surrounding region; the conditions that could develop by the turn of the century, if the current rate of desiccation continues, would be catastrophic.

While debt-for-nature swaps have great potential advantages, they present a number of potential limitations to both the donor and the recipient (Patterson 1990; Gullison and Losos 1993). Debt swaps will not necessarily change the underlying problems that led to environmental degradation in the first place. Farmers will still need land to farm, timber industries will continue to log, and cash-hungry Third

World countries will still have a motivation to exploit the environment for profit. Also, spending money on conservation programs might divert money away from other necessary domestic programs, such as medical care and schools. In addition, protection of nature reserves can be economically burdensome to financially strapped governments. If the overall government budget is increased to pay for conservation activities, it might contribute to inflation. There is also the public perception that land is being "sold" to foreign concerns. If the public feels that their government is giving up control of the country to foreign governments or organizations, the people will be more likely to ignore park regulations and encroach on protected lands.

Despite these concerns, debt-for-nature swaps appear to be one of the most innovative new mechanisms for encouraging conservation activities in the developing world. Now that these swaps appear to be feasible in countries like Costa Rica, other countries and organizations are becoming encouraged to try them. Debt swaps are now being incorporated into major foreign assistance programs, such as the Enterprise for the Americas, and are influencing the establishment of new funding mechanisms, such as the Global Environment Facility.

Summary

1. Conservation groups and governments in the developed countries are increasing funding to protect biological diversity in developing tropical countries. United States institutions increased their funding for the tropics by 68% between 1987 and 1989, with private foundations showing the biggest increase. U.S. funding of biodiversity activities has been concentrated in Latin America. While the increased levels of funding are welcome, the amount of money is still inadequate to deal with the loss of biological diversity that is taking place. The Global Environment Facility may be an important new source of funds.

2. International development banks, including the World Bank, have often funded massive projects that cause widespread environmental damage. The most notable such projects in Brazil have involved the construction of roads and mining operations in remote areas and have led to widespread deforestation. The World Bank is now attempting to be more environmentally responsible in its lending policies

3. Massive environmental damage has occurred in the planned economies of the former Soviet Union, a prime example being the destruction of the Aral Sea by poorly conceived irrigation projects.

4. An innovative new approach for preserving biodiversity involves debt-for-nature swaps in which foreign debt obligations of a government are traded for increased conservation funding. Already $100 million of foreign debt has been exchanged in swaps. Even though this amount is impressive, it is only about 0.01% of Third World debt.

Suggested Readings

Abramovitz, J. N. 1991. *Investing in Biological Diversity: U.S. Research and Conservation Efforts in Developing Countries.* WRI, Washington, DC. Unique presentation of biodiversity funding, with a list of projects, research institutions, and sources of funding.

Anderson, A. B. (ed.). 1990. *Alternatives to Deforestation.* Columbia University Press, Irvington, NY. The factors in Amazonia's deforestation are reviewed.

Cavalcanti, C. 1991. Government policy and ecological concerns: Some lessons from the Brazilian experience. *In* R. Costanza (ed.), *Ecological Economics: The Science and Management of Sustainability*, pp. 474–485. Columbia University Press, New York. The long history of environmental destruction in the Amazon.

Dogsé, P. and B. von Droste. 1990. *Debt-For-Nature Exchanges and Biosphere Reserves.* UNESCO, Paris. Excellent summary of debt-for-nature swaps, including case histories.

Fearnside, P. M. 1987. Deforestation and international economic development projects in Brazilian Amazonia. *Conservation Biology* 1: 214–221. An angry but well-presented critique of major development projects in the Amazon.

Goodland, R. J. A. 1992. Environmental priorities for financing institutions. *Environmental Conservation.* 19: 9–22. The new World Bank environmental policy is explained.

Gullison, R. E. and E. C. Losos. 1993. The role of foreign debt in deforestation in Latin America. *Conservation Biology* 7: 140–147. Linkages among tropical deforestation, poverty, and foreign debt are examined.

McNeely, J. A. 1987. How dams and wildlife can coexist: Natural habitats, agriculture, and major water resource development projects in tropical Asia. *Conservation Biology* 1: 228–238. Examples from Asia show how habitat protection can benefit nearby development projects.

Rich, B. 1990. Multilateral development banks and tropical deforestation. *In* S. Head and R. Heinzman (eds.), *Lessons of the Rainforest.* Sierra Club Books, San Francisco. A critical look at international banking policies.

Shaffer, M. L. and K. A. Saterson. 1987. The biological diversity program of the U.S. Agency for International Development. *Conservation Biology* 1: 280–283. The protection of biological diversity becomes part of U.S. foreign policy.

CHAPTER 22

An Agenda for the Future

There is no mystery as to why biological diversity is showing a rapid, worldwide decline. Biological communities are destroyed and species driven to extinction either because someone is making money by doing it, or because people are simply using up resources in order to survive. The money may be made by people who are indigenous to the region, landless people from outside the region, banks, local business interests, large businesses in urban centers, business interests in other countries, and governments. In order for conservation policies to work, people at all levels of the society must see that it is in their own interest to work for conservation. If conservationists can demonstrate that the protection of biological diversity has more value than the destruction of biological diversity, then people and their governments might be more willing to preserve biological diversity.

There is a consensus among conservation biologists that there are a number of major problems involved in preserving biological diversity, and that certain changes in policies and practices are needed.

Problem: Protecting biological diversity is difficult when most of the world's species remain undescribed by scientists.

Response: More scientists need to be trained to identify and classify species, and funding should be increased in this area (Raven and Wilson 1992). Enthusiastic nonscientists can often play an important role in this effort once they are given some training and guidance by scientists.

Problem: Many conservation issues are global in scope, involving many countries.

Response: Countries are increasingly willing to discuss international conservation issues, as shown by the 1992 Earth Summit, as well as to sign treaties such as the recent Biodiversity Convention and CITES. International conservation efforts are expanding, but these positive efforts are countered by the political instability and war affecting many countries in the world. One positive development is the trend toward establishing binational "Peace Parks" that straddle borders; such parks are good for wildlife as well as serving as national security buffer zones.

Problem: Developed countries of the world place a greater emphasis on the preservation of biological diversity than the poorer Third World countries that have the most biological diversity.

Response: Developed countries and international conservation organizations should be willing to offer financial support to developing countries that establish and maintain national parks and other protected areas (Katzman and Cale 1990). This seems fair and reasonable when the developed countries have the funds to pay for these parks and are able to utilize the protected biological resources in their agriculture, industry, research programs, and educational systems.

Problem: Governments in the developing world are forced to exploit their natural resources to pay off foreign debt and to pay for social programs.

Response: Concerned citizens in the developed world should pressure their governments to reduce or forgive foreign debt payments. Reducing debt payments will reduce the pressure on developing countries to exploit their resources in a nonsustainable manner. Also, debt forgiveness can be directly tied to conservation efforts, as in debt-for-nature swaps (Dogsé and von Droste 1990).

Problem: Economic analyses often paint a falsely encouraging picture of development projects that are environmentally damaging.

Response: New types of cost–benefit analyses must be used that include environmental and human costs, such as the costs of soil erosion, water pollution, loss of natural products, loss of traditional knowledge with potential economic value, loss of tourist potential, loss of species of possible future value, and loss of home sites (Daly and Cobb 1989). In particular, the effects of large projects on indigenous people have typically been ignored in economic analyses and should be given more attention. Environmental impact analyses also need to include comparative studies of similar projects completed elsewhere, and the probabilities and costs of possible worst-case scenarios.

Problem: Ecosystem services do not receive the recognition they deserve in economic activities.

Response: Economic activities should be linked with the maintenance of ecosystem services through fees, penalties, and land acquisition. The "polluter pays" principle must be adopted, in which polluters pay for cleaning up the environmental damage their activities have caused. Factories and human settlements must become morally and financially responsible for the pollution they cause. A step in this direction is the recent initiative by electric power companies to plant trees in the tropics to absorb the excess carbon dioxide that their factories produce. As further examples, hydroelectric projects and dam projects need to be linked to watershed protection and acquisition. Coastal fishing industries must become involved in the protection of coastal wetlands, mangroves, and estuaries that are important feeding grounds and nurseries for commercial species.

Problem: Much of the destruction of the world's biological diversity is caused by people who are desperately poor and are simply trying to survive.

Response: Conservation biologists need to assist local people to organize and develop sustainable economic activities. Because the funds of conservationists are limited and problems are urgent, such conservation efforts should be concentrated on people who are affecting areas of major biological importance. Conservation biologists may be most effective when they work in cooperation with charitable and humanitarian organizations in these regions.

Problem: Decisions on land conversion and the establishment of protected areas are often made by central governments with little input from people in the region being affected. At the extreme, local people may only find out about a new national park when they are ordered off their traditional lands. Consequently, local people sometimes feel alienated from conservation projects and do not support them.

Response: Local people have to believe that they will benefit from the project and that their involvement is important. To achieve this goal, environmental impact statements and other project information should be publicly available to encourage open discussion. Local people should be provided with whatever assistance they may need in order to understand and evaluate the implications of the project being presented to them (Clay 1991).

Problem: Revenues, business activities, and scientific research associated with national parks do not directly benefit surrounding communities.

Response: Whenever possible, local people should be trained and employed in parks as a way of utilizing local knowledge and providing local income (Lewis et al. 1990). Also, a portion of park revenues can be used to fund community projects such as schools, clinics, roads, and community businesses—an infrastructure that benefits the whole village or region. Biologists working in national parks should periodically explain the purpose and results of their work to nearby communities and school groups, and listen to what the local people have to say.

Problem: National parks and conservation areas in developing countries have inadequate budgets to pay for conservation activities. Revenues that they collect are often returned to government treasuries.

Response: Funds for park management can often be raised from foreign tourists and scientists by charging them for admission, lodging, or meals. Making sure that these revenues and profits remain at the park and in the surrounding area is important. One possibility is an "international rate" charged to foreign visitors, with a lower rate charged to local citizens, so that local people can afford to visit the park. When the number of foreign visitors becomes too high for the capacity of the park, a simple solution is to keep raising the fees so that park revenue remains constant but visitor pressure is reduced. At Chichén Itzá in Mexico, local residents visit the park on Sunday when admission is free, while foreign tourists predominate during the week when admission is relatively expensive. Zoos and conservation organizations in the developed world can make direct contributions to conservation efforts in developing countries, strengthening the most significant programs.

Problem: People cut down tropical forests and graze grasslands to establish title to the land, even when lands are not suitable for agriculture.

Response: Change the laws so that people can obtain titles to harvest trees and use grassland on a selective basis as long as the health of the biological community is maintained.

Problem: Timber companies that lease forests and ranchers that rent grassland from the government often damage the land and reduce its productive capacity in pursuit of short-term profits; they do not own the land and typically have only a short-term lease, so have no interest in the long-range productivity of the land.

Response: Grant long-term leases that are contingent on the productivity and biological diversity of the land being maintained.

Problem: In some countries, governments are inefficient, slow-moving, and bound by excessive regulation, and consequently are ineffective at protecting land and biological communities.

Response. In these countries, nongovernment conservation organizations are often the most effective agents for dealing with conservation issues. Local conservation organizations should be encouraged and supported politically, scientifically, and financially. New foundations should be started by individuals, organizations, and businesses to support conservation efforts financially.

Problem: Many businesses, banks, and governments are uninterested in and unresponsive to conservation issues.

Response: Lobbying efforts may be effective at changing the policies of institutions that want to avoid bad publicity. Petitions, rallies, letter-writing campaigns, press releases, and economic boycotts all have their place when reasonable requests for change are ignored. The key point is that conservationists have to be realistic and have a long-term perspective. But unrealistic demands for immediate change sometimes also have value in eliciting realistic counterproposals from previously unresponsive institutions. In many situations, radical environmental groups dominate media attention with dramatic, publicity-grabbing actions, while mainstream conservation organizations follow behind to negotiate a compromise.

The Role of Conservation Biologists in Achieving the Agenda

Conservation biology as a discipline is different from many other bodies of knowledge in that it has developed an active role in the preservation of biological diversity in all forms: the preservation of species, genetic variability within species, biological communities, and ecosystem functions. Members of all of the diverse disciplines that contribute to conservation biology share the common goal of protecting biological diversity. Their different perspectives and wide range of experiences foster an inclusive atmosphere that permits people with diverse interests to find a common ground (Norton 1991).

The ideas and theories of conservation biology are increasingly being incorporated into decisions about park management and species protection. At the same time, botanical gardens, museums, nature centers, zoos, national parks, and aquaria are reorienting their programs to meet the challenges of protecting biological diversity. The need for large parks and the need to protect large populations of endangered species are two particular topics that have received widespread attention in both academic and popular literature. The vulnerability of small populations to local extinction, even when they are carefully protected and managed, and the alarming rates of species extinction and destruction of unique biological communities world-wide have also been highly publicized. As a result of this publicity, the need to protect biological diversity is entering the political debate and has been targeted as a priority for government programs.

One of the most serious challenges facing conservation biology is reconciling the needs of people and the need to preserve biological diversity (MacNeil 1989; McKinnon et al. 1992). How can poor people, particularly in the developing world, be convinced to respect nature reserves and biological diversity when they are desperate to obtain the food, fuel, and other natural products that they need for their daily survival? Park managers need to find compromises, such as the concept of biosphere reserves, that allow people to obtain the natural resources that they need to support their families in a way that does not damage the natural communities of the park. At national and international levels, the world's resources must be distributed more fairly to end the inequities that exist today. Effective programs must be established to stabilize the world's human population. At the same time, the destruction of natural resources by industries must be controlled, so that the short-term quest for profits does not lead to a long-term ecological catastrophe (Goodland 1992). Management strategies to preserve biological diversity also need to be developed for the 95% of the terrestrial environment that will remain outside of protected areas, as well for the vast, largely unexplored marine environment.

If these challenges are to be met successfully, conservation biologists must take on several active roles. First, they must become more effective as *educators*, in the public forum as well as in the classroom. Conservation biologists need to educate as broad a range of people as possible about the problems that stem from loss of biological diversity. While conservation biologists often teach college students and write technical papers addressing these issues, they reach only a limited audience in this way; conservation biologists need to reach a wider range of people through speaking in villages, schools, parks, and neighborhood gatherings. Also, the themes of conservation need to be more widely incorporated into public discussions. Conservation biologists must spend a greater percentage of their time writing pieces for newspapers and magazines, as well as appearing on radio, television, and other mass media (Crystal 1989). Remember that only a few thousand people read most scientific papers. In contrast, millions of adults saw the movies *Medicine Man* and *The Emerald Forest*, while tens of millions of children have watched *Ferngully: The Last Rainforest* and *Seabert the Seal*, all of which have very powerful conservation themes.

Second, conservation biologists must become *politically active.* Involvement in the political process allows conservation biologists to influence the passage of new laws to support the preservation of biological diversity, or, alternatively, to argue against legislation that would prove harmful to species or ecosystems (Caldwell 1989). Though the political process can be time-consuming and tedious, it is often the only way to accomplish major conservation goals, such as acquiring new land for reserves or preventing overexploitation of unique habitats. Conservation biologists need to master the language and methods of the legal process and form effective alliances with environmental lawyers, citizen groups, and politicians.

Third, conservation biologists need to become *organizers* within the biological community. Many professional biologists in universities, museums, high schools, and government agencies tend to concentrate their energies on the specialized needs of their own professional niche. These people may not realize that the world's biological diversity is under imminent threat of destruction, and that their contributions are urgently needed to save it. By stimulating interest in this problem among their colleagues, conservation biologists can increase the ranks of trained professional advocates fighting the destruction of natural resources; these professional biologists may also find their involvement to be personally and professionally beneficial, as their new interests may result in heightened scientific creativity and more inspired teaching.

Fourth, conservation biologists need to become *motivators,* convincing a range of people to support conservation efforts. At a local level, conservation programs have to be created and presented in ways that provide incentives for local people to support them. This approach may sometimes mean empowering local communities to develop their own conservation programs; at the least, such programs must provide practical benefits to the community. Discussions, educational efforts, and publicity need to be a major part of any such program. Careful attention must be devoted to convincing business leaders and politicians to support conservation efforts. Many of these people will support conservation efforts when they are presented in the right way; sometimes conservation is perceived to have good publicity value, or supporting it is perceived to be better than a confrontation that may otherwise result. National leaders may be among the most difficult people to convince, since they must respond to a diversity of interests; however, whether it is by reason, pressure, demonstrations, or petitions, once converted to the conservation perspective these leaders may be in a position to make a major contribution to the cause.

Finally and most importantly, conservation biologists need to become effective *managers* and *practitioners* of conservation projects. They must be willing to walk on the ground to find out what is really happening, to get dirty, to talk with local people, to knock on doors, and to take risks. Conservation biologists must learn everything they can about the species and communities that they are trying to protect, and then make that knowledge available to others. If conservation biologists are willing to put their ideas into practice, and to work with park managers, land-use planners, politicians, and local people, then progress will follow. Getting the right mixture of models, new theories, innovative approaches, and practical examples will be the key to the success of the discipline. Once this balance is found, conservation biologists working with an energized citizenry will be in a position to protect the world's biological diversity during this unprecedented era of change.

Suggested Readings

Allen, W. 1988. Biocultural restoration of a tropical forest: Architects of Costa Rica's emerging Guanacaste National Park plan to make it an integral part of local culture. *BioScience* 38: 156–161. Outstanding account of the restoration of a tropical ecosystem that highlights the need for public education and personal dedication.

Conservation Biology, published by Blackwell Scientific Publications, Cambridge, MA. The cutting-edge journal in the field. Browsing back issues is well worthwhile. Excellent articles can also be found in *BioScience* and *Biological Conservation*.

Giono, J. 1989. *The Man Who Planted Trees*. Collins Dove, Melbourne, Australia. Simple, beautiful story about how one person can make a change in society.

Gore, A. 1992. *Earth in the Balance*. The U.S. Vice President passionately and intelligently argues the need to give greater effort to environmental protection.

Grumbine, R. E. 1993. *Ghost Bears: Exploring the Biodiversity Crisis*. Island Press, Washington, D.C. This popular book weaves together elements of conservation biology, law, policy, and activism.

McNeely, J. A., K. Miller, W. Reid, R. Mittermeier and T. Werner. 1990. *Conserving the World's Biological Diversity*. IUCN, Gland, Switzerland. Strategies and priorities for conservation.

Miller, K. and L. Tangley. 1991. *Trees of Life: Saving Tropical Forests and Their Biological Wealth*. Beacon Press, Boston. Excellent summary of ideas and source of information.

Mishra, H. R. 1984. A delicate balance: Tigers, rhinoceros, tourists, and park management vs. the needs of local people in Royal Chitwan National Park, Nepal. *In* J. A. McNeely and K. R. Miller (eds.), *National Parks, Conservation, and Development: The Role of Protected Areas in Sustaining Society*, pp. 197–205. Smithsonian Institution Press, Washington, D.C. A complex variety of interests must be satisfied in order to protect biological diversity.

Naess, A. 1989. *Ecology, Community, and Lifestyle.* Cambridge University Press, Cambridge. A leader of the Deep Ecology movement argues for a personal commitment to political activism and changes in lifestyle.

Rolston, H., III. 1988. *Environmental Ethics: Values In and Duties To the Natural World.* Temple University Press, Philadelphia. This and other writings by Rolston illuminate the ethical arguments for valuing and protecting all species and biological communities.

Wilson, E. O. 1992. *The Diversity of Life.* The Belknap Press of Harvard University Press, Cambridge, MA. The best popular book on the biodiversity crisis, and eminently suitable to recommend or give to interested people.

WRI/IUCN/UNEP. 1992. *Global Biodiversity Strategy: Guidelines for Action to Save, Study and Use Earth's Biotic Wealth Sustainably and Equitably.* World Resources Institute, Washington, D.C. Current views on needed policy changes, with a long list of key people in the field.

APPENDIX

Selected Environmental Organizations and Sources of Information

The best single reference on conservation activities is the *Conservation Directory*, updated each year by the National Wildlife Federation, 1400 Sixteenth Street N.W., Washington, D.C. 20036. This directory lists thousands of local, national, and international conservation organizations, conservation publications, and leaders in the field of conservation. Another publication of interest is *The New Complete Guide to Environmental Careers* (1993), published by Island Press, 1718 Connecticut Avenue N.W., Washington, D.C. 20009.

The following lists some major organizations and resources.

Center for Marine Conservation
1725 De Sales St. N.W., Suite 500
Washington, D.C. 20036 U.S.A.
  Focus on marine wildlife and ocean and coastal habitats.

Center for Plant Conservation and Missouri Botanical Garden
P.O. Box 299
St. Louis, MO 63166 U.S.A.
  Major centers for worldwide plant conservation activities.

CITES Secretariat, UNEP
15 Chemin des Anemones
Case Postale 356
1219 Chatelaine
Geneva, Switzerland
  The official body that regulates international trade in endangered species.

Conservation International
1015 18th St. N.W., Suite 100
Washington, D.C. 20036 U.S.A.
  Active in conservation efforts and working for sustainable development, particularly in Latin America. Strong emphasis on working with local people.

Earthwatch
P.O. Box 403N
Mt. Auburn St.
Watertown, MA 02272 U.S.A.
  Clearinghouse for international projects in which volunteers can work with scientists.

Environmental Defense Fund
257 Park Avenue South
New York, N.Y. 10010 U.S.A.
  Leading environmental organization; involved in scientific, legal, and economic issues.

Friends of the Earth
218 "D" St. S.E.
Washington, D.C. 20003 U.S.A.
  International environmental organization working to improve public policy.

Greenpeace U.S.A., Inc.
1436 "U" St. N.W.
Washington, D.C. 20009 U.S.A.
  An activist organization, known for grass-roots efforts and dramatic protests against environmental damage.

**Institute for Economic Botany and
New York Botanical Gardens**
Bronx, N.Y. 10458 U.S.A.
  Active in research and conservation programs
  involving plants that are useful to people.

International Council for Bird Preservation
32 Cambridge Road, Girton
Cambridge CB3 0PJ United Kingdom
  Determines conservation status and priorities
  for birds throughout the world.

International Council of Environmental Law
Adenaueralle 214
D-5300 Bonn 1, Germany
  An international center for environmental
  law.

**International Union for the Conservation of
Nature and Natural Resources (IUCN)**
Avenue de Mont Blanc
CH-1196 Gland, Switzerland
  Also known as the World Conservation
  Union. This is the premier coordinating body
  for international conservation efforts. Pro-
  duces directories of specialists who are
  knowledgable about captive breeding pro-
  grams and other aspects of conservation.

International Union of Biological Sciences
51 Boulevard de Montmorency
75016 Paris, France
  Coordinates international scientific research;
  publishes on major scientific topics.

National Audubon Society
950 Third Avenue
New York. N.Y. 10022 U.S.A.
  Their extensive program includes wildlife
  conservation, public education, research, and
  political lobbying.

National Wildlife Federation
1400 Sixteenth St. N.W.
Washington, D.C. 20036 U.S.A.
  Advocates for wildlife conservation. Publishes
  the *Conservation Directory*, as well as such
  outstanding children's publications as *Ranger
  Rick* and *Your Big Backyard*.

The Nature Conservancy
1815 North Lynn St.
Arlington, VA 22209 U.S.A.
  Emphasis on land preservation. Maintains extensive records on rare species distribution in
  the Americas, particularly North America.

Rain Forest Action Network
301 Broadway, Suite "A"
San Francisco, CA 94133 U.S.A.
  Works actively for rain forest conservation.

Royal Botanical Garden, Kew
Richmond
Surrey TW9 3AE, United Kingdom
  The famous "Kew Gardens" are home to a
  leading botanical research institute.

Sierra Club
730 Polk St.
San Francisco, CA 94109 U.S.A.
  Leading advocate for the preservation of wilderness and open space.

**Smithsonian Institution
and National Zoological Park**
1000 Jefferson Drive S.W.
Washington, D.C. 20560 U.S.A.
  The National Zoo and the nearby U.S. National Museum of Natural History represent a
  vast resource of literature, biological materials, and skilled people.

Society for Conservation Biology
c/o Blackwell Scientific Publications, Inc.
238 Main Street
Cambridge, MA 02142 U.S.A.
  Leading scientific society for the field. Develops and publicizes new ideas and scientific
  results through the journal *Conservation Biology*.

United Nations Development Program (UNDP)
U.N. Plaza
New York, N.Y. 10017 U.S.A.
  Funds and coordinates international economic
  development activities, particularly those that
  use natural resources in a responsible way.

United Nations Environment Program (UNEP)
P.O. Box 30552
Nairobi, Kenya
or
1899 "F" St. N.W.
Washington, D.C. 20006 U.S.A.
  International program of research and management relating to major environmental
  problems.

United States Fish and Wildlife Service Washington, D.C. 20240 U.S.A.

The leading U.S. government agency in the conservation of endangered species, with a vast research and management network. Major activities also take place within other federal government units, such as the National Marine Fisheries Service and the U.S. Forest Service. The Agency for International Development is active in many developing nations. Individual state governments have comparable units, with National Heritage programs being especially relevant. The *Conservation Directory* helps to show how these units are organized.

Wildlife Conservation International and New York Zoological Society
Bronx Zoo
185th St. & Southern Blvd.
Bronx, N.Y. 10460 U.S.A.

Leaders in wildlife conservation and research.

World Bank
1818 "H" St. N.W.
Washington, D.C. 20433 U.S.A.

A multinational bank involved in economic development; increasingly concerned with environmental issues.

World Conservation Monitoring Centre
219 Huntingdon Road
Cambridge CB3 0DL, United Kingdom

Monitors global wildlife trade, the status of endangered species, natural resource use, and protected areas.

World Resources Institute (WRI)
1709 New York Ave. N.W.
Washington, D.C. 20006 U.S.A.

Research center producing excellent original position papers on environmental, conservation, and development topics. An extensive listing of projects, funders, and institutions can be found in the WRI publication *Investing in Biological Diversity* by J. Abramovitz (1991).

World Wildlife Fund (WWF)
1250 24th St. N.W.
Washington, D.C. 20037 U.S.A.

Also known as the Worldwide Fund for Nature. Major conservation organization, with branches throughout the world. Active in both research and in the management of national parks.

Xerces Society
10 Ash St. S.W.
Portland, OR 97204 U.S.A.

Focuses on the conservation of insects and other invertebrates.

Zoological Society of London
Regents Park
London NW1 4RY, United Kingdom

Center for worldwide activities to preserve nature.

Bibliography

Abele, L. G. and E. F. Connor. 1979. Application of island biogeography theory to refuge design: Making the right decision for the wrong reasons. *In* M. Linn (ed.), *Proceedings of the First Conference on Scientific Research in the National Parks*, New Orleans, Louisiana, November 9–12, 1976. Volume I, pp. 89–94. National Parks Service, U.S. Department of the Interior, Washington, D.C.

Abramovitz, J. N. 1991. *Investing in Biological Diversity: U.S. Research and Conservation Efforts in Developing Countries*. World Resources Institute, Washington, D.C.

Ackerman, D. 1992. Last refuge of the monk seal. *National Geographic* 181(January): 128–144.

Adams, D. and M. Carwardine. 1990. *Last Chance to See*. Harmony Books, New York.

Agency for International Development (AID). 1985. *U.S. Strategy on the Conservation of Biological Diversity: An Interagency Task Force Report to Congress*. U.S. Government Printing Office, Washington, D.C.

Akçakaya, H. R. 1990. Bald ibis *Geronticus eremita* population in Turkey: An evaluation of the captive breeding project for reintroduction. *Biological Conservation* 51: 225–237.

Al-Thukair, A. A. and S. Golubic. 1991. Five new *Hyella* species from the Arabian Gulf. *Algological Studies* 64: 167–197.

Alcorn, J. B. 1984. Development policy, forests, and peasant farms: Reflections on Huastec-managed forests' contributions to commercial production and resource conservation. *Economic Botany* 38: 389–406.

Alcorn, J. B. 1991. Ethics, economies, and conservation. *In* M. L. Oldfield and J. B. Alcorn (eds.), *Biodiversity: Culture, Conservation and Ecodevelopment*, pp. 317–349. Westview Press, Boulder, CO.

Alcorn, J. B. 1993. Indigenous peoples and conservation. *Conservation Biology* 7: 424–426.

Allen, W. H. 1988. Biocultural restoration of a tropical forest: Architects of Costa Rica's emerging Guanacaste National Park plan to make it an integral part of local culture. *BioScience* 38: 156–161.

Allendorf, F. W. and R. F. Leary. 1986. Heterozygosity and fitness in natural populations of animals. *In* M. E. Soulé (ed.), *Conservation Biology: The Science of Scarcity and Diversity*, pp. 57–76. Sinauer Associates, Sunderland, MA.

Allendorf, F. W. and C. Servheen. 1986. Genetics and the conservation of grizzly bears. *Trends in Ecology and Evolution* 1: 88–89.

Altaba, C. R. 1990. The last known population of the freshwater mussel *Margaritifera auricularia* (Bivalvia, Unionoida): A conservation priority. *Biological Conservation* 52: 271–286.

Altieri, M. A. and M. K. Anderson. 1992. Peasant farming systems, agricultural modernization, and the conservation of crop genetic resources in Latin America. *In* P. L. Fiedler and S. K. Jain (eds.), *Conservation Biology: The Theory and Practice of Nature Conservation, Preservation and Management* pp. 49–64. Chapman and Hall, New York.

Altieri, M. A. and L. C. Merrick. 1988. Agroecology and in situ conservation of native crop diversity in the Third World. *In* E. O. Wilson and F. M. Peter (eds.), *Biodiversity*, pp. 361–369. National Academy Press, Washington, D.C.

Altieri, M. A., M. K. Anderson and L. C. Merrick. 1987. Peasant agriculture and the conservation of crop and wild plant resources. *Conservation Biology* 1: 49–58.

Alverson, W. S., D. M. Waller and S. L. Solheim. 1988. Forests too deer: Edge effects in northern Wisconsin. *Conservation Biology* 2: 348–358.

Alyanak, L. 1992. The Road From Rio. *The New Road* 24: 3–5.

Ames, M. H. 1991. Saving some cetaceans may require breeding in captivity. *BioScience* 41: 746–749.

Anderson, A. B. 1990. Deforestation in Amazonia: Dynamics, causes, and alternatives. *In* A. B. Anderson (ed.), *Alternatives to Deforestation*, pp. 3–23. Columbia University Press, New York.

Anderson, A. B. (ed.). 1990. *Alternatives to Deforestation*. Columbia University Press, New York.

Anderson, D. and R. H. Grove. 1987. *Conservation in Africa: Peoples, Policies and Practice*. Cambridge University Press, Cambridge.

Anderson, R. M. 1982. Epidemiology. *In* F. E. G. Cox (ed.), *Modern Parasitology*, pp. 204–251. Blackwell Scientific Publications, Oxford.

Anderson, R. M. and R. M. May. 1980. Infectious diseases and population cycles of forest insects. *Science* 210: 658–661.

Anonymous. 1986. How to have your iguanas and eat 'em too. *Discover* 7:7.

Anonymous. 1989. *1988 Annual Report on the Status of California's State Listed, Threatened, and Endangered Plants and Animals*. State of California Department of Fish and Game.

Anonymous. 1990. On to the past. *Scientific American* 262: 18.

Antonovics, J. 1976. The nature of limits to natural selection. *Annals of the Missouri Botanical Garden* 63: 224–247.

515

Antonovics, J., A. D. Bradshaw and R. G. Turner. 1971. Heavy metal tolerance in plants. *Advances in Ecological Research* 7: 1–85.

Arita, H. T., J. G. Robinson and K. H. Redford. 1990. Rarity in Neotropical forest mammals and its ecological correlates. *Conservation Biology* 4: 181–192.

Ashley, M. V., D. J. Melnick and D. Western. 1990. Conservation genetics of the black rhinoceros (*Diceros bicornis*). I. Evidence from the mitochondrial DNA of three populations. *Conservation Biology* 4: 71–77.

Ashton, P. S. 1984. Botanic gardens and experimental grounds. *In* V. H. Heywood and S. M. Moore (eds.), *Current Concepts in Plant Taxonomy*, pp. 39–48. Academic Press, London.

Ashton, P. S. 1988. Conservation of biological diversity in botanical gardens. *In* E. O. Wilson and F. M. Peter (eds.), *Biodiversity*, pp. 269–278. National Academy Press, Washington, D.C.

Ayres, J. M., R. E. Bodmer and R. A. Mittermeier. 1991. Financial considerations of reserve design in countries with high primate density. *Conservation Biology* 5: 109–114.

Babbel, G. R. and R. K. Selander. 1974. Genetic variability in edaphically restricted and widespread plant species. *Evolution* 28: 619–630.

Badiner, A. H. 1990. *Dharma Gaia: A Harvest of Essays in Buddhism and Ecology*. Parallax Press, Berkeley, CA.

Bailey, R. G. and H. C. Hogg. 1986. A world ecoregions map for resource reporting. *Environmental Conservation* 13: 195–202.

Baillie, S. R. 1991. Monitoring terrestrial breeding bird populations. *In* B. Goldsmith (ed.), *Monitoring for Conservation and Ecology*, pp. 112–132. Chapman and Hall, New York.

Balick, M. J. and R. Mendelsohn. 1992. Assessing the economic value of traditional medicines from tropical rain forests. *Conservation Biology* 6: 128–130.

Baltz, D. M. 1991. Introduced fishes in marine systems and inland seas. *Biological Conservation* 56: 151–177.

Barrett, S. C. H. and J. R. Kohn. 1991. Genetic and evolutionary consequences of small population size in plants: Implications for conservation. *In* D. A. Falk and K. E. Holsinger (eds.), *Genetics and Conservation of Rare Plants*, pp. 3–30. Oxford University Press, New York.

Barrie, L. A. and J. M. Hales. 1984. The spatial distributions of precipitation acidity and major ion wet deposition in North America during 1980. *Tellus* 368: 333–355.

Bartley, D., M. Bagley, G. Gall and B. Bentley. 1992. Use of linkage disequilibrium data to estimate effective size of hatchery and natural fish populations. *Conservation Biology* 6: 365–375.

Batisse, M. 1986. Developing and focusing the biosphere reserve concept. *Nature and Resources* 22: 1–10.

Bawa, K. S. 1990. Plant–pollinator interactions in tropical rain forests. *Annual Review of Ecology and Systematics* 21: 399–422.

Bawa, K. S. 1992. Mating systems, genetic differentiation and speciation in tropical rain forest plants. *Biotropica* (special issue) 24: 250–255.

Bazzaz, F. A. and E. D. Fajer. 1992. Plant life in a CO_2-rich world. *Scientific American* (January): 68–74.

Beamish, R. J., et al. 1975. Long-term acidification of a lake and the resulting effect on fishes. *Ambio* 4: 98–102.

Bean, M. J. 1983. *The Evolution of National Wildlife Law*, 2nd ed. Praeger Publishers, New York.

Beck, B. B., D. G. Kleiman, J. M. Dietz, I. Castro, C. Carvalho, A. Martins and B. Rettberg-Beck. 1991. Losses and reproduction in reintroduced golden lion tamarins *Leontopithecus rosalia*. *Dodo: Journal of the Jersey Wildlife Preserve Trust* 27: 50–61.

Bedigian, D. 1991. Genetic diversity of traditional sesame cultivars and cultural diversity in Sudan. *In* M. L. Oldfield and J. B. Alcorn (eds.), *Biodiversity: Culture, Conservation and Ecodevelopment*, pp. 25–36. Westview Press, Boulder, CO.

Beebee, T. J. C., et al. 1990. Decline of the Natterjack Toad *Bufo calamita* in Britain: Palaeoecological, documentary and experimental evidence for breeding site acidification. *Biological Conservation* 53: 1–20.

Beehler, B. 1985. Conservation of New Guinea rainforest birds. *In* J. W. Diamond and T. E. Lovejoy (eds.), *Conservation of Tropical Forest Birds*, pp. 233–247. Technical Publication No. 4, International Council for Bird Preservation, Cambridge, England.

Begley, S., L. Wilson, M. Hager and P. Annin. 1991. Return of the wolf. *Newsweek* 118: 44–50.

Begon, M., J. L. Harper and C. R. Townsend. 1990. *Ecology: Individuals, Populations, and Communities*, 2nd ed. Blackwell Scientific Publications, Oxford.

Benirschke, K. 1983. The impact of research on the propagation of endangered species in zoos. *In* C. M. Schonewald-Cox, S. M. Chambers, B. MacBryde and L. Thomas (eds.), *Genetics and Conservation: A Reference for Managing Wild Animal and Plant Populations*, pp. 402–413. Benjamin/Cummings, Menlo Park, CA.

Berger, J. 1990. Persistence of different-sized populations: An empirical assessment of rapid extinctions in bighorn sheep. *Conservation Biology* 4: 91–98.

Best, P. B. 1988. Right whales *Eubalaena australis* at Tristan da Cunha: A clue to the 'non-recovery' of depleted stocks? *Biological Conservation* 46: 23–51.

Bibby, C. J., M. J. Crosby, M. F. Heath, T. H. Johnson, T. H. Long, A. J. Sattersfield and S. J. Thirgood. 1992. *Putting Biodiversity on the Map: Global Priorities for Conservation*. International Council for Bird Preservation, Cambridge, England.

Bickford, W. E. 1991. Massachusetts' landscape-level land protection. *In* D. J. Decker, M. E. Krasny, G. R. Goff, C. R. Smith and D. W. Gross (eds.), *Challenges in the Con-*

servation of Biological Resources: A Practitioner's Guide, pp. 183–196. Westview Press, Boulder, CO.

Bierregaard, R. O., T. E. Lovejoy, V. Kapos, A. A. Dos Santos and R. W. Hutchings. 1992. The biological dynamics of tropical rainforest fragments. *BioScience* 42: 859–866.

Billington, H. L. 1991. Effect of population size on genetic variation in a dioecious conifer. *Conservation Biology* 5: 115–119.

Bishop, R. C. 1987. Economic values defined. In D. J. Decker and G. R. Goff (eds.), *Valuing Wildlife: Economic and Social Perspectives*, pp. 24–33. Westview Press, Boulder, CO.

Bleich, V. C., J. D. Wehausen and S. A. Holl. 1990. Desert-dwelling mountain sheep: Conservation implications of a naturally fragmented distribution. *Conservation Biology* 4: 383–389.

Blockhus, J. M., M. Dillenbeck, J. A. Sayer and P. Wegge (eds.). 1992. *Conserving Biological Diversity in Managed Tropical Forests*. IUCN, Gland, Switzerland.

Blondin, A. R. 1989. The owl complex. *Journal of Forestry* 87(8): 37–40.

Bond, W. 1983. On alpha diversity and the richness of the Cape flora: A study in southern Cape Fynbos. In F. J. Kruger, D. T. Mitchell and J. U. M. Jarvis (eds.), *Mediterranean-type Ecosystems: The Role of Nutrients*, pp. 337–356. Springer-Verlag, Berlin.

Bonnell, M. L. and R. K. Selander. 1974. Elephant seals: Genetic variation and near-extinction. *Science* 184: 908–909.

Booth, W. 1987. Combing the earth for cures to cancer, AIDS. *Science* 237: 969–970.

Booth, W. 1988. The otter-urchin-kelp scenario. *Science* 241: 157.

Bormann, F. H. 1976. An inseparable linkage: Conservation of natural ecosystems and conservation of fossil energy. *BioScience* 26:759.

Bormann, F. 1982. The effects of air pollution on the New England landscape. *Ambio* 11: 338–346.

Bormann, F. H. and S. R. Kellert (eds.). 1991. *Ecology, Economics, Ethics: The Broken Circle*. Yale University Press, New Haven, CT.

Botanic Gardens Conservation Secretariat. 1987. *The International Transfer Format for Botanic Gardens Plant Records*. Hunt Institute for Botanical Documentation, Carnegie Mellon University, Pittsburgh, PA.

Botkin, D. B. 1990. *Discordant Harmonies: A New Ecology for the Twenty-First Century*. Oxford University Press, New York.

Boyce, M. S. 1992. Population viability analysis. *Annual Review of Ecology and Systematics* 23: 481–506.

Boyce, M. S. and R. S. Miller. 1985. Ten-year periodicity in whooping crane census. *Auk* 105: 658–660.

Boyle, K. J. and R. C. Bishop. 1986. The economic valuation of endangered species of wildlife. *Transactions of the North American Wildlife and Natural Resources Conference* 51: 153–161.

Bradshaw, A. D. 1983. The reconstruction of ecosystems. *Journal of Applied Ecology* 20: 1–17.

Bradshaw, A. D. 1990. The reclamation of derelict land and the ecology of ecosystems. In W. R. Jordan III, M. E. Gilpin and J. D. Aber (eds.), *Restoration Ecology: A Synthetic Approach to Ecological Research*. pp. 53–74. Cambridge University Press, Cambridge.

Bradshaw, A. D. and M. J. Chadwick. 1980. *The Restoration of Land*. Blackwell Scientific Publications, Oxford.

Brass, D. (no date). The Nature Conservancy letter to investors. The Nature Conservancy.

Bratton, S. P. 1985. Effects of disturbance by visitors on two woodland orchid species in Great Smoky Mountains National Park, USA. *Biological Conservation* 31: 211–227.

Breman, H. 1992. Desertification control, the West African case: Prevention is better than cure. *Biotropica* (special issue) 24: 328–334.

Brenan, J. P. M. 1978. Some aspects of the phytogeography of tropical Africa. *Annals of the Missouri Botanical Garden*. 65: 437–478.

Briggs, J. C. 1974. *Marine Zoogeography*. McGraw-Hill, New York.

Briscoe, D. A., J. M. Malpica, A. Robertson, G. J. Smith, R. Frankham, R. G. Banks and J. S. F. Barker. 1992. Rapid loss of genetic variation in large captive populations of *Drosophila* flies: Implications for the genetic management of captive populations. *Conservation Biology* 6: 416–425.

Brosius, J. P. 1990. Penan hunter-gatherers of Sarawak, East Malaysia. *AnthroQuest* 42: 1–7.

Browder, J. O. 1988. The social cost of rain forest destruction: A critique and economic analysis of the "hamburger debate." *Interciencia* 13: 115–120.

Browder, J. O. 1990. Extractive reserves will not save the tropics. *BioScience* 40: 626.

Brown, B. E. and J. C. Ogden. 1993. Coral bleaching. *Scientific American* 268: 64–70.

Brown, G. M. and J. H. Goldstein. 1984. A model for valuing endangered species. *Journal of Environmental Economics and Management* 11: 303–309.

Brown, J. H. and A. C. Gibson. 1983. *Biogeography*. C. V. Mosby Company, St. Louis, MO.

Brown, L. and D. Amadon. 1989. *Eagles, Hawks, and Falcons of the World*. Wellfeet Press, NJ.

Brown, P. J., P. T. Manders, D. P. Bands, F. J. Kruger and R. H. Andrag. 1991. Prescribed burning as a conservation management practice: A case history from the Cederberg Mountains, Cape Province, South Africa. *Biological Conservation* 56: 133–150.

Browning, J. A. 1991. Conserving crop plant–pathogen coevolutionary processes in situ. In M. L. Oldfield and J. B. Alcorn (eds.), *Biodiversity: Culture, Conservation and Ecodevelopment*, pp. 59–85. Westview Press, Boulder, CO.

Brush, S. B. 1989. Rethinking crop genetic resource conservation. *Conservation Biology* 3: 19–29.

Budd, J. T. C. 1991. Remote sensing techniques for monitoring land-cover. In B. Goldsmith (ed.), *Monitoring* for *Conservation and Ecology*, pp. 33–60. Chapman and Hall, New York.

Burbidge, A. A. and N. L. McKenzie. 1989. Patterns in the modern decline of western Australia's vertebrate fauna: Causes and conservation implications. *Biological Conservation* 50: 143–198.

Burdick, D. M., D. Cushman, R. Hamilton and J. G. Gosselink. 1989. Faunal changes and bottomland hardwood forest loss in the Tensas watershed, Louisiana. *Conservation Biology* 3: 282–292.

Burke, R. L. 1989. Florida gopher tortoise relocation: Overview and case study. *Biological Conservation* 48: 295–309.

Burke, R. L. 1990. Conservation of the world's rarest tortoise. *Conservation Biology* 4: 122–124.

Burley, F. W. 1988. Monitoring biological diversity for setting priorities in conservation. *In* E. O. Wilson and F. M. Peter (eds.), *Biodiversity*, pp. 227–230. National Academy Press, Washington, D.C.

Burrill, A., I. Douglas-Hamilton and J. Mackinnon. 1986. Protected areas as refuges for elephants. *In* J. Mackinnon and K. Mackinnon (eds.), *Protected Areas Systems Review of the Afrotropical Realm*. IUCN, Gland, Switzerland.

Bustamante, R. H. and J. C. Castilla. 1990. Impact of human exploitation on populations of the intertidal southern bull-kelp *Durvilaea antarctica* (Phaeophya, Durvileales) in Central Chile. *Biological Conservation* 52: 205–220.

Buttrick, S. C. 1992. Habitat management: A decision-making process. *Rhodora* 94: 258–286.

Buzas, M. A. and S. J. Culver. 1991. Species diversity and dispersal of benthic Foraminifera. *BioScience* 41: 483–489.

Cade, T. J. 1983. Hybridization and gene exchange among birds in relation to conservation. *In* C. M. Schonewald-Cox, S. M. Chambers, B. MacBryde and L. Thomas (eds.), *Genetics and Conservation: A Reference for Managing Wild Animal and Plant Populations*, pp. 288–309. Benjamin/Cummings, Menlo Park, CA.

Cade, T. J. 1988. Using science and technology to reestablish species lost in nature. *In* E. O. Wilson and F. M. Peter (eds.), *Biodiversity*, pp. 279–288. National Academy Press, Washington, D.C.

Cade, T. J. and D. M. Bird. 1990. Peregrine falcons, *Falco peregrinus*, nesting in an urban environment: A review. *The Canadian Field-Naturalist* 104: 209–218.

Cade, T. J. and C. G. Jones. 1993. Progress in restoration of the Mauritius kestrel. *Conservation Biology* 7: 169–175.

Cade, T. J., J. H. Enderson, C. G. Thelander and C. M. White (eds.). 1988. *Peregrine Falcon Populations: Their Management and Recovery*. The Peregrine Fund, Inc. Boise, ID.

Cahn, R. 1979a. The God committee. *Audubon* 81: 10–12.

Cahn, R. 1979b. The triumph of wrong. *Audubon* 81: 5–6.

Cahn, R. and P. Cahn. 1985. Saved but threatened. *Audubon* 87: 48–51.

Cairns, J., Jr. 1986. Restoration, reclamation, and regeneration of degraded or destroyed ecosystems. *In* M. E. Soulé (ed.), *Conservation Biology: The Science of Scarcity and Diversity*, pp. 153–181. Sinauer Associates, Sunderland, MA.

Caldecott, J. 1988. *Hunting and Wildlife Management in Sarawak*. IUCN, Gland, Switzerland.

Caldwell, L. 1985. Science will not save the Biosphere but politics might. *Environmental Conservation* 12: 195–197.

Callicott, J. B. 1986. On the intrinsic value of nonhuman species. *In* B. G. Norton (ed.), *The Preservation of Species: The Value of Biological Diversity*, pp. 138–172. Princeton University Press, Princeton, NJ.

Callicott, J. B. 1989. *In Defense of the Land Ethic: Essays in Environmental Philosophy*. State University of New York Press, Albany.

Callicott, J. B. 1990. Whither conservation ethics? *Conservation Biology* 4: 15–20.

Campbell, S. 1980. Is reintroduction a realistic goal? *In* M. E. Soulé and B. A. Wilcox (eds.), *Conservation Biology: An Evolutionary-Ecological Perspective*, pp. 263–269. Sinauer Associates, Sunderland, MA.

Canby, T. Y. 1991. The Persian Gulf after the storm. *National Geographic* 180(August): 2–35.

Carlton, J. T. 1989. Man's role in changing the face of the ocean: Biological invasions and implications for conservation of near shore environments. *Conservation Biology* 3: 265–273.

Carrier, J. 1991. The Colorado: A river drained dry. *National Geographic* 179(June): 4–35.

Carroll, C. R. 1992. Ecological management of sensitive natural areas. *In* P. L. Fiedler and S. K. Jain (eds.), *Conservation Biology: The Theory and Practice of Nature Conservation, Preservation and Management*, pp. 347–372. Chapman and Hall, New York.

Carroll, J. B. 1984. The conservation and wild status of the Rodrigues fruit bat *Pteropus rodricensis*. *Myotis* 21–22: 148–154.

Carson, H. L. 1983. The genetics of the founder effect. *In* C. M. Schonewald-Cox, S. M. Chambers, B. MacBryde and L. Thomas (eds.), *Genetics and Conservation: A Reference for Managing Wild Animal and Plant Populations*, pp. 189–200. Benjamin/Cummings, Menlo Park, CA.

Carson, H. L. and A. R. Templeton. 1984. Genetic revolutions in relation to speciation phenomena: The founding of new populations. *Annual Review of Ecology and Systematics* 15: 97–131

Carson, R. 1962. *Silent Spring*. Reprinted 1982 by Penguin, Harmondsworth, UK.

Carter, J. A. 1988. Survival training for chimps. *Smithsonian* 12: 90–101.

Case, T. J., D. T. Bolger and A. D. Richman. 1992. Reptilian extinctions: The last ten thousand years. *In* P. L. Fiedler and S. K. Jain (eds.), *Conservation Biology: The Theory and Practice of Nature Conservation,*

Preservation and Management pp. 91–126. Chapman and Hall, New York.

Caswell, H. 1989. *Matrix Population Models: Construction, Analysis and Interpretation*. Sinauer Associates, Sunderland, MA.

Caufield, C. 1985. *In the Rainforest*. Alfred A. Knopf, New York.

Caughley, G., H. Dublin and I. Parker. 1990. Projected decline of the African elephant. *Biological Conservation* 54: 157–164.

Cavalcanti, C. 1991. Government policy and ecological concerns: Some lessons from the Brazilian experience. *In* R. Costanza (ed.), *Ecological Economics: The Science and Management of Sustainability*, pp. 474–485. Columbia University Press, New York.

Ceballos-Lascuráin, H. (ed.). 1993. *Tourism and Protected Areas*. IUCN, Gland, Switzerland.

Center for Plant Conservation. 1991. Genetic sampling guidelines for conservation collections of endangered plants. *In* D. A. Falk and K. E. Holsinger (eds.), *Genetics and Conservation of Rare Plants*, pp. 224–238. Oxford University Press, New York.

Chadwick, D. H. 1991. Elephants: Out of time, out of space. *National Geographic* 179(May): 2–49.

Chaney, W. R. and M. Basbous. 1978. The cedars of Lebanon, witnesses of history. *Economic Botany* 32: 118–123.

Charlesworth, D. and B. Charlesworth. 1987. Inbreeding depression and its evolutionary consequences. *Annual Review of Ecology and Systematics* 18: 237–268.

Chase, A. 1986. *Playing God in Yellowstone: The Destruction of America's First National Park*. Atlantic Monthly Press, Boston.

Cheke, A. S. and J. F. Dahl. 1981. The status of bats on western Indian Ocean islands, with special reference to *Pteropus*. *Mammalia* 45: 205–238.

Cherfas, J. 1991. Disappearing mushrooms: Another mass extinction? *Science* 254: 1458.

Cherfas, J. 1993. Backgarden biodiversity. *Conservation Biology* 7: 6–7.

Chernela, J. 1987. Endangered ideologies: Tukano fishing taboos. *Cultural Survival Quarterly* 11: 50–52.

Christensen, N. L. 1985. Shrubland fire regimes and their evolutionary consequences. *In* S. T. A. Pickett and P. S. White (eds.), *The Ecology of Natural Disturbance and Patch Dynamics*, pp. 85–100. Academic Press, New York.

Clark, T. W. 1987. Black-footed ferret recovery: A progress report. *Conservation Biology* 1: 8–13.

Clark, W. C. 1989. Managing planet Earth. *Scientific American* 261: 47–54.

Clay, J. 1991. Cultural survival and conservation: Lessons from the past twenty years. *In* M. L. Oldfield and J. B. Alcorn (eds.), *Biodiversity: Culture, Conservation and Ecodevelopment*, pp. 248–273. Westview Press, Boulder, CO.

Clay, J. 1992. Some general principles and strategies for developing markets in America and Europe for non-timber forest products: Lessons from Cultural Enterprises, 1989–1990. *In* D. C. Nepstad and S. Schwartzman (eds.), *Non-Timber Products from Tropical Forests: Evaluation of a Conservation and Development Strategy*. The New York Botanical Garden, Bronx, NY.

Coblentz, B. E. 1990. Exotic organisms: A dilemma for conservation biology. *Conservation Biology* 4: 261–265.

Cody, M. L. 1986. Diversity, rarity, and conservation in Mediterranean-climate regions. *In* M. E. Soulé (ed.), *Conservation Biology: The Science of Scarcity and Diversity*, pp. 123–152. Sinauer Associates, Inc., Sunderland, MA.

Cohn, J. P. 1985. Duke primate center fosters research. *BioScience* 35: 691–695.

Cohn, J. P. 1987. Red wolf in the wilderness. *BioScience* 37: 313–316.

Cohn, J. P. 1988a. Halting the rhino's demise. *BioScience* 38: 740–744.

Cohn, J. P. 1988b. Culture and conservation. *BioScience* 38: 450–453.

Cohn, J. P. 1988c. Captive breeding for conservation. *BioScience* 38: 312–316.

Cohn, J. P. 1989. Iguana conservation and economic development. *BioScience* 39: 359–363.

Cohn, J. P. 1990a. Endangered wolf population increases. *BioScience* 40: 628–632.

Cohn, J. P. 1990b. Elephants: Remarkable and endangered. *BioScience* 40: 10–14.

Cohn, J. P. 1990c. An island for conservation. *BioScience* 40: 342–345.

Cohn, J. P. 1991a. Ferrets return from near-extinction. *BioScience* 41: 132–135.

Cohn, J. P. 1991b. New focus on wildlife health. *BioScience* 41: 448–450.

Collar, N. J. and S. N. Stuart. 1985. *Threatened Birds of Africa and Related Islands*, 3rd ed. ICBP/IUCN, Cambridge, England.

Collett, M. 1989. Bolivia blazes trail. . .to where? *Christian Science Monitor*, 10 July 1989.

Colwell, R. K. 1973. Competition and coexistence in a simple tropical community. *American Naturalist* 107: 737–760.

Colwell, R. K. 1986. Community biology and sexual selection: Lessons from hummingbird flower mites. *In* T. J. Case and J. R. Diamond (eds.), *Ecological Communities*, pp. 406–424. Harper & Row, New York.

Colwell, R. K. 1992. Making sense of ecological complexity: A personal and conceptual retrospective. *Biotropica* Special Issue 24: 226–232.

Committee of Scientists. 1979. Final report. Federal Register 44(88): 26599–26657.

Commoner, B. 1971. *The Closing Circle*. Knopf, New York.

Commoner, B. 1990. *Making Peace with the Planet*. Gollancz, London.

Conant, R. 1958. *A Field Guide to Reptiles and Amphibians*. Houghton Mifflin, Boston.

Conant, S. 1988. Saving endangered species by translocation. *BioScience* 38: 254–257.

Condit, R., S. P. Hubbel and R. B. Foster. 1992. Short-term dynamics of a Neotropical forest. *BioScience* 42: 822–828.

Connell, J. H. and E. Orias. 1964. The ecological regulation of species diversity. *American Naturalist* 98: 399–404.

Connor, E. F. and E. D. McCoy. 1979. The statistics and biology of the species–area relationship. *American Naturalist* 13: 791–833.

Conservation International. 1990. *The Rain Forest Imperative.* Conservation International, Washington, D.C.

Conway, W. G. 1980. An overview of captive propagation. *In* M. E. Soulé and B. A. Wilcox (eds.), *Conservation Biology: An Evolutionary-Ecological Perspective*, pp. 199–208. Sinauer Associates, Sunderland, MA.

Conway, W. G. 1988. Can technology aid species preservation? *In* E. O. Wilson and F. M. Peter (eds.), *Biodiversity*, pp. 263–268. National Academy Press, Washington, D.C.

Cooley, J. L. and J. H. Cooley (eds.). 1984. *Natural Diversity in Forest Ecosystems.* Proceedings of the Workshop, 1982 November 29–December 1, Athens GA. University of Georgia, Institute of Ecology, Athens.

Costa, V. R. 1989. Xingu: Hidrelétricas coroam quatro séculos de agressões. (Interview with anthropologists Leinad Santos and Lucia Andrade). *Ciencia Hoje* 59: 74–75.

Costanza, R. (ed.). 1991. *Ecological Economics: The Science and Management of Sustainability.* Columbia University Press, New York.

Cottam, G. 1990. Community dynamics on an artifical prairie. *In* W. R. Jordan III, M. E. Gilpin and J. D. Aber (eds.), *Restoration Ecology: A Synthetic Approach to Ecological Research.* pp. 257–270. Cambridge University Press, Cambridge.

Council on Environmental Quality and the Department of State. 1980–1981. *The Global 2000 Report to the President: Entering the Twenty-First Century.* U.S. Government Printing Office, Washington, D.C.

Courtenay, W. R. Jr. and J. R. Staufer Jr. (eds.). 1984. *Distribution, Biology, and Management of Exotic Fishes.* The Johns Hopkins University Press, Baltimore.

Cox, G. W. 1993. *Conservation Ecology.* W. C. Brown, Dubuque, IA.

Cox, P. A., T. Elmquist, E. D. Pierson and W. E. Rainey. 1991. Flying foxes as strong interactors in South Pacific island ecosystems: A conservation hypothesis. *Conservation Biology* 5: 448–454.

Crawford, M. 1986. Condor recovery effort hurt by strategy debate. *Science* 231: 213–214.

Crow, J. F. and N. E. Morton. 1955. Measurement of gene frequency drift in small populations. *Evolution* 9: 202–214.

Crumpacker, D. W., S. W. Hodge, D. Friedley and W. P. Gregg, Jr. 1988. A preliminary assessment of the status of major terrestrial and wetland ecosystems on federal and Indian lands in the United States. *Conservation Biology* 2: 103–115.

Crystal, L. 1989. American broadcast journalism: Its coverage of conservation crises. *In* D. Western and M. Pearl (eds.), *Conservation for the Twenty-First Century*, pp. 289–293. Oxford University Press, New York.

Currie, D. J. 1991. Energy and large-scale patterns of animal- and plant-species richness. *American Naturalist* 137: 27–49.

Curtis, J. T. 1956. A prairie continuum in Wisconsin. *Ecology* 36: 558–566.

Dahl, A. L. 1986. *Review of the Protected Areas System in Oceania.* IUCN/UNEP, Gland, Switzerland.

Dallmeier, F. (ed.). 1992. *Long-term Monitoring of Biological Diversity in Tropical Forest Areas.* MAB Digest No. 11. UNESCO, Paris.

Daly, D. 1992. Tree of life. *Audubon* 70: 76–85

Daly, H. E. and J. B. Cobb, Jr. 1989. *For the Common Good: Redirecting the Economy Toward Community, the Environment, and a Sustainable Future.* Beacon Press, Boston.

Daniel, J. G. and A. Kulasingam. 1974. Problems arising from large-scale forest clearing for agricultural use. *Malaysian Forester* 37: 152–160.

Danielson, R. M. 1985. Mycorrhizae and reclamation of stressed terrestrial environments. *In* R. L. Tate and D. A. Klein (eds.), *Soil Reclamation Processes*, pp. 173–201. Marcel Dekker, New York.

Darling, J. D. 1988. Working with whales. *National Geographic* 174 : 886–908.

Darwin, C. R. 1859. *On the Origin of Species.* John Murray, London.

Darwin, C. R. 1868. *The Variation of Animals and Plants Under Domestication.* John Murray, London.

Darwin, C. R. 1876. *The Effects of Cross and Self-Fertilization in the Vegetable Kingdom.* John Murray, London.

Dasmann, R. F. 1973. A system for defining and classifying natural regions for purposes of conservation. Occasional Paper 7, pp. 1–47. IUCN, Gland, Switzerland.

Dasmann, R. F. 1987. World parks, people, and land use. *In* R. Hermann and T. B. Craig (eds.), *Conference on Science in National Parks: The Fourth Triennial Conference on Research in National Parks and Equivalent Reserves*, pp. 122–127. The George Wright Society and the U.S. National Park Service.

Dasmann, R. F. 1991. The importance of cultural and biological diversity. *In* M. L. Oldfield and J. B. Alcorn (eds.), *Biodiversity: Culture, Conservation and Ecodevelopment*, pp. 7–15. Westview Press, Boulder, CO.

Dasmann, R. F., J. P. Milton and P. H. Freeman. 1973. *Ecological Principles for Economic Development.* John Wiley, London.

Daughery, C. H., A. Cree, J. M. Hay and M. B. Thompson. 1990. Neglected taxonomy and continuing extinctions of tuatara (*Sphenodon*). *Science* 347: 177–179.

Davies, S. (ed.). 1987. *Tree of Life: Buddhism and Protection of Nature: With a Declaration on Environmental Ethics by His Holiness the Dalai Lama.* Buddhist Perception of Nature, Geneva, Switzerland.

Davis, M. B. 1990. Climatic change and the survival of forest species. *In* G. M. Woodwell (ed.), *The Earth in Transition: Patterns and Processes of Biotic Impoverishment.* Cambridge University Press, Cambridge.

Davis, M. B. and C. Zabinski. 1992. Changes in geographical range resulting from greenhouse warming: Effects on biodiversity in forests. *In* R. Peters and T. E. Lovejoy (eds.), *Global Warming and Biological Diversity*, pp. 297–308. Yale University Press, New Haven, CT.

Davis, S. D., et al. 1986. *Plants In Danger: What Do We Know?* IUCN, Gland, Switzerland.

Davy, A. J. and R. L. Jefferies. 1981. Approaches to monitoring of rare plant populations. *In* H. Synge (ed.), *The Biological Aspects of Rare Plant Conservation.* John Wiley, Chichester, U.K.

Dawson, W. R., J. D. Ligon, J. R. Murphy, J. P. Myers, D. Simberloff and J. Verner. 1987. Report of the scientific advisory panel on the spotted owl. *Condor* 89: 205–229.

de Klemm, C. 1990. *Wild Plant Conservation and the Law.* IUCN, Gland, Switzerland.

de Klemm, C. 1993. *Guidelines for CITES Implementation Legislation.* IUCN, Gland, Switzerland.

De Mauro, M. 1989. Aspects of the reproductive biology of the endangered *Hymenoxys acaulis* var. *glabra*: Implications for conservation. M.Sc. Thesis, University of Illinois, Chicago.

De Wolf, R., R. Goossens, J. MacKinnon and W. S. Cai. 1988. Remote sensing for wildlife managment: Giant panda habitat mapping from Landsat MSS images. *Geocarto International* 1: 41–49.

Debt-for-Development Coalition. 1990. *A Guide for Debt-for-Development: Making the International Debt Crisis Work for Development.* Revised edition, February 1990. Debt-for-Development Coalition, U.S. Overseas Cooperative Development Committee, Volunteers in Overseas Cooperative Assistance, Washington, D.C.

Decker, D. J., M. E. Krasny, G. R. Goff, C. R. Smith and D. W. Gross (eds.). 1991. *Challenges in the Conservation of Biological Resources: A Practitioner's Guide.* Westview Press, Boulder, CO.

Del Tredici, P. 1991. Ginkgos and people: A thousand years of interaction. *Arnoldia* 51: 2–15.

Del Tredici, P., Hsieh Ling and Guang Yang. 1992. The gingkos of Tian Mu Shan. *Conservation Biology* 6: 202–209.

Dennis, B. 1981. Extinction and waiting times in birth–death processes: Applications to endangered species and insect pest control. *In* C. Jaillie, G. P. Patil and B. A. Baldessari (eds.), *Statistical Distributions in Scientific Work*, Vol. 6, pp. 298–301. Reidel, Dordrecht, Germany.

Dennis, B., P. L. Munholland and J. M. Scott. 1991. Estimation of growth and extinction parameters for endangered species. *Ecological Monographs* 61: 115–143.

Denslow, J. S. and C. Padoch (eds.). 1988. *People of the Tropical Rain Forest.* University of California Press, Berkeley.

DeSteven, D. 1991. Experiments on mechanisms of tree establishment in old-field succession: Seedling emergence. *Ecology* 72: 1066–1075.

Devall, B. and G. Sessions. 1985. *Deep Ecology: Living as if Nature Mattered.* Gibbs Smith Publisher, Salt Lake City.

Diamond, A. W. 1985. The selection of critical areas and current conservation efforts in tropical forest birds. *In* A. W. Diamond and T. E. Lovejoy (eds.), *Conservation of Tropical Forest Birds*, pp. 33- 48. Technical Publication No. 4, International Council for Bird Preservation, Cambridge, England.

Diamond, J. M. 1975. The island dilemma: Lessons of modern biogeographic studies for the design of natural reserves. *Biological Conservation* 7: 129–146.

Diamond, J. M. 1984. "Normal" extinctions of isolated populations. *In* M. H. Nitecki (ed.), *Extinctions*, pp. 191–245. University of Chicago Press, Chicago.

Diamond, J. M. 1986. The design of a nature reserve system for Indonesian New Guinea. *In* M. E. Soulé (ed.), *Conservation Biology: The Science of Scarcity and Diversity*, pp. 485–503. Sinauer Associates, Sunderland, MA.

Diamond, J. M. 1987. Extant unless proven extinct? Or, extinct unless proven extant? *Conservation Biology* 1: 77–81.

Diamond, J. M. 1988a. Factors controlling species diversity: Overview and synthesis. *Annals of the Missouri Botanical Gardens* 75: 117–129.

Diamond, J. M. 1988b. Red books or green lists? *Nature* 332: 304–305.

Diamond, J. M. 1990. Reflections on goals and on the relationship between theory and practice. *In* W. R. Jordan III, M. E. Gilpin and J. D. Aber (eds.), *Restoration Ecology: A Synthetic Approach to Ecological Research.* pp. 329–336. Cambridge University Press, Cambridge.

Diamond, J. M., K. D. Bishop and S. van Balen. 1987. Bird survival in an isolated Javan woodland: Island or mirror? *Conservation Biology* 1: 132–142.

Dinerstein, E. and G. F. McCracken. 1990. Endangered greater one-horned rhinoceros carry high levels of genetic variation. *Conservation Biology* 4: 417–422.

Dinerstein, E. and E. Wikramanayake. 1992. Beyond 'Hotspots': How to prioritize investments in biodiversity in the Indo-Pacific Region. *Conservation Biology* 7: 39–52.

Dirig, R. 1988. Nabokov's blue snowflakes. *Natural History* 97: 69.

Di Silvestro, R. L. 1987. Saga of AC-9, the last free condor. *Audubon* 89: 12–14.

Doak, D. 1989. Spotted owls and old growth logging in the Pacific Northwest. *Conservation Biology* 3: 389–396.

Dobson, A. P. and A. M. Lyles. 1989. The population dynamics and conservation of primate populations. *Conservation Biology* 3: 362–380.

Dobson, A. P. and J. H. Poole. 1992. Ivory: Why the ban must stay! *Conservation Biology* 6: 149–151.

Docherty, D. E. and R. I. Romaine. 1983. Inclusion body disease of cranes: A serological follow-up to the 1978 die-off. *Avian Diseases* 27: 830–835.

Dodd, C. K. Jr. 1990. Effects of habitat fragmentation on a stream-dwelling species, the flattened musk turtle *Stenotherus depressus. Biological Conservation* 54: 33–45.

Dogsé, P. and B. von Droste. 1990. *Debt-for-Nature Exchanges and Biosphere Reserves.* UNESCO, Paris.

Dolnick, E. 1989. Panda paradox. *Discover* 10: 70–74.

Donázar, J. A. and C. Fernández. 1990. Population trends of the griffon vulture *Gyps fulvus* in northern Spain between 1969 and 1989 in relation to conservation measures. *Biological Conservation* 53: 83–91.

Donovan, G. P. 1986. Thirty-eighth annual meeting of the International Whaling Commission, June 1986. *Polar Record* 23: 437–441.

Doughty, R. W. 1989. *Return of the Whooping Crane.* University of Texas Press, Austin.

Dowling, T. E. and M. R. Childs. 1992. Impact of hybridization on a threatened trout of the southwestern United States. *Conservation Biology* 6: 355–364.

Downing, T. E., S. B. Hecht, H. A. Pearson and C. Garcia-Downing. 1992. *Development or Destruction: The Conversion of Tropical Forest to Pasture in Latin America.* Westview Press, Boulder, CO.

Drake, J. A., et al. (eds.). 1989. *Biological Invasions: A Global Perspective.* SCOPE Report No. 37. John Wiley, New York.

Drayton, B. 1993. Changes in the Flora of the Middlesex Fells, 1894–1993. Masters Thesis, Boston University, Boston.

Dregné, H. E. 1983. *Desertification of Arid Lands.* Academic Press, New York.

Dresser, B. L. 1988. Cryobiology, embryo transfer, and artificial insemination in ex situ animal conservation programs. *In* E. O. Wilson and F. M. Peter (eds.), *Biodiversity,* pp. 296–308. National Academy Press, Washington, D.C.

Dudley, L. C. C. 1992. The Chimane conservation program in Beni, Bolivia: An effort for local participation. *In* K. H. Redford and C. Padoch (eds.), *Conservation of Neotropical Rainforests,* pp. 228–244. Columbia University Press, New York.

Duffey, E. 1971. The management of Woodwalton Fen: A multidisciplinary approach. *In* E. Duffey and A. S. Watts (eds.), *The Scientific Management of Animal and Plant Communities for Conservation,* pp. 581–597. Blackwell Scientific Publications, Oxford.

Duffey, E. and A. S. Watts (eds.). 1971. *The Scientific Management of Animal and Plant Communities for Conservation.* Blackwell Scientific Publications, Oxford.

Duffey, E., N. G. Morris, J. Sheail, L. K. Ward, D. A. Wells and T. C. E. Wells. 1974. *Grassland Ecology and Wildlife Management.* Chapman and Hall, London.

Duffus, D. A. and P. Dearden. 1990. Non-consumptive wildlife-oriented recreation: A conceptual framework. *Biological Conservation* 53: 213–231.

Dufour, D. L. 1990. Use of tropical rainforest by native Amazonians. *BioScience* 40: 652–659.

Dwivedi, O. P. and B. Tiwari. 1987. *Environmental Crisis and Hindu Religion.* Gatanjali Publishing House, New Delhi.

Dyer, M. I. and M. M. Holland. 1991. The biosphere reserve concept: Needs for a network design. *BioScience* 41: 319–325.

Eberhardt, L. L. and J. M. Thomas. 1991. Designing environmental field studies. *Ecological Monographs* 61: 53–73.

Economic and Social Commission for Asia and the Pacific. 1985. *Marine Environmental Problems and Issues in the ESCAP Region.* United Nations, ESCAP, Bangkok, Thailand.

Ehrenfeld, D. W. 1970. *Biological Conservation.* Holt, Rinehart and Winston, New York.

Ehrenfeld, D. W. 1981. *The Arrogance of Humanism.* Oxford University Press, New York.

Ehrenfeld, D. W. 1988. Why put a value on biodiversity? *In* E. O. Wilson and F. M. Peter (eds.), *Biodiversity,* pp. 212–216. National Academy Press, Washington, D.C.

Ehrenfeld, D. W. 1989. Hard times for diversity. *In* D. Western and M. Pearl (eds.), *Conservation for the Twenty-First Century,* pp. 247–250. Oxford University Press, New York.

Ehrenfeld, D. W. and P. J. Bentley. 1985. Judaism and the practice of stewardship. *Judaism* 34: 301–311.

Ehrlich, P. R. 1988. The loss of diversity: Causes and consequences. *In* E. O. Wilson and F. M. Peter (eds.), *Biodiversity,* pp. 21–27. National Academy Press, Washington, D.C.

Ehrlich, P. R. and A. H. Ehrlich. 1968. *The Population Bomb.* Amereon, Mattituck, NY.

Ehrlich, P. R. and A. H. Ehrlich. 1981. *Extinction: The Causes and Consequences of the Disappearance of Species.* Random House, New York.

Ehrlich, P. R. and H. A. Mooney. 1983. Extinction, substitution and ecosystem services. *BioScience* 33: 248–254.

Ehrlich, P. R. and D. D. Murphy. 1987. Conservation lessons from long-term studies of checkerspot butterflies. *Conservation Biology* 1: 122–131.

Ehrlich, P. R. and P. H. Raven. 1964. Butterflies and

plants: A study in coevolution. *Evolution* 18: 586–608.

Eisenberg, J. F. and L. D. Harris. 1989. Conservation: A consideration of evolution, population, and life history. In D. Western and M. Pearl (eds.), *Conservation for the Twenty-First Century*, pp. 99–107. Oxford University Press, New York.

Eisner, T. 1991. Chemical prospecting: A proposal for action. In F. H. Bormann and S. R. Kellert (eds.), *Ecology, Economics, Ethics: The Broken Circle*, pp. 196–202. Yale University Press, New Haven, CT.

El Serafy, S. 1989. The proper calculation of income from depletable natural resources. In *Environmental Accounting for Sustainable Development*. World Bank/UNEP Symposium. The World Bank, Washington, D.C.

Elfring, C. 1989. Preserving land through local land trusts. *BioScience* 39: 71–74.

Ellenberg, H. 1978. *Vegetation mitteleuropas mit den alpen in ökologischer sicht.* 2 Auft. Ulmer, Stuttgart, Germany.

Ellis, R. 1992. Whale kill begins anew. *Audubon* 94: 20–22.

Ellis, W. S. 1990. A Soviet sea lies dying. *National Geographic* 177(February): 73–93.

Ellstrand, N. C. 1992. Gene flow by pollen: Implications for plant conservation genetics. *Oikos* 63: 77–86.

Elton, C. S. 1958. *The Ecology of Invasions.* John Wiley, New York.

Emerson, R. W. 1836. *Nature.* James Monroe and Co., Boston, MA.

Endangered Species Coalition. 1992. *The Endangered Species Act: A Commitment Worth Keeping.* The Wilderness Society, Washington, D.C.

Erdelon, W. 1988. Forest ecosystems and nature conservation in Sri Lanka. *Biological Conservation* 43: 115–135.

Erwin, T. L. 1982. Tropical forests: Their richness in Coleoptera and other arthropod species. *Coleopterists Bulletin* 36: 74–75.

Erwin, T. L. 1983. Beetles and other insects of tropical forest canopies at Manaus, Brazil, sampled by insecticidal fogging. In S. L. Sutton, T. C. Whitmore, and A. C. Chadwick (eds.), *Tropical Rain Forest: Ecology and Management*, pp. 59–75. Blackwell Scientific Publications, Edinburgh.

Erwin, T. L. 1991. How many species are there? Revisited. *Conservation Biology* 5: 330–333.

Estes, J. A., D. O. Duggins and G. B. Rathbun. 1989. The ecology of extinctions in kelp forest communities. *Conservation Biology* 3: 252–264.

Etter, R. J. and J. F. Grassle. 1992. Patterns of species diversity in the deep sea as a function of sediment particle size diversity. *Nature* 360: 575–578.

Eudey, A. A. 1987. *Action Plan for Asian Primate Conservation: 1987–1991.* IUCN Species Survival Commission Primate Specialist Group. IUCN, Gland, Switzerland.

Faanes, C. 1990. Cuban sandhills still declining. *International Crane Foundation Bugle* 16: 2.

Facelli, J. M and S. T. Pickett. 1991. Plant litter: Light interceptions and effects on an old-field plant community. *Ecology* 72: 1024–1031.

Fackelmann, K. 1984. The zoo ark: Charting a new course. *BioScience* 34: 606–612.

Falconer, D. S. 1981. *Introduction to Quantitative Genetics*, 2nd ed. Longman, New York.

Falk, D. A. 1991. Joining biological and economic models for conserving plant genetic diversity. In D. A. Falk and K. E. Holsinger (eds.), *Genetics and Conservation of Rare Plants*, pp. 209–224. Oxford University Press, New York.

Falk, D. A. 1992. From conservation biology to conservation practice: Strategies for protecting plant diversity. In P. L. Fiedler and S. K. Jain (eds.), *Conservation Biology: The Theory and Practice of Nature Conservation, Preservation and Management* pp. 397–432. Chapman and Hall, New York.

Falk, D. A. and K. E. Holsinger. 1991. *Genetics and Conservation of Rare Plants.* Oxford University Press, New York.

Falk, D. A. and P. Olwell. 1992. Scientific and policy considerations in restoration and reintroduction of endangered species. *Rhodora* 94: 287–315.

Farnsworth, N. R. 1988. Screening plants for new medicines. In E. O. Wilson and F. M. Peter (eds.), *Biodiversity*, pp. 83–97. National Academy Press, Washington, D.C.

Favre, D. S. 1989. *International Trade in Endangered Species: A Guide to CITES.* Klumer Academic, Dordrecht, Netherlands.

Fearnside, P. M. 1987. Deforestation and international economic development projects in Brazilian Amazonia. *Conservation Biology* 1: 214–221.

Fearnside, P. M. 1988. O carvão de Carajás. *Ciencia Hoje* 48: 17–21.

Fearnside, P. M. 1989. Extractive reserves in Brazilian Amazonia. *BioScience* 39: 387–393

Fearnside, P. M. 1990. Predominant land uses in Brazilian Amazonia. In A. Anderson (ed.), *Alternatives to Deforestation: Steps Toward Sustainable Use of the Amazon Rain Forest*, pp. 233–251. Columbia University Press, New York.

Feschbach, M. and A. Friendly, Jr. 1991. *Ecocide in the USSR: Health and Nature Under Seige.* Basic Books, New York.

Fiedler, P. L. and S. K. Jain (eds.). 1992. *Conservation Biology: The Theory and Practice of Nature Conservation, Preservation and Management.* Chapman and Hall, New York.

Fillon, F. L., A. Jacquemot and R. Reid. 1985. *The Importance of Wildlife to Canadians.* Canadian Wildlife Service, Ottawa.

Fischer, A. G. 1960. Latitudinal variations in organic diversity. *Evolution* 14: 64–81.

Fisher, C. R., Jr. 1990. Marine invertebrates and their chemolithoautotrophic symbionts. *Review of Aquatic Sciences* 2: 399–436.

Fitter, R. and M. Fitter. 1987. *The Road to Extinction.* IUCN, Gland, Switzerland.

Fitzgerald, S. 1989. *International Wildlife Trade: Whose Business Is It?* World Wildlife Fund, Washington, D.C.

Folk, M. J. and T. C. Tacha. 1990. Crane roost sites. *Journal of Wildife Management* 54: 480–489.

Food and Agriculture Organization of the United Nations. 1982. *Management and Utilization of Mangroves in Asia and the Pacific.* Environment Paper 3. FAO, Rome.

Food and Agriculture Organization of the United Nations. 1985. *Mangrove Management in Thailand, Malaysia, and Indonesia.* Environment Paper 4. FAO, Rome.

Food and Agriculture Organization of the United Nations. 1988. *Current Fisheries Statistics.* FAO, Rome.

Foose, T. J. 1983. The relevance of captive populations to the conservation of biotic diversity. *In* C. M. Schonewald-Cox, S. M. Chambers, B. MacBryde and L. Thomas (eds.), *Genetics and Conservation: A Reference for Managing Wild Animal and Plant Populations,* pp. 374–401. Benjamin/Cummings, Menlo Park, CA.

Forman, R. T. and M. Godron. 1981. Patches and structural components for a landscape ecology. *BioScience* 31: 733–740.

Forman, R. T. and M. Godron. 1986. *Landscape Ecology.* John Wiley, New York.

Forrester, D. J. 1971. Bighorn sheep lungworm-pneumonia complex. *In* J. W. Davis and R. C. Anderson (eds.), *Parasitic Diseases of Wild Mammals,* pp. 158–173. Iowa State University Press, Ames.

Fowler, S. V., et al. 1989. Survey and management proposals for a tropical deciduous forest reserve at Ankarana in northern Madagascar. *Biological Conservation* 47: 297–313.

France, R. L. and N. C. Collins. 1993. Extirpation of crayfish in a lake affected by long-range anthropogenic acidification. *Conservation Biology* 7: 184–188.

Frankel, O. H. and M. E. Soulé. 1981. *Conservation and Evolution.* Cambridge University Press, Cambridge.

Franklin, I. R. 1980. Evolutionary change in small populations. *In* M. E. Soulé and B. A. Wilcox (eds.), *Conservation Biology: An Evolutionary-Ecological Perspective,* pp. 135–149. Sinauer Associates, Sunderland, MA.

Franklin, J. F. 1985. Design of natural area preserves in Hawai'i. *In* C. P. Stone and J. M. Scott (eds.). *Hawai'i's Terrestrial Ecosystems: Preservation and Management,* pp. 459–474. Cooperative National Park Resources Studies Unit, University of Hawaii, Manoa.

Franklin, J. F. 1989. Towards a new forestry. *American Forester* Nov/Dec: 37–44.

Franklin, J. F. and R. T. T. Forman. 1987. Creating landscape patterns by forest cutting: Ecological consequences and principles. *Landscape Ecology* 1: 5–18.

Frazer, N. B. 1992. Sea turtle conservation and halfway technology. *Conservation Biology* 6: 179–184.

Freeland, W. J. and W. J. Boulton. 1992. Coevolution of food webs: Parasites, predators and plant secondary compounds. *Biotropica* (Special Issue) 24: 309–327.

Fricke, H. 1988. Coelacanths: The fish that time forgot. *National Geographic* 173: 824–838.

Fricke, H. and K. Hissmann. 1990. Natural habitat of the coelacanths. *Nature* 346: 323–4.

Fujita, M. S. and M. D. Tuttle. 1991. Flying foxes (Chiroptera: Pteropodidae): Threatened animals of key ecological and economic importance. *Conservation Biology* 5: 455 – 463.

Futuyma, D. J. 1986. *Evolutionary Biology,* 2nd ed. Sinauer Associates, Sunderland, MA.

Gadgil, M. and R. Guha. 1992. *This Fissured Land: An Ecological History of India.* Oxford University Press, Oxford.

Gage, J. D. and P. A. Tyler. 1991. *Deep-Sea Biology: A Natural History of Organisms at the Deep Seafloor.* Cambridge University Press, Cambridge.

Gagné, W. C. 1988. Conservation priorities in Hawaiian natural systems. *BioScience* 38: 264–271.

Game, M. and G. F. Peterken. 1984. Nature reserve selection strategies in the woodlands of central Lincolnshire, England. *Biological Conservation* 29: 157–181.

Gaston, K. J. 1991. The magnitude of global insect species richness. *Conservation Biology* 5: 283–296.

Gates, D. M. 1993. *Climate Change and Its Biological Consequences.* Sinauer Associates, Sunderland, MA.

Gavin, T. A. 1991. New approaches in managing biodiversity: A matter of scale. *In* D. J. Decker, M. E. Krasny, G. R. Goff, C. R. Smith and D. W. Gross (eds.), *Challenges in the Conservation of Biological Resources: A Practitioner's Guide,* pp. 97–103. Westview Press, Boulder, CO.

Gentry, A. H. 1982. Neotropical floristic diversity: phytogeographical connections between Central and South America, Pleistocene climatic fluctuations, or an accident of the Andean orogeny? *Annals of the Missouri Botanical Garden* 69: 557–593.

Gentry, A. H. 1986. Endemism in tropical versus temperate plant communities. *In* M. E. Soulé (ed.), *Conservation Biology: The Science of Scarcity and Diversity,* pp. 153–181. Sinauer Associates, Sunderland, MA.

Gerrodette, T. and W. G. Gilmartin. 1990. Demographic consequences of changing pupping and hauling sites of the Hawaiian monk seal. *Conservation Biology* 4: 423–430.

Gersh, J. and R. Pickert. 1991. Land-use modeling: Accommodating growth while conserving biological resources in Dutchess County, New York. *In* D. J. Decker, M. E. Krasny, G. R. Goff, C. R. Smith and D. W. Gross (eds.), *Challenges in the Conservation of*

Biological Resources: A Practitioner's Guide, pp. 233–242. Westview Press, Boulder, CO.

Getz, W. M. and R. G. Haight. 1989. *Population Harvesting: Demographic Models of Fish, Forest, and Animal Resources*. Princeton University Press, Princeton, NJ.

Gibbons, A. 1992. Conservation biology in the fast lane. *Science* 255: 20–22.

Gilbert, L. E. 1980. Food web organization and the conservation of Neotropical diversity. *In* M. E. Soulé and B. A. Wilcox (eds.), *Conservation Biology: An Evolutionary-Ecological Perspective*, pp. 11–33. Sinauer Associates, Sunderland, MA.

Gillis, A. M. 1990. The new forestry. *BioScience* 40: 558–562.

Gillis, M. 1986. *Non-wood Forest Products in Indonesia*. Department of Forestry, University of North Carolina, Chapel Hill, NC.

Gillis, M. 1991. Economics, ecology, and ethics: Mending the broken circle for tropical forests. *In* F. H. Bormann and S. R. Kellert (eds.), *Ecology, Economics, Ethics: The Broken Circle*, pp. 155–179. Yale University Press, New Haven, CT.

Gilpin, M. E. 1990. Experimental community assembly: competition, community structure, and the order of species introductions. *In* W. R. Jordan III, M. E. Gilpin and J. D. Aber (eds.), *Restoration Ecology: A Synthetic Approach to Ecological Research*, pp. 151–161. Cambridge University Press, Cambridge.

Gilpin, M. E. and M. E. Soulé. 1986. Minimum viable populations: Processes of species extinction. *In* M. E. Soulé (ed.), *Conservation Biology: The Science of Scarcity and Diversity*, pp. 19–34. Sinauer Associates, Sunderland, MA.

Giono, J. 1989. *The Man Who Planted Trees*. Collins Dove, Melbourne, Australia.

Giovannoni, S. J., T. B. Britschgi, C. L. Moyer and K. G. Field. 1990. Genetic diversity in Sargasso Sea bacterioplankton. *Nature* 345: 60–63.

Gipps, J. H. W. (ed.). 1991. *Beyond Captive Breeding: Reintroducing Endangered Species through Captive Breeding*. Zoological Society of London Symposia, No. 62. Clarendon Press, Oxford.

Girdner, B. 1992. The condor will soar anew: zoo program brings California birds back from the brink of extinction. *The Boston Globe*. January 6, 1992.

Gliessman, S. R. 1991. Ecological basis of traditional management of wetlands in tropical Mexico: Learning from agroecosystem models. *In* M. L. Oldfield and J. B. Alcorn (eds.), *Biodiversity: Culture, Conservation and Ecodevelopment*, pp. 211–229. Westview Press, Boulder, CO.

Gold, Thomas. 1992. The deep, hot biosphere. *Proceedings of the National Academy of Science* 89: 6045–6049.

Goldblatt, P. 1978. An analysis of the flora of Southern Africa: Its characteristics, relationships and origins. *Annals of the Missouri Botanical Garden* 65: 369–436.

Goldman, B. and F. H. Talbot. 1976. Aspects of the ecology of coral reef fishes. *In* O. A. Jones and R. Endean (eds.), *Biology and Geology of Coral Reefs*, Vol. 3, pp. 125–154. Academic Press, New York.

Goldsmith, B. (ed.). 1991. *Monitoring for Conservation and Ecology*. Chapman and Hall, New York.

Goleman, D. 1991. Shamans and their lore may vanish with the forests. *New York Times* Vol. CXL No. 48,628. June 11, 1991, pp. C1/C13.

Gomez-Pompa, A. and A. Kaus. 1988. Conservation by traditional cultures in the tropics. *In* V. Martin (ed.), *For the Conservation of Earth*. Fulcrum Inc., Golden, CO.

Gomez-Pompa, A. and A. Kaus. 1992. Taming the wilderness myth. *BioScience* 42: 271–279.

Goodland, R. J. A. 1987. The World Bank's wildlands policy: A major new means of financing conservation. *Conservation Biology* 1: 210–213.

Goodland, R. J. A. 1989. The environmental implications of major projects in Third World development. *In* P. Morris (ed.), *Major Projects and the Environment*, pp. 9–34. Major Projects Association, Oxford, England.

Goodland, R. J. A. (ed.). 1990a. *Race to Save the Tropics: Ecology and Economics for a Sustainable Future*. Island Press, Washington, D.C.

Goodland, R. J. A. 1990b. The World Bank's new environmental policy for dams and reservoirs. *Water Resources Development* 6: 226–239.

Goodland, R. J. A. 1991. *Tropical Deforestation: Solutions, Ethics and Religions*. Environment Working Paper No. 43. The World Bank, Washington, D.C.

Goodland, R. J. A. 1992. Environmental priorities for financing institutions. *Environmental Conservation* 19: 9–22.

Goodland, R., H. Daly and S. El Serafy. 1992. The urgent need for a rapid transition to global environmental sustainability. Paper presented at Environment Canada, Environmental Sustainability Seminar Series, Hull, Ottawa, Canada, 11 December 1992.

Goodland, R., H. Daly and S. El Serafy (eds.). 1992. *Population, Technology and Lifestyle: The Transition to Sustainability*. Island Press, Washington, D.C.

Goodland, R., A. Juras and R. Pachauri. 1992. Can hydroreservoirs in tropical moist forests be made environmentally acceptable? *Energy Policy* (June): 507–515.

Gordon, R. E. 1993. *Conservation Directory*. National Wildlife Federation, Washington, D.C.

Gore, A. 1992. *Earth in the Balance: Ecology and the Human Spirit*. Houghton Mifflin, New York.

Gorr, T., T. Kleinschmidt and H. Fricke. 1991. Close tetrapod relationship of the coelacanth *Latimeria* indicated by haemoglobin sequences. *Nature* 351: 394–396.

Gosselink, J. G., et al. 1990. Landscape conservation in a forested wetland watershed. *BioScience* 40(8): 588–600.

Gradwohl, J. and R. Greenberg. 1988. *Saving the Tropical Forests*. Earthscan Ltd., London.

Graedel, T. E. and P. J. Crutzen. 1989. The changing atmosphere. *Scientific American* 261: 58–68.

Graham, F. 1990a. Ban bankrupts the Far East's ivory carvers. *Audubon* 92: 64–65.

Graham, F. 1990b. Kite vs. stork. *Audubon* 92: 104–111.

Graham, R. L., M. G. Turner and V. H. Dale. 1990. How increasing CO_2 and climate change affect forests. *BioScience* 40(8): 575–587.

Grant, P. R. and B. R. Grant. 1992. Darwin's finches: Genetically effective population sizes. *Ecology* 73: 766–784.

Grassle, J. F. 1985. Hydrothermal vent animals: Distribution and biology. *Science* 229: 713–717.

Grassle, J. F. 1991. Deep-sea benthic biodiversity. *BioScience* 41: 464–469.

Grassle, J. F. and N. J. Maciolek. 1992. Deep-sea species richness: Regional and local diversity estimates from quantitative bottom samples. *American Naturalist* 139: 313–341.

Grassle, J. F., P. Lasserre, A. D. McIntyre and G. C. Ray. 1991. Marine biodiversity and ecosystem function. *Biology International* Special Issue 23: i–iv, 1–19.

Green, B. H. 1989. Conservation in cultural landscapes. *In* D. Western and M. Pearl (eds.), *Conservation for the Twenty-First Century*, pp. 182–198. Oxford University Press, New York.

Green, C. H. and S. M. Tunstall. 1991. Is the economic evaluation of environmental resources possible? *Journal of Environmental Management* 33: 123–141.

Greeson, P. E., J. R. Clark and J. E. Clark (eds.). 1979. *Wetland Functions and Values: The State of Our Understanding.* American Water Resources Association, Minneapolis.

Gregg, W. P., Jr. 1991. MAB Biosphere Reserves and conservation of traditional land use systems. *In* M. L. Oldfield and J. B. Alcorn (eds.), *Biodiversity: Culture, Conservation and Ecodevelopment*, pp. 274–294. Westview Press, Boulder, CO.

Griffith, B., J. M. Scott, J. W. Carpenter and C. Reed. 1989. Translocation as a species conservation tool: Status and strategy. *Science* 245: 477–480.

Grigg, G. 1989. Kangaroo harvesting and the conservation of arid and semi-arid rangelands. *Conservation Biology* 3: 194–197.

Grigg, R. W. and D. Epp. 1989. Critical depth for the survival of coral islands: Effects on the Hawaiian archipelago. *Science* 243: 638–641.

Groom, M. J. and N. Schumaker. 1993. Evaluating landscape change: Patterns of worldwide deforestation and local fragmentation. *In* P. M. Kareiva, J. G. Kingsolver, and R. B. Huey (eds.), *Biotic Interactions and Global Change*, pp. 24–44. Sinauer Associates, Sunderland, MA.

Gross, D. W., B. T. Wilkins, R. R. Quinn and A. E. Zepp. 1991. Local land protection and planning efforts. *In* D. J. Decker, M. E. Krasny, G. R. Goff, C. R. Smith and D. W. Gross (eds.), *Challenges in the Conservation of Biological Resources: A Practitioner's Guide*, pp. 355–366. Westview Press, Boulder, CO.

Grove, N. 1988. Quietly conserving nature. *National Geographic* 174(January): 818–844.

Grove, R. H. 1990. Colonial conservation, ecological hegemony, and popular resistance: Towards a global synthesis. *In* J. M. MacKenzie (ed.), *Imperialism and the Natural World.* University of Manchester Press, Manchester.

Grove, R. H. 1992. Origins of Western environmentalism. *Scientific American* 267: 42–47.

Grove, R. H. and J. J. Burdon (eds.). 1986. *Ecology of Biological Invasions.* Cambridge University Press, Cambridge.

Grumbine, R. E. 1993. *Ghost Bears: Exploring the Biodiversity Crisis.* Island Press, Washington, D.C.

Guerrant, E. O. 1992. Genetic and demographic considerations in the sampling and reintroduction of rare plants. In P. L. Fiedler and S. K. Jain (eds.), *Conservation Biology: The Theory and Practice of Nature Conservation, Preservation, and Management*, pp. 321–344. Chapman & Hall, New York.

Guha, R. 1989a. Radical American environmentalists and wilderness preservation: A third world critique. *Environmental Ethics* 11: 71–83.

Guha, R. 1989b. *The Unquiet Woods: Ecological Change and Peasant Resistance in the Himalaya.* Oxford University Press, Delhi.

Gulick, P., C. Hershey and J. Esquinas Alcazar. 1983. *Genetic Resources of Cassava and Wild Relatives.* International Board for Plant Genetic Resources 82/111, Rome.

Gullison, R. E. and E. C. Losos. 1993. The role of foreign debt in deforestation in Latin America. *Conservation Biology* 7: 140–147.

Gup, T. 1992. The stealth secretary. *Time* 139(21): 57–59.

Gupta, T. A. and A. Guleria. 1982. *Non-Wood Forest Products from India.* IBH Publishing Co., New Delhi.

Guzmán, H. M. 1991. Restoration of coral reefs in Pacific Costa Rica. *Conservation Biology* 5: 189–195.

Haas, P. M., M. A. Levy and E. A. Parson. 1992. Appraising the Earth Summit: How should we judge UNCED's success? *Environment* 34(8): 7–35.

Hafernik, J. E., Jr. 1992. Threats to invertebrate biodiversity: Implications for conservation strategies. *In* P. L. Fiedler and S. K. Jain (eds.), *Conservation Biology: The Theory and Practice of Nature Conservation, Preservation and Management* pp. 171–195. Chapman and Hall, New York.

Hagan, J. (ed.). 1992. *Neotropical Migrants.* Smithsonian Institution Press, Washington, D.C.

Haig, S. M., J. D. Ballou and S. R. Derrickson. 1990. Management options for preserving genetic diversity: Reintroduction of Guam rails to the wild. *Conservation Biology* 4: 290–300.

Hair, J. D. 1988. The economics of conserving wetlands: A widening circle. Paper presented at Workshop in Economics, IUCN General Assembly, 4–5 February 1988, Costa Rica.

Hair, J. D. and G. A. Pomerantz. 1987. The educational value of wildlife. *In* D. J. Decker and G. R. Goff (eds.), *Valuing Wildlife: Economic and Social Perspectives*, pp. 197–207. Westview Press, Boulder, CO.

Hales, D. F. 1984. The World Heritage Convention: Status and directions. *In* J. A. McNeely and K. R. Miller (eds.), *National Parks, Conservation, and Development: The Role of Protected Areas in Sustaining Society*, pp. 744–750. Smithsonian Institution Press, Washington, D.C.

Hall, A. V., B. de Winter, S. P. Fourie and T. H. Arnold. 1984. Threatened plants in Southern Africa. *Biological Conservation* 28: 5–20.

Hall, L. A. 1987. Transplantation of sensitive plants as mitigation for environmental impacts. *In* T. S. Elias (ed.), *Conservation and Management of Rare and Endangered Plants*, pp. 413–420. California Native Plant Society, Sacramento, CA.

Hammack, J. and G. M. Brown, Jr. 1974. *Waterfowl and Wetlands: Toward Bioeconomic Analysis*. Johns Hopkins University Press, Baltimore.

Hamrick, J. L and M. J. W. Godt. 1989. Allozyme diversity in plant species. *In* A. H. D. Brown, M. T. Clegg, A. L. Kahler and B. S. Weir (eds.), *Plant Population Genetics, Breeding, and Genetic Resources*, pp. 43–63. Sinauer Associates, Sunderland, MA.

Hamrick, J. L, M. J. W. Godt, D. A. Murawski and M. D. Loveless. 1991. Correlations between species traits and allozyme diversity: Implications for conservation biology. *In* D. A. Falk and K. E. Holsinger (eds.), *Genetics and Conservation of Rare Plants*, pp. 75–86. Oxford University Press, New York.

Hansen, A. J., T. A. Spies, F. J. Swanson and J. L. Ohmann. 1991. Conserving biodiversity in managed forests. *BioScience* 41: 382–392.

Hansen, S. 1989. Debt for nature swaps: Overview and discussion of key issues. *Ecological Economics* 1: 77–93.

Hanski, I. 1989. Metapopulation dynamics: Does it help to have more of the same? *Trends in Ecology and Evolution* 4: 113–114.

Hardin, G. 1968. The tragedy of the commons. *Science* 162: 1243–1248.

Hardin, G. 1985. *Filters against Folly: How to Survive Despite Economists, Ecologists, and the Merely Eloquent*. Viking Press, New York.

Harding, P. T. National species distribution surveys. *In* B. Goldsmith (ed.), *Monitoring for Conservation and Ecology*. Chapman and Hall, New York.

Hargrove, E. C. 1986a. Environmental ethics and Asian and comparative philosophy. *Environmental Professional* 8: 291–292.

Hargrove, E. C. (ed.) 1986b. *Religion and the Environmental Crisis*. University of Georgia Press, Athens.

Hargrove, E. C. 1989. *Foundations of Environmental Ethics*. Prentice-Hall, Englewood Cliffs, NJ.

Harper, J. L. 1977. *Population Biology of Plants*. Academic Press, New York.

Harris, L. D. 1984. *The Fragmented Forest: Island Biogeographic Theory and the Preservation of Biotic Diversity*. University of Chicago Press, Chicago.

Harris, R. B. and F. W. Allendorf. 1989. Genetically effective population size of large mammals: An assessment of estimators. *Conservation Biology* 3: 181–91.

Harrison, J. L. 1968. The effect of forest clearance on small mammals. In *Conservation in Tropical Southeast Asia*. IUCN, Morges.

Harrison, S. and J. F. Quinn. 1989. Correlated environments and the persistence of metapopulations. *Oikos* 56: 293–298.

Hart, J. A. 1978. From subsistence to market: a case study of the Mbuti net-hunters. *Human Ecology* 6: 32–53.

Hartell, K. E. 1992. Non-native fishes known from Massachusetts fresh waters. *Occasional Reports of the Museum of Comparative Zoology Fish Department* 2: 1–9.

Harwood, J. 1990. Are scientific quotas needed for the assessment of whale stocks? *Mammal Review* 20: 13–16.

Hawkes, J. G. 1983. *The Diversity of Crop Plants*. Harvard University Press, Cambridge, MA.

Hawkins, A. F. A., P. Chapman, J. U. Ganzhorn, Q. M. C. Bloxham, S. C. Barlow and S. J. Tonge. 1990. Vertebrate conservation in Ankarana Special Reserve, northern Madagascar. *Biological Conservation* 54: 83–110.

Hawksworth, D. L. 1990. The long-term effects of air pollutants on lichen communities in Europe and North America. *In* G. M. Woodwell (ed.), *The Earth in Transition: Patterns and Processes of Biotic Impoverishment*, pp. 45–64. Cambridge University Press, Cambridge.

Hawksworth, D. L. 1991a. The fungal dimension of biodiversity: Magnitude, significance, and conservation. *Mycological Research* 95: 641–655.

Hawksworth, D. L. (ed.). 1991b. *The Biodiversity of Microorganisms and Invertebrates: Its Role in Sustainable Agriculture*. CAB International, Wallingford, U.K.

Hawksworth, D. L. 1992. Biodiversity in microorganisms and its role in ecosystem function. *In* O. T. Solbrig, H. M. van Emden and P. G. W. J. van Oordt (eds.), *Biodiversity and Global Change*, pp. 83–94. International Union of Biological Sciences, Paris.

Hayden, B. P., G. C. Ray and R. Dolan. 1984. Classification of coastal and marine environments. *Environmental Conservation* 11: 199–207.

Hecht, S. B. and A. Cockburn. 1989. *The Fate of the Forest: Developers, Destroyers, and Defenders of the Amazon*. Verso, London.

Hecht, S. B., R. B. Norgaard and G. Possio. 1988. The economics of cattle ranching in eastern Amazonia. *Interciencia* 13: 233–240.

Hellawell, J. M. 1986. *Biological Indicators of Freshwater Pollution and Environmental Management*. Elsevier Applied Science Publisher, London.

Hellawell, J. M. 1991. Development of a rationale for monitoring. *In* B. Goldsmith (ed.), *Monitoring for Conservation and Ecology*, pp. 1–14. Chapman and Hall, New York.

Hinrichsen, D. 1987. The forest decline enigma. *BioScience* 37(8): 542–546.

Hobbs, R. J. 1989. The nature and effects of disturbance relative to invasions. *In* J. A. Drake and H. A. Mooney (eds.), *Biological Invasions: A Global Perspective*, pp. 389–405. Thomson Press, New Dehli, India.

Hodgson, J. G. 1986. Commonness and rarity in plants with special reference to the Sheffield flora. Part I. The identity, distribution and habitat characteristics of the common and rare species. Part III. Taxonomic and evolutionary aspects. *Biological Conservation* 36: 199–252, 275–296.

Holdren, C. 1991. Endangered languages. *Science* 251: 159.

Homan, T. (ed.). 1991. *A Yearning Toward Wildness: Environmentalism and Inspiration from Henry David Thoreau.* Peachtree Publishers, Atlanta.

Homer-Dixon, T. F., J. H. Boutwell and G. W. Rathjens. 1993. Environmental change and violent conflict. *Scientific American* 268: 38–45.

Hoose, P. M. 1981. *Building an Ark: Tools for the Preservation of Natural Diversity through Land Protection.* Island Press, Covelo, CA.

Horgan, J. 1992. It came from within. *Scientific American* 267: 20.

Hornocker, M. G. 1992. Learning to live with mountain lions. *National Geographic* 182(July): 52 – 65.

Horton, T. 1992. The Endangered Species Act: Too tough, too weak, or too late? *Audubon* (March/April): 68–74.

Hourigan, T. F. and E. S. Reese. 1987. Mid-ocean isolation and the evolution of Hawaiian reef fishes. *Trends in Ecology and Evolution* 2: 187–191.

Houseal, B., C. MacFarland, G. Archibold and A. Chiaria. 1985. Indigenous cultures and protected areas in Central America. *Cultural Survival Quarterly* 9: 10–20.

Howarth, F. G. 1973. The cavericolous fauna of Hawaiian lava tubes. I. Introduction. *Pacific Insects Monographs* 15: 139–151.

Howarth, F. G. 1990. Hawaiian terrestrial arthropods: An overview. *Bishop Museum Occasional Papers* 30: 4–26.

Howarth, F. G. 1991. Environmental impacts of classical biological control. *Annual Review of Entomology* 36: 485–509.

Howarth, F. G. and G. W. Ramsay. 1991. The conservation of island insects and their habitats. *In* N. M. Collins and J. A. Thomas (eds.), *The Conservation of Insects and Their Habitats*, pp. 71–107. Academic Press, New York.

Howarth, F. G., S. H. Sohmer and W. D. Duckworth. 1988. Hawaiian natural history and conservation efforts. *BioScience* 38: 232–253.

Howe, H. F. 1984. Implications of seed dispersal by animals for tropical reserve management. *Biological Conservation* 30: 261–281.

Hoyt, E. 1988. *Conserving the Wild Relatives of Crops.* IBPGR/IUCN/WWF, Rome.

Hufschmidt, M. M. and R. Srivardhana. 1986. The Nam Pong water resources project in Thailand. *In* J. A. Dixon and M. Hufschmidt (eds.), *Economic Valuation Techniques for the Environment: A Case Study Workbook*, pp. 141–162. Johns Hopkins University Press, Baltimore, MD.

Hughes, N. F. 1986. Changes in the feeding biology of the Nile perch (*Lates nilotica* L.)(Pisces: Centropomidae) in Lake Victoria, East Africa, since its introduction in 1960, and its impact on the native fish community of the Nyanza Gulf. *Journal of Fish Biology* 29: 541–548.

Hunt, H. E. and R. D. Slack. 1989. Winter diets of whooping and sandhill cranes in south Texas. *Journal of Wildlife Management* 53: 1150–1154.

Huston, M. A. 1985. Patterns of species diversity in relation to depth at Discovery Bay, Jamaica. *Bulletin of Marine Science* 37: 928–935.

Hutchings, M. J. 1987a. The population biology of the early spider orchid, *Ophrys sphegodes* Mill. I. A demographic study from 1975–1984. *Journal of Ecology* 75: 711–727.

Hutchings, M. J. 1987b. The population biology of the early spider orchid, *Ophrys sphegodes* Mill. II. Temporal patterns in behaviour. *Journal of Ecology* 75: 729–742.

Hutchings, M. J. 1991. Monitoring plant populations: census as an aid to conservation. *In* B. Goldsmith (ed.), *Monitoring for Conservation and Ecology*, pp. 61–76. Chapman and Hall, New York.

Hutchins, M. and C. Wemmer. 1986. Wildlife conservation and animal rights: Are they compatible? *In* M. W. Fox and L. D. Mickley (eds.), *Advances in Animal Welfare Science 1986/87*, pp. 111–137. Humane Society of the U.S., Washington, D.C.

Hvengaard, G. T., J. R. Butler and D. K. Krystofiak. 1989. Economic values of bird watching at Point Pelee National Park, Canada. *Wildlife Society Bulletin* 17: 526–531.

Iltis, H. H. 1988. Serendipity in the exploration of biodiversity: What good are weedy tomatoes? *In* E. O. Wilson and F. M. Peter (eds.), *Biodiversity*, pp. 98–105. National Academy Press, Washington, D.C.

Isliker, H. and B. Schurch (eds.) 1981. *The Impact of Malnutrition on Immune Defense in Parasitic Infection.* Nestlé Foundation Publication Series Volume 2, Switzerland.

IUCN (International Union for Conservation of Nature and Natural Resources). 1980. *World Conservation Strategy: Living Resource Conservation for Sustainable Development.* IUCN/UNEP/WWF. Gland, Switzerland.

IUCN. 1984. Categories, objectives and criteria for protected areas. *In* J. A. McNeely and K. R. Miller (eds.), *National Parks, Conservation and Development*, pp. 47–53. Smithsonian Institution Press, Washington, D.C.

IUCN. 1985. *United Nations List of National Parks and Protected Areas.* IUCN, Gland, Switzerland.

IUCN. 1988. *1988 IUCN Red List of Threatened Animals.* IUCN, Gland, Switzerland.

IUCN. 1990. *Lemurs of Madagascar and the Comoros: The IUCN Red Data Book.* IUCN, Gland, Switzerland.

IUCN/UNEP (IUCN/United Nations Environment Program). 1986a. *Review of the Protected Areas System in the Indo–Malayan Realm.* IUCN, Gland, Switzerland.

IUCN/UNEP. 1986b. *Review of the Protected Areas System in the Afrotropical Realm.* IUCN, Gland, Switzerland.

IUCN/UNEP. 1986c. *Review of the Protected Areas System in the Oceania.* IUCN, Gland, Switzerland.

IUCN/UNEP. 1986d. *Managing Protected Areas in the Tropics.* IUCN, Gland, Switzerland.

IUCN/UNEP. 1988. *Coral Reefs of the World.* 3 Volumes. IUCN, Gland, Switzerland.

IUCN/UNEP/WWF. (IUCN/UNEP/Worldwide Fund for Nature). 1980. *World Conservation Strategy: Living Resource Conservation for Sustainable Development.* IUCN, Gland, Switzerland.

IUCN/UNEP/WWF. 1991. *Caring for the Earth: A Strategy for Sustainable Living.* IUCN, Gland, Switzerland.

IUCN/WWF. 1989. *The Botanic Gardens Conservation Strategy.* IUCN, Gland, Switzerland.

Iverson, G. C., P. A. Vohs and T. C. Tacha. 1987. Habitat use by mid-continent sandhill cranes during spring migration. *Journal of Wildife Management* 51: 448–458.

Jackson, J. B. C. 1991. Adaptation and diversity of reef corals. *BioScience* 41: 475–482.

Jackson, L. L. 1992. The role of ecological restoration in conservation biology. *In* P. L. Fiedler and S. K. Jain (eds.), *Conservation Biology: The Theory and Practice of Nature Conservation, Preservation and Management* pp. 433–452. Chapman and Hall, New York.

Jacobson, G. L. Jr., H. Almquist-Jacobson and J. C. Winne. 1991. Conservation of rare plant habitat: Insights from the recent history of vegetation and fire at Crystal Fen, northern Maine, USA. *Biological Conservation* 57: 287–314.

Jacobson, S. K. 1990. Graduate education in conservation biology. *Conservation Biology* 4: 431–440.

Jannasch, H. W. and M. J. Motti. 1985. Geomicrobiology of deep-sea hydrothermal vents. *Science* 229(4715): 717–720.

Janos, D. P. 1980. Vesicular-arbuscular mycorrhizae affect lowland tropical rain forest plant growth. *Ecology* 61: 151–162.

Janson, C. H. and L. H. Emmons. 1990. Ecological structure of the nonflying mammal community at Cocha Cashu Biological Station, Manu National Park, Peru. *In* A. H. Gentry (ed.), *Four Neotropical Forests*, pp. 314–338. Yale University Press, New Haven, CT.

Janzen, D. H. 1983. No park is an island: Increase in interference from outside as park size decreases. *Oikos* 41: 402–410.

Janzen, D. H. 1986. The eternal external threat. *In* M. E. Soulé (ed.), *Conservation Biology: The Science of Scarcity and Diversity*, pp. 286–303. Sinauer Associates, Sunderland, MA.

Janzen, D. H. 1988a. Tropical dry forests: The most endangered major tropical ecosystem. *In* E. O. Wilson and F. M. Peter (eds.), *Biodiversity*, pp. 130–137. National Academy Press, Washington, D.C.

Janzen, D. H. 1988b. Tropical ecological and biocultural restoration. *Science* 239: 243–244.

Järvinen, O. 1979. Geographical gradients of stability in European land bird communities. *Oecologia* 38: 51–69.

Jeffery, D. 1989. Yellowstone: The great fires of 1988. *National Geographic* 175: 255–273.

Jenkins, M. D. (ed.) 1987. *Madagascar: An Environmental Profile.* IUCN, Gland, Switzerland.

Jenkins, R. E. Jr. 1988. Information management for the conservation of biodiversity. *In* E. O. Wilson and F. M. Peter (eds.), *Biodiversity*, pp. 231–239. National Academy Press, Washington, D.C.

Johannes, R. E. 1978. Traditional marine conservation methods in Oceania and their demise. *Annual Review of Ecology and Systematics* 9: 49–64.

Johannes, R. E. 1982. Traditional conservation methods and protected marine areas in Oceania. *Ambio* 11: 258–261.

Johns, A. D. 1985. Selective logging and wildlife conservation in tropical rain forest: Problems and recommendations. *Biological Conservation* 31: 355–375.

Johns, A. D. 1987. The use of primary and selectively logged rainforest by Malaysian hornbills (Bucerotidae) and implications for their conservation. *Biological Conservation* 40: 179–190.

Johns, A. D. 1988. Economic development and wildlife conservation in Brazilian Amazonia. *Ambio* 17: 302–306.

Johns, D. M. 1990. The relevance of deep ecology to the Third World: Some preliminary comments. *Environmental Ethics* 12: 233–252.

Johnsgard, P. A. 1983. The Platte: A river of birds. *Nature Conservancy News* 33(5): 6–11.

Johnsgard, P. A. 1990. *Hawks, Eagles, and Falcons of North America.* Smithsonian Institution Press, Washington, D.C.

Johnsgard, P. A. 1991. *Crane Music: A Natural History of American Cranes.* Smithsonian Institution Press, Washington, D. C.

Johnson, K., G. Schaller and H. Jinchu. 1988. Responses of giant pandas to a bamboo die-off. *National Geographic Research* 4: 161–177.

Johnson, N. In press. *What to Save First? Setting Biodiversity Conservation Priorities in a Crowded World.* Biodiversity Support Program, Washington, D.C.

Jones, H. L. and J. M. Diamond. 1976. Short-time-base studies of turnover in breeding birds of the California Channel Islands. *Condor* 76: 526–549.

Jones, P. D. and T. M. L. Wigley. 1990. Global warming trends. *Scientific American* 263: 84–91.

Jones, R. F. 1990. Farewell to Africa. *Audubon* 92: 50–104.

Jordan, W. R. III. 1988. Ecological restoration: Reflections on a half-century of experience at the University of Wisconsin–Madison Arboretum. *In* E. O. Wilson and F. M. Peter (eds.), *Biodiversity*, pp. 311–316. National Academy Press, Washington, D.C.

Jordan, W. R. III, M. E. Gilpin and J. D. Aber (eds.). 1990. *Restoration Ecology: A Synthetic Approach to Ecological Research.* Cambridge University Press, Cambridge.

Jordan, W. R. III, M. E. Gilpin and J. D. Aber. 1990. Restoration ecology: ecological restoration as a technique for basic research. *In* W. R. Jordan III, M. E. Gilpin and J. D. Aber (eds.), *Restoration Ecology: A Synthetic Approach to Ecological Research.* pp. 3–21. Cambridge University Press, Cambridge.

Joyce, C. 1993. Taxol: Search for a cancer drug. *BioScience* 43: 133–136.

Julien, M. H. (ed.). 1987. *Biological Control of Weeds: A World Catalog of Agents and Their Target Weeds.* CAB CIB Contr., Slough, London.

Kapos, V. 1989. Effects of isolation on the water status of forest patches in the Brazilian Amazon. *Journal of Tropical Ecology* 5: 173–185.

Karron, J. D. 1987. A comparison of levels of genetic polymorphisms and self-compatibility in geographically restricted and widespread plant congeners. *Evolutionary Ecology* 1: 47–58.

Katzman, M. T. and W. G. Cale, Jr. 1990. Tropical forest preservation using economic incentives. *BioScience* 40: 827–832.

Kaufman, L. 1986. Why the ark is sinking. *In* L. Kaufman and K. Mallory (eds.), *The Last Extinction*, pp. 1–41. MIT Press, Cambridge, MA.

Kaufman, L. 1988. Caught between a reef and a hard place: Why aquaria must invest in captive propagation. Proceedings of the National Meeting of the American Association of Zoological Parks and Aquaria 1988, pp. 365–382.

Kaufman, L. 1992. Catastrophic change in a species-rich freshwater ecosystem: Lessons from Lake Victoria. *BioScience* 42: 846–858.

Kaufman, L. and K. Mallory (eds.). 1986. *The Last Extinction.* MIT Press, Cambridge, MA.

Kavanagh, M., A. A. Rahim and C. J. Hails. 1989. *Rainforest Conservation in Sarawak: An International Policy for WWF.* WWF Malaysia, Kuala Lumpur.

Kay, J. 1988. Concepts of nature in the Hebrew bible. *Environmental Ethics* 10: 307–328.

Keen, W. H. and J. Gagliardi. 1981. Effect of brown bullheads on release of phosphorus in sediment and water systems. *Progressive Fish-Culturalist* 43: 183–185.

Kenchington, R. A. and M. T. Agardy. 1990. Achieving marine conservation through biosphere reserve planning and management. *Environmental Conservation* 17: 39–44.

Kennedy, D. M. 1987. What's new at the zoo? *Technology Review* 90: 66–73.

Kepler, C. B. and J. M. Scott. 1985. Conservation of island ecosystems. *In* P. O. Moors (ed.), *Conserva-*

tion of Island Birds, pp. 255–271. International Council for Bird Preservation, Cambridge, England.

Keyfitz, N. 1989. The growing human population. *Scientific American* 261: 119–126.

Khush, G. S. and K. C. Ling. 1974. Inheritance of resistance to grassy stunt virus and its vector in rice. *Journal of Heredity* 65: 134–136.

Kiew, R. 1991. *The State of Nature Conservation in Malaysia.* Malayan Nature Society, Kuala Lumpur.

Kiff, L. 1990. To the brink and back: The battle to save the California condor. *Terra* 28: 6–18.

Kimmins, J. P. 1987. *Forest Ecology.* Macmillan, New York.

Kimura, M. and J. F. Crow. 1963. The measurement of effective population numbers. *Evolution* 17: 279–288.

King, W. B. 1985. Island birds: Will the future repeat the past? *In* P. J. Moors (ed.), *Conservation of Island Birds*, pp. 3–15. International Council for Bird Preservation, Cambridge, England.

Kinnaird, M. F. and T. G. O'Brien. 1991. Viable populations for an endangered forest primate, the Tana River crested mangabey (*Cercocebus galeritus galeritus*). *Conservation Biology* 5: 203–213.

Kirkpatrick, R. C. 1992. Roundtable: Ecology, government legitimacy, and a changing world order. *BioScience* 42: 867–869.

Kleiman, D. G. 1989. Reintroduction of captive mammals for conservation. *BioScience* 39:152–161.

Kleiman, D. G., B. B. Beck, A. J. Baker, J. D. Ballou, L. Dietz and J. M. Dietz. 1990. The conservation program for the golden lion tamarin, *Leontopithecus rosalia. Endangered Species Update* 8: 82–84.

Klein, B. C. 1989. Effects of forest fragmentation on dung and carrion beetle communities in central Amazonia. *Ecology* 70: 1715–1725.

Kline, V. M. and E. A. Howell. 1990. Prairies. *In* W. R. Jordan III, M. E. Gilpin and J. D. Aber (eds.), *Restoration Ecology: A Synthetic Approach to Ecological Research.* pp. 75–84. Cambridge University Press, Cambridge.

Klinowska, M. 1991. *Dolphins, Porpoises, and Whales of the World: The IUCN Red Data Book.* IUCN, Gland, Switzerland.

Kloppenburg, J. and D. L. Kleinman. 1987. The plant germplasm controversy *BioScience* 37: 190–198.

Knight, R. R. and L. L. Eberhardt. 1985. Population dynamics of Yellowstone grizzly bears. *Ecology* 66: 323–334.

Koechlin, J., J. L. Guillaumet and P. Morat. 1974. *Flore et Vegetation de Madagascar.* J. Cramer, Vaduz, Liechtenstein.

Koester, V. 1989. *The RAMSAR Convention on the Conservation of Wetlands: A Legal Analysis of the Adoption and Implementation of the Convention in Denmark.* IUCN, Gland, Switzerland.

Kornfield, I. and K. E. Carpenter. 1984. Cyprinids of Lake Lanao, Philippines: Taxonomic validity, evolutionary rates and speciation scenarios. *In* A. A. Echelle and I. Kornfield (eds.), *Evolution of Species Flocks*, pp. 69–84. University of Maine Press, Orono.

Kraus, S. D. 1990. Rates and potential cause of mortality in North Atlantic right whales (*Eubalaena glacialis*). *Marine Mammal Science* 6: 278–291.

Kristensen, R. M. 1983. Loricifera, a new phylum with *Aschelminthes* characters from the meiobenthos. *Zeitschrift fur Zoologische Systematik* 21: 163–180.

Kruger, F. J. 1977. Ecological reserves in the Cape fynbos: Toward a strategy for conservation. *South African Journal of Science* 73: 81–85.

Kruger, F. J. and R. C. Bigalke. 1984. Fire in fynbos. *In* P. de V. Booysen and N. M. Tainton (eds.), *Ecological Effects of Fire in South African Ecosystems*, pp. 67–114. Ecological Studies 48, Springer-Verlag, Berlin.

Krutilla, J. V. and A. C. Fisher. 1975. *The Economics of Natural Environments: Studies in the Valuation of Commodity and Amenities Resources*. Johns Hopkins University Press, Baltimore, MD.

Küchler, A. W. 1964. Potential natural vegetation of the conterminous United States. Special Publication Number 36 (map and manual), American Geographical Society, New York.

Kusler, J. A. and M. E. Kentula (eds.) 1990. *Wetland Creation and Restoration: The Status of the Science*. Island Press, Washington, D.C.

Lacy, R. C. 1987. Loss of genetic diversity from managed populations: Interacting effects of drift, mutation, immigration, selection, and population subdivision. *Conservation Biology* 1: 143–158.

Lamberson, R. H., R. McElvey, B. R. Noon and C. Voss. 1992. A dynamic analysis of northern spotted owl viability in a fragmented forest landscape. *Conservation Biology* 6: 505–512.

Lambert, F. 1991. The conservation of fig-eating birds in Malaysia. *Biological Conservation* 58: 31–40.

Lamprey, H. F. 1974. Management of flora and fauna in national parks. *In* H. Elliott (ed.), *Second World Conference on National Parks*, pp. 237–248. IUCN, Morges, Switzerland.

Land Trust Exchange. 1988. *Land Trust Standard Practices*. Land Trust Exchange, Alexandria, VA.

Lande, R. 1988a. Genetics and demography in biological conservation. *Science* 241: 1455–1460.

Lande, R. 1988b. Demographic models of the northern spotted owl (*Strix occidentalis caurina*). *Oecologia* 75: 601–7.

Lande, R. and G. F. Barrowclough. 1987. Effective population size, genetic variation, and their use in population management. *In* M. E. Soulé (ed.), *Viable Populations for Management*, pp. 87–124. Cambridge University Press, Cambridge.

Lasiak, T. 1991. The susceptibility and/or resilience of rocky littoral molluscs to stock depletion by the indigenous coastal people of Transkei, southern Africa. *Biological Conservation* 56: 245–264.

Lasiak, T. and A. Dye. 1989. The ecology of the brown mussel *Perna perna* in Transkei, southern Africa: Implications for the management of a traditional food resource. *Biological Conservation* 47: 245–257.

Laurance, W. F. 1991a. Ecological correlates of extinction proneness in Australian tropical rain forest mammals. *Conservation Biology* 5: 79–89.

Laurance, W. F. 1991b. Edge effects in tropical forest fragments: Application of a model for the design of nature reserves. *Biological Conservation* 57: 205–219.

Le Houérou, H. N. and H. Gillet. 1986. Desertization in African arid lands. *In* M. E. Soulé (ed.), *Conservation Biology: The Science of Scarcity and Diversity*, pp. 444–461. Sinauer Associates, Sunderland, MA.

Ledec, G. 1989. A proposed strategy for the World Bank to promote increased conservation of biological diversity. World Bank, Washington, D.C. Manuscript.

Ledig, F. T. 1988. The conservation of diversity in forest trees. *BioScience* 38: 471–479.

Lehmkuhl, J. F., R. K. Upreti and U. R. Sharma. 1988. National parks and local development: Grasses and people in Royal Chitwan National Park, Nepal. *Environmental Conservation* 15: 143–148.

Leigh, J. H., J. D. Briggs and W. Hartley. 1982. The conservation status of Australian plants. *In* R. H. Groves and W. D. L. Ride (eds.), *Species at Risk: Research in Australia*, pp 13–25. Springer-Verlag, New York.

Leighton, M. and N. Wirawan. 1986. Catastrophic drought and fire in Borneo tropical rain forest associated with the 1982–1983 El Niño southern oscillation event. *In* G. T. Prance (ed.), *Tropical Rain Forests and World Atmosphere*, pp. 75–102. Westview Press, Boulder, CO.

Leopold, A. 1933. *Game Management*. Charles Scribners Sons, New York.

Leopold, A. 1939a. A biotic view of land. *Journal of Forestry* 37: 113–116.

Leopold, A. 1939b. The farmer as a conservationist. *American Forests* 45: 294–299, 316, 323.

Leopold, A. 1949. *A Sand County Almanac, and Sketches Here and There*. Oxford University Press, New York.

Leopold, A. 1953. *Round River*. Oxford University Press, Oxford.

Lesica, P. and F. W. Allendorf. 1992. Are small populations of plants worth preserving? *Conservation Biology* 6: 135–139.

Lesica, P., R. F. Leary, F. W. Allendorf and D. E. Bilderback. 1988. Lack of genic diversity within and among populations of an endangered plant, *Howellia aquatilis*. *Conservation Biology* 2: 275–282.

Levins, R. 1970. Extinction. *In Some Mathematical Questions in Biology*, Vol. II, pp. 75–108. American Mathematical Society, Providence, RI.

Lewis, D., G. B. Kaweche and A. Mwenya. 1990. Wildlife conservation outside protected areas: Lessons from an experiment in Zambia. *Conservation Biology* 4: 171–180.

Lewis, W. H. and M. P. F. Elvin-Lewis. 1977. *Medical Botany*. John Wiley, New York.

Likens, G. E. 1991. Toxic winds: Whose responsibility? *In* F. H. Bormann and S. R. Kellert (eds.), *Ecology, Economics, Ethics: The Broken Circle*, pp. 136–152. Yale University Press, New Haven, CT.

Likens, G. E., F. H. Bormann, R. S. Pierce, J. S. Eaton and N. M. Johnson. 1977. *Biogeochemistry of a Forested Ecosystem*. Springer-Verlag, New York.

Lin, S. C. and L. P. Yuan. 1980. Hybrid rice breeding in China. Pages 35–51 in *Innovative Approaches to Rice Breeding*. IRRI, Manila, Philippines.

Lindberg, K. 1991. *Policies for Maximizing Nature Tourism's Ecological and Economic Benefits*. World Resources Institute, Washington, D.C.

Line, L. 1993. Silence of the songbirds. *National Geographic* 183(June): 62–91.

Lipske, M. 1991. Big hopes for bold beasts: Can grizzlies and wolves be reintroduced safely into old haunts? *National Wildlife* 29: 44–53.

Locke, J. 1690. Second Treatise on Government. Reprinted in J. M. Porter (ed.). 1989. *Classics in Political Philosophy*. Scarborough, Ontario.

Loiselle, B. A. and J. G. Blake. 1992. Population variation in a tropical bird community. *BioScience* 42: 838–845.

Loope, L. L., O. Hamann and C. P. Stone. 1988. Comparative conservation biology of oceanic archipelagoes: Hawaii and the Galápagos. *BioScience* 38: 272–282.

Lovejoy, T. 1984. Aid debtor nation's ecology. *New York Times* 4 October 1984.

Lovejoy, T. E., et al. 1986. Edge and other effects of isolation on Amazon forest fragments. *In* M. E. Soulé (ed.), *Conservation Biology: The Science of Scarcity and Diversity*, pp. 257–285. Sinauer Associates, Sunderland, MA.

Lovelock, J. 1988. *The Ages of Gaia*. Norton, New York.

Lowenthal, D. 1991. Environmental conflict. *National Geographic Research and Exploration* 7: 266–275.

Lubchenco, J., et al. 1991. The sustainable biosphere initiative: An ecological research agenda. *Ecology* 72: 371–412.

Ludwig, D., R. Hilborn and C. Walters. 1993. Uncertainty, resource exploitation, and conservation: Lessons from history. *Science* 260: 17, 36.

Luoma, J. R. 1992. Born to be wild. *Audubon* 94: 50–61.

Lutz, R. A. 1991. The biology of deep-sea vents and seeps. *Oceanus* 34: 75–83.

Lynch, J. F. and D. F. Whigham. 1984. Effects of forest fragmentation on breeding bird communities in Maryland, USA. *Biological Conservation* 28: 287–324.

MacArthur, R. H. 1955. Fluctuations of animal populations, and a measure of community stability. *Ecology* 36: 533–536.

MacArthur, R. H. and E. O. Wilson. 1967. *The Theory of Island Biogeography*. Princeton University Press, Princeton, NJ.

Mace, G. M. and R. Lande. 1991. Assessing extinction threats: Towards a reevaluation of IUCN threatened species categories. *Conservation Biology* 5: 148–157.

Machlis, G. and K. Johnson. 1987. Panda outposts. *National Parks* 61(9–10): 14–16.

Machlis, G. E. and D. L. Tichnell. 1985. *The State of the World's Parks: An International Assessment of Resource Management, Policy, and Research*. Westview Press, Boulder, CO.

MacKenzie, J. J. and M. T. El-Ashry. 1988. *Ill Winds: Airborne Pollutions's Toll on Trees and Crops*. World Resources Institute, Washington, D.C.

MacKinnon, J. 1983. Irrigation and watershed protection in Indonesia. Report to IBRD Regional Office, Jakarta.

MacKinnon, J. and K. MacKinnon. 1986a. *Review of the Protected Areas System in the Indo-Malayan Realm*. IUCN/UNEP, Gland, Switzerland.

MacKinnon, J. and K. MacKinnon. 1986b. *Review of the Protected Areas System in the Afro-tropical Realm*. IUCN/UNEP, Gland, Switzerland.

MacKinnon, J., K. MacKinnon, G. Child and J. Thorsell. 1992. *Managing Protected Areas in the Tropics*. IUCN, Gland, Switzerland.

Mackintosh, G. (ed.). 1990. *Preserving Communities and Corridors*. Defenders of Wildlife, Washington, D.C.

MacNeill, J. 1989. Strategies for sustainable economic development. *Scientific American* 261: 155–165.

Maehr, D. S. 1990. The Florida panther and private lands. *Conservation Biology* 4: 167–170.

Magnuson, J. J. 1990. Long-term ecological research and the invisible present. *BioScience* 40: 495–501.

Maguire, L. A. and C. Servheen. 1992. Integrating biological and sociological concerns in endangered species management: Augmentation of grizzly bear populations. *Conservation Biology* 6: 426–434.

Mahar, D. 1988. *Government Policies and Deforestation in Brazil's Amazon Region*. World Bank Environment Department Working Paper No. 7. Washington, D.C.

Makarewicz, J. C. and P. Bertram. 1991. Evidence for the restoration of the Lake Erie ecosystem. *BioScience* 41: 216–223.

Maltby, E. 1988. Wetland resources and future prospects: an international perspective. *In* J. Zelazny and J. S. Feierabend (eds.), *Wetlands: Increasing Our Wetland Resources*, pp. 3–14. National Wildlife Federation, Washington, D.C.

Manire, C. A. and S. H. Gruber. 1990. Many sharks may be headed toward extinction. *Conservation Biology* 4: 10–11.

Mares, M. A. 1992. Neotropical mammals and the myth of Amazonian biodiversity. *Science* 255: 976–979.

Marsh, G. P. 1864. *Man and Nature; or, Physical Geography as Modified by Human Action*. Reprinted in 1965, D. Lowenthal (ed.). Harvard University Press, Cambridge, MA.

Martin, P. S. 1973. The discovery of America. *Science* 179: 969–974.

Martin, P. S. 1986. Refuting late Pleistocene extinction models. *In* D. K. Elliott (ed.), *Dynamics of Extinction*, pp. 107–130. John Wiley, New York.

Martin, P. S. and R. G. Klein (eds.). 1984. *Quaternary Extinctions: A Prehistoric Revolution.* University of Arizona Press, Tucson.

Martin, V. (ed.). 1988. *For the Conservation of Earth.* Fulcrum, Golden, CO.

Master, L. L. 1991. Assessing threats and setting priorities for conservation. *Conservation Biology* 5: 559–563.

Mathews, A. 1992. *Where the Buffalo Roam.* Grove Weidenfeld, New York.

Matusek, J. E. and G. L. Beggs. 1988. Fish species richness in relation to lake area, pH, and other abiotic factors in Ontario lakes. *Canadian Journal of Fisheries and Aquatic Sciences* 45: 1931–1941.

May, R. M. 1988a. Conservation and disease. *Conservation Biology* 2: 28–30.

May, R. M. 1988b. How many species are there on Earth? *Science* 241: 1441–1449.

May, R. M. 1992. How many species inhabit the Earth? *Scientific American* 267: 42–48.

McCloskey, J. M. and H. Spalding. 1989. A reconnaissance-level inventory of the amount of wilderness remaining in the world. *Ambio* 18: 221–227.

McDonald, K. A. and J. H. Brown. 1992. Using montane mammals to model extinctions due to global change. *Conservation Biology* 6: 409–415.

McGoodwin, J. R. 1990. *Crisis in the World's Fisheries: People, Problems, and Politics.* Stanford University Press, Stanford, CA.

McIntosh, R. P. 1985. *The Background of Ecology: Concept and Theory.* Cambridge University Press, Cambridge.

McMinn, J. W. No date. Biological diversity research: An analysis. U.S. Department of Agriculture Forest Service, Southeastern Forest Experiment Station. General Technical Report SE-71.

McNamee, T. 1986. Yellowstone's missing element. *Audubon* 88: 12–19.

McNaughton, S. J. 1989. Ecosystems and conservation in the twenty-first century. *In* D. Western and M. Pearl (eds.), *Conservation for the Twenty-First Century*, pp. 109–120. Oxford University Press, New York.

McNeely, J. A. 1987. How dams and wildlife can coexist: Natural habitats, agriculture, and major water resource development projects in tropical Asia. *Conservation Biology* 1: 228–238.

McNeely, J. A. 1988. *Economics and Biological Diversity: Developing and Using Economic Incentives to Conserve Biological Resources.* IUCN, Gland, Switzerland.

McNeely, J. A. 1989. Protected areas and human ecology: How national parks can contribute to sustaining societies of the twenty-first century. *In* D. Western and M. Pearl (eds.), *Conservation for the Twenty-First Century*, pp. 150–165. Oxford University Press, New York.

McNeely, J. A. 1990. The future of national parks. *Environment* 32: 16–41.

McNeely, J. A. (ed.). 1993a. *Protected Areas and Modern Societies: Regional Reviews of Conservation Issues.* IUCN, Gland, Switzerland.

McNeely, J. A. (ed.). 1993b. *Building Partnerships for Conservation.* IUCN, Gland, Switzerland.

McNeely, J. A. and K. R. Miller (eds.). 1984. *National Parks, Conservation and Development: The Role of Protected Areas in Sustaining Society.* Smithsonian Institution Press, Washington, D.C.

McNeely, J. A., et al. 1990. *Conserving the World's Biological Diversity.* IUCN/WRI/CI/WWF-US/World Bank, Gland, Switzerland.

McPhee, J. 1971. *Encounters with the Archdruid.* Farrar, Straus, and Giroux, New York.

Meagher, T. R., J. Antonovics and R. Primack. 1978. Experimental ecological genetics in *Plantago*. III. Genetic variation and demography in relation to survival of *Plantago cordata*, a rare species. *Biological Conservation* 17: 243–257.

Medina, E. and O. Huber. 1992. The role of biodiversity in the function of savannah ecosystems. *In* O. T. Solbrig, H. M. van Emden and P. G. W. J. van Oordt (eds.), *Biodiversity and Global Change*, pp. 139–158. International Union of Biological Sciences, Paris.

Meffe, G. K., A. H. Ehrlich and D. Ehrenfeld. 1993. Human population control: The missing agenda. *Conservation Biology* 7: 1–3.

Menges, E. S. 1986. Predicting the future of rare plant populations: Demographic monitoring and modeling. *Natural Areas Journal* 6: 13–25.

Menges, E. S. 1990. Population viability analysis for an endangered plant. *Conservation Biology* 4: 52–62.

Menges, E. S. 1991. The application of minimum viable population theory to plants. *In* D. A. Falk and K. E. Holsinger (eds.), *Genetics and Conservation of Rare Plants*, pp. 45–61. Oxford University Press, New York.

Menges, E. S. 1992. Stochastic modeling of extinction in plant populations. *In* P. L. Fiedler and S. K. Jain (eds.), *Conservation Biology: The Theory and Practice of Nature Conservation, Preservation and Management* pp. 253–275. Chapman and Hall, New York.

Merrill, E. D. [1948] 1991. *Metasequoia*, another "living fossil". *Arnoldia* 51: 12–16.

Mickleburgh, S. P., P. A. Racey and A. M. Hutson. 1992. *Old World Fruit Bats.* IUCN, Gland, Switzerland.

Mies, M. 1991. *Consumption Patterns of the North: The Cause of Environmental Destruction and Poverty in the South.* UNCED/UNICEF/UNFPA, Geneva, Switzerland.

Miller, J. A. 1984. Hatching a plan for iguana stew. *Science News* 126:87.

Miller, K. R. and L. Tangley. 1991. *Trees of Life: Saving Tropical Forests and Their Biological Wealth.* Beacon Press, Boston.

Miller, R. M. 1990. Mycorrhizae and succession. *In* W. R. Jordan III, M. E. Gilpin and J. D. Aber (eds.), *Restoration Ecology: A Synthetic Approach to Ecological Research.* pp. 205–220. Cambridge University Press, Cambridge.

Mills, K. H. and D. W. Schindler. 1986. Biological indicators of lake acidification. *Water, Air, and Soil Pollution* 30: 779–89.

Milner-Gulland, E. J. and R. Mace. 1991. The impact of the ivory trade on the African elephant *Loxodonta africana* population as assessed by data from the trade. *Biological Conservation* 55: 215–229.

Ministry for Population and Environment. 1991. *Biodiversity Action Plan for Indonesia*. Ministry of State for Population and the Environment, Jakarta.

Mishra, H. 1984. A delicate balance: Tigers, rhinoceros, tourists and park management vs. the needs of the local people in Royal Chitwan National Park, Nepal. *In* J. A. McNeely and K. R. Miller (eds.), *National Parks, Conservation, and Development: The Role of Protected Areas in Sustaining Society*, pp. 197–205. Smithsonian Institution Press, Washington, D.C.

Mishra, H., C. Wemmer, J. L. D. Smith and P. Wegge. 1992. Biopolitics of saving Asian mammals in the wild: Balancing conservation with human needs in Nepal. Noragic Technical Series No. 11, NLH, Norway.

Mitchell, J. G. 1992. Our disappearing wetlands. *National Geographic* 182(October): 3–45.

Mittermeier, R. A. 1988. Primate diversity and the tropical forest: Case studies from Brazil and Madagascar and the importance of the megadiversity countries. *In* E. O. Wilson and F. M. Peter (eds.), *Biodiversity*, pp. 145–154. National Academy Press, Washington, D.C.

Mittermeier R. A. and T. B. Werner. 1990. Wealth of plants and animals unites "megadiversity" countries. *Tropicus* 4: 1, 4–5.

Mlot, C. 1989a. The science of saving endangered species. *BioScience* 39: 68–70.

Mlot, C. 1989b. Blueprint for conserving plant diversity. *BioScience* 39: 364–368.

Mlot, C. 1989c. Global risk assessment. *BioScience* 39: 428–430.

Mlot, C. 1992. Botanists sue Forest Service to preserve biodiversity. *Science* 257: 1618–1619.

Mohsin, A. K. M. and M. A. Ambak. 1983. *Freshwater Fishes of Peninsular Malaysia*. University Pertanian Malaysia Press, Kuala Lumpur, Malaysia.

Mooney, H. A. and J. A. Drake (eds.). 1986. *Ecology of Biological Invasions of North America and Hawaii*. Ecological Studies 58. Springer-Verlag, New York.

Moore, N. 1987. *The Bird of Time: Science and the Politics of Nature Conservation*. Cambridge University Press, New York.

Moorhouse, R. J. and R. G. Powlesland. 1991. Aspects of the ecology of kakapo *Strigops habroptilus* liberated on Little Barrier Island (Hauturu), New Zealand. *Biological Conservation* 56: 349–365.

Morris, M. G. 1971. The management of grassland for the conservation of invertebrate animals. *In* E. Duffey and A. S. Watt (eds.), *The Scientific Management of Animal and Plant Communities for Conservation*, pp. 527–552. Blackwell Scientific, London.

Moyle, P. B. and R. A. Leidy. 1992. Loss of biodiversity in aquatic ecosystems: Evidence from fish faunas. *In* P. L. Fiedler and S. K. Jain (eds.), *Conservation Biology: The Theory and Practice of Nature Conservation*, pp. 127–169. Chapman and Hall, New York.

Muir, J. 1901. *Our National Parks*. Houghton Mifflin, Boston, MA.

Muir, J. 1916. *A Thousand Mile Walk to the Gulf*. Houghton Mifflin, Boston.

Murphy, D. D. and S. B. Weiss. 1988. Ecological studies and the conservation of the Bay Checkerspot Butterfly, *Euphydryas editha bayensis*. *Biological Conservation* 46: 183–200.

Murphy, D. D., K. E. Freas and S. B. Weiss. 1990. An environment–metapopulation approach to population viability analysis for a threatened invertebrate. *Conservation Biology* 4: 41–51.

Murphy, P. G. and A. E. Lugo. 1986. Ecology of tropical dry forest. *Annual Review of Ecology and Systematics* 17: 67–88.

Mwalyosi, R. B. B. 1991. Ecological evaluation for wildlife corridors and buffer zones for Lake Manyara National Park, Tanzania, and its immediate environment. *Biological Conservation* 57: 171–186.

Myers, F. W. and A. Anderson. 1992. Microbes from 20,000 feet under the sea. *Science* 255: 28–29.

Myers, J. G. 1934. The arthropod fauna of a rice-ship, trading from Burma to the West Indies. *Journal of Animal Ecology* 3: 146–149.

Myers, K. 1986. Introduced vertebrates in Australia, with emphasis on the mammals. *In* R. H. Grove and J. J. Burdon (eds.), *Ecology of Biological Invasions*, pp. 120–136. Cambridge University Press, Cambridge.

Myers, N. 1979. *The Sinking Ark: A New Look at the Problem of Disappearing Species*. Pergamon Press, New York.

Myers, N. 1980. *Conversion of Tropical Moist Forests*. National Academy of Sciences, Washington, D.C.

Myers, N. 1983. *A Wealth of Wild Species*. Westview Press, Boulder, CO.

Myers, N. 1984. *The Primary Source: Tropical Forests and Our Future*. Norton, New York.

Myers, N. 1986. Tropical deforestation and a mega-extinction spasm. *In* M. E. Soulé (ed.), *Conservation Biology: The Science of Scarcity and Diversity*, pp. 394–409. Sinauer Associates, Sunderland, MA.

Myers, N. 1987. The extinction spasm impending: Synergisms at work. *Conservation Biology* 1: 14–21.

Myers, N. 1988a. Threatened biotas: "Hotspots" in tropical forests. *Environmentalist* 8: 1–20.

Myers, N. 1988b. Tropical forests: Much more than stocks of wood. *Journal of Tropical Ecology* 4: 209–221.

Myers, N. 1991a. The biodiversity challenge: Expanded "hotspots" analysis. *Environmentalist* 10: 243–256.

Myers, N. 1991b. Tropical deforestation: The latest situation. *BioScience* 41: 282.

Nabhan, G. P. 1985. Native crop diversity in Aridoamerica: Conservation of regional gene pools. *Economic Botany* 39: 387–399.

Nabhan, G. P. 1986. Native American crop diversity, genetic resource conservation and the policy of neglect. *Agriculture and Human Values* 2: 14–17.

Nabhan, G. P., A. M. Rea, K. L. Reichhardt, E. Mellink and C. F. Hutchinson. 1982. Papago influences on habitat and biotic diversity: Quitovac oasis ethnoecology. *Journal of Ethnobiology* 2: 124–143.

Nabhan, G. P., D. House, H. Suzan A., W. Hodgson, L. Hernandez and G. Malda. 1991. Conservation and use of rare plants by traditional cultures of the U.S./Mexico borderlands. *In* M. L. Oldfield and J. B. Alcorn (eds.), *Biodiversity: Culture, Conservation and Ecodevelopment*, pp. 127–146. Westview Press, Boulder, CO.

Naess, A. 1986. Intrinsic value: Will the defenders of nature please rise? *In* M. E. Soulé (ed.), *Conservation Biology: The Science of Scarcity and Diversity*, pp. 153–181. Sinauer Associates, Sunderland, MA.

Naess, A. 1989. *Ecology, Community and Lifestyle*. Cambridge University Press, Cambridge.

Nagasaki, F. 1990. The case for scientific whaling. *Nature* 344: 189–90.

Nash, R. 1982. *Wilderness and the American Mind*. Yale University Press, New Haven, CT.

Nash, R. 1989. *The Rights of Nature*. University of Wisconsin Press, Madison.

Nash, S. 1991. What price nature? *BioScience* 41: 677–680.

National Academy of Sciences (NAS). 1980. *Research Priorities in Tropical Biology*. Committee on Research Priorities in Tropical Biology, National Academy of Sciences, Washington, D.C.

National Academy of Sciences/National Research Council (NAS/NRC). 1972. *Genetic Vulnerability of Major Crop Plants*. Washington, D.C.

National Wildlife Federation (NWF). 1992. *Conservation Directory 1992*. National Wildlife Federation, Washington, D.C.

Nations, J. D. 1988. Deep ecology meets the developing world. *In* E. O. Wilson and F. M. Peter (eds.), *Biodiversity*, pp. 79–82. National Academy Press, Washington, D.C.

Naveh, Z. and A. S. Lieberman. 1984. *Landscape Ecology: Theory and Application*. Springer-Verlag, New York.

Nedelman, J. and J. A. Thompson. 1987. The statistical demography of whooping cranes. *Ecology* 68: 1401–1411.

Needham, J. 1962. *Science and Civilization in China*. Cambridge University Press, Cambridge.

Nei, M., T. Maruyama and R. Chakraborty. 1975. The bottleneck effect and genetic variability in populations. *Evolution* 29: 1–10.

Nepstad, D. C. and S. Schwartzman (eds.). 1992. *Non-Timber Products from Tropical Forests: Evaluation of a Conservation and Development Strategy*. New York Botanical Gardens, Bronx, NY.

Nepstad, D.C., F. Brown, L. Luz, A. Alechandra and V. Viana. 1992. Biotic impoverishment of Amazonian forests by rubber tappers and cattle ranchers. *In* D.

C. Nepstad and S. Schwartzman (eds.), *Non-Timber Products from Tropical Forests: Evaluation of a Conservation and Development Strategy*. New York Botanical Garden, Bronx, NY.

Netherlands National Committee for IUCN/Steering Group for World Conservation Strategy. 1988. *The Netherlands and the World Ecology: Towards a National Conservation Strategy in and by the Netherlands*. Netherlands National Committee, Amsterdam.

New England Wild Flower Society. 1992. New England plant conservation program. *Wild Flower Notes* 7.

Newmark, W. D. 1985. Legal and biotic boundaries of western North American national parks: A problem of congruence. *Biological Conservation* 33: 197–208.

Newton, I. 1979. *Population Ecology of Raptors*. Buteo Books, Vermillion, SD.

Nilsson, G. 1983. *The Endangered Species Handbook*. Animal Welfare Institute, Washington, D.C.

Noonan, P. F. and M. D. Zagata 1982. Wildlife in the marketplace: Using the profit motive to maintain wildlife habitat. *Wildlife Society Bulletin* 10: 46–49.

Norman, C. 1981. Snail darter's status threatened. *Science* 212: 761.

Norris, K. S. 1992. Dolphins in crisis. *National Geographic* 182(September): 2–35.

Norse, E. A. 1987. International lending and the loss of biological diversity. *Conservation Biology* 1: 259–260.

Norse, E. A., et al. 1986. *Conserving Biological Diversity in Our National Forests*. The Wilderness Society, Washington, D.C.

Norton, B. G. (ed.). 1986a. *The Preservation of Species: The Value of Biological Diversity*. Princeton University Press, Princeton, NJ.

Norton, B. G. 1986b. On the inherent danger of undervaluing species. *In* B. G. Norton (ed.), *The Preservation of Species: The Value of Biological Diversity*, pp. 110–137. Princeton University Press, Princeton, NJ.

Norton, B. G. 1988. Commodity, amenity and morality: the limits of quantification in valuing biodiversity. *In* E. O. Wilson and F. M. Peter (eds.), *Biodiversity*, pp. 200–205. *Biodiversity*. National Academy Press, Washington, D.C.

Norton, B. G. 1991. *Toward Unity Among Environmentalists*. Oxford University Press, New York.

Noss, R. F. 1983. A regional landscape approach to maintain diversity. *BioScience* 33: 700–706.

Noss, R. F. 1987. From plant communities to landscapes in conservation inventories: A look at The Nature Conservancy (USA). *Biological Conservation* 41: 11–37.

Noss, R. F. 1992. Essay: Issues of scale in conservation biology. *In* P. L. Fiedler and S. K. Jain (eds.), *Conservation Biology: The Theory and Practice of Nature Conservation, Preservation and Management* pp. 239–250. Chapman and Hall, New York.

O'Brien, S. J. and J. F. Evermann. 1988. Interactive influence of infectious disease and genetic diversity in natural populations. *Trends in Ecology and Evolution* 3: 254–259.

O'Brien, S. J., et al. 1985a. The cheetah is depauperate in genetic variation. *Science* 221: 459–462.

O'Brien, S. J., et al. 1985b. Genetic basis for species vulnerability in the cheetah. *Science* 227: 1428–1434.

Ochumba , P. B. O. and D. I. Kibara. 1989. Observations on blue-green algal blooms in the open waters of Lake Victoria, Kenya. *African Journal of Ecology* 27: 23–34.

Odum, E. P. 1993. *Ecology and Our Endangered Life-Support Systems*, 2nd ed. Sinauer Associates, Sunderland, MA.

Oedekoven, K. 1980. The vanishing forest. *Environmental Policy and Laws* 6: 184–185.

Office of Technology Assessment of the U.S. Congress (OTA). 1987. *Technologies to Maintain Biological Diversity*. OTA-F-330. U.S. Government Printing Office, Washington, D.C.

Ogutu-Ohwayo, R. 1990. The decline of the native fishes of Lakes Victoria and Kyoga (East Africa) and the impact of introduced species, especially the Nile perch, *Lates niloticus* and the Nile tilapia, *Oreochromis niloticus*. *Environmental Biology of Fishes* 27: 81–90.

Oldfield, M. L. 1984. *The Value of Conserving Genetic Resources*. Sinauer Associates, Sunderland, MA.

Oldfield, M. L. and J. B. Alcorn. 1987. Conservation of traditional agroecosystems. *BioScience* 37: 199–208.

Oldfield, M. L. and J. B. Alcorn. (eds.). 1991. *Biodiversity: Culture, Conservation and Ecodevelopment*. Westview Press, Boulder, CO.

Olivieri, I., D. Couvet and P. H. Gouyon. 1990. The genetics of transient populations: Research at the metapopulation level. *Trends in Ecology and Evolution* 5: 207–210.

Olson, S. L. 1989. Extinction on islands: Man as a catastrophe. *In* D. Western and M. Pearl (eds.), *Conservation for the Twenty-First Century*, pp. 50–53. Oxford University Press, Oxford.

Olson, S. L. and H. F. James. 1982. Fossil birds from the Hawaiian Islands: Evidence for wholesale extinction by man before western contact. *Science* 217: 633–635.

Ono, R. D., J. D. Williams and A. Wagner. 1983. *Vanishing Fishes of North America*. Stone Wall Press, Washington, D.C.

Oren, D. C. 1987. Grande Carajás, international financing agencies, and biological diversity in southeastern Brazilian Amazonia. *Conservation Biology* 1: 222–227.

Orians, G. H. 1980. Habitat selection: General theory and applications to human behavior. *In* J. S. Lockard (ed.), *The Evolution of Human Social Behavior*, pp. 49–66. Elsevier/North Holland, New York.

Orians, G. H., G. M. Brown, W. E. Kunin and J. E. Sweirzbinski (eds.). 1990. *The Preservation and Valuation of Biological Resources*. University of Washington Press, Seattle.

Osborn, F. 1948. *Our Plundered Planet*. Little, Brown, Boston.

Ottochilo, W. K. 1987. The causes of the recent heavy elephant mortality in the Tsavo ecosystem, Kenya, 1975–1980. *Biological Conservation* 41: 279–289.

Packer, C. 1992. Captives in the wild. *National Geographic* 181(April): 122–136.

Packer, C., A. E. Pusey, H. Rowley, D. A. Gilbert, J. Martenson and S. J. O'Brien. 1991. Case study of a population bottleneck: Lions of the Ngorongoro Crater. *Conservation Biology* 5: 219–230.

Pagel, M. D., R. M. May and A. R. Collie. 1991. Ecological aspects of the geographical distribution and diversity of mammalian species. *American Naturalist* 137: 791–815.

Paine, R. T. 1966. Food web complexity and species diversity. *American Naturalist* 100: 65–75.

Palmer, M. E. 1987. A critical look at rare plant monitoring in the United States. *Biological Conservation* 39: 113–127.

Panayotou, T. and P. S. Ashton. 1992. *Not by Timber Alone: Economics and Ecology for Sustaining Tropical Forests*. Island Press, Washington, D.C.

Pandit, A. K. 1991. Conservation of wildlife resources in wetland ecosystems of Kashmir, India. *Journal of Environmental Management* 33: 143–154.

Panwar, H. S. 1987. Project Tiger: The reserves, the tigers, and their future. *In* R. L. Tilson and U. S. Seal (eds.). *Tigers of the World: The Biology, Biopolitics, Management and Conservation of an Endangered Species*, pp. 100–117. Noyes Publications, Park Ridge, NJ.

Parenti, L. 1984. Biogeography of the Andean killifish genus *Orestias* with comments on the species flock concept. *In* A. A. Echelle and I. Kornfield (eds.), *Evolution of Species Flocks*, pp. 85–92. University of Maine Press, Orono.

Parikh, J. and K. Parikh. 1991. *Consumption Patterns: The Driving Force of Environmental Stress*. UNCED, Geneva, Switzerland.

Parson, E. A., P. M. Haas and M. A. Levy. 1992. A summary of the major documents signed at the Earth Summit and the Global Forum. *Environment* 34: 12–15, 34–36.

Pascua, M. P. 1991. Ozette: A Makah village in 1491. *National Geographic* 180(October): 38–53.

Paterson, R. and P. Paterson. 1989. The status of the recovering stock of Humpback whales *Megaptera novaeangliae* in East Australian waters. *Biological Conservation* 47: 33–48.

Patterson, A. 1990. Debt for nature swaps and the need for alternatives. *Environment* 32: 5–32.

Pavlik, B. M. and M. G. Barbour. 1988. Demographic monitoring of endemics and dune plants, Eureka Valley, California. *Biological Conservation* 46: 217–42.

Pearce, D. W. 1987. The sustainable use of natural resources in developing countries. *In* R. K. Turner (ed.), *Sustainable Environmental Management: Principles and Practice*, pp. 102–118. F. Pinter, London.

Pearce, D., A. Markandya and E. B. Barbier. 1989. *Blueprint for a Green Economy*. Earthscan, London.

Pearce, F. 1990. Bolivian Indians march to save their homeland. *New Scientist* 25 August 1990.

Pearl, M. C. 1989. How the developed world can promote conservation in emerging nations. *In* D. Western and M. Pearl (eds.), *Conservation for the Twenty-First Century*, pp. 274–283. Oxford University Press, New York.

Pearson, D. L. and F. Cassola. 1992. World-wide species richness patterns of tiger beetles (Coleoptera: Cicindelidae): Indicator taxon for biodiversity and conservation studies. *Conservation Biology* 6: 376–391.

Pechmann, J. H. K., D. E. Scott, R. D. Semlitsch, J. P. Caldwell, L. J. Vitt and J. W. Gibbons. 1991. Declining amphibian populations: The problem of separating human impacts from natural fluctuations. *Science* 253: 892–895.

Peet, R. K. 1974. The measurement of species diversity. *Annual Review of Ecology and Systematics* 5: 285–307.

Peeters, J. P. and J. T. Williams. 1984. Towards better use of gene banks with special reference to information. *Plant Genetic Resource News (FAO)* 60: 22–32.

Peluso, N. L. 1992. The Ironwood problem: (Mis)management and development of an extractive rainforest product. *Conservation Biology* 6: 210–219.

Peres, C. A. 1990. Effects of hunting on western Amazonian primate communities. *Biological Conservation* 54: 47–59.

Perrin, W. F. 1991. Why are there so many kinds of whales and dolphins? *BioScience* 41: 460–461.

Peterken, G. F. 1982. *Woodland Conservation and Management*. Chapman and Hall, New York.

Peters, C. M., A. H. Gentry and R. Mendelsohn. 1989. Valuation of a tropical forest in Peruvian Amazonia. *Nature* 339: 655–656.

Peters, R. L. II. 1988. The effect of global climatic change on natural communities. *In* E. O. Wilson and F. M. Peter (eds.), *Biodiversity*, pp. 450–461. National Academy Press, Washington, D.C.

Peters, R. L. and J. D. S. Darling. 1985. The greenhouse effect and nature reserves. *BioScience* 35: 707–717.

Peters, R. L. and T. E. Lovejoy (eds.). 1992. *Global Warming and Biological Diversity*. Yale University Press, New Haven, CT.

Peterson, G. L and A. Randall. 1984. *Valuation of Wildlife Resource Benefits*. Westview Press, Boulder, CO.

Phillips, B. 1992. Belize nature reserves. *Belize Today* 6(8): 32–33.

Phillips, K. 1990. Where have all the frogs and toads gone? *BioScience* 40: 422–424.

Phillips, M. K. 1990a. The red wolf: Recovery of an endangered species. *Endangered Species Update* 8: 79–81.

Phillips, M. K. 1990b. Measures of the value and success of a reintroduction projects: Red wolf reintroduction in Alligator River National Wildlife Refuge. *Endangered Species Update* 8: 24–26.

Phillips, M. K. and V. G. Henry. 1992. Comments on red wolf taxonomy. *Conservation Biology* 6: 596–599.

Pianka, E. 1966. Latitudinal gradients in species diversity: A review of the concepts. *American Naturalist* 100: 33–46.

Pielou, E. C. 1969. *An Introduction to Mathematical Ecology*. Wiley-Interscience, New York.

Pielou, E. C. 1979. *Biogeography*. John Wiley, New York.

Pimm, S. L. 1982. *Food Webs*. Chapman and Hall, London.

Pimm, S. L. 1984. The complexity and stability of ecosystems. *Nature* 307: 321–326.

Pimm, S. L. 1991. *The Balance of Nature: Ecological Issues in the Conservation of Species and Communities*. University of Chicago Press, Chicago.

Pimm, S. L., H. L. Jones and J. Diamond. 1988. On the risk of extinction. *American Naturalist* 132: 757–785.

Pinchot, G. 1947. *Breaking New Ground*. Harcourt, Brace & Co., New York.

Pinedo-Vasquez, M., D. Zarin and P. Jipp. 1992. Community forest and lake reserves in the [Peruvian Amazon]: A local alternative for sustainable development of tropical forests. *In* D. C. Nepstad and S. Schwartzman (eds.), *Non-Timber Products from Tropical Forests: Evaluation of a Conservation and Development Strategy*. New York Botanical Gardens, Bronx, NY.

Pinto, L. F. 1990. A Cor do Verde. *Jornal do Commercio*, p. 4 (Caderno Cidades.) Recife, Brazil.

Pister, E. P. 1985. Desert pupfishes: Reflections on reality, desirability, and conscience. *Environmental Biology of Fishes* 12: 3–11.

Plotkin, M. J. 1988. The outlook for new agricultural and industrial products from the tropics. *In* E. O. Wilson and F. M. Peter (eds.), *Biodiversity*, pp. 106–116. National Academy Press, Washington, D.C.

Plotkin, M. J. and L. Famolare. 1992. *Sustainable Harvest and Marketing of Rain Forest Products*. Island Press, Washington, D.C.

Plucknett, D. L., N. J. H. Smith, J. T. Williams and N. M. Anishetty. 1987. *Gene Banks and the World's Food*. Princeton University Press, Princeton, NJ.

Poffenberger, M. (ed.). 1990. *Keepers of the Forest*. Kumarian, West Hartford, CT.

Pollard, E. 1991. Monitoring butterfly numbers. *In* B. Goldsmith (ed.), *Monitoring for Conservation and Ecology*, pp. 87–111. Chapman and Hall, New York.

Poore, D. and J. Sayer. 1991. *The Management of Tropical Moist Forest Lands*. IUCN, Gland, Switzerland.

Poore, D., P. Burgess, J. Palmer, S. Rietbergen and T. Synnott. 1989. *No Timber Without Trees: Sustainability in the Tropical Forest*. Earthscan Publications, London.

Popper, F. J. and D. E. Popper. 1991. The reinvention of the American frontier. *Amicus Journal* (Summer): 4–7.

Porteous, P. L. 1992. Eagles on the rise. *National Geographic* 182(November): 42–55.

Posey, D. A. 1983. Indigenous knowledge and development: An ideological bridge to the future. *Ciencia e Cultura* 35: 877–894.

Posey, D. A. 1992. Traditional knowledge, conservation, and "the rain forest harvest". *In* M. Plotkin and L. Famolare (eds.), *Sustainable Harvest and Marketing of Rain Forest Products*, pp. 46–50. Island Press, Washington, D.C.

Poten, C. J. 1991. A shameful harvest: America's illegal wildlife trade. *National Geographic* 180(September): 106–132.

Powell, A. N. and F. J. Cuthbert. 1993. Augmenting small populations of plovers: An assessment of cross-fostering and captive-rearing. *Conservation Biology* 7: 160–168.

Power, T. M. 1991. Ecosystem preservation and the economy in the Greater Yellowstone area. *Conservation Biology* 5: 395–404.

Prance, G. T. 1982. *Biological Diversification in the Tropics*. Columbia University Press, New York.

Prance, G. T., W. Balée, B. M. Boom and R. L. Carneiro. 1987. Quantitative ethnobotany and the case for conservation in Amazonia. *Conservation Biology* 1: 296–310.

Prescott-Allen, C. and R. Prescott-Allen. 1986. *The First Resource: Wild Species in the North American Economy*. Yale University Press, New Haven, CT.

Prescott-Allen, R. 1986. *National Conservation Strategies and Biological Diversity*. Report to the IUCN, Gland, Switzerland.

Prescott-Allen, R. and C. Prescott-Allen. 1978. Sourcebook for a world conservation strategy: Threatened vertebrates. General Assembly Paper GA.78/10 Addendum 6. IUCN, Gland, Switzerland.

Prescott-Allen, R. and C. Prescott-Allen. 1982. *What's Wildlife Worth? Economic Contributions of Wild Plants and Animals to Developing Countries*. Earthscan, London.

Price, P. W. 1992. The resource-based organization of communities. *Biotropica* (Special Issue) 24: 273–282.

Primack, R. B. 1988. Forestry in Fujian Province (People's Republic of China) during the Cultural Revolution. *Arnoldia* 48: 26–29.

Primack, R. B. 1991a. Logging, conservation and native rights in Sarawak forests. *Conservation Biology* 5: 126–130.

Primack, R. B. 1991b. Response to Salzman. *Conservation Biology* 5: 266–7.

Primack, R. B. 1992. Tropical community dynamics and conservation biology. *BioScience* 42: 818–820.

Primack, R. B. and P. Hall. 1992. Biodiversity and forest change in Malaysian Borneo. *BioScience* 42: 829–837.

Primack, R. B. and S. L. Miao. 1992. Dispersal can limit local plant distribution. *Conservation Biology* 6: 513–519.

Primack, R. B., E. Hendry and P. Del Tredici. 1986. Current status of *Magnolia virginiana* in Massachusetts. *Rhodora* 88: 357–365.

Prins, H. H. T. 1987. Nature conservation as an integral part of optimal land use in East Africa: The case of the Masai ecosystem of northern Tanzania. *Biological Conservation* 40: 141–161.

Pritchard, P. C. 1991. "The best idea America ever had": The National Parks Service turns 75. *National Geographic* 180(August): 36–59.

Proceedings of the U.S. Strategy Conference on Biological Diversity, November 16–18, 1981. 1982. Department of State Publication 9262. 0-377-963. U.S. Government Printing Office, Washington, D.C.

Pullin, A. S. and S. R. J. Woodell. 1987. Response of the fen violet, *Viola persicifolia* Schreber, to different management regimes at Woodwalton Fen National Nature Reserve, Cambridgeshire, England. *Biological Conservation* 41: 203–217.

Putwain, P. D. and D. A. Gilham. 1990. The significance of the dormant viable seed bank in the restoration of heathlands. *Biological Conservation* 52: 1–16.

Quinn, J. F. and S. P. Harrison. 1988. Effects of habitat fragmentation and isolation on species richness: Evidence from biogeographic patterns. *Oecologia* 75: 132–140.

Rabinowitz, D., S. Cairnes and T. Dillon. 1986. Seven forms of rarity and their frequency in the flora of the British Isles. *In* M. E. Soulé (ed.), *Conservation Biology: The Science of Scarcity and Diversity*, pp. 182–204. Sinauer Associates, Sunderland, MA.

Ralls, K. and J. Ballou. 1983. Extinction: Lessons from zoos. *In* C. M. Schonewald-Cox, S. M. Chambers, B. MacBryde and L. Thomas (eds.), *Genetics and Conservation: A Reference for Managing Wild Animal and Plant Populations*, pp. 164–184. Benjamin/Cummings, Menlo Park, CA.

Ralls, K. and J. Ballou. 1987. Captive breeding programs for populations with a small number of founders. *Trends in Ecology and Evolution* 1: 19–22.

Ralls, K. and R. L. Brownell. 1989. Protected species: Research permits and the value of basic research. *BioScience* 39: 394–396.

Ralls, K., P. H. Harvey and A. M. Lyles. 1986. Inbreeding in natural populations of birds and mammals. *In* M. E. Soulé (ed.), *Conservation Biology: The Science of Scarcity and Diversity*, pp. 35–56. Sinauer Associates, Sunderland, MA.

Ralls, K., J. D. Ballou and A. Templeton. 1988. Estimates of lethal equivalents and the cost of inbreeding in mammals. *Conservation Biology* 2: 185–193.

Rand, S. A. and H. W. Greene. 1982. Latitude and climate in the phenology of reproduction in the green iguana. *In Iguanas of the World: Their Behavior, Ecology, and Conservation*. Noges Publications, Park Ridge, NJ.

Randall, A. 1986. Human preferences, economics, and

the undervaluing of species. *In* B. G. Norton (ed.), *The Preservation of Species*, pp. 79–109. Princeton University Press, Princeton, NJ.

Randall, A. 1987. *Resource Economics*, 2nd ed. John Wiley, New York.

Ratcliffe, D. A. 1984. *Nature Conservation in Great Britain*. Nature Conservancy Council, Shrewsbury.

Rauh, W. 1979. Problems of biological conservation in Madagascar. *In* D. Bramwell (ed.), *Plants and Islands*, pp. 405–421. Academic Press, New York.

Raup, D. M. 1978. Cohort analysis of generic survivorship. *Paleobiology* 4: 1–15.

Raup D. M. 1979. Size of the Permo-Triassic bottleneck and its evolutionary implications. *Science* 206: 217–218.

Raup, D. M. 1988. Diversity crises in the geological past. *In* E. O. Wilson and F. M. Peter (eds.), *Biodiversity*, pp. 51–57. National Academy Press, Washington, D.C.

Raup, D. M. and J. J. Sepkoski, Jr. 1982. Mass extinctions in the marine fossil record. *Science* 215: 1501–1503.

Raup, D. M. and S. M. Stanley. 1978. *Principles of Paleontology*, 2nd ed. W.H. Freeman, San Francisco, CA.

Raven, P. H. 1976. Ethics and attitudes. *In* J. B. Simmons (ed.), *Conservation of Threatened Plants*, pp. 155–179. Plenum, New York.

Raven, P. H. 1981. Research in botanical gardens. *Botanische Jahrbuecher fur Systematik Pflanzengeschichte und Pflanzengeographie* 102: 53–72.

Raven, P. H. and E. O. Wilson. 1992. A fifty-year plan for biodiversity surveys. *Science* 258: 1099–1100.

Ravenscroft, N. O. M. 1990. The ecology and conservation of the silver-studded blue butterfly *Plejebus argus* L. on the sandlings of East Anglia, England. *Biological Conservation* 53: 21–36.

Ray, G. C. 1991. Coastal-zone biodiversity patterns. *BioScience* 41: 490–498.

Ray, G. C. and J. F. Grassle. 1991. Marine biological diversity. *BioScience* 41: 453–457.

Ray, G. C. and W. P. Gregg, Jr. 1991. Establishing biosphere reserves for coastal barrier ecosystems. *BioScience* 41: 301–309.

Rebelo, A. G. and W. R. Siegfried. 1990. Protection of fynbos vegetation: Ideal and real-world options. *Biological Conservation* 54: 15–31.

Redford, K. H. 1992. The empty forest. *BioScience* 42: 412–422.

Redford, K. H. and C. Padoch (eds.). 1992. *Conservation of Neotropical Rainforests: Working from Traditional Resource Use*. Columbia University Press, New York.

Redford, K. H. and J. G. Robinson. 1991. Subsistence and commercial uses of wildlife in Latin America. *In* J. G. Robinson and K. H. Redford (eds.), *Neotropical Wildlife Use and Conservation*, pp. 6–23. University of Chicago Press, Chicago.

Reid, D. G., et al. 1989. Giant panda *Ailuropoda malanoleuca* behaviour and carrying capacity following a bamboo die-off. *Biological Conservation* 49: 85–104.

Reid, D. G., Hu Jinchu and H. Yan. 1991. Ecology of the red panda *Ailurus fulgens* in the Wolong Reserve, Sichuan, China. *Journal of Applied Ecology* 28: 229–43.

Reid, W. V. 1992. *The United States Needs a National Biodiversity Policy*. Issues and Ideas Brief. World Resources Institute, Washington, D.C.

Reid, W. V. and K. R. Miller. 1989. *Keeping Options Alive: The Scientific Basis for Conserving Biodiversity*. World Resources Institute, Washington, D.C.

Reinthal, P. N. and M. L. J. Stiassny. 1991. The freshwater fishes of Madagascar: A study of endangered fauna with recommendations for a conservation strategy. *Conservation Biology* 5: 231–243.

Repetto, R. 1990a. Deforestation in the tropics. *Scientific American* 262: 36–42.

Repetto, R. 1990b. *Promoting Environmentally Sound Economic Progress: What the North Can Do*. World Resources Institute, Washington, D.C.

Repetto, R. 1992. Accounting for environmental assets. *Scientific American* 266(June): 94–100.

Repetto, R., et al. 1989. *Wasting Assets: Natural Resources in the National Income Accounts*. World Resources Institute, Washington, D.C.

Rhoades, R. E. 1991. World's food supply at risk. *National Geographic* 179(April): 74–105.

Rich, B. 1990. Multilateral development banks and tropical deforestation. *In* S. Head and R. Heinzman (eds.), *Lessons of the Rainforest*. Sierra Club Books, San Francisco.

Richter-Dyn, N. and N. S. Goel. 1972. On the extinction of a colonizing species. *Population Biology* 3: 406–433.

Ricklefs, R. E. 1993. *The Economy of Nature*. W. H. Freeman and Co., New York.

Roberts, L. 1988. Conservationists in Panda-monium. *Science* 241: 529–531.

Robins, C. R. 1991. Regional diversity among Caribbean fish species. *BioScience* 41: 458–459.

Robinson, J. G. 1993. The limits of caring: Sustainable living and the loss of biodiversity. *Conservation Biology* 7: 20–28.

Robinson, M. H. 1988. Bioscience education through bioparks. *BioScience* 38: 630–634.

Robinson, M. H. 1992. Global change, the future of biodiversity and the future of zoos. *Biotropica* (Special Issue) 24: 345–352.

Rohlf, D. L. 1989. *The Endangered Species Act: A Guide to its Protections and Implementation*. Stanford Environmental Law Society, Stanford, CA.

Rohlf, D. L. 1991. Six biological reasons why the Endangered Species Act doesn't work—and what to do about it. *Conservation Biology* 5: 273–282.

Rojas, M. 1992. The species problem and conservation: What are we protecting? *Conservation Biology* 6: 170–178.

Rolston, H. III. 1981. Values in nature. *Environmental Ethics* 3: 113–128.

Rolston, H. III. 1985a. Valuing wildlands. *Environmental Ethics* 7: 23–48.

Rolston, H. III. 1985b. Duties to endangered species. *BioScience* 35: 718–726.

Rolston, H. III. 1987. On behalf of bioexuberance. *Garden* 11: 2–4, 31–32.

Rolston, H. III. 1988a. *Enviromental Ethics: Values In and Duties To the Natural World.* Temple University Press, Philadelphia.

Rolston, H. III. 1988b. In defense of ecosystems. *Garden* 12: 2–5, 32.

Rolston, H. III. 1989a. *Philosophy Gone Wild: Essays on Environmental Ethics.* Prometheus Books, Buffalo, NY.

Rolston, H. III. 1989b. Biology without conservation: An environmental misfit and contradiction in terms. *In* D. Western and M. Pearl (eds.), *Conservation for the Twenty-First Century,* pp. 232–240. Oxford University Press, New York.

Rolston, H. III. In press. Duties to Endangered Species. *Encyclopedia of Environmental Biology.* Harcourt/Academic Press, New York.

Rowell, G. 1991. Falcon rescue. *National Geographic* 179(June): 106–115.

Ruckelshaus, W. D. 1989. Toward a sustainable world. *Scientific American* 261(September): 166–175.

Safina, C. 1993. Bluefin tuna in the West Atlantic: Negligent management and the making of an endangered species. *Conservation Biology* 7: 229–234.

Salafsky, N., B. L. Dugelby and J. W. Terborgh. 1993. Can extractive reserves save the rain forest? An ecological and socioeconomic comparison of nontimber forest product extraction systems in Petén, Guatemala, and West Kalimantan, Indonesia. *Conservation Biology* 7: 39–52.

Sale, J. B. 1981. *The Importance and Values of Wild Plants and Animals in Africa.* IUCN, Gland, Switzerland.

Salm, R. and J. Clark. 1984. *Marine and Coastal Protected Areas: A Guide for Planners and Managers.* IUCN, Gland, Switzerland.

Salwasser, H. 1991a. Roles for land and resource managers in conserving biological diversity. *In* D. J. Decker, M. E. Krasny, G. R. Goff, C. R. Smith and D. W. Gross (eds.), *Challenges in the Conservation of Biological Resources: A Practitioner's Guide,* pp. 11–31. Westview Press, Boulder, CO.

Salwasser, H. 1991b. New perspectives for sustaining diversity in the U.S. National Forest ecosystems. *Conservation Biology* 5: 567–569.

Salwasser, H., C. M. Schonewald-Cox and R. Baker. 1987. The role of interagency cooperation in managing for viable populations. *In* M. E. Soulé (ed.), *Viable Populations for Conservation,* pp. 159–173. Cambridge University Press, Cambridge.

Samour, H. J., D. M. J. Spratt, M. G. Hart, B. Savage and C. M. Hawkey. 1987. A survey of the Aldabra giant tortoise population introduced on Curieuse Island, Seychelles. *Biological Conservation* 41: 147–158.

Saunders, D., G. Arnold, A. Burbidge and A. Hopkins. 1987. The role of remnants of native vegetation in nature conservation: Future directions. *In* D. Saunders, G. Arnold, A. Burbidge and A. Hopkins, (eds.), *Nature Conservation: The Role of Remnants of Native Vegetation,* pp. 259–268. Surrey Beatty and Sons, Chipping Norton, N.S.W., Australia.

Savidge, J. A., 1987. Extinction of an island forest avifauna by an introduced snake. *Ecology* 68: 660–668.

Sayer, J. A. 1991. *Rainforest Buffer Zones: Guidelines for Protected Area Management.* IUCN, Gland, Switzerland.

Sayer, J. A. and S. Stuart. 1988. Biological diversity and tropical forests. *Environmental Conservation.* 15: 193–194.

Sayer, J. A. and T. C. Whitmore. 1991. Tropical moist forests: Destruction and species extinction. *Biological Conservation* 55:199–213.

Schaller, G., T. Qitao, P. Wenshi, Q. Zisheng, W. Xiaoming, H. Jinchu and S. Heming. 1989. The feeding ecology of giant panda and asiatic black bear in the Tangjiahe Reserve, China. *In* J. Gittleman (ed.), *Carnivore Behavior, Ecology, and Evolution,* pp. 212–241. Cornell University Press, Ithaca, NY.

Scheffer, V. B. 1991. *The Shaping of Environmentalism in America.* University of Washington Press, Seattle.

Schindler, D. W. 1988. Effects of acid rain on freshwater ecosystems. *Science* 239: 149–157.

Schlesinger, W. H., et al. 1990. Biological feedbacks in global desertification. *Science* 247: 1043–1048.

Schmidt, K. F. 1992. Rebirth in the Prairie State. *U.S. News and World Report* May 18 1992.

Schneider, S. H. 1989. The changing climate. *Scientific American* 261(September): 70–79.

Schofield, C. L. 1988. Lake acidification in wilderness areas: An evaluation of impacts and options for rehabilitation. *In* J. K. Agee and D. R. Johnson (eds.), *Ecosystem Management for Parks and Wilderness,* pp. 135–144. University of Washington Press, Seattle.

Schofield, E. K. 1989. Effects of introduced plants and animals on island vegetation: Examples from the Galápagos Archipelago. *Conservation Biology* 3: 227–238.

Schonewald-Cox, C. M. 1983. Guidelines to management: A beginning attempt. *In* C. M. Schonewald-Cox, S. M. Chambers, B. MacBryde and L. Thomas (eds.), *Genetics and Conservation: A Reference for Managing Wild Animal and Plant Populations,* pp. 414–445. Benjamin/Cummings, Menlo Park, CA.

Schonewald-Cox, C. M. and M. Buechner. 1992. Park protection and public roads. *In* P. L. Fiedler and S. K. Jain (eds.), *Conservation Biology: The Theory and Practice of Nature Conservation, Preservation and Management* pp. 373–396. Chapman and Hall, New York.

Schonewald-Cox, C. M., S. M. Chambers, B. MacBryde and L. Thomas (eds.). 1983. *Genetics and Conservation: A Reference for Managing Wild Animal and*

Plant Populations. Benjamin/Cummings, Menlo Park, CA.

Schultes, R. E. and R. F. Raffauf. 1990. *The Healing Forest: Medicinal and Toxic Plants of the Northwest Amazonia*. Dioscorides Press, Portland, OR.

Schumacher, E. F. 1973. *Small is Beautiful: Economics as if People Mattered*. Harper & Row, New York.

Schwartzman, S. 1992. Land distribution and the social costs of frontier development in Brazil: Social and historical context of extractive reserves. *In* D. C. Nepstad and S. Schwartzman (eds.), *Non-Timber Products from Tropical Forests: Evaluation of a Conservation and Development Strategy*. New York Botanical Garden, Bronx, NY.

Scott, J. M., B. Csuti, J. D. Jacobi and J. E. Estes. 1987. Species richness. *BioScience* 37: 782–788.

Scott, J. M., C. B. Kepler, C. van Riper III and S. I. Fefer. 1988. Conservation of Hawaii's vanishing avifauna. *BioScience* 38: 232–253.

Scott, J. M., B. Csuti, J. E. Estes and H. Anderson. 1989. Status assessment of biodiversity protection. *Conservation Biology* 3: 85–87.

Scott, J. M., B. Csuti and S. Ciacco. 1991a. Gap analysis: Assessing protection needs. *In* W. E. Hudson (ed.). *Landscape Linkages and Biodiversity*. Island Press, Washington, D.C.

Scott, J. M., B. Csuti and F. Davis. 1991b. Gap analysis: An application of Geographic Information Systems for wildlife species. *In* D. J. Decker, M. E. Krasny, G. R. Goff, C. R. Smith and D. W. Gross (eds.), *Challenges in the Conservation of Biological Resources: A Practitioner's Guide*, pp. 167–179. Westview Press, Boulder, CO.

Scott, M. E. 1988. The impact of infection and disease on animal populations: Implications for conservation biology. *Conservation Biology* 2: 40–56.

Seal, U. S. 1988. Intensive technology in the care of ex situ populations of vanishing species. *In* E. O. Wilson and F. M. Peter (eds.), *Biodiversity*, pp. 289–295. National Academy Press, Washington, D.C.

Seal, U. S., E. T. Thorne, M. A. Bogan and S. H. Anderson (eds.). 1989. *Conservation Biology and the Black-Footed Ferret*. Yale University Press, New Haven, CT.

Selander, R. K. 1983. Evolutionary consequences of inbreeding. *In* C. M. Schonewald-Cox, S. M. Chambers, B. MacBryde and L. Thomas (eds.), *Genetics and Conservation: A Reference for Managing Wild Animal and Plant Populations*, pp. 201–215. Benjamin/Cummings, Menlo Park, CA.

Sepkoski, J. J., Jr. and D. M. Raup. 1986. Periodicity in marine extinction events. *In* D. K. Elliott (ed.), *Dynamics of Extinction*, pp. 3–36. John Wiley, New York.

Sessions, G. 1987. The deep ecology movement: A review. *Environmental Review* 11: 105–125.

Shafer, C. L. 1990. *Nature Reserves: Island Theory and Conservation Practice*. Smithsonian Institution Press, Washington, D.C.

Shaffer, M. L. 1981. Minimum population sizes for species conservation. *BioScience* 31: 131–134.

Shaffer, M. L. 1987. Minimum viable populations: Coping with uncertainty. *In* M. E. Soulé (ed.), *Viable Populations for Conservation*, pp. 69–86. Cambridge University Press, Cambridge.

Shaffer, M. L. 1990. Population viability analysis. *Conservation Biology* 4: 39–40.

Shaffer, M. L. 1991. Population viability analysis. *In* D. J. Decker, M. E. Krasny, G. R. Goff, C. R. Smith and D. W. Gross (eds.), *Challenges in the Conservation of Biological Resources: A Practitioner's Guide*, pp. 107–118. Westview Press, Boulder, CO.

Shaffer, M. L and F. B. Samson. 1985. Population size and extinction: a note on determining critical population size. *American Naturalist* 125: 144–152.

Shaffer, M. L. and K. A. Saterson. 1987. The biological diversity program of the U.S. Agency for International Development. *Conservation Biology* 1: 280–283.

Shapiro, J., B. Forsberg, V. LaMarra, G. Lindmark, M. Lynch, E. Smeltzer and G. Zoto. 1982. *Experiments and Experiences in Biomanipulation*. Interim Report No. 19. University of Minnesota Press, Minneapolis.

Shaw, W. W. and W. R. Mangun. 1984a. *Nonconsumptive Use of Wildlife in the United States*. U.S. Fish and Wildlife Service Resource Publication 154, Washington, D.C.

Shaw, W. W. and W. R. Mangun. 1984b. Tourism and nonconsumptive uses of wildlife in the western states. *Proceedings of the Western Association of Fish and Wildlife Agencies and the Western Division of the American Fisheries Society* 64: 171–180.

Sherwin, W. B., N. D. Murray, J. M. Graves and P. R. Brown. 1991. Measurement of genetic variation in endangered populations: Bandicoots (Marsupialia: Peramelidae) as an example. *Conservation Biology* 5: 103–108.

Shinn, E. A. 1989. What is really killing the corals. *Sea Frontiers* (March/April): 72–81.

Shulman, S. 1986. Seeds of controversy. *BioScience* 36: 647–651.

Siderits, K. and R. E. Radtke. 1977. Enhancing forest wildlife habitat through diversity. *Transactions of the North American Wildlife and Natural Resources Conference* 42: 425–434.

Silvertown, J. 1991. Dorothy's dilemma and the unification of plant population biology. *Trends in Ecology and Evolution* 6: 346–348.

Simberloff, D. 1986a. Are we on the verge of a mass extinction in tropical rain forests? *In* D. K. Elliott (ed.), *Dynamics of Extinction*, pp. 165–180. John Wiley, New York.

Simberloff, D. 1986b. Introduced insects: A biogeographic and systematic perspective. *In* H. A. Mooney and J. A. Drake (eds.), *Ecology of Biological Invasions of North America and Hawaii*, pp. 3–26. Ecological Studies 58. Springer-Verlag, New York.

Simberloff, D. 1987. The spotted owl fracas: Mixing academic, applied and political ecology. *Ecology* 68: 766–772.

Simberloff, D. 1988. The contribution of population and community biology to conservation science. *Annual Review of Ecology and Systematics* 19: 473–511.

Simberloff, D. S. and L. G. Abele. 1976. Island biogeography theory and conservation practice. *Science* 191: 285–286.

Simberloff, D. S. and L. G. Abele. 1982. Refuge design and island biogeographic theory: effects of fragmentation. *American Naturalist* 120: 41–50.

Simberloff, D. and J. Cox. 1987. Consequences and costs of conservation corridors. *Conservation Biology* 1: 63–71.

Simberloff, D. and N. Gotelli. 1984. Effects of insularization on plant species richness in the prairie - forest ecotone. *Biological Conservation* 29: 27–46.

Simberloff, D., J. A. Farr, J. Cox and D. W. Mehlman. 1992. Movement corridors: Conservation bargains or poor investments? *Conservation Biology* 6: 493–505.

Simon, D. J. (ed.). 1988. *Our Common Lands: Defending the National Parks*. Island Press, Washington, D.C.

Simpson, B. B. and M. Conner-Ogorzaly. 1986. *Economic Botany: Plants in Our World*. McGraw-Hill, New York.

Sinden, J. and A. Worrell. 1979. *Unpriced Values: Decisions Without Market Prices*. John Wiley, New York.

Singer, P. 1979. Not for humans only. In K. E. Goodpaster and K. M. Sayre (eds.), *Ethics and Problems of the Twenty-First Century*, pp. 191–206. University of Notre Dame, Notre Dame, IN.

Slatyer, R. O. 1983. The origin and evolution of the World Heritage Convention. *Ambio* 12: 138–145.

Smith, J. B. and D. A. Tirpack (eds.). 1988. *The Potential Effects of Global Climate Changes on the United States*, Vol. 2. U.S. Environmental Protection Agency, Washington, D.C.

Snyder, B., J. Thisted, B. Burgess and M. Richard. 1985. Pigeon herpesvirus mortalities in foster reared Mauritius pink pigeons. In *Proceedings of the American Association of Zoo Veterinarians*, pp. 69–70. Scottsdale, Arizona.

Snyder, H. A. and N. F. Snyder. 1990. The comeback of the California Condor. *Birds International* 2: 10–23.

Society for Ecological Restoration. 1991. Program and abstracts, 3rd Annual Conference, Orlando, FL 18–23 May 1991.

Solbrig, O. T., H. M. van Emden and P. G. W. J. van Oordt (eds.). 1992. *Biodiversity and Global Change*. International Union of Biological Sciences, Paris.

Soulé, M. E. 1980. Thresholds for survival: Maintaining fitness and evolutionary potential. *In* M. E. Soulé and B. A. Wilcox (eds.), *Conservation Biology: An Evolutionary-Ecological Perspective*, pp. 151–170. Sinauer Associates, Sunderland, MA.

Soulé, M. E. 1983. What do we really know about extinction? *In* C. M. Schonewald-Cox, S. M. Chambers, B. MacBryde and L. Thomas (eds.), *Genetics and Conservation: A Reference for Managing Wild Animal and Plant Populations*, pp. 111–124. Benjamin/Cummings, Menlo Park, CA.

Soulé, M. E. 1985. What is conservation biology? *BioScience* 35: 727–734.

Soulé, M. E. (ed.). 1986. *Conservation Biology: The Science of Scarcity and Diversity*. Sinauer Associates, Sunderland, MA.

Soulé, M. E. (ed.). 1987. *Viable Populations for Conservation*. Cambridge University Press, Cambridge.

Soulé, M. E. 1990. The onslaught of alien species, and other challenges in the coming decades. *Conservation Biology* 4: 233–239.

Soulé, M. E. and D. Simberloff. 1986. What do genetics and ecology tell us about the design of nature reserves? *Biological Conservation* 35: 19–40.

Soulé, M. E. and B. A. Wilcox (eds.). 1980. *Conservation Biology: An Evolutionary-Ecological Perspective*. Sinauer Associates, Sunderland, MA.

Species Survival Commission. 1990. *Membership Directory*. IUCN, Gland, Switzerland.

Spencer, C. N., B. R. McClelland and J. A. Stanford. 1991. Shrimp stocking, salmon collapse, and eagle displacement. *BioScience* 41: 14–21.

Standley, L. A. 1992. Taxonomic issues in rare species protection. *Rhodora* 94: 218–242.

Stanley Price, M. R. 1989. *Animal Reintroductions: The Arabian Oryx in Oman*. Cambridge University Press, Cambridge.

Steele, R. C. and R. C. Welch (eds.). 1973. *Monks Wood: A Nature Reserve Record*. The Nature Conservancy, Monks Wood Experimental Station, Huntington, England.

Stehli, F. G. and J. W. Wells. 1971. Diversity and age patterns in hermatypic corals. *Systematic Zoology* 20: 115–125.

Stevens, W. K. 1992. U.S. moves to impose limits on killing of coastal sharks. *New York Times*, December 22, p. C4.

Stevens, T. P. 1988. *California State Mussel Watch Marine Water Quality Monitoring Program 1986–1987*. California State Water Resource Control Board, Sacramento.

Stewart, M. M. and C. Ricci. 1988. Dearth of the blues. *Natural History* 97: 67–71.

St. John, H. 1973. *List and Summary of the Flowering Plants in the Hawaiian Islands*. Pacific Tropical Botanical Garden Memoir No. 1. Cathay Press, Hong Kong.

Stolzenburg, W. 1992. The mussels' message. *Nature Conservancy* 42(Nov/Dec): 16–23.

Strong, D. H. 1988. *Dreamers and Defenders: American Conservationists*. University of Nebraska Press, Lincoln.

Stuart, S. N. 1987. *Why We Need Action Plans: Species*. Newsletter #8, February. IUCN Species Survival Commission, Gland, Switzerland.

Swanson, F. J. and R. E. Sparks. 1990. Long-term ecological research and the invisible place. *BioScience* 40: 502–508.

Szafer, W. 1968. The ure-ox, extinct in Europe since the

seventeenth century: An attempt at conservation that failed. *Biological Conservation* 1: 45–47.

Takegawa, J. E. and S. R. Beissinger. 1989. Cyclic drought, dispersal, and the conservation of the snail kite in Florida: Lessons in critical habitat. *Conservation Biology* 3: 302–311.

Tamarin, R. H. 1993. *Principles of Genetics*, 4th ed. Wm. C. Brown, Dubuque, IA.

Tangley, L. 1986. Saving tropical forests. *BioScience* 36: 4–15.

Tangley, L. 1988a. Beyond national parks. *BioScience* 38: 146–161.

Tangley, L. 1988b. Studying (and saving) the tropics. *BioScience* 38: 375–385.

Tangley, L. 1988c. Research priorities for conservation. *BioScience* 38: 444–448.

Tangley, L. 1990. Cataloging Costa Rica's diversity. *BioScience* 40: 633–636.

Tansley, S. A. 1988. The status of threatened Proteaceae in the Cape flora, South Africa, and the implications for their conservation. *Biological Conservation* 43: 227–239.

Tattersall, I. 1993. Madagascar's lemurs. *Scientific American* 268: 110–117.

Taylor, P. W. 1986. *Respect for Nature: A Theory of Environmental Ethics*. Princeton University Press, Princeton, NJ.

Temple, S. A. 1990. The nasty necessity: Eradicating exotics. *Conservation Biology* 4: 113–115.

Temple, S. A. 1991. Conservation biology: New goals and new partners for managers of biological resources. *In* D. J. Decker, M. E. Krasny, G. R. Goff, C. R. Smith and D. W. Gross (eds.), *Challenges in the Conservation of Biological Resources: A Practitioner's Guide*, pp. 45–54. Westview Press, Boulder, CO.

Templeton, A. R. 1986. Coadaptation and outbreeding depression. *In* M. E. Soulé (ed.), *Conservation Biology: The Science of Scarcity and Diversity*, pp. 105–116. Sinauer Associates, Sunderland, MA.

Terborgh, J. 1974. Preservation of natural diversity: The problem of extinction-prone species. *BioScience* 24: 715–722.

Terborgh, J. 1976. Island biogeography and conservation: Strategy and limitations. *Science* 193: 1029–1030.

Terborgh, J. 1986. Keystone plant resources in the tropical forest. *In* M. E. Soulé (ed.), *Conservation Biology: The Science of Scarcity and Diversity*, pp. 330–344. Sinauer Associates, Sunderland, MA.

Terborgh, J. 1989. *Where Have All the Birds Gone? Essays on the Biology and Conservation of Birds That Migrate to the American Tropics*. Princeton University Press, Princeton, NJ.

Terborgh, J. 1992a. Why American songbirds are vanishing. *Scientific American* 264: 98–104.

Terborgh, J. 1992b. Maintenance of diversity in tropical forests. *Biotropica* (Special Issue) 24: 283–292.

Terborgh, J. and B. Winter. 1980. Some causes of extinction. *In* M. E. Soulé and B. A. Wilcox (eds.), *Conservation Biology: An Evolutionary-Ecological Perspective*, pp. 119–133. Sinauer Associates, Sunderland, MA.

Terborgh, J. and B. Winter. 1983. A method for siting parks and reserves with special reference to Columbia and Ecuador. *Biological Conservation* 27: 45–58.

Thiollay, J. M. 1986. Structure comparée du peuplement avien dans trois sites de forêt primaire en Guyane. *Revue d'Ecologie et de Biolgie du Sol* 41: 59–105.

Thiollay, J. M. 1989. Area requirements for the conservation of rain forest raptors and game birds in French Guiana. *Conservation Biology* 3: 128–137.

Thiollay, J. M. 1992. Influence of selective logging on bird species diversity in a Guianan rain forest. *Conservation Biology* 6: 47–63.

Thiollay, J. and B. U. Meyburg. 1988. Forest fragmentation and the conservation of raptors: A survey on the island of Java. *Biological Conservation* 44: 229–250.

Thomas, C. D. 1990. What do real population dynamics tell us about minimum viable population sizes? *Conservation Biology* 4: 324–327.

Thomas, D. W. 1982. The ecology of an African savanna fruit bat community: Resource partitioning and role in seed dispersal. Ph.D. Dissertation, University of Aberdeen, Scotland.

Thomas, J. W. and H. Salwasser. 1989. Bringing conservation biology into a position of influence in natural resource management. *Conservation Biology* 3: 123–127.

Thomas, J. W., et al. 1990. A conservation strategy for the northern spotted owl: Report of the interagency scientific committee to address the conservation of the northern spotted owl. Portland, OR.

Thomas, K. S. 1991. *Living Fossil: The Story of the Coelacanth*. Norton, New York.

Thoreau, H. D. 1854. *Walden*. Ticknor and Fields, Boston, MA.

Thoreau, H. D. 1863. *Excursions*. Ticknor and Fields, Boston, MA.

Thorne, E. T. and B. Oakleaf. 1991. Species rescue for captive breeding: Black-footed ferret as an example. *Symposia of the Zoological Society of London* 62: 241–261.

Thorne, E. T. and E. S. Williams. 1988. Disease and endangered species: the black-footed ferret as a recent example. *Conservation Biology* 2: 66–74.

Thorne, R. F. 1967. A flora of Santa Catalina Island, California. *Aliso* 6:1–77.

Thorsell, J. and J. Sawyer. 1992. *World Heritage: The First Twenty Years*. IUCN, Gland, Switzerland.

Tobin, R. 1990. *The Expendable Future: U.S. Politics and the Protection of Biological Diversity*. Duke University Press, Durham, NC.

Toledo, V. M. 1988. La diversidad biológica de México. In *Ciencia y Desarollo*. Conacyt, Mexico City.

Toledo, V. M. 1991. Patzcuaro's lesson: Nature, production, and culture in an indigenous region of Mexico. *In* M. L. Oldfield and J. B. Alcorn (eds.), *Biodiversity:*

Culture, Conservation and Ecodevelopment, pp. 147–171. Westview Press, Boulder, CO.

Toufexis, A. 1992. A new endangered species: human protectors of the planet put their lives on the line. *Time* 139(17): 48–50.

Tremaine, R. 1993. Valuing tropical rainforests. *Conservation Biology* 7: 7–8.

Triggs, S. J., R. G. Powlesland and C. H. Daugherty. 1989. Genetic variation and conservation of kakapo (*Strigops habroptilus*: Psittaciformes). *Conservation Biology* 3: 92–96.

Tunnicliffe, V. 1992. Hydrothermal vent communities of the deep sea. *American Scientist* 80: 336–349.

Turner, B. L. II. 1976. Prehistoric population density in the Maya lowlands: New evidence from old approaches. *Geographical Review* 66: 73–82.

Udvardy, M. D. F. 1975. A classification of the biogographical provinces of the world. Occasional Paper 18. IUCN, Gland, Switzerland.

United Nations. 1993a. *Agenda 21: Rio Declaration and Forest Principles.* Post-Rio Edition. United Nations Publications, New York.

United Nations. 1993b. *The Global Partnership for Environment and Development.* United Nations Publications, New York.

UNESCO (United Nations Educational, Scientific, and Cultural Organization). 1974. *Task Force on Criteria and Guidelines for the Choice and Establishment of Biosphere Reserves.* Final Report. MAB Report Series No. 22. UNESCO, Paris.

UNESCO. 1980. The mighty minnow. *UNESCO Courier* 33: 19.

UNESCO. 1984. Action plan for the biosphere reserves. *Natural Resources* 20: 1–12.

UNESCO. 1985. Action plan for biosphere reserves. *Environmental Conservation* 12: 17–27.

Untermaier, J. 1991. *Legal Aspects of the Conservation of Wetlands.* IUCN, Gland, Switzerland.

Urban, D. L., R. V. O'Neill and H. H. Shugart, Jr. 1987. Landscape ecology. *BioScience* 37: 119–127.

U.S. Congress. 1973. Sec. 2(a) in *Endangered Species Act* 87 STAT. 884 (Public Law 93-205).

USDA Forest Service. 1988. Final supplement to the environmental impact statement for an amendment to the Pacific Northwest regional guide, Vols. 1 and 2. USDA Forest Service, Portland, OR.

Usher, M. B. 1975. *Biological Management and Conservation: Ecological Theory, Application and Planning.* Chapman and Hall, London.

Usher, M. B. 1991. Scientific requirements of a monitoring programme. *In* B. Goldsmith (ed.), *Monitoring for Conservation and Ecology*, pp. 15–32. Chapman and Hall, New York.

Valdéz, J. 1992. Defenders of the reef. *Belize Today* 6(8): 16–18.

Van Riper, C., III, S. G. Van Riper, M. L. Goff and M. Laird. 1986. The epizootiology and ecological significance of malaria in Hawaiian land birds. *Ecological Monographs* 56: 327–344.

Van Swaay, C.A.M. 1990. An assessment of the changes in butterfly abundance in the Netherlands during the twentieth century. *Biological Conservation* 52: 287–302.

Van Tighem, K. V. 1986. Have our national parks failed us? *Park News* 31–33.

Van Waerbeck, K. and J. C. Reyes. 1990. Catch of small cetaceans at Pucusana Port, central Peru, during 1987. *Biological Conservation* 51: 15–22.

Vaughan, D. A. and L. A. Sitch. 1991. Gene flow from the jungle to farmers. *BioScience* 41: 22–28.

Vaux, P. D. and C. R. Goldman. 1990. Dams and development in the tropics: The role of applied ecology. *In* R. Goodland (ed.), *Race to Save the Tropics: Ecology and Economics for a Sustainable Future*, pp. 101–124. Island Press, Washington, D.C.

Vedder, A. 1989. In the hall of the mountain king. *Animal Kingdom* 92: 31–43.

Verboom, J., A. Schotman, P. Opdam and J. A. J. Metz. 1991. European nuthatch metapopulations in a fragmented agricultural landscape. *Oikos* 61: 149–156.

Vernon, J. E. N. 1986. *Corals of Australia and the Indo-Pacific.* Angus and Robertson, London.

Vessey, S. H. 1964. Effects of grouping on levels of circulating antibodies in mice. *Proceedings of the Society for Experimental Biology and Medicine* 115: 252–255.

Vietmeyer, N. 1989. Iguana mama. *International Wildlife* 19: 24–27.

Vindevogel, H., H. Debruyne and P. P. Patoret. 1985. Observation of pigeon *herpesvirus 1* re-excretion during the reproductive period in conventionally reared homing pigeons. *Journal of Comparative Pathology* 95: 105–112.

Vitousek, P. M., P. R. Ehrlich, A. H. Ehrlich and P. A. Matson. 1986. Human appropriation of the products of photosynthesis. *BioScience* 36: 368–373.

von Furer-Haimendorf, C. 1964. *The Sherpas of Nepal.* John Murray, London.

Waley, A. 1934. *The Way and Its Power: A Study of Tao Te Ching and Its Place in Chinese Thought.* Allen & Unwin, London.

Walkinshaw, L. 1949. The sandhill cranes. Bulletin 29, Cranbrook Institute of Science, Bloomfield Hills, MI.

Waller, D. M., D. M. O'Malley and S. C. Gawler. 1988. Genetic variation in the extreme endemic *Pedicularis furbishiae.* *Conservation Biology* 1: 335–340.

Walsh, R. G., J. B. Loomis and R. A. Gillman. 1984. Valuing option, existence, and bequest demands for wilderness. *Land Economics* 60: 14–19.

Walsh, R. G., R. D. Bjonback, R. A. Aiken and D. H. Rosenthal. 1990. Estimating the public benefits of protecting forest quality. *Journal of Environmental Management* 30: 175–189.

Waples, R. S. and D. J. Teel. 1990. Conservation genetics of Pacific salmon. I. Temporal changes in allele frequency. *Conservation Biology* 4: 144–156.

Ward, D. M., R. Weller and M. M. Bateson. 1990. 16S rRNA sequences reveal numerous uncultured microorganisms in a natural community. *Nature* 345: 63–65.

Ward, G. C. 1992. India's wildlife dilemma. *National Geographic* 181(May): 2–29.

Warford, J. 1987. Nature resource management and economic development. *In* P. Jacobs and D. Munro (eds.), *Conservation with Equity: Strategies for Sustainable Development*, pp. 71–85. IUCN, Gland, Switzerland.

Warren, M. S. 1990. The successful conservation of an endangered species, the heath fritillary butterfly *Mellicta athalia*, in Britain. *Biological Conservation* 55: 37–56.

Waser, N. M. and M. V. Price. 1989. Optimal outcrossing in *Ipomopsis aggregata*: Seed set and offspring fitness. *Evolution* 43: 1097–1109.

Waser, N. M., M. V. Price, A. M. Montalvo and R. N. Gray. 1987. Female mate choice in a perennial herbaceous wildflower, *Delphinium nelsonii*. *Evolutionary Trends in Plants* 1: 29–33.

Waters, T. 1992. Sympathy for the devil. *Discover* 13: 62.

Wayne, R. K., et al. 1991. Conservation genetics of the endangered Isle Royale gray wolf. *Conservation Biology* 5: 41–51.

Webster, R. E. 1987. Habitat conservation plans under the Endangered Species Act. *San Diego Law Review* 24: 243–271.

Weiss, S. B., et al. 1991. Forest canopy structure at overwintering monarch butterfly sites: Measurements with hemispherical photography. *Conservation Biology* 5: 165–175.

Welch, E. B. and G. D. Cooke. 1990. Lakes. *In* W. R. Jordan III, M. E. Gilpin and J. D. Aber (eds.), *Restoration Ecology: A Synthetic Approach to Ecological Research*. pp. 109–129. Cambridge University Press, Cambridge.

Wells, S. M., C. Shappard and M. D. Jenkins. 1988a. *Coral Reefs of the World, Volume 1: Atlantic and Eastern Pacific*. IUCN, Gland, Switzerland.

Wells, S. M., C. Shappard and M. D. Jenkins. 1988b. *Coral Reefs of the World, Volume 2: Indian Ocean, Red Sea and Gulf*. IUCN, Gland, Switzerland.

Wells, S. M., C. Shappard and M. D. Jenkins. 1989. *Coral Reefs of the World, Volume 3: Central and Western Pacific*. IUCN, Gland, Switzerland.

Werner, D. I. 1991. The rational use of green iguanas. *In* J. G. Robinson and K. H. Redford (eds.), *Neotropical Wildlife Use and Conservation*. University of Chicago Press, Chicago.

Werner, D. I. and T. J. Miller. 1984. Artificial nests for female green iguanas. *Herpetological Review* 15: 57–58.

Western, D. 1985. Conservation-based rural development. *In* F. R. Thibodeau and H. Field (eds.), *Sustaining Tomorrow*. University Press of New England, Hanover, NH.

Western, D. 1989. Conservation without parks: Wildlife in the rural landscape. *In* D. Western and M. Pearl (eds.), *Conservation for the Twenty-First Century*, pp. 158–165. Oxford University Press, New York.

Western, D. and W. Henry. 1979. Economics and conservation in Third World national parks. *BioScience* 29: 414–418.

Western, D. and M. Pearl (eds.). 1989. *Conservation for the Twenty-First Century*. Oxford University Press, New York.

Western, D. and J. Ssemakula. 1981. The future of the savannah ecosystem: Ecological islands or faunal enclaves? *African Journal of Ecology* 19: 7–19.

Whitcomb, R. F., C. S. Robbins, J. F. Lynch, B. L. Whitcomb, M. K. Klimkiewicz and D. Bystrak. 1981. Effects of forest fragmentation on avifauna of the eastern deciduous fores. *In* R. L. Burgess and D. M. Sharpe (eds.), *Forest Island Dynamics in Man-Dominated Landscapes*, pp. 125–205. Springer-Verlag, New York.

White, F. 1983. *The Vegetation of Africa: A Descriptive Memoir to Accompany the Unesco/AITFAT/UNSO Vegetation Map of Africa*. UNESCO, Paris.

Whitmore, T. C. 1990. *An Introduction to Tropical Rain Forests*. Clarendon Press, Oxford.

Whittaker, R. H. 1975. *Communities and Ecosystems*, 2nd ed. Macmillan, New York.

Whitten, A. J. 1987. Indonesia's transmigration program and its role in the loss of tropical rain forests. *Conservation Biology* 1: 239–246.

Whitten, A. J., K. D. Bishop, S. V. Nash and Lynn Clayton. 1987. One or more extinctions from Sulawesi, Indonesia? *Conservation Biology* 1: 42–48.

Wijnstekers, W. 1992. *The Evolution of CITES*. CITES Secretariat, Geneva, Switzerland.

Wikramanayake, E. D. 1990. Conservation of endemic rain forest fishes of Sri Lanka: Results of a translocation experiment. *Conservation Biology* 4: 32–37.

Wilcove, D. S. 1985. Nest predation in forest tracts and the decline of migratory songbirds. *Ecology* 66: 1211–1214.

Wilcove, D. S. and R. M. May. 1986. National park boundaries and ecological realities. *Nature* 324: 206–207.

Wilcove, D. and D. Murphy. 1991. The spotted owl controversy and conservation biology. *Conservation Biology* 5: 261–262.

Wilcove, D. S., C. H. McLellan and A. P. Dobson. 1986. Habitat fragmentation in the temperate zone. *In* M. E. Soulé (ed.), *Conservation Biology: The Science of Scarcity and Diversity*, pp. 237–256. Sinauer Associates, Sunderland, MA.

Wilcove, D. S., M. McMillan and K. C. Winston. 1993. What exactly is an endangered species? An analysis of the U.S. Endangered Species List, 1985–1991. *Conservation Biology* 7: 87–93.

Wildt, D. E. 1992. Genetic resource banks for conserving

wildlife species: Justification, examples, and becoming organized on a global scale. *Animal Reproduction Science* 28: 247–257.

Wilkes, G. 1977. The world's crop plant germplasm—an endangered resource. *The Bulletin of the Atomic Scientists* 33: 8–16.

Wilkes, G. 1987. Plant genetic resources: Why privatize a public good? *BioScience* 37: 215–217.

Wilkes, G. 1989. Germplasm preservation: Objectives and needs. *In* L. Knutson and A. K. Stoner (eds.), *Biotic Diversity and Germplasm Preservation, Global Imperatives*, pp. 13–41. Kluwer Academic, Boston.

Wilkes, G. 1991. In situ conservation of agricultural systems. *In* M. L. Oldfield and J. B. Alcorn (eds.), *Biodiversity: Culture, Conservation and Ecodevelopment*, pp. 86–101. Westview Press, Boulder, CO.

Wilkie, D. S., J. G. Sidle and G. C. Boundzanga. 1992. Mechanized logging, market hunting, and a bank loan in the Congo. *Conservation Biology* 6: 570–580.

Williams, J. D. and R. M. Nowak. 1986. Vanishing species in our own backyard: Extinct fish and wildlife of the United States and Canada. *In* L. Kaufman and K. Mallory (eds.), *The Last Extinction*, pp. 107–140. MIT Press, Cambridge, MA.

Williams, M. (ed.). 1990. *Wetlands: A Threatened Landscape*. Basil Blackwell, Oxford.

Williams, S. B. 1984. Protection of plant varieties and parts as intellectual property. *Science* 225: 18–23.

Williams, T. 1986. The final ferret fiasco. *Audubon* 88: 110–119.

Willis, E. O. 1979. The composition of avian communities in remanescent woodlots in southern Brazil. *Papeis Avulsos Zoologicas* 33: 1–25.

Willwerth, J. 1992. The $25 million bird. *Time* Magazine, January 27.

Wilson, E. O. 1984. *Biophilia*. Harvard University Press, Cambridge, MA.

Wilson, E. O. 1985. The biological diversity crisis. *BioScience* 35: 700–705.

Wilson, E. O. 1987. The little things that run the world: The importance and conservation of invertebrates. *Conservation Biology* 1: 344–346.

Wilson, E. O. 1989. Threats to biodiversity. *Scientific American* 261(September): 108–116.

Wilson, E. O. 1991. Rain forest canopy: The high frontier. *National Geographic* 180(December): 78–107.

Wilson, E. O. 1992. *The Diversity of Life*. The Belknap Press of Harvard University Press, Cambridge, MA.

Wilson, E. O. and F. M. Peter (eds.). 1988. *Biodiversity*. National Academy Press, Washington, D.C.

Witte, F., T. Goldschmidt, J. H. Wanink, M. J. P. van Oijen, P. C. Goudswaard, E. L. M. Witte-Maas and N. Bouton. 1992. The destruction of an endemic species flock: Quantitative data on the decline of the haplochromine species from the Mwanza Gulf of Lake Victoria. *Environmental Biology of Fishes* 34: 1–28.

Wong, M 1985. Understory birds as indicators of regeneration in a patch of selectively logged west Malaysian rainforest. *In* A. W. Diamond and T. E. Lovejoy (eds.), *Conservation of Tropical Forest Birds*, pp. 249–263. Technical Publication No. 4. International Council for Bird Preservation, Cambridge, England.

Woodwell, G. M. 1990. *The Earth in Transition*. Cambridge University Press, Cambridge.

Woolf, N. B. 1986. New hope for exotic species. *BioScience* 36: 594–597.

Woolf, N. B. 1990. Biotechnologies sow seeds for the future. *BioScience* 40: 346–348.

World Commission on Environment and Development (WCED). 1987. *Our Common Future*. Oxford University Press, Oxford.

World Conservation Monitoring Centre. 1992. *Global Biodiversity: Status of the Earth's Living Resources*. Compiled by the World Conservation Monitoring Centre, Cambridge, UK. Chapman and Hall, London.

World Conservation Monitoring Centre/IUCN. 1992a. *Protected Areas of the World, Volume 1: Indomalaya, Oceania, Australia, and Antarctic*. IUCN, Gland, Switzerland.

World Conservation Monitoring Centre/IUCN. 1992b. *Protected Areas of the World, Volume 2: Palearctic*. IUCN, Gland, Switzerland.

World Conservation Monitoring Centre/IUCN. 1992c. *Protected Areas of the World, Volume 3: Afrotropical*. IUCN, Gland, Switzerland.

World Conservation Monitoring Centre/IUCN. 1992d. *Protected Areas of the World, Volume 4: Nearctic and Neotropical*. IUCN, Gland, Switzerland.

World Resources Institute (WRI). 1993 *Biodiversity Prospecting: Using Genetic Resources for Sustainable Development*. World Resources Institute, Washington, D.C.

World Resources Institute/International Institute for Environment and Development (WRI/IIED). 1986. *World Resources 1986*. Basic Books, New York.

WRI/IIED. 1987. *World Resources 1987*. Basic Books, New York.

WRI/IIED. 1988. *World Resources 1988*. Basic Books, New York.

WRI/UNEP/UNDP. 1992. *World Resources 1992–1993*. Oxford University Press, New York.

WRI/IUCN/UNEP. 1992. *Global Biodiversity Strategy: Guidelines for Action to Save, Study, and Use Earth's Biotic Wealth Sustainably and Equitably*. World Resources Institute, Washington, D.C.

Wright, S. 1931. Evolution in Mendelian populations. *Genetics* 16: 97–159.

Wyman, R. L. 1990. What's happening to the amphibians? *Conservation Biology* 4: 350–352.

Yager, J. 1981. Remipedia, a new class of Crustacea from a marine cave in the Bahamas. *Journal of Crustacean Biology* 1: 328–333.

Yahner, R. H. 1988. Changes in wildlife communities near edges. *Conservation Biology* 2: 333–339.

Yoakum, J. and W. P. Dasmann. 1971. Habitat manipulation practices. *In* R. H. Giles (ed.), *Wildlife Management Techniques*, pp. 173–231. Wildlife Society, Washington, D.C.

Yonzon, P. B. and M. L. Hunter, Jr. 1991a. Cheese, tourists, and red pandas in the Nepal Himalayas. *Conservation Biology* 5: 196–202.

Yonzon, P. B. and M. L. Hunter, Jr. 1991b. Conservation of the red panda *Ailurus fulgens*. *Biological Conservation* 57: 1–11.

Yost, J. and P. Kelley. 1983. Shotguns, blowguns, and spears: The analysis of technological efficiency. *In* R. B. Hames and W. T. Vickers (eds.), *Adaptive Responses of Native Amazonians*, pp. 189–224. Academic Press, New York.

Young, R. A, D. J. P. Swift, T. L. Clarke, G. R. Harvey and P. R. Betzer. 1985. Dispersal pathways for particle-associated pollutants. *Science* 229: 431–435.

Zaidi, I. H. 1986. On the ethics of man's interaction with the environment: an Islamic approach. *In* E. C. Hargrove (ed.), *Religion and the Environmental Crisis*, pp. 107–126. University of Georgia Press, Athens.

Zonneveld, I. and R. T. Forman. 1990. *Changing Landscapes: An Ecological Perspective*. Springer-Verlag, New York.

Zwank, P. J. and C. D. Wilson. 1987. Survival of captive, parent-reared Mississippi sandhill cranes released on a refuge. *Conservation Biology* 1: 165–168.

Index